Lecture Notes in Computer Sc

T0238520

Commenced Publication in 1973
Founding and Former Series Editors:
Gerhard Goos, Juris Hartmanis, and Jan van Leeuwen

Vladimir P. Gerdt   Wolfram Koepf
Ernst W. Mayr   Evgenii V. Vorozhtsov (Eds.)

Computer Algebra in Scientific Computing

14th International Workshop, CASC 2012
Maribor, Slovenia, September 3-6, 2012
Proceedings

 Springer

Volume Editors

Vladimir P. Gerdt
Joint Institute for Nuclear Research (JINR)
Laboratory of Information Technologies (LIT)
141980 Dubna, Russia
E-mail: gerdt@jinr.ru

Wolfram Koepf
Universität Kassel
Institut für Mathematik
Heinrich-Plett-Straße 40, 34132 Kassel, Germany
E-mail: koepf@mathematik.uni-kassel.de

Ernst W. Mayr
Technische Universität München
Institut für Informatik
Lehrstuhl für Effiziente Algorithmen
Boltzmannstraße 3, 85748 Garching, Germany
E-mail: mayr@in.tum.de

Evgenii V. Vorozhtsov
Russian Academy of Sciences
Institute of Theoretical and Applied Mechanics
630090 Novosibirsk, Russia
E-mail: vorozh@itam.nsc.ru

ISSN 0302-9743   e-ISSN 1611-3349
ISBN 978-3-642-32972-2   e-ISBN 978-3-642-32973-9
DOI 10.1007/978-3-642-32973-9
Springer Heidelberg Dordrecht London New York

Library of Congress Control Number: 2012944896

CR Subject Classification (1998): F.2, G.2, E.1, I.1, I.3.5, G.1, F.1

LNCS Sublibrary: SL 1 – Theoretical Computer Science and General Issues

Typesetting: Camera-ready by author, data conversion by Scientific Publishing Services, Chennai, India

Printed on acid-free paper

Springer is part of Springer Science+Business Media (www.springer.com)

Preface

One of the directions of research at the Center of Applied Mathematics and Theoretical Physics, University of Maribor (CAMTP), is the application of methods and algorithms of computer algebra to studying some long-standing problems of the theory of differential equations, such as the Poincaré center problem and Hilbert's 16th problem. In the work of the group, led by Valery Romanovski, efficient computational approaches to studying the center problem and the closely related isochronicity problem have been developed. It allowed the group to completely solve the problems for many classes of polynomial systems of ODEs. In recent work (with V. Levandovskyy, D.S. Shafer, and others), they also developed a powerful algorithmic method to obtain some bounds on the number of small limit cycles bifurcating from elementary singular points of polynomial systems of ODEs, i.e., to evaluate algorithmically the cyclicity of the elementary center and focus. Research on applications of computer algebra to differential equations and dynamical systems at CAMTP is carried out in collaboration with colleagues worldwide working in similar directions; among them we can mention X. Chen, M. Han, W. Huang, Y.-R. Liu and W. Zhang (China), V. Edneral (Russia), J. Giné (Spain), and A. Mahdi and D.S. Shafer (USA). Some goals and features of the approaches mentioned above are described in a recent book [V.G. Romanovski, D.S. Shafer. The center and cyclicity problems: a computational algebra approach. Boston, Basel–Berlin: Birkhäuser, 2009; ISBN 978-0-8176-4726-1].

In 2010, CAMTP, in collaboration with the Institute of Mathematics, Physics, and Mechanics (IMFM), the Faculty of Natural Science and Mathematics of the University of Maribor, and with the support of the Slovenian Research Agency, organized the conference "Symbolic Computation and Its Applications" (SCA). The concept of this meeting was to bring together researchers from various areas of natural sciences, who employ and/or develop symbolic techniques, and to provide a platform for discussions and exchange of ideas. Following the success of the meeting, a second conference was organized in May 2012 at RWTH Aachen University, thus turning SCA into a series of conferences.

In connection with the above, it was decided to hold the 14th CASC Workshop in Maribor. The 13 earlier CASC conferences, CASC 1998, CASC 1999, CASC 2000, CASC 2001, CASC 2002, CASC 2003, CASC 2004, CASC 2005, CASC 2006, CASC 2007, CASC 2009, CASC 2010, and CASC 2011 were held, respectively, in St. Petersburg (Russia), in Munich (Germany), in Samarkand (Uzbekistan), in Konstanz (Germany), in Yalta (Ukraine), in Passau (Germany), in St. Petersburg (Russia), in Kalamata (Greece), in Chişinău (Moldova), in Bonn (Germany), in Kobe (Japan), in Tsakhkadzor (Armenia), and in Kassel (Germany), and they all proved to be very successful.

This volume contains 28 full papers submitted to the workshop by the participants and accepted by the Program Committee after a thorough reviewing process. Additionally, the volume includes two abstracts of invited talks.

One of the main themes of the CASC workshop series, namely, polynomial algebra, is represented by contributions devoted to new algorithms for computing comprehensive Gröbner and involutive systems, parallelization of the Gröbner bases computation, the study of quasi-stable polynomial ideals, new algorithms to compute the Jacobson form of a matrix of Ore polynomials, a recursive Leverrier algorithm for inversion of dense matrices whose entries are monic polynomials, root isolation of zero-dimensional triangular polynomial systems, optimal computation of the third power of a long integer, investigation of the complexity of solving systems with few independent monomials, the study of ill-conditioned polynomial systems, a method for polynomial root-finding via eigen-solving and randomization, an algorithm for fast dense polynomial multiplication with Java using the new opaque typed method, and sparse polynomial powering using heaps.

The invited talk by K. Yokoyama deals with the usage of modular techniques for efficient computation of ideal operations. The following applications of modular techniques are considered: Gröbner bases computation and computation of minimal polynomials. The methods for recovering the true result from the results of modular computations are also discussed.

Several papers are devoted to using computer algebra for the investigation of various mathematical and applied topics related to ordinary differential equations (ODEs): algebraic methods for investigating the qualitative behavior of bio-chemical reaction networks, algorithms for detecting the Hopf bifurcation in high-dimensional chemical reaction networks, the solution of linear ODEs with rational coefficients, also known as D-finite (or holonomic) series, the calculation of normal forms and the first integrals of the Euler–Poisson equations, conditions for the first integral of the cubic Lotka–Volterra system in a neighborhood of the origin, and the analysis of the asymptotic stabilizability of planar switched linear ODE systems.

Two papers deal with applications of symbolic computation in mechanics: the investigation of stability of equilibrium positions in the spatial circular restricted four-body problem of celestial mechanics, and the investigation of stability of a gyroscopic system with four degrees of freedom and with three parameters.

New symbolic-numeric algorithms presented in this volume deal with the solution of the boundary-value problem for the Schrödinger equation in cylindrical coordinates and the solution of the Navier–Stokes equations for the three-dimensional viscous incompressible fluid flows.

Other applications of computer algebra systems presented in this volume include the investigation of the questions of existence of polynomial solutions for linear partial differential equations and (q-)difference equations, new algorithms for rational reparameterization of any plane curve, Maple-based algorithms for determining the intersection multiplicity of two plane curves, and the reduction of the solution of the combinatorial problem of rainbow connectivity to the solution of a certain system of polynomial equations.

The invariant theory, which is at the crossroads of several mathematical disciplines, is surveyed in the invited talk by G. Kemper. Some examples are given, in which invariant theory is applied to graph theory, computer vision, and coding theory. The talk also gives an overview of the state of the art of algorithmic invariant theory.

The CASC 2012 workshop was supported financially by the Slovenian Research Agency and CAMTP. Our particular thanks are due to the members of the CASC 2012 local Organizing Committee in Slovenia: M. Robnik and V. Romanovski (CAMTP, Maribor) and M. Petkovšek (University of Ljubljana), who ably handled local arrangements in Maribor. Furthermore, we want to thank the Program Committee for their thorough work. Finally, we are grateful to W. Meixner for his technical help in the preparation of the camera-ready manuscript for this volume.

July 2012   V.P. Gerdt
                                                             W. Koepf
                                                             E.W. Mayr
                                                        E.V. Vorozhtsov

Organization

CASC 2012 has been organized jointly by the Department of Informatics at the Technische Universität München, Germany, and the Center for Applied Mathematics and Theoretical Physics at the University of Maribor, Slovenia.

Workshop General Chairs

Vladimir P. Gerdt (JINR, Dubna) Ernst W. Mayr (TU München)

Program Committee Chairs

Wolfram Koepf (Kassel) Evgenii V. Vorozhtsov (Novosibirsk)

Program Committee

Sergei Abramov (Moscow) Valery Romanovski (Maribor)
François Boulier (Lille) Markus Rosenkranz (Canterbury)
Hans-Joachim Bungartz (Munich) Mohab Safey El Din (Paris)
Victor F. Edneral (Moscow) Yosuke Sato (Tokyo)
Ioannis Z. Emiris (Athens) Werner M. Seiler (Kassel)
Jaime Gutierrez (Santander) Doru Stefanescu (Bucharest)
Victor Levandovskyy (Aachen) Thomas Sturm (Saarbrücken)
Marc Moreno Maza (London, CAN) Agnes Szanto (Raleigh)
Alexander Prokopenya (Warsaw) Stephen M. Watt (W. Ontario, CAN)
Eugenio Roanes-Lozano (Madrid) Andreas Weber (Bonn)

Local Organization

Valery Romanovski (Maribor) Marko Petkovšek (Ljubljana)
Marko Robnik (Maribor)

Website

http://wwwmayr.in.tum.de/CASC2012/

Table of Contents

On Polynomial Solutions of Linear Partial Differential and (q-)Difference Equations

S.A. Abramov[1,*], M. Petkovšek[2,**]

[1] Computing Centre of the Russian Academy of Sciences, Vavilova,
40, Moscow 119333, Russia
sergeyabramov@mail.ru
[2] University of Ljubljana, Faculty of Mathematics and Physics, Jadranska 19,
SI-1000, Ljubljana, Slovenia
Marko.Petkovsek@fmf.uni-lj.si

Abstract. We prove that the question of whether a given linear partial differential or difference equation with polynomial coefficients has non-zero polynomial solutions is algorithmically undecidable. However, for equations with constant coefficients this question can be decided very easily since such an equation has a non-zero polynomial solution iff its constant term is zero. We give a simple combinatorial proof of the fact that in this case the equation has polynomial solutions of *all* degrees. For linear partial q-difference equations with polynomial coefficients, the question of decidability of existence of non-zero polynomial solutions remains open. Nevertheless, for such equations with constant coefficients we show that the space of polynomial solutions can be described algorithmically. We present examples which demonstrate that, in contrast with the differential and difference cases where the dimension of this space is either infinite or zero, in the q-difference case it can also be finite and non-zero.

1 Introduction

Polynomial solutions of linear differential and (q-)difference equations often serve as a building block in algorithms for finding other types of closed-form solutions. Computer algebra algorithms for finding polynomial (see, for example, [4]) and rational (see [1,2,7,10,8] etc.) solutions of linear *ordinary* differential and difference equations with polynomial coefficients are well known. Note, however, that relatively few results about rational solutions of *partial* linear differential and (q-)difference equations can be found in the literature. Only recently, M. Kauers and C. Schneider [11,12] have started work on the algorithmic aspects of finding universal denominators for rational solutions in the difference case. Once such a denominator is obtained, one needs to find polynomial solutions of the equation satisfied by the numerators of the rational solutions of the original equation. This is our motivation for considering polynomial solutions of linear

---

* Supported by RFBR grant 10-01-00249-a.
** Supported by MVZT RS grant P1-0294.

V.P. Gerdt et al. (Eds.): CASC 2012, LNCS 7442, pp. 1–11, 2012.

partial differential and (q-)difference equations with polynomial coefficients in the present paper.

Let K be a field of characteristic 0, and let $x_1, \ldots, x_m$ be independent variables where $m \geq 2$. In Section 2, using an argument similar to the one given in [9, Thm. 4.11], we show that there is no algorithm which, for an arbitrary linear differential or difference operator L with coefficients from $K[x_1, \ldots, x_m]$, determines whether or not there is a non-zero polynomial $y \in K[x_1, \ldots, x_m]$ such that $L(y) = 0$ (Theorem 1). The proof is based on the Davis-Matiyasevich–Putnam–Robinson theorem (DMPR) which states that the problem of solvability of Diophantine equations is algorithmically undecidable, i.e., that there is no algorithm which, for an arbitrary polynomial $P(t_1, \ldots, t_m)$ with integral coefficients, determines whether or not the equation $P(t_1, \ldots, t_m) = 0$ has an integral solution [14,17]. In fact, we use the equivalent form which states that existence of *non-negative* integral solutions of $P(t_1, \ldots, t_m) = 0$ is undecidable as well.

Of course, by limiting the class of operators considered, the corresponding problem may become decidable. For example, it is well known that a partial linear differential or difference operator L with coefficients in K (a.k.a. an operator with *constant* coefficients) has a non-zero polynomial solution iff $L(1) = 0$ (see, for example, [20, Lemma 2.3]). In addition, in Section 3 we show that in this case, the equation $L(y) = 0$ has polynomial solutions of degree d *for all* $d \in \mathbb{N}$ (Theorem 2). This is contrasted with the univariate case $m = 1$, where the degree of a polynomial solution cannot exceed ord L (but note that, when a univariate L is considered to be m-variate with $m \geq 2$, and $L(1) = 0$, equation $L(y) = 0$ does have solutions of all degrees). In the differential case, when the affine algebraic variety defined by $\sigma(L) = 0$ (where $\sigma : K[\partial/\partial x_1, \ldots, \partial/\partial x_n] \to K[x_1, \ldots, x_n]$ is the ring homomorphism given by $\sigma|_K = \mathrm{id}_K$, $\sigma(\partial/\partial x_j) = x_j$) is not singular at 0, and for d large enough, Theorem 2 follows from [20, Prop. 3.3(e)]. Here we present a short direct proof based on a simple counting argument. For a given $d \in \mathbb{N}$, all solutions of degree d of such an equation can be found, e.g., by the method of undetermined coefficients. Of course, there exist more efficient ways to do that: in [19], the application of Janet bases to the computation of (formal) power series and polynomial solutions is considered; in [19, Ex. 4.6], the command `PolySol` for computing polynomial solutions from the *Janet* Maple package is illustrated. Computing polynomial solutions using Gröbner bases is described in [21, Sect. 10.3, 10.4] and [19, Sect. 10.8]. The more general problem of finding polynomial solutions of holonomic systems with polynomial coefficients (if they exist) is treated in [16,22], and the resulting algorithms are implemented in Macaulay2 [13].

Our attention was drawn to these problems by M. Kauers. In a letter to the first author he presented a proof of undecidability of existence of non-zero polynomial solutions of partial differential equations with polynomial coefficients, and attributed it to mathematical folklore. In our paper, a simple common proof for the differential and difference cases is proposed. The situation when coefficients are constant is clarified as well.

In Section 4 we consider the q-difference case, assuming that $K = k(q)$ where k is a subfield of K and q is transcendental over k (q-calculus, as well as the theory and algorithms for q-difference equations, are of interest in combinatorics, especially in the theory of partitions [5, Sect. 8.4], [6]). The question of decidability of existence of non-zero polynomial solutions of an arbitrary q-difference equation with polynomial coefficients is still open. As for the equations with constant coefficients, we formulate and prove a necessary condition for existence of a non-zero polynomial solution: if $L(1) = p(q) \in K[q]$, then $p(1) = 0$, or, more succinctly: $(L(1))(1) = 0$. We also show that the dimension of the space of polynomial solutions of a linear q-difference equation with constant coefficients can be, in contrast with the differential and difference cases, not only zero or infinite, but also finite positive. An explicit description of this space can be obtained algorithmically. We consider this as one of the first steps in the program to find wider classes of closed-form solutions of multivariate q-difference equations.

Terminology and notation. We write $x = (x_1, \ldots, x_m)$ for the variables, $D = (D_1, \ldots, D_m)$ for partial derivatives ($D_i = \frac{\partial}{\partial x_i}$), and $\Delta = (\Delta_1, \ldots, \Delta_m)$ for partial differences ($\Delta_i = E_i - 1$ where $E_i f(x) = f(x_1, \ldots, x_i + 1, \ldots, x_m)$). Multiindices from $\mathbb{N}^m$ (where $\mathbb{N} = \{0, 1, 2, \ldots\}$) are denoted by lower-case Greek letters, so that a partial linear operator of order at most r with polynomial coefficients is written as

$$L = \sum_{|\mu| \le r} a_\mu(x) D^\mu \tag{1}$$

in the differential case, and

$$L = \sum_{|\mu| \le r} a_\mu(x) \Delta^\mu \tag{2}$$

in the difference case, with $a_\mu(x) \in K[x_1, \ldots, x_m]$ in both cases. We denote the dot product of multiindices $\mu, \alpha \in \mathbb{N}^m$ by $\mu \cdot \alpha = \mu_1 \alpha_1 + \cdots + \mu_m \alpha_m$.

We call $y(x) \in K[x_1, \ldots, x_m]$ a *solution* of L if $L(y) = 0$.

Let $c \in K \setminus \{0\}$. As usual, we define

$$\deg_{x_i}(c x_1^{n_1} \cdots x_m^{n_m}) = n_i$$

for $i = 1, \ldots, m$, and

$$\deg(c x_1^{n_1} \cdots x_m^{n_m}) = n_1 + \cdots + n_m.$$

For $p(x) \in K[x_1, \ldots, x_m] \setminus \{0\}$ we set $\deg_{x_i} p(x)$ for $i = 1, \ldots, m$ to be equal to $\max \deg_{x_i} t$, and $\deg p(x)$ to be equal to $\max \deg t$ where the maximum is taken over all the terms t of the polynomial $p(x)$. We define $\deg_{x_i} 0 = \deg 0 = -\infty$ for $i = 1, \ldots, m$.

We denote the rising factorial by

$$a^{\overline{n}} = \prod_{i=0}^{n-1} (a + i).$$

2   Equations with Polynomial Coefficients

Theorem 1. *There is no algorithm to decide whether an arbitrary linear partial differential resp. difference operator L with polynomial coefficients in an arbitrary number m of variables, of the form (1) resp. (2), has a non-zero polynomial solution.*

Proof. Let $P(t_1, \ldots, t_m) \in \mathbb{Z}[t_1, \ldots, t_m]$ be arbitrary. For $i = 1, \ldots, m$ write $\theta_i = x_i D_i$ and $\sigma_i = x_i \Delta_i$. Then

$$\theta_i(x_1^{n_1} \cdots x_m^{n_m}) \;=\; n_i x_1^{n_1} \cdots x_m^{n_m} \tag{3}$$

and

$$\sigma_i(x_1^{\overline{n}_1} \cdots x_m^{\overline{n}_m}) \;=\; n_i x_1^{\overline{n}_1} \cdots x_m^{\overline{n}_m}, \tag{4}$$

for $i = 1, \ldots, m$. Define an operator L of the form (1) resp. (2) by setting $L = P(\theta_1, \ldots, \theta_m)$ in the differential case, and $L = P(\sigma_1, \ldots, \sigma_m)$ in the difference case. Let $f(x_1, \ldots, x_m) \in K[x_1, \ldots, x_m]$ be a polynomial over K. From (3) and (4) it follows that L annihilates f iff it annihilates each term of f separately, so L has a non-zero polynomial solution iff it has a monomial solution (where in the difference case we assume that the polynomial f is expanded in terms of the rising factorial basis). But we have

$$L(x_1^{n_1} \cdots x_m^{n_m}) = P(n_1, \ldots, n_m) x_1^{n_1} \cdots x_m^{n_m}$$

in the differential case, and

$$L(x_1^{\overline{n}_1} \cdots x_m^{\overline{n}_m}) = P(n_1, \ldots, n_m) x_1^{\overline{n}_1} \cdots x_m^{\overline{n}_m}$$

in the difference case. So L has a monomial solution iff there exist $n_1, \ldots, n_m \in \mathbb{N}$ such that $P(n_1, \ldots, n_m) = 0$. Hence an algorithm for deciding existence of non-zero polynomial solutions of linear partial differential or difference operators with polynomial coefficients would give rise to an algorithm for deciding existence of non-negative integral solutions of polynomial equations with integral coefficients, in contradiction to the DMPR theorem.

Remark 1. *In [9, Thm. 4.11], it is shown that there is no algorithm for deciding existence of formal power series solutions of an inhomogeneous partial differential equations with polynomial coefficients and right-hand side equal to 1 (see also Problem 13 in [15, p. 62] and Problem 3 in [18, p. 27]). Even though the same polynomial P in θ_i is used in the proof of Theorem 1 as in the proof of [9, Thm. 4.11], it is not at all clear whether the former follows from the latter.*

Remark 2. *Since the DMPR theorem holds for any fixed number $m \geq 9$ of variables as well (cf. [17]), the same is true of Theorem 1.*

3   Equations with Constant Coefficients

In this section we assume that L is an operator of the form (1), (2) with coefficients $a_\mu \in K$.

For $i = 1, \ldots, m$, let

$$\delta_i = \begin{cases} D_i, & \text{in the differential case,} \\ \Delta_i, & \text{in the difference case.} \end{cases}$$

Lemma 1. *Let $L \in K[\delta_1, \ldots, \delta_m]$ and let the equation*

$$L(y) = 0 \tag{5}$$

have a polynomial solution of degree $k \geq 0$. Then this equation has a polynomial solution of degree j for $j = 0, 1, \ldots, k$.

Proof. By induction on j from k down to 0.

$j = k$: This holds by assumption.

$0 \leq j \leq k-1$: By inductive hypothesis, equation (5) has a polynomial solution $y(x) = p(x_1, \ldots, x_m)$ of degree $j + 1$. Let $t = cx_1^{n_1} \cdots x_m^{n_m}$ be a term of the polynomial p such that $\deg t = j + 1$, and let $i \in \{1, \ldots, m\}$ be such that $\deg_{x_i} t > 0$. Then $\delta_i(p)$ has the desired properties. Indeed, $\deg \delta_i(p) = \deg p - 1 = j$ and, since operators with constant coefficients commute, $L(\delta_i(p)) = \delta_i(L(p)) = \delta_i(0) = 0$.

Theorem 2. *Let $m \geq 2$, and let $L \in K[\delta_1, \ldots, \delta_m]$ be a linear partial differential or difference operator with constant coefficients. The following assertions are equivalent:*

(a) For each $k \in \mathbb{N}$, L has a polynomial solution of degree k.
(b) L has a non-zero polynomial solution.
(c) $L(1) = 0$.

Proof. (a) $\Rightarrow$ (b): Obvious.

(b) $\Rightarrow$ (c): Assume that L has a non-zero polynomial solution $p(x)$. Then $\deg p \geq 0$, and by Lemma 1, L has a solution of degree 0. Hence $L(1) = 0$ as well.

(c) $\Rightarrow$ (a): It is well known that, in m variables, the number of monomials of degree d is $\binom{d+m-1}{m-1}$, and the number of monomials of degree at most d is $\binom{d+m}{m}$. Set

$$d = \binom{k+1}{2}$$

and denote by $\mathcal{M}$ the set of all monomials in the variables $x_1, \ldots, x_m$ of degrees $k, k+1, \ldots, d$. Then

$$|\mathcal{M}| = \binom{d+m}{m} - \binom{k-1+m}{m}.$$

Let $\mathcal{P} = L(\mathcal{M})$. From (c) it follows that the free term c_0 of L is equal to 0, hence $\deg L(t) < \deg t$ for any $t \in \mathcal{M}$, and so the degrees of polynomials in $\mathcal{P}$ do not exceed $d - 1$.

If $\mathcal{M}$ contains two distinct monomials m_1 and m_2 such that $L(m_1) = L(m_2)$ then $p = m_1 - m_2$ is a non-zero polynomial solution of L of degree at least k.

Otherwise, L is injective on $\mathcal{M}$, and so $|\mathcal{P}| = |\mathcal{M}|$. From $d + 1 > k(k+1)/2$, $d \geq k$ and $m \geq 2$ it follows that

$$
\begin{aligned}
(d+1)^{\overline{m}} - d^{\overline{m}} &= m(d+1)^{\overline{m-1}} \\
&= m(d+1)\,(d+2)^{\overline{m-2}} \\
&> m\,\frac{k(k+1)}{2}\,(k+2)^{\overline{m-2}} \\
&\geq k^{\overline{m}},
\end{aligned}
$$

hence $(d+1)^{\overline{m}} - k^{\overline{m}} > d^{\overline{m}}$. Dividing this by $m!$ we see that

$$
\begin{aligned}
|\mathcal{P}| \;=\; |\mathcal{M}| &= \binom{d+m}{m} - \binom{k-1+m}{m} \\
&> \binom{d-1+m}{m}.
\end{aligned}
$$

Since the dimension of the space of polynomials of degrees at most $d - 1$ is $\binom{d-1+m}{m}$, it follows that the set $\mathcal{P}$ is linearly dependent. Hence there is a non-trivial linear combination p of the monomials in $\mathcal{M}$ such that $L(p) = 0$. Clearly, p is a non-zero polynomial solution of L of degree at least k.

In either case (if L is injective on $\mathcal{M}$ or not) we have obtained a non-zero polynomial solution of L of degree at least k. By Lemma 1 it follows that L has a non-zero polynomial solution of degree k.

4 q-Difference Equations with Constant Coefficients

The question of decidability of the existence of non-zero polynomial solutions of an arbitrary q-difference equation with polynomial coefficients is still open. In this section we consider equations with coefficients from K, assuming that $K = k(q)$ where k is a subfield of K and q is transcendental over k.

We write $Q = (Q_1, \ldots, Q_m)$ for partial q-shift operators where

$$
Q_i f(x) = f(x_1, \ldots, qx_i, \ldots, x_m),
$$

so that a partial linear q-difference operator with constant coefficients of order at most r is written as

$$
L = \sum_{|\mu| \leq r} a_\mu Q^\mu \tag{6}
$$

with $a_\mu \in K$. Clearly, for multiindices μ and α,

$$Q^\mu x^\alpha = Q_1^{\mu_1} \cdots Q_m^{\mu_m} x_1^{\alpha_1} \cdots x_m^{\alpha_m}$$
$$= Q_1^{\mu_1} x_1^{\alpha_1} \cdots Q_m^{\mu_m} x_m^{\alpha_m}$$
$$= (q^{\mu_1} x_1)^{\alpha_1} \cdots (q^{\mu_m} x_m)^{\alpha_m}$$
$$= q^{\mu_1 \alpha_1 + \cdots + \mu_m \alpha_m} x_1^{\alpha_1} \cdots x_m^{\alpha_m}$$
$$= q^{\mu \cdot \alpha} x^\alpha. \tag{7}$$

Lemma 2. *An operator L of the form (6) has a nonzero polynomial solution iff it has a monomial solution.*

Proof. If L has a monomial solution x^α, then x^α is also a non-zero polynomial solution of L.

Conversely, assume that $p(x) \in K[x]$ is a non-zero polynomial solution of L. Write

$$p(x) = \sum_\alpha c_\alpha x^\alpha$$

where only finitely many c_α are non-zero, and define its support by

$$\operatorname{supp} p = \{\alpha \in \mathbb{N}^m; \ c_\alpha \neq 0\}.$$

Then

$$L(p) = \sum_\mu a_\mu \sum_\alpha c_\alpha Q^\mu x^\alpha$$
$$= \sum_\mu a_\mu \sum_\alpha c_\alpha q^{\mu \cdot \alpha} x^\alpha \quad \text{(by (7))}$$
$$= \sum_\alpha c_\alpha \left(\sum_\mu a_\mu q^{\mu \cdot \alpha} \right) x^\alpha,$$

hence from $L(p) = 0$ it follows that

$$\sum_\mu a_\mu q^{\mu \cdot \alpha} = 0$$

whenever $c_\alpha \neq 0$. Therefore, by (7),

$$L(x^\alpha) = \sum_\mu a_\mu Q^\mu x^\alpha = \sum_\mu a_\mu q^{\mu \cdot \alpha} x^\alpha = 0$$

for all such α, so x^α is a monomial solution of L for each $\alpha \in \operatorname{supp} p$.

By clearing denominators in the equation $L(y) = 0$, we can assume that the coefficients of L are in $k[q]$, hence we can rewrite

$$L = \sum_\mu \sum_i a_{\mu,i} q^i Q^\mu \tag{8}$$

where only finitely many $a_{\mu,i} \in k$ are non-zero. Define

$$\operatorname{supp} L = \{(\mu, i) \in \mathbb{N}^{m+1}; \ a_{\mu,i} \neq 0\}.$$

Let P be a partition of $\operatorname{supp} L$. We call such a partition *balanced* if

$$\sum_{(\mu,i)\in B} a_{\mu,i} = 0$$

for every block $B \in P$. To any $\alpha \in \mathbb{N}^m$ we assign the partition $P_{L,\alpha}$ of $\operatorname{supp} L$ induced by the equivalence relation

$$(\mu, i) \sim (\nu, j) \quad \text{iff} \quad \mu \cdot \alpha + i = \nu \cdot \alpha + j.$$

Lemma 3. $L(x^\alpha) = 0$ *iff* $P_{L,\alpha}$ *is balanced.*

Proof.

$$L(x^\alpha) = \sum_{(\mu,i)\in\operatorname{supp} L} a_{\mu,i} q^i Q^\mu x^\alpha$$

$$= \sum_{(\mu,i)\in\operatorname{supp} L} a_{\mu,i} q^{\mu\cdot\alpha+i} x^\alpha,$$

hence $L(x^\alpha) = 0$ iff $\sum_{(\mu,i)\in\operatorname{supp} L} a_{\mu,i} q^{\mu\cdot\alpha+i} = 0$. Since q is transcendental over k, the latter equality holds iff $\sum_{(\mu,i)\in B} a_{\mu,i} = 0$ for every block $B \in P_{L,\alpha}$, i.e., iff $P_{L,\alpha}$ is balanced.

Corollary 1. L *in (8) has a non-zero polynomial solution iff there is an* $\alpha \in \mathbb{N}^m$ *such that* $P_{L,\alpha}$ *is balanced.*

Proof. This follows from Lemmas 2 and 3.

Corollary 2. *If* L *in (8) has a non-zero polynomial solution then* $\sum_\mu a_\mu = 0$.

Proof. This follows from Corollary 1 since if $P_{L,\alpha}$ is balanced then $\sum_\mu a_\mu = 0$.

¿From Corollary 1 we obtain the following algorithm for deciding existence of non-zero polynomial solutions of L in (8):

> **for** each balanced partition P of $\operatorname{supp} L$ **do**
> let S be the system of $|\operatorname{supp} L|$ linear equations
>
> $$\mu \cdot \alpha + i = v_B, \quad (\mu, i) \in B \in P$$
>
> for the unknown vectors α and $v = (v_B)_{B\in P}$
> **if** S has a solution (α, v) with $\alpha \in \mathbb{N}^m$ **then**
> return "yes" and stop
> return "no".

Corollary 3. *The problem of existence of non-zero polynomial solutions of partial linear q-difference operators with constant coefficients is decidable.*

Note that one can convert the above decision algorithm into a procedure for providing a finite description of a (possibly infinite) basis for the space of all polynomial solutions of equation $L(y) = 0$.

The following simple examples demonstrate that, in contrast with the differential and difference cases, there are partial linear q-difference equations with constant coefficients such that the dimension of their space of polynomial solutions is: a) infinite, b) finite positive, c) zero.

Example 1. *Let $L_1 = Q_1^2 Q_2 + q Q_1 Q_2^2 - 2q^2 Q_2^3$. Then*

$$L_1(x_1^{\alpha_1} x_2^{\alpha_2}) = (q^{2\alpha_1 + \alpha_2} + q^{\alpha_1 + 2\alpha_2 + 1} - 2q^{3\alpha_2 + 2}) x_1^{\alpha_1} x_2^{\alpha_2}$$

and supp $L_1 = \{(2,1,0),(1,2,1),(0,3,2)\}$. The only balanced partition of this set is the single-block partition $P = \{\text{supp } L_1\}$, and we obtain the system of linear equations

$$2\alpha_1 + \alpha_2 = \alpha_1 + 2\alpha_2 + 1 = 3\alpha_2 + 2$$

for α_1 and α_2. This system has infinitely many non-negative integer solutions of the form $\alpha_1 = t + 1$, $\alpha_2 = t$ where $t \in \mathbb{N}$. Therefore, every non-zero linear combination of monomials of the form $x_1^{t+1} x_2^t$ where $t \in \mathbb{N}$, is a non-zero polynomial solution of the operator L_1.

Example 2. *Let $L_2 = Q_1^4 Q_2 + Q_1^2 Q_2^3 - 2q^2 Q_1^3$. Then*

$$L_2(x_1^{\alpha_1} x_2^{\alpha_2}) = (q^{4\alpha_1 + \alpha_2} + q^{2\alpha_1 + 3\alpha_2} - 2q^{3\alpha_1 + 2}) x_1^{\alpha_1} x_2^{\alpha_2}$$

and supp $L_2 = \{(4,1,0),(2,3,0),(3,0,2)\}$. Again the only balanced partition of this set is the single-block partition, and we obtain the system of linear equations

$$4\alpha_1 + \alpha_2 = 2\alpha_1 + 3\alpha_2 = 3\alpha_1 + 2$$

for α_1 and α_2. The only solution of this system is $\alpha_1 = \alpha_2 = 1$, so the operator L_2 has a 1-dimensional space of polynomial solutions spanned by $x_1 x_2$.

Example 3. *Let $L_3 = Q_1^2 Q_2 + Q_1 Q_2^2 - 2q Q_2^3$. Then*

$$L_3(x_1^{\alpha_1} x_2^{\alpha_2}) = (q^{2\alpha_1 + \alpha_2} + q^{\alpha_1 + 2\alpha_2} - 2q^{3\alpha_2 + 1}) x_1^{\alpha_1} x_2^{\alpha_2}$$

and supp $L_3 = \{(2,1,0),(1,2,0),(0,3,1)\}$. Once again the only balanced partition of this set is the single-block partition, and we obtain the system of linear equations

$$2\alpha_1 + \alpha_2 = \alpha_1 + 2\alpha_2 = 3\alpha_2 + 1$$

for α_1 and α_2. Since this system has no solution, the operator L_3 has no non-zero polynomial solution.

5   Conclusion

In this paper, we have investigated the computational problem of existence of non-zero polynomial solutions of linear partial differential and difference equations with polynomial coefficients. We have shown that the problem is algorithmically undecidable. This means that there is no hope of having a general algorithm for deciding existence of such solutions in a computer algebra system now or ever in the future.

However, we have shown that the existence problem is decidable in the case of partial linear differential or difference equations with constant coefficients: such an equation $L(y) = 0$ has non-zero polynomial solutions iff $L(1) = 0$. Moreover, when the latter condition is satisfied, this equation has polynomial solutions of any desired degree. A number of methods exist to search for such solutions efficiently (see, e.g., [19,21]).

For partial equations with constant coefficients in the q-difference case which is of interest in combinatorics, we have formulated and proved a necessary condition for existence of non-zero polynomial solutions: $(L(1))(1) = 0$ (note that $L(1)$ is a polynomial in q). We have also shown that when the latter condition is satisfied, the dimension of the space of polynomial solutions in some particular cases can be finite and even zero (then no non-zero polynomial solutions exist). An explicit description of this space can be obtained algorithmically, and the corresponding algorithm is straightforward to implement in any computer algebra system.

The following interesting problems remain open:

1. (Un)decidability of existence of non-zero polynomial solutions of a given linear partial differential or difference equation with polynomial coefficients when the number of variables m is between 2 and 8.

2. (Un)decidability of existence of non-zero polynomial solutions of a given linear partial q-difference equation with polynomial coefficients (both the general problem when the number m of variables is arbitrary, and the problems related to particular numbers of variables).

Problem 1 seems to be very hard since the problem of solvability of Diophantine equations in m variables with m between 2 and 8 is still open (cf. [17]). Concerning Problem 2, note that in the ordinary case ($m = 1$), certain existence problems in the q-difference case are decidable although the analogous problems in the differential and difference cases are not (see, e.g., [3]). An example of an open problem which might be easier than Problems 1 or 2 is the existence problem of non-zero polynomial solutions for q-*differential* equations. We will continue to pursue this line of inquiry.

Acknowledgements. The authors are grateful to M. Kauers for kindly providing a version of the proof of Theorem 1 in the differential case (as mentioned in the Introduction), to S. P. Tsarev for interesting and helpful discussions, and to several anonymous referees for their valuable remarks and references to the literature.

References

1. Abramov, S.A.: Rational solutions of linear difference and differential equations with polynomial coefficients. U.S.S.R. Comput. Math. and Math. Phys. 29(6), 7–12 (1989)
2. Abramov, S.A.: Rational solutions of linear difference and q-difference equations with polynomial coefficients. Programming and Comput. Software 21(6), 273–278 (1995)
3. Abramov, S.A.: On some decidable and undecidable problems related to q-difference equations with parameters. In: Proc. ISSAC 2010, pp. 311–317 (2010)
4. Abramov, S.A., Bronstein, M., Petkovšek, M.: On polynomial solutions of linear operator equations. In: Proc. ISSAC 1995, pp. 290–296 (1995)
5. Andrews, G.E.: The Theory of Partitions. Encyclopedia of Mathematics and its Applications. Addison-Wesley, Reading Mass. (1976)
6. Andrews, G.E.: q-Series: Their Development and Application in Analysis, Number Theory, Combinatorics, Physics, and Computer Algebra. CBMS Regional Conference Series, vol. 66. AMS, Providence (1986)
7. Barkatou, M.A.: A fast algorithm to compute the rational solutions of systems of linear differential equations. RR 973-M– Mars 1997, IMAG–LMC, Grenoble (1997)
8. Barkatou, M.A.: Rational solutions of systems of linear difference equations. J. Symbolic Comput. 28(4-5), 547–567 (1999)
9. Denef, J., Lipshitz, L.: Power series solutions of algebraic differential equations. Math. Ann. 267(2), 213–238 (1984)
10. van Hoeij, M.: Rational solutions of linear difference equations. In: Proc. ISSAC 1998, pp. 120–123 (1998)
11. Kauers, M., Schneider, C.: Partial denominator bounds for partial linear difference equations. In: Proc. ISSAC 2010, pp. 211–218 (2010)
12. Kauers, M., Schneider, C.: A refined denominator bounding algorithm for multivariate linear difference equations. In: Proc. ISSAC 2011, pp. 201–208 (2011)
13. Leykin, A.: D-modules for Macaulay 2. In: Mathematical Software, Beijing, pp. 169–179. World Sci. Publ., River Edge (2002)
14. Matiyasevich, Y.V.: Hilbert's Tenth Problem. MIT Press, Cambridge (1993)
15. Matiyasevich, Y.V.: On Hilbert's tenth problem. PIMS Distinguished Chair Lectures (2000), http://www.mathtube.org/lecture/notes/hilberts-tenth-problem
16. Oaku, T., Takayama, N., Tsai, H.: Polynomial and rational solutions of holonomic systems. J. Pure Appl. Algebra 164(1-2), 199–220 (2001)
17. Pheidas, T., Zahidi, K.: Undecidability of existential theories of rings and fields: A survey. In: Hilbert's Tenth Problem: Relations with Arithmetic and Algebraic Geometry. Contemp. Math., vol. 270, pp. 49–105 (2000)
18. Sadovnikov, A.: Undecidable problems about polynomials: Around Hilbert's 10th problem. Lecture notes (2007), http://www14.informatik.tu-muenchen.de/konferenzen/Jass07/courses/1/Sadovnikov/Sadovnikov_Paper.pdf
19. Seiler, W.M.: Involution. The formal theory of differential equations and its applications in computer algebra. In: Algorithms and Computation in Mathematics, vol. 24. Springer, Berlin (2010)
20. Smith, S.P.: Polynomial solutions to constant coefficient differential equations. Trans. Amer. Math. Soc. 329(2), 551–569 (1992)
21. Sturmfels, B.: Solving Systems of Polynomial Equations. CBMS Regional Conferences Series, vol. 97. Amer. Math. Soc., Providence (2002)
22. Tsai, H., Walther, U.: Computing homomorphisms between holonomic D-modules. J. Symbolic Comput. 32(6), 597–617 (2001)

An Algebraic Characterization
of Rainbow Connectivity

Prabhanjan Ananth and Ambedkar Dukkipati

Department of Computer Science and Automation
Indian Institute of Science, Bangalore 560012, India
{prabhanjan,ambedkar}@csa.iisc.ernet.in

Abstract. The use of algebraic techniques to solve combinatorial problems is studied in this paper. We formulate the rainbow connectivity problem as a system of polynomial equations. We first consider the case of two colors for which the problem is known to be hard and we then extend the approach to the general case. We also present a formulation of the rainbow connectivity problem as an ideal membership problem.

Keywords: Graphs, NulLA alogirithm, ideal membership.

1 Introduction

The use of algebraic concepts to solve combinatorial optimization problems has been a fascinating field of study explored by many researchers in theoretical computer science. The combinatorial method introduced by Noga Alon [1] offered a new direction in obtaining structural results in graph theory. Lovász [2], De Loera [3] and others formulated popular graph problems like vertex coloring, independent set as a system of polynomial equations in such a way that solving the system of equations is equivalent to solving the combinatorial problem. This formulation ensured the fact that the system has a solution if and only if the corresponding instance has a "yes" answer.

Solving system of polynomial equations is a well studied problem with a wealth of literature on this topic. It is well known that solving system of equations is a notoriously hard problem. De Loera et al. [4] proposed the NulLA approach (Nullstellensatz Linear Algebra) which used Hilbert's Nullstellensatz to determine the feasibility among a system of equations. This approach was further used to characterize some classes of graphs based on degrees of the Nullstellensatz certificate.

In this work, we study the algebraic characterization of a relatively new concept in graph theory termed as rainbow connectivity. We first show how to model the rainbow connectivity problem as an ideal membership problem and then using a result from [3], we propose an algorithm to solve the rainbow connectivity problem. We then show how to encode the k-rainbow connectivity problem as a system of polynomial equations for the case when $k = 2$. We then show how to extend this for any constant k.

V.P. Gerdt et al. (Eds.): CASC 2012, LNCS 7442, pp. 12–21, 2012.

In Section 2, we review the basics of encoding of combinatorial problems as systems of polynomial equations. Further, we describe NulLA along with the preliminaries of rainbow connectivity. In Section 3, we propose a formulation of the rainbow connectivity problem as an ideal membership problem. We then present encodings of the rainbow connectivity problem as a system of polynomial equations in Section 4.

2   Background and Preliminaries

The encoding of well known combinatorial problems as system of polynomial equations is described in this section. The encoding schemes of the vertex coloring and the independent set problem is presented. Encoding schemes of well known problems like Hamiltonian cycle problem, MAXCUT, SAT and others can be found in [5]. The term encoding is formally defined as follows:

Definition 1. *Given a language L, if there exists a polynomial-time algorithm A that takes an input string I, and produces as output a system of polynomial equations such that the system has a solution if and only if $I \in L$, then we say that the system of polynomial equations encodes I.*

It is a necessity that the algorithm that transforms an instance into a system of polynomial equations has a polynomial running time in the size of the instance I. Else, the problem can be solved by brute force and trivial equations $0 = 0$ ("yes" instance) or $1 = 0$ ("no" instance) can be output. Further since the algorithm runs in polynomial time, the size of the output system of polynomial equations is bounded above by a polynomial in the size of I. The encodings of vertex coloring and stable set problems are presented next.

We use the following notation throughout this paper. Unless otherwise mentioned all the graphs $G = (V, E)$ have the vertex set $V = \{v_1, \ldots, v_n\}$ and the edge set $E = \{e_1, \ldots, e_m\}$. The notation $v_{i_1} - v_{i_2} - \cdots - v_{i_s}$ is used to denote a path $\mathcal{P}$ in G, where $e_{i_1} = (v_{i_1}, v_{i_2}), \ldots, e_{i_{s-1}} = (v_{i_{s-1}}, v_{i_s}) \in E$. The path $\mathcal{P}$ is also denoted by $v_{i_1} - e_{i_1} - \cdots - e_{i_{s-1}} - v_{i_s}$ and $v_{i_1} - \mathcal{P} - v_{i_s}$.

2.1 k-Vertex Coloring and Stable Set Problem

The vertex coloring problem is one of the most popular problems in graph theory. The minimum number of colors required to color the vertices of the graph such that no two adjacent vertices get the same color is termed as the vertex coloring problem. We consider the decision version of the vertex coloring problem. The k-vertex coloring problem is defined as follows: Given a graph G, does there exist a vertex coloring of G with k colors such that no two adjacent vertices get the same color. There are a quite a few encodings known for the k-vertex colorability problem. We present one such encoding given by Bayer [6]. The polynomial ring under consideration is $\Bbbk[x_1, \ldots, x_n]$.

Theorem 1. *A graph $G = (V, E)$ is k-colorable if and only if the following zero-dimensional system of equations has a solution:*

$$x_i^k - 1 = 0, \ \forall v_i \in V,$$

$$\sum_{d=0}^{k-1} x_i^{k-1-d} x_j^d = 0, \ \forall (v_i, v_j) \in E.$$

Proof Idea. If the graph G is k-colorable, then there exists a proper k-coloring of graph G. Denote these set of k colors by k^{th} roots of unity. Consider a point $p \in \Bbbk^n$ such that i^{th} co-ordinate of p (denoted by $p^{(i)}$) is the same as the color assigned to the vertex x_i. The equations corresponding to each vertex (of the form $x_i^k - 1 = 0$) are satisfied at point p. The equations corresponding to the edges can be rewritten as

$$\frac{x_i^k - x_j^k}{x_i - x_j} = 0.$$

Since $x_i^k = x_j^k = 1$ and $x_i \neq x_j$, even the edge equation is satisfied at p.

Assume that the system of equations have a solution p. It can be seen that p cannot have more than k distinct co-ordinates. We color the vertices of the graph G as follows: color the vertex v_i with the value $p^{(i)}$. It can be shown that if the system is satisfied then in the edge equations, x_i and x_j need to take different values. In other words, if (v_i, v_j) is an edge then $p^{(i)}$ and $p^{(j)}$ are different. Hence, the vertex coloring of G is a proper coloring. □

A stable set (independent set) in a graph is a subset of vertices such that no two vertices in the subset are adjacent. The stable set problem is defined as the problem of finding the maximum stable set in the graph. The cardinality of the largest stable set in the graph is termed as the independence number of G. The encoding of the decision version of the stable set problem is presented. The decision version of the stable set problem deals with determining whether a graph G has a stable set of size at least k. The following result is due to Lovász [2].

Lemma 1. *A graph $G = (V, E)$ has an independent set of size $\geq k$ if and only if the following zero-dimensional system of equations has a solution*

$$x_i^2 - x_i = 0, \ \forall i \in V,$$

$$x_i x_j = 0, \ \forall \{i, j\} \in E,$$

$$\sum_{i=1}^{n} x_i - k = 0 .$$

The number of solutions equals the number of distinct independent sets of size k.

The proof of the above result can be found in [5].

2.2   NulLA Algorithm

De Loera et al. [4] proposed the Nullstellensatz Linear Algebra Algorithm (NulLA) which is an approach to ascertain whether the polynomial system has a solution or not. Their method relies on the one of the most important theorems in algebraic geometry, namely the Hilbert Nullstellensatz. The Hilbert Nullstellensatz theorem states that the variety of an ideal is empty over an algebraically closed field iff the element 1 belongs to the ideal. More formally,

Theorem 2. *[7] Let $\mathfrak{a}$ be a proper ideal of $\Bbbk[x_1,\ldots,x_n]$. If $\Bbbk$ is algebraically closed, then there exists $(a_1,\ldots,a_n) \in \Bbbk^n$ such that $f(a_1,\ldots,a_n) = 0$ for all $f \in \mathfrak{a}$.*

Thus, to determine whether a system of equations $f_1 = 0, \ldots, f_s = 0$ has a solution or not is the same as determining whether there exists polynomials h_i where $i \in \{1,\ldots,s\}$ such that $\sum_{i=1}^{s} h_i f_i = 1$. Denote the quantity $\max_{1 \leq i \leq s}(\deg(f_i))$ by d. A result by Kollár [8] shows that the degree of the coefficient polynomials h_i can be bounded above by $\{\max(3,d)\}^n$ where n is the number of indeterminates. Hence, each h_i can be expressed as a sum of monomials of degree at most $\{\max(3,d)\}^n$, with unknown coefficients. By expanding the summation $\sum_{i=1}^{s} h_i f_i$, a system of linear equations is obtained with the unknown coefficients being the variables. Solving this system of linear equations will yield us the polynomials h_i such that $\sum_{i=1}^{s} h_i f_i = 1$. The equation $\sum_{i=1}^{s} h_i f_i = 1$ is known as Nullstellensatz certificate and is said to be of degree d if $\max_{1 \leq i \leq s}\{\deg(h_i)\} = d$. There have been efforts to determine the bounds on the degree of the Nullstellensatz certificate which in turn has an impact on the running time of NulLA algorithm. The description of the NulLA algorithm can be found in [5]. The running time of the algorithm depends on the degree bounds on the polynomials in the Nullstellensatz certificate. It was shown in [9] that if $f_1 = 0, \ldots, f_s = 0$ is an infeasible system of equations then there exists polynomials $h_1, \ldots, h_s$ such that $\sum_{i=1}^{s} h_i f_i = 1$ and $\deg(h_i) \leq n(d-1)$ where $d = \max\{\deg(f_i)\}$. Thus with this bound, the running time of the above algorithm in the worst case is exponential in $n(d-1)$. Even though this is still far being practical, for some special cases of polynomial systems this approach seems to be promising. More specifically this proved to be beneficial for the system of polynomial equations arising from combinatorial optimization problems [5]. Also using NulLA, polynomial-time procedures were designed to solve the combinatorial problems for some special class of graphs [10].

2.3   Rainbow Connectivity

The concept of rainbow connectivity was introduced by Chartrand et. al. [11] as a measure of strengthening connectivity. Consider an edge colored graph G. A rainbow path is a path consisting of distinctly colored edges. The graph G is said to be rainbow connected if between every two vertices there exists a rainbow path. The least number of colors required to edge color the graph G such that G is rainbow connected is called the rainbow connection number of the graph, denoted by $rc(G)$. The problem of determining $rc(G)$ for a graph G is termed as the rainbow connectivity problem. The corresponding decision version, termed as the k-rainbow connectivity problem is defined as follows: Given a graph G, decide whether $rc(G) \leq k$. The k-rainbow connectivity problem is NP-complete even for the case $k = 2$.

3   Rainbow Connectivity as an Ideal Membership Problem

Combinatorial optimization problems like vertex coloring [3,12] were formulated as a membership problem in polynomial ideals. The general approach is to associate a polynomial to each graph and then consider an ideal which contains all and only those graph polynomials that have some property (for example, chromatic number of the corresponding graph is less than or equal to k). To test whether the graph has a required property, we just need to check whether the corresponding graph polynomial belongs to the ideal. In this section, we describe a procedure of solving the k-rainbow connectivity problem by formulating it as an ideal membership problem. By this, we mean that a solution to the ideal membership problem yields a solution to the k-rainbow connectivity problem. We restrict our attention to the case when $k = 2$.

In order to formulate the 2-rainbow connectivity problem as a membership problem, we first consider an ideal $I_{m,3} \subset \mathbb{Q}[x_{e_1}, \ldots, x_{e_m}]$. Then the problem of deciding whether the given graph G can be rainbow connected with 2 colors or not is reduced to the problem of deciding whether a polynomial f_G belongs to the ideal $I_{m,3}$ or not. The ideal $I_{m,3}$ is defined as the ideal vanishing on $V_{m,3}$, where $V_{m,3}$ is defined as the set of all points which have at most 2 distinct coordinates. The following theorem was proved by De Loera [3]:

Theorem 3. *The set of polynomials*

$$\mathcal{G}_{m,3} = \{ \prod_{1 \leq r < s \leq 3} (x_{e_{i_r}} - x_{e_{i_s}}) \mid 1 \leq i_1 < i_2 < i_3 \leq m \}$$

is a universal Gröbner basis[1] for the ideal $I_{m,3}$.

We now associate a polynomial f_G to each graph G such that f_G belongs to the ideal $I_{m,3}$ if and only if the rainbow connection number of the graph G is at least 3. Assume that the diameter of G is at most 2, because if not we have $rc(G) \geq 3$. We first define the path polynomials for every pair of vertices $(v_i, v_j) \in V \times V$ as follows: If v_i and v_j are adjacent then $P_{i,j} = 1$, else

$$P_{i,j} = \sum_{e_a, e_b \in E:\ v_i - e_a - e_b - v_j \in G} (x_{e_a} - x_{e_b})^2 .$$

---

[1] A set of generators of an ideal is said to be a universal Gröbner basis if it is a Gröbner basis with respect to every term order.

The polynomial f_G is nothing but the product of path polynomials between any pair of vertices. Formally, f_G is defined as follows:

$$f_G = \prod_{v_i, v_j \in V; \ i < j} P_{i,j} .$$

Note that f_G can be computed in polynomial time.

Theorem 4. *The polynomial $f_G \in I_{m,3}$ if and only if $rc(G) \geq 3$.*

Proof. To prove the theorem, it is enough to show that $\forall p \in V_{m,3}$, $f_G(p) = 0$ if and only if rainbow connection number of G is at least 3. Assume that the rainbow connection number of G is at most 2. This means that there exists an edge coloring of the graph with two colors such that the graph is 2-rainbow connected. We can visualize this coloring of edges as a tuple $(c_1, \ldots, c_m)$ where $c_i \in \mathbb{Q}$ and the edge e_i is given the color c_i. It can be seen that the point $p = (c_1, \ldots, c_m)$ belongs to $V_{m,3}$. We claim that $f_G(p) \neq 0$. For that, we show that $P_{i,j}(p) \neq 0$ for all $(v_i, v_j) \in V \times V$. Assume that v_i and v_j are not adjacent (this is because $P_{i,j}(p) \neq 0$ for adjacent pair of vertices (v_i, v_j)). Since G is rainbow connected, there is a rainbow path from v_i to v_j and let e_a, e_b be the two edges in this path. Correspondingly, $(c_a - c_b)$ is non-zero and hence $(c_a - c_b)^2$ is positive. This implies that $P_{i,j}(p) \neq 0$ for every pair of vertices (v_i, v_j). Hence, $f_G(p)$ is non-zero.
Assume that $f_G(p) \neq 0$ for some $p = (c_1, \ldots, c_m) \in V_{m,3}$. First, we consider the case when p has no distinct coordinates. In this case, it can be seen that $P_{i,j}$ has to be 1 for every $i < j$ and $i, j \in \{1, \ldots, n\}$. This further means that the graph is a complete graph in which case a single color suffices to rainbow connect the graph. Henceforth, we restrict our attention to the case when p has exactly two distinct coordinates. Using p, we color the edges of the graph G with two colors such that G is rainbow connected. Assume without loss of generality that b and r are the only two values taken by the entries in p. Color the edges of G as follows: If $c_i = b$ then color the edge e_i with blue else color the edge e_i with red. Since, $f_G(p) \neq 0$ we have $P_{i,j}(p) \neq 0$ for all $i, j \in \{1, \ldots, n\}$. Consider a non adjacent pair of vertices (v_i, v_j). This implies that there exists a and b such that $(x_{e_a} - x_{e_b})^2$ is in the support of $P_{i,j}$ and $(c_a - c_b)^2$ is non-zero. Correspondingly, the path from v_i to v_j containing the edges e_a and e_b is a rainbow path since e_a and e_b are colored distinctly. Thus, G is rainbow connected which implies that $rc(G) \leq 2$.   □

The above characterization gives us a computational algebraic procedure to decide whether the rainbow connection of a graph is at most 2 or not.

1. Given a graph G, find its corresponding polynomial f_G.
2. Divide f_G by $\mathcal{G}_{m,3}$.
3. If the division algorithm gives a non-zero remainder then the rainbow connection number of the graph is at most 2 else $rc(G) \geq 3$.

4   Encoding of Rainbow Connectivity

Consider the polynomial ring $\mathbb{F}_2[x_{e_1}, \ldots, x_{e_m}]$. As before, assume that the diameter of G is at most 2. We present an encoding of the 2-rainbow connectivity problem as a system of polynomial equations S defined as follows:

$$\prod_{e_a, e_b \in E : v_i - e_a - e_b - v_j \in G} (x_{e_a} + x_{e_b} + 1) = 0; \ \forall i, j \in \{1, \ldots, n\}, i < j, \ (v_i, v_j) \notin E$$

If all pairs of vertices are adjacent (as in the case of clique), we have the trivial system $0 = 0$.

Proposition 1. *The rainbow connection number of G is at most 2 if and only if S has a solution in $\mathbb{F}_2^m$.*

Proof. Let $p = (c_1, \ldots, c_m) \in \mathbb{F}_2^m$ be a solution to S. Consider the edge coloring $\chi : E \to \{\text{blue, red}\}$ defined as follows: $\chi(e_i) = $ blue if $c_i = 1$ else $\chi(e_i) = $ red. Now, consider a pair of vertices $(v_i, v_j) \notin E$. Since the equation corresponding to (i, j) is satisfied at p, there exists a and b such that e_a and e_b are edges in the path from v_i to v_j and $c_a + c_b + 1 = 0$. This implies that c_a and c_b have different values and hence the edges e_a and e_b are colored differently. In other words there is a rainbow path between v_i and v_j. Since, this is true for any pair of vertices, the graph G is rainbow connected.

Assume that $rc(G) \leq 2$. Then, let $\chi : E \to \{\text{blue,red}\}$ be an edge coloring of G such that G is rainbow connected. Let $p = (c_1, \ldots, c_m)$ be a point in $\mathbb{F}_2^m$ such that $c_i = 1$ if $\chi(e_i) = $ blue else $c_i = 0$. The claim is that p is a solution for the system of polynomial equations S. Consider a pair of non adjacent vertices (v_i, v_j) in G. Since G is rainbow connected there exists a rainbow path from v_i to v_j. Let e_a and e_b be the edges on this path. Since these two edges have distinct colors, correspondingly the expression $c_a + c_b + 1$ has the value zero. In other words, the point p satisfies the equation corresponding to i, j. Since this is true for any pair of vertices the point p satisfies S. $\square$

Example. Consider a graph $G_n = (V, E)$ such that $V = \{a, v_1, \ldots, v_n\}$ and $E = \{(a, v_i) \mid i \in \{1, \ldots, n\}\}$. We denote the edge (a, v_i) by e_i. It can be easily seen that the rainbow connection number of the graph G_n, for $n \geq 3$, is at least 3. We show this by using the system of equations denoted by S as follows. The system of equations S for G_n, for $n \geq 3$, is given by:

$$x_{e_i} + x_{e_j} + 1 = 0, \qquad \forall i, j \in \{1, \ldots, n\}, i < j .$$

Since $(x_{e_1} + x_{e_2} + 1) + (x_{e_2} + x_{e_3} + 1) + (x_{e_1} + x_{e_3} + 1) = 1$, we have the fact that 1 belongs to the ideal $\mathfrak{a} = \langle x_{e_i} + x_{e_j} + 1 : \forall i, j \in \{1, \ldots, n\}, i < j \rangle$. This means that the solution set of $\mathfrak{a}$ is empty which further implies that the system of equations S defined for G_n, for $n \geq 3$, has no solution. From the above proposition, we have the result that the rainbow connection number of G_n is at least 3.

We now generalize the encoding for the 2-rainbow connectivity problem to the k-rainbow connectivity problem. We will only consider graphs of diameter at most k. This encoding is similar to the one described for the k-vertex coloring problem. The polynomial ring under consideration is $\mathbb{C}[x_{e_1}, \ldots, x_{e_m}]$.

Theorem 5. *The rainbow connection number of a graph $G = (V, E)$ is $\leq k$ if and only if the following zero-dimensional system of equations has a solution:*

$$x_{e_i}^k - 1 = 0, \ \forall e_i \in E$$

$$\prod_{v_i - \mathcal{P} - v_j} \left(\sum_{e_a, e_b \in \mathcal{P}} \left(\sum_{d=0}^{k-1} x_{e_a}^{k-1-d} x_{e_b}^d \right)^k \right) = 0, \ \forall (v_i, v_j) \notin E$$

Proof. Assume that the system of polynomial equations has a solution p. We color the edges of the graph as follows: Color the edge e_i with $p^{(i)}$ (i^{th} coordinate of p). Consider a pair of non adjacent vertices $(v_i, v_j) \in V \times V$. Corresponding to this pair, there is an equation in the system which is satisfied at p. This implies that for some path $\mathcal{P}$ between v_i and v_j, the polynomial

$$\sum_{e_a, e_b \in \mathcal{P}} \left(\sum_{d=0}^{k-1} x_{e_a}^{k-1-d} x_{e_b}^d \right)^k$$

vanishes to zero at point p. This further implies that

$$\left(\sum_{d=0}^{k-1} x_{e_a}^{k-1-d} x_{e_b}^d \right)^k$$

is zero for any pair of edges e_a, e_b on the path $\mathcal{P}$. This can happen only when $p^{(a)}$ is different from $p^{(b)}$. Correspondingly any two edges e_a and e_b on the path $\mathcal{P}$ are assigned different colors. Thus the path $\mathcal{P}$ between vertices v_i and v_j is a rainbow path. This is true for all pairs of vertices and hence the graph is rainbow connected. Since the point p has at most k distinct coordinates (this is because p satisfies equations of the form $x_{e_i}^k - 1 = 0$), we have the rainbow connection number of G to be at most k.

Let the rainbow connection number of graph G be at most k. We find a point p belonging to the solution set of the given system of polynomial equations. As in the case of proof of Theorem 1, denote the k colors by k^{th} roots of unity. Let $p \in \mathbb{C}^m$ such that the entry $p^{(i)}$ of p is equal to the color assigned to the edge e_i. The set of equations $x_{e_i}^k - 1 = 0$ are satisfied at p. Consider a pair of vertices $(v_i, v_j) \notin E$ in graph G. Since graph G is k-rainbow connected, there is a rainbow path $\mathcal{P}$ between v_i and v_j. Consider any two edges e_a and e_b on the path $\mathcal{P}$. Since e_a and e_b are colored differently, the indeterminates x_{e_a} and x_{e_b} are given different values. This further implies that the expression $\sum_{d=0}^{k-1} x_{e_a}^{k-1-d} x_{e_b}^d$ is zero. Thus, for a rainbow path $\mathcal{P}$ between v_i and v_j, the summation

$$\sum_{e_a, e_b \in \mathcal{P}} \left(\sum_{d=0}^{k-1} x_{e_a}^{k-1-d} x_{e_b}^d \right)^k$$

is zero and hence, the equation corresponding to the pair of vertices (v_i, v_j) is satisfied at point p. Since this is true for any pair of vertices, the point p satisfies the given system of polynomial equations.   □

The above given formulation of the k-rainbow connectivity problem, for any k, as a system of polynomial equations is not a valid encoding since the encoding procedure does not run in time polynomial in n. However, if k is a constant then we have a polynomial time algorithm to exhaust all the paths of length at most k between every pair of vertices. Using this, we can transform the graph instance into a system of polynomial equations in time polynomial in n. Hence if k is a constant, Theorem 5 gives a valid polynomial time encoding of the k-rainbow connectivity problem.

5 Conclusion

In this paper, we reviewed methods to solve graph theoretic problems algebraically. One of the most popular being formulation of the combinatorial problems as a system of polynomial equations. Using this formulation, an approach to determine the infeasibility of the system of polynomial equations, namely NulLA, is described. We solve the rainbow connectivity problem in two ways. We formulate the problem as a system of polynomial equations and using NulLA this will give a solution to our original problem. We also formulate the problem as an ideal membership problem such that determination of whether the graph can be colored with some number of colors is equivalent to determining whether a specific polynomial belongs to a given ideal or not.

An interesting future direction might be to analyze the special cases for which the rainbow connectivity problem is tractable using the above characterization (the rainbow connectivity problem is NP-hard for the general case). In order to achieve this, it would be interesting to get some bounds on the degree of the Nullstellensatz certificate for the polynomial system corresponding to the rainbow connectivity problem.

References

1. Alon, N.: Combinatorial Nullstellensatz. Combinatorics, Probability and Computing 8(1&2), 7–29 (1999)
2. Lovász, L.: Stable sets and polynomials. Discrete Mathematics 124(1-3), 137–153 (1994)
3. De Loera, J.: Gröbner bases and graph colorings. Beiträge Algebra Geom. 36(1), 89–96 (1995)
4. De Loera, J., Lee, J., Malkin, P., Margulies, S.: Hilbert's Nullstellensatz and an algorithm for proving combinatorial infeasibility. In: ISSAC 2008: Proceedings of the Twenty-first International Symposium on Symbolic and Algebraic Computation, pp. 197–206. ACM (2008)
5. Margulies, S.: Computer algebra, combinatorics, and complexity: Hilberts Nullstellensatz and NP-complete problems. PhD thesis, University of California (2008)

6. Bayer, D.: The division algorithm and the Hilbert scheme. PhD thesis, Harvard University (1982)
7. Cox, D.A., Little, J., O'Shea, D.: Ideals, Varieties, and Algorithms, 3rd edn. Undergraduate Texts in Mathematics. Springer (2007)
8. Kollár, J.: Sharp effective Nullstellensatz. American Mathematical Society 1(4) (1988)
9. Brownawell, W.: Bounds for the degrees in the Nullstellensatz. The Annals of Mathematics 126(3), 577–591 (1987)
10. Loera, J., Lee, J., Margulies, S., Onn, S.: Expressing combinatorial problems by systems of polynomial equations and hilberts nullstellensatz. Combinatorics, Probability and Computing 18(04), 551–582 (2009)
11. Chartrand, G., Johns, G., McKeon, K., Zhang, P.: Rainbow connection in graphs. Math. Bohem 133(1), 85–98 (2008)
12. Alon, N., Tarsi, M.: A note on graph colorings and graph polynomials. Journal of Combinatorial Theory Series B 70, 197–201 (1997)

Application of the Method of Asymptotic Solution to One Multi-Parameter Problem*

Alexander Batkhin

Keldysh Institute of Applied Mathematics,
Miusskaya sq. 4, Moscow, 125047, Russia

Abstract. We propose software implementation of the method of computation of asymptotic expansions (see [1,2]) of branches of the set of zeros of a polynomial in three variables near a singular point at which this polynomial is annulled with its partial derivatives. We apply this method for investigation of the set of stability of some gyroscopic system with 4 degrees of freedom and with 3 parameters. It is also possible to compute the set of stability with the help of this method for more general system with 5 parameters.

1 Introduction

We consider a mechanical system in gravitational field which consists of a massive thin disk rigidly connected with vertically positioned rotor of an engine with the help of two massless bars. These bars are pivotally connected with each other and with the uniformly rotating rotor by elastic Hooke joints. Each Hooke joint provides 3 degrees of freedom. Such a system is statically unstable and is shown in Fig. 1.

The equations of motion of such a system in linear approximation can be reduced to linear Hamiltonian system with 4 degrees of freedom of the form

$$\dot{X} = JA(P)X, \quad X = (Y, Z)^{\mathrm{T}}, \quad Y, Z \in \mathbb{R}^4, \tag{1}$$

where J is simplectic unit matrix, P is the vector of parameters, and A is symmetric constant matrix.

We investigate the set of stability of stationary point $X = 0$.

Definition 1. *The set of stability Σ for the system (1) is the set of parameters P from the parameters space $\Pi = \mathbb{R}^n$ for which the stationary solution $X = 0$ to system (1) is Lyapunov stable.*

For special case $K = 2$, $k = 1$ (see Fig. 1), the set of stability was analytically described in [3]. The general case was investigated by the author in [4] by elimination theory methods. The goal of this work is to show how to solve the problem of computing the set of stability by methods of Power Geometry applied to computing the exact solutions of an algebraic equation.

---

* This work was supported by RFBR, Grant No. 11-01-00023.

V.P. Gerdt et al. (Eds.): CASC 2012, LNCS 7442, pp. 22–33, 2012.

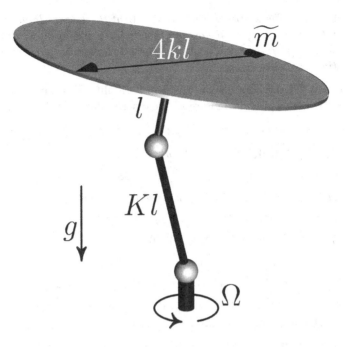

Fig. 1.

The vector of parameters P becomes three-dimensional in special case and is denoted by $Q = (x, y, z)$. The symmetric matrix $A(Q)$ is

$$
A = \begin{pmatrix}
a_{11} & 0 & a_{13} & 0 & 0 & -1 & 0 & 0 \\
0 & a_{11} & 0 & a_{13} & 1 & 0 & 0 & 0 \\
a_{13} & 0 & a_{33} & 0 & 0 & -1/2 & 0 & 0 \\
0 & a_{13} & 0 & a_{33} & 1/2 & 0 & 0 & 0 \\
0 & 1 & 0 & 1/2 & 1/2 & 0 & -1/2 & 0 \\
-1 & 0 & -1/2 & 0 & 0 & 1/2 & 0 & -1/2 \\
0 & 0 & 0 & 0 & -1/2 & 0 & 1 & 0 \\
0 & 0 & 0 & 0 & 0 & -1/2 & 0 & 1
\end{pmatrix}
$$

where $a_{11} = x + y - 2z + 4$, $a_{33} = y - z + 1$, $a_{13} = 2 - y$.

The characteristic polynomial of the matrix $JA(Q)$ includes only even powers of λ, i. e. it is a polynomial in $\mu = \lambda^2$ and further it is called *semi-characteristic*:

$$
f(\mu) = \sum_{k=0}^{m} f_k(Q)\mu^k, \quad f_m = 1. \tag{2}
$$

The conditions of stationary point stability are stated by the following

Theorem 1 ([5]). *The stationary point $X = 0$ of the linear Hamiltonian system (1) is stable if and only if all nonzero roots $\mu_1, \ldots, \mu_m$ of the semi-characteristic polynomial (2) are real and negative, and all elementary divisors of matrix JA for multiple roots are simple.*

As was shown in [3], the boundary of the set of stability Σ is part of the set $\mathcal{G} = \{Q : g(Q) = 0\}$, where $g(Q)$ is a factor of the discriminant of the polynomial (2):

$$
\begin{aligned}
g(\mathbf{Q}) = {} & 512z^6 - 4352z^5y - 768z^5x + 14848z^4y^2 + 5376z^4yx + 512z^4x^2 - \\
& - 25408z^3y^3 - 14656z^3y^2x - 2752z^3yx^2 - 192z^3x^3 + \\
& + 21800z^2y^4 + 19168z^2y^3x + 5360z^2y^2x^2 + 736z^2yx^3 + 40z^2x^4 - \\
& - 7500zy^5 - 11700zy^4x - 4376zy^3x^2 - 904zy^2x^3 - 92zyx^4 - 4zx^5 + \\
& + 2500y^5x + 1200y^4x^2 + 344y^3x^3 + 48y^2x^4 + 4yx^5 - \\
& - 256z^5 + 2880z^4y + 1344z^4x - 14976z^3y^2 - 6720z^3yx - \\
& - 1344z^3x^2 + 37928z^2y^3 + 13816z^2y^2x + 5144z^2yx^2 + 456z^2x^3 - \\
& - 45120zy^4 - 14464zy^3x - 6784zy^2x^2 - 1152zyx^3 - 64zx^4 + \\
& + 20250y^5 + 6490y^4x + 3156y^3x^2 + 740y^2x^3 + 82yx^4 + 2x^5 + \\
& + 1872z^4 + 2016z^3y - 5088z^3x - 35496z^2y^2 + 15888z^2yx + \\
& + 2200z^2x^2 + 67608zy^3 - 12936zy^2x - 5176zyx^2 - 344zx^3 - \\
& - 37827y^4 + 828y^3x + 2782y^2x^2 + 412yx^3 + 13x^4 - \\
& - 13824z^3 + 62208z^2y + 6912z^2x - 93312zy^2 - 20736zyx - \\
& - 1152zx^2 + 46656y^3 + 15552y^2x + 1728yx^2 + 64x^3.
\end{aligned}
\tag{3}
$$

The structure of the set $\mathcal{G}$ was computed in [3] with the help of Gröbner basis. Here we show how to compute the set $\mathcal{G}$ using Power Geometry algorithms given in [2].

The main step in investigation of the set $\mathcal{G}$ is to provide the significant simplification of the polynomial $g(Q)$. In order to do that we find the singular points of the set $\mathcal{G}$.

Definition 2. *Let $\varphi(P)$ be a polynomial. Point $P = P_0$ of the set $\varphi(P) = 0$ is called a singular point of the order k if the polynomial $\varphi(P)$ and all its partial derivatives up to the order k vanish at this point but at least one derivative of order $k + 1$ does not vanish.*

We have one a priori singular point of the set $\mathcal{G}$ at infinity. Below we apply the algorithm of asymptotic solution of equation (3) described in [2]. The definition of used objects of Power Geometry can be found ibidem.

2   Asymptotic Analysis of Set $\mathcal{G}$ at Infinity

The Newton polyhedron $\Gamma(g)$ of polynomial (3) is shown in Fig. 2 from the side of large power exponents. It has only one two-dimensional face with positive outward normal. This is the pentagonal face $\Gamma_{01}^{(2)}$ with the normal $N_{01} = (1, 1, 1)$. It is associated with the truncated polynomial

$$\hat{g}_{01}^{(2)}(\mathbf{Q}) = 4 \left(2z^2 - zx + xy - 3zy\right) \left(x^2 + 6xy + 25y^2 - 4zx - 28zy + 8z^2\right)^2. \quad (4)$$

The truncations corresponding to the edges of the face $\Gamma_{01}^{(2)}$ have no multiple roots.

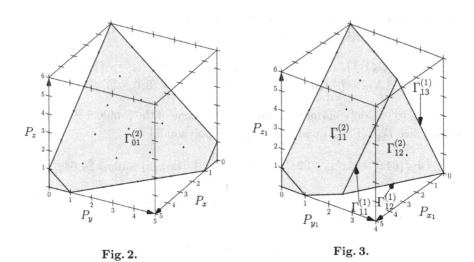

Fig. 2.   Fig. 3.

The zero of the first form in (4) defines a conic surface $\mathcal{C}_0$ with the center at the origin. This surface approximates (in the first approximation) behavior of the two-dimensional component of the set $\mathcal{G}$ at infinity.

Roots of the second quadratic form in (4) form the real straight line $\mathcal{B} = \{x = y, z = 2y\}$ with the directing vector $\tau_\mathcal{B} = (1, 1, 2)$ lying in the intersection of the two complex planes on which this quadratic form vanishes. This line $\mathcal{B}$, in the first approximation, specifies asymptotic direction of branches of the set $\mathcal{G}$. In order to find a parametric expansion of these branches, we use the procedure described in [2]. New variables of the kth cycle will have index k. This index will be the first index in the notation of other objects: faces, their normals, corresponding truncated polynomials, normal cones, and cones of the problem.

Let us find the next approximation of the set $\mathcal{G}$ near this line $\mathcal{B}$. To this end, we go to the local coordinates along the straight line $\mathcal{B}$:

$$x = x_1 + y_1, \quad y = y_1, \quad z = z_1 + 2y_1. \quad (5)$$

In this change of variables, y_1 is the line parameter, and x_1 and z_1 are local coordinates.

The Newton polyhedron $\Gamma(g_1(\mathbf{Q}_1))$ of the transformed polynomial $g(\mathbf{Q})$ is shown in Fig. 3 in the variables $\mathbf{Q}_1 = (x_1, y_1, z_1)$. In accordance with formula (1.9) from [2], the cone of the problem is

$$\mathbf{K}_1 = \{S = \mu_1 N_{01} + \lambda_1(-1,0,0) + \lambda_2(0,0,-1)\},$$

where $\mu_1 \geqslant 0$ and $\lambda_{1,2} > 0$. In the coordinate representation, the components of vector S from the cone of the problem $\mathbf{K}_1$ are written as $s_1 = \mu_1 - \lambda_1$, $s_2 = \mu_1$, and $s_3 = \mu_1 - \lambda_2$. Following Remark (f) in [2], we obtain the system of inequalities

$$\mu_1 = s_2 \geqslant 0, \lambda_1 = s_2 - s_1 > 0, \lambda_2 = s_2 - s_3 > 0, \tag{6}$$

which efficiently selects vectors from $\mathbf{K}_1$. The outward normals to the faces of the polyhedron $\Gamma(g(\mathbf{Q}_1))$ are given by

$$N_{11} = (1,1,1), \ N_{12} = (1,2,1), \ N_{13} = (0,-1,0), N_{14} = -(1,1,1),$$
$$N_{15} = (-1,0,0), \ N_{16} = (1,0,0), \ N_{17} = (0,0,-1).$$

According to (6), only normal N_{12} falls into the cone of the problem $\mathbf{K}_1$. To this normal, face $\Gamma_{12}^{(2)}$ (Fig. 3) and the truncated polynomial

$$\hat{g}_{12}^{(2)} = 4y_1^2(64z_1^4 - 64z_1^3 x_1 + 32z_1^2 x_1^2 - 8z_1 x_1^3 + x_1^4 + 64z_1 y_1 x_1 - 16y_1 x_1^2 + 64y_1^2), \tag{7}$$

correspond. The three edges $\Gamma_{11}^{(1)}$, $\Gamma_{12}^{(1)}$, and $\Gamma_{13}^{(1)}$ of the face $\Gamma_{12}^{(2)}$ are associated with the truncated polynomials

$$\hat{g}_{11}^{(1)} = 4y_1^2\left(x_1^2 - 4x_1 z_1 + 8z_1^2\right)^2, \tag{8}$$

$$\hat{g}_{12}^{(1)} = 4y_1^2\left(x_1^2 - 8y_1\right)^2, \tag{9}$$

$$\hat{g}_{13}^{(1)} = 4y_1^2(64z_1^4 + 64y_1^2). \tag{10}$$

The discriminant of the parenthesized polynomial in (8) is negative; i. e., this polynomial has no nonzero real roots. The parenthesized polynomial in (10) also has no nonzero real roots (since it is a sum of squares). The roots of polynomial (9) are $y_1 = x_1^2/2$ and $z_1 = 0$; they fall into case (a) of step 4 in [2] and will be studied later as case (2).

The discriminant of the second multiplier of polynomial (7) with respect to variable y_1 is equal to $-4z_1^2(x_1 - 2z_1)^2$. Since the degree of polynomial (7) in y_1 is even, it can have real roots only in the following two cases:

(1) if $x_1 - 2z_1 = 0$, then $y_1 = -z_1^2/2$;
(2) if $z_1 = 0$, then $y_1 = x_1^2/8$ is a root of polynomial (9).

Consider cases (1) and (2) separately.

2.1   Expansion of the Family $\mathcal{P}_1$ of Singular Points

In case (1), we make the change of variables

$$x_1 = x_2 + 2z_2, \quad y_1 = y_2 - z_2^2/2, \quad z_1 = z_2, \tag{11}$$

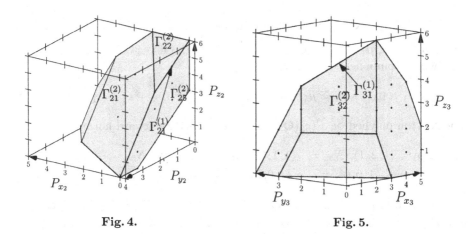

Fig. 4.   Fig. 5.

making thus x_2 and y_2 to be local variables. The Newton polyhedron of the polynomial $g_2(\mathbf{Q}_2)$ is shown in Fig. 4 in the variables $\mathbf{Q}_2 = (x_2, y_2, z_2)$.

The new cone of the problem is

$$\mathbf{K}_2 = \{S = \mu_1 N_{12} + \lambda_1(-1, 0, 0) + \lambda_2(0, -1, 0)\}, \tag{12}$$

where $\mu_1 \geqslant 0$, and $\lambda_{1,2} > 0$. Then, the components of vector S falling into the cone of the problem (12) must satisfy the following inequalities:

$$s_3 = \mu_1 \geqslant 0, s_3 - s_1 = \lambda_1 > 0, 2s_3 - s_2 = \lambda_2 > 0. \tag{13}$$

The faces of the Newton polyhedron $\Gamma(g_2(\mathbf{Q}_2))$ has the following outward normals:

$$N_{21} = (1, 2, 1), \; N_{22} = (0, 1, 1), \; N_{23} = (0, -1, 0), \; N_{24} = -(1, 1, 1),$$
$$N_{25} = (-1, 0, 0), \; N_{26} = (1, 0, 0), \; N_{27} = (0, 0, -1), \; N_{28} = (2, 2, 1).$$

According to (13), only normal N_{22} falls into the cone of the problem (12), and the truncated polynomial corresponding to the face $\Gamma_{22}^{(2)}$ is $\hat{g}_{22}^{(2)} = 16z_2^6 x_2^2 + 64z_2^4(2z_2 + y_2)^2$. All its roots lie on the straight line $x_2 = 0$, $2z_2 + y_2 = 0$ and are roots of the truncation $\hat{g}_{21}^{(1)}$ corresponding to the edge $\Gamma_{21}^{(1)}$ of the face $\Gamma_{22}^{(2)}$, which defines the following change of variables:

$$x_2 = x_3, \quad y_2 = y_3 - 2z_3, \quad z_2 = z_3. \tag{14}$$

The truncations corresponding to the other edges of the face $\Gamma_{22}^{(2)}$ are not meaningful according to Remark (d) in [2]. The edge $\Gamma_{21}^{(1)}$ is a common edge of the faces $\Gamma_{22}^{(2)}$ and $\Gamma_{25}^{(2)}$; hence, the new problem cone is

$$\mathbf{K}_3 = \{S = \mu_1 N_{22} + \mu_2 N_{25} + \lambda_1(0, -1, 0)\}, \tag{15}$$

where $\mu_{1,2} \geqslant 0$, $\mu_1 + \mu_2 > 0$, and $\lambda_1 > 0$. Only those vectors S belong to this cone whose components satisfy the inequalities

$$s_3 = \mu_1 \geqslant 0, s_3 - s_1 = \mu_1 + \mu_2 > 0, s_1 = -\mu_2 \leqslant 0, s_3 - s_2 = \lambda_1 > 0. \tag{16}$$

The Newton polyhedron $\Gamma(g_3(\mathbf{Q}_3))$ has the following outward normals:

$$N_{31} = (1, 2, 1), \ N_{32} = -(1, 1, 0), \ N_{33} = (0, -1, 0), \ N_{34} = -(2, 2, 1),$$
$$N_{35} = (-1, 0, 0), \ N_{36} = (1, 0, 0), \ N_{37} = (0, 0, -1), \ N_{28} = (2, 2, 1).$$

According to (16), only normal N_{32} falls into the problem cone (15). Since this face $\Gamma_{32}^{(2)}$ is parallel to the OZ axis (Fig. 5), we got a "hole" in the polyhedron; i. e., according to Remark (e) in [2], the expansion has been terminated.

Collecting substitutions (5), (11), and (14) together, we obtain the resulting expansion

$$x = -z_3^2/2, y = -2z_3 - z_3^2/2, z = -3z_3 - z_3^2, \tag{17}$$

which defines the one-parameter family $\mathcal{P}_1$ of singular points of order 1 of the set $\mathcal{G}$. Moreover, the polynomial $f(\mu)$ has the root of multiplicity 3 along the family $\mathcal{P}_1$. Direct computations show that there are two singular points of order 2 $Q_0 = (0, 0, 0)$ and $Q_1 = (-2, 2, 2)$ in the family $\mathcal{P}_1$. At the point Q_1, the polynomial $f(\mu)$ has the root of multiplicity 4.

2.2 Expansion of the Family $\mathcal{P}_2$ of Singular Points

In case (2), we make the substitution

$$x_1 = x_4, \quad y_1 = y_4 + x_4^2/8, \quad z_1 = z_4. \tag{18}$$

The faces adjacent to the edge $\Gamma_{12}^{(1)}$, corresponding to polynomial (9), have normals $N_{12} = (1, 2, 1)$ and $N_{17} = (0, 0, -1)$. The new cone of the problem is given by

$$\mathbf{K}_4 = \{S = \mu_1 N_{12} + \mu_2 N_{17} + \lambda_1(0, -1, 0)\}, \tag{19}$$

where $\mu_{1,2} \geqslant 0$, $\mu_1 + \mu_2 > 0$, and $\lambda_1 > 0$. Only those vectors S belong to this cone whose components satisfy the following system of inequalities:

$$s_1 = \mu_1 \geqslant 0, 2s_1 - s_3 = \mu_1 + \mu_2 > 0, s_1 - s_3 = \mu_2 \geqslant 0, \lambda_1 = 2s_1 - s_2 > 0. \tag{20}$$

The Newton polyhedron $\Gamma(g_4(\mathbf{Q}_4))$ is shown in Fig. 6.

It has the following outward normals:

$$N_{41} = (1, 2, 1), \ N_{42} = (1, 1, 0), \ N_{43} = (0, -1, 0), \ N_{44} = -(1, 1, 1),$$
$$N_{45} = (-1, 0, 0), \ N_{46} = (1, 2, 2), \ N_{47} = (0, 0, -1).$$

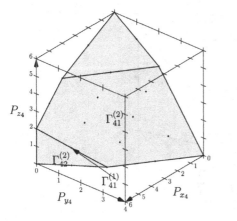

Fig. 6.

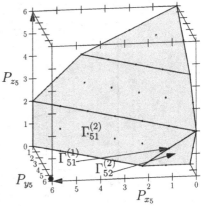

Fig. 7.

According to (20), only normal N_{42} falls into the cone of the problem (19). The truncation of the original polynomial corresponding to the face $\Gamma_{42}^{(2)}$ is $\hat{g}_{42}^{(2)} = 2x_4^4(z_4^2 x_4^2 + 2z_4 y_4 x_4 + 2y_4^2 - 2y_4 x_4 + x_4^2)$. The factor in the parentheses is written in the form of sum of two squares as $(z_4 x_4 + y_4)^2 + (y_4 - x_4)^2$, which vanishes only under the condition $z_4 = -1$, $x_4 = y_4$, which defines the following change of variables:

$$x_4 = x_5 + y_5, \quad y_4 = y_5, \quad z_4 = -1 + z_5. \tag{21}$$

The new cone of the problem is given by

$$\mathbf{K}_5 = \{S = \mu_1 N_{42} + \lambda_1(-1,0,0) + \lambda_2(0,0,-1)\}, \tag{22}$$

where $\mu_1 \geqslant 0$ and $\lambda_{1,2} > 0$. Appropriate vectors S should satisfy the system of inequalities $s_2 = \mu_1 \geqslant 0$, $s_2 - s_1 = \lambda_1 > 0$, $s_3 = -\lambda_2 < 0$. The new Newton polyhedron $\Gamma(g_5(\mathbf{Q}_5))$ shown in Fig. 7 has the following outward normals:

$$N_{51} = (1,1,0), \ N_{52} = (0,1,-1), \ N_{53} = (0,-1,0), \ N_{54} = (1,1,1),$$
$$N_{55} = (-1,0,0), \ N_{46} = (1,1,2), \ N_{57} = (0,0,-1),$$

and only the normal N_{52} falls into the problem cone (22).

The truncation of the polynomial $g_5(\mathbf{Q}_5)$ corresponding to this face is the polynomial $\hat{g}_{52}^{(2)} = y_5^2(z_5 y_5(z_5 y_5 - 2x_5 - 7) + (2x_5 + 7)^2)$, which vanishes when $x_5 = -7/2$ and $z_5 = 0$. After substitution $x_5 = x_6 - 7/2$, $y_5 = y_6$, $z_5 = z_6$, we obtain the face $\Gamma_{61}^{(2)}$ with the normal vector $N_{61} = (-1,0,-1)$ (Fig. 8) parallel to the OY axis, i. e., a "hole" in the polyhedron. According to Remark (e) in [2], the expansion has been terminated.

Collecting substitutions (5), (18), and (21) together, we obtain the following expansion

$$x = (-63 + 36y_5 + 4y_5^2)/32, \ y = (49 + 4y_5 + 4y_5^2)/32, \ z = (33 + 4y_5 + 4y_5^2)/16, \tag{23}$$

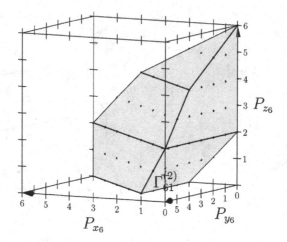

Fig. 8.

which gives the second family $\mathcal{P}_2$ of singular points of order 1 of the set $\mathcal{G}$. Along this family, the polynomial $f(\mu)$ has pair of roots of multiplicity 2. There are two singular points of order 2 laying in the family $\mathcal{P}_2$: $Q_2 = (7/2, 7/2, 6)$, $Q_3 = (-5/2, 3/2, 2)$.

3   Structure of the Set of Stability $\mathcal{G}$

We form new basis in the parameter space Π which consists of the following vectors: $\overrightarrow{Q_0 Q_1}/2$, $\overrightarrow{Q_2 Q_3}/2$, $\overrightarrow{Q_4 Q_5}/2$, where $Q_4 = (-1/2, 3/2, 2)$ and $Q_5 = (0, 2, 3)$ are vertices of parabolas (17) and (23), respectively.

In new coordinates $\widetilde{Q} = (U, V, W)$, the polynomial g takes the simplest form

$$\begin{aligned} g(\widetilde{Q}) =&\, 64\,U^6 + 192\,U^4 V^2 - 4\,U^4 W^2 + 192\,U^2 V^4 - 8\,U^2 V^2 W^2 + 64\,V^6 - \\ &- 4\,V^4 W^2 + 72\,U^4 W - 4\,U^2 W^3 - 72\,V^4 W + 4\,V^2 W^3 + 60\,U^4 - \\ &- 312\,U^2 V^2 + 20\,U^2 W^2 + 60\,V^4 + 20\,V^2 W^2 - W^4 + 36\,U^2 W - \\ &- 36\,V^2 W + 12\,U^2 + 12\,V^2 + 2\,W^2 - 1. \end{aligned} \tag{24}$$

It can easily be shown that the set $\mathcal{G}$ consists of two one-dimensional components which are the families $\mathcal{P}_1$ and $\mathcal{P}_2$ and ruled surface $\widetilde{\mathcal{G}}$ with parametrization

$$U = u \sin \varphi, \quad V = (u+1) \cos \varphi, \quad W = 4u + 2 \cos^2 \varphi + 1. \tag{25}$$

The surface $\widetilde{\mathcal{G}}$ has two parabolic segments of self-intersection, namely the segment of parabola $\mathcal{P}_1$ between points Q_0 and Q_1 and the segment of parabola $\mathcal{P}_2$ between points Q_2 and Q_3. The boundary of the set of stability Σ together with other objects are shown in Fig. 9.

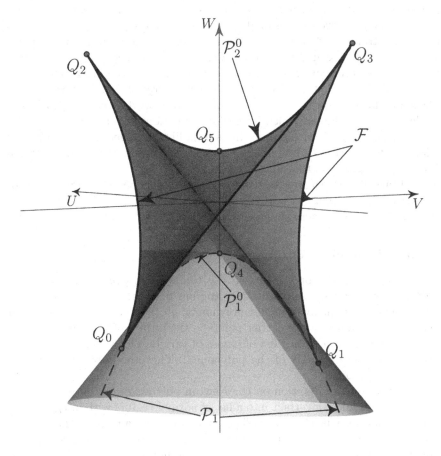

Fig. 9.

4 Software Implementation

Implementation of the algorithms described in [2] required the use of several software tools. All calculations related to polynomials and plane algebraic curves were carried out with the help of the computer algebra system **Maple**. For this system, the library of procedures **PGeomlib** was written, which implements spatial Power Geometry algorithms. To work with plane algebraic curves, package **algcurves** was used. By means of this package, the kind of the curve was determined (function **genus**), and rational parameterization of curves of kind 0 was calculated (function **parametrization**). It also provided procedures for work with elliptic and hyperelliptic curves. To calculate singular points of the set $\mathcal{G}$, it is required to solve systems of algebraic equations, which were solved by means of the **Groebner** package.

Basic objects of spatial Power Geometry were calculated with the use of program **qconvex** from freely distributed software package **Qhull**. Given a support

$\mathbf{S}(g)$ in the form of a list of point coordinates, this program computes the Newton polyhedron, its two-dimensional faces, and normals to them. The objects obtained are transferred to the Maple environment, where all other computations are performed. Currently, the data exchange interface between the computer algebra system Maple and program qconvex from the Qhull package is implemented on the file level. Note that the procedures of library PGeomlib are implemented in such a way that it is possible to work with polynomials in three variables the coefficients of which are rational functions of other parameters.

Below is a description of the procedures from library PGeomlib, which can be divided into two – basic and auxiliary – groups. The names of all objects stored in the library begin with the capital letters PG.

The library PGeomlib was tested for different versions of CAS Maple for operating systems Win32/64 and MacOSX, and it is available for download at https://www.dropbox.com/sh/epanm7gzz5xyqt7/uyqCztx9Lk.

4.1 Basic Procedures of Library PGeomlib

PGsave computes and stores in a text file the support $\mathbf{S}(g)$ of a polynomial $g(Q)$ for subsequent processing by program qconvex. The procedure has two obligatory input parameters: polynomial $g(Q)$ and a name of the file for storing coordinates of the carrier points in the format of program qconvex. An optional parameter is a list of names of variables for which it is required to construct the support of the polynomial. The procedure uses auxiliary procedure PGsupp.

PGgetnormals gets information on the Newton polyhedron, its faces, and normals to them and converts it into a list of normal vectors with integer coefficients. The procedure has one obligatory parameter – a name of the file with results of operation of program qconvex – and returns list of support planes of the Newton polyhedron determined by the normal vector and shift. The procedure uses auxiliary procedure PGnormpriv.

PGtruncface, PGtruncedge, and PGtruncfwe are three variants of the procedure calculating truncated polynomials corresponding to a face, an edge, or a face and all adjacent edges. In the second variant, the edge is given by two adjacent faces. The procedure uses auxiliary procedures PGsupp and PGgetneighbours.

PGfitnormal selects normals from the list of normals of the Newton polyhedron that fall into the problem cone $\mathbf{K}$ given by a list of linear inequalities.

PGplot is a procedure for the visualization of the Newton polyhedron $\Gamma(g)$ and support $\mathbf{S}(g)$ of a polynomial $g(Q)$.

4.2 Auxiliary Procedures of Library PGeomlib

PGsupp returns the support $\mathbf{S}(g)$ of a polynomial $g(Q)$ in the form of a list of vector power exponents of monomials.

PGnormpriv converts a list of vectors with commensurable floating-point coordinates into a list of collinear vectors with integer coordinates. This procedure is required because program `qconvex` stores normals to faces of the Newton polyhedron in the floating-point format, whereas all operations on vector power exponents are to be performed in integer arithmetic.

`PGneighbours` for each vertex of the Newton polyhedron, calculates the numbers of faces adjacent to it.

4.3   Scheme of Using Library `PGeomlib`

Let us describe schematically the order of work with library `PGeomlib`.

For a polynomial $g(Q)$, by means of procedure `PGsave`, support $\mathbf{S}(g)$ is calculated and stored for subsequent processing by program `qconvex`, which computes the Newton polyhedron $\Gamma(g)$ and its support faces. These objects are obtained by means of procedure `PGgetnormals`. Then, the cone of problem $\mathbf{K}$ is specified in the form of a list of inequalities, and appropriate faces are selected by means of procedure `PGfitnormals`. For them, the truncated polynomials are calculated by means of one of the procedures `PGtruncface`, `PGtruncedge`, or `PGtruncfwe`.

All computations in the previous example were carried out in accordance with the above-specified scheme.

References

1. Bruno, A.D., Batkhin, A.B.: Asymptotic solution of an algebraic equation. Doklady Mathematics 84(2), 634–639 (2011)
2. Bruno, A.D., Batkhin, A.B.: Resolution of an algebraic singularity by power geometry algorithms. Programming and Computer Software 38(2), 57–72 (2012)
3. Batkhin, A.B., Bruno, A.D., Varin, V.P.: Sets of stability of Mmulti-parameter Hamiltonian problems. J. Appl. Math. and Mech. 76(1), 56–92 (2012)
4. Batkhin, A.B.: Stability of Certain Multiparameter Hamiltonian System. Preprint No. 69, Keldysh Inst. Appl. Math., Moscow (2011) (in Russian)
5. Malkin, I.G.: Theory of Stability of Motion. U.S. Atomic Energy Commission, Office of Technical Information, Oak Bridge (1958)

A New Algorithm
for Long Integer Cube Computation
with Some Insight into Higher Powers

Marco Bodrato[1] and Alberto Zanoni[2]

[1] mambaSoft
Via S.Marino, 118 – 10137 Torino, Italy
bodrato@mail.dm.unipi.it
[2] Dipartimento di Scienze Statistiche
Università "Sapienza", P.le Aldo Moro 5 – 00185 Roma, Italy
zanoni@volterra.uniroma2.it

Abstract. A new approach for the computation of long integer cube (third power) based on a splitting-in-two divide et impera approach and on a modified Toom-Cook-3 unbalanced method is presented, showing that the "classical" square-and-multiply algorithm is not (always) optimal. The new algorithm is used as a new basic tool to improve long integer exponentiation: different techniques combining binary and ternary exponent expansion are shown. Effective implementations by using the GMP library are tested, and performance comparisons are presented.

AMS Subject Classification: 11N64, 11A25, 13B25

Keywords and phrases: Toom-Cook, cube, third power, long integers.

1 Introduction

Fast long integer arithmetic is at the very core of many computer algebra systems. Starting with the works of Karatsuba [1], Toom [2] and Cook [3], who found methods to lower asymptotic complexity for multiplication and squaring from $O(n^2)$ to $O(n^e)$, with $1 < e \leqslant \log_2 3$, many efforts have been done to find optimized implementations in arithmetic software packages.

The family of Toom-Cook (Toom, for short) methods is an infinite set of polynomial algorithms (Toom-3, Toom-4, etc. – Karatsuba may be identified with Toom-2). The original family was generalized by Bodrato and Zanoni in [4] considering unbalanced operands – polynomials with different degrees – with the so-called Toom-$(k + 1/2)$ methods (Toom-2.5, Toom-3.5, etc.) and with the unbalanced use of classic methods as well.

Each of them may be viewed as solving a polynomial interpolation problem, with base points not specified a priori, from which a matrix to be inverted arises. In a software implementation, a set of basic operations (typically additions, subtractions, bit shiftings, multiplications and divisions by small numbers, etc.)

V.P. Gerdt et al. (Eds.): CASC 2012, LNCS 7442, pp. 34–46, 2012.

is given. Practically, this is a set of very efficiently implemented basic functions, and the idea is to use them to evaluate factors in the base points, invert the resulting matrix step by step and recompose data to obtain the final result.

Asymptotically speaking, the best theoretical result is Fürer's [5], but it has not been used in practice yet. The actual competitors with Toom methods are Schönhage-Strassen and FFT-based methods [6], eventually faster for very large numbers, but there is a wide range where Toom beats any other algorithm.

To the best of our knowledge, although long integer product, squaring and generic power raising have been extensively studied, the particular problem of computing the third power u^3 has not been deeply treated. Usually, a generic power is obtained by a clever sequence of squarings and multiplications, and cube is simply treated as a particular instance of this general procedure.

In this paper a new perspective for cube computation based on unbalanced Toom-3 method is presented in full details, showing that a non trivial algorithm exists, faster (in a certain range) than computing first $U = u^2$ and then the product $U \cdot u$. A practical performance comparison with respect to GMP library [7] is reported as well, beside some empirical studies on extending the saving to larger exponents by using cube in addition to squarings and products.

2   Mathematical Setting

For simplicity, we study only *positive* long integers. Consider base expansion representation: for $0 < u \in \mathbb{N}$, we fix a base $1 < \mathcal{B} \in \mathbb{N}$ and consider the polynomial determined by the representation of u in base $\mathcal{B}$ with degree $d = \lfloor \log_{\mathcal{B}}(u) \rfloor$ and having as coefficients the digits of u, each smaller than $\mathcal{B}$. In computer science, common choices are $\mathcal{B} = 2, 2^8, 2^{16}, 2^{32}, 2^{64}$: in particular, GMP library digits are named *limbs*. Following the *divide et impera* approach, we set $n = \lceil (d+1)/2 \rceil$, $t = d + 1 - n \in \{n-1, n\}$ and $x = y^n$, so that one has

$$u = \sum_{i=0}^{\lfloor \log_{\mathcal{B}}(u) \rfloor} \hat{a}_i \mathcal{B}^i \implies \hat{a}(y) = \sum_{i=0}^{d} \hat{a}_i y^i = \left(\sum_{i=n}^{d} \hat{a}_i y^{i-n} \right) x + \left(\sum_{i=0}^{n-1} \hat{a}_i y^i \right) \quad (1)$$

$$= a_1 x + a_0 = a(x)$$

Let $f(x) = a(x)^3$: we can compute u^3 as $f(x)|_{x=\mathcal{B}^n}$. The core idea of the new cube algorithm is based on considering not directly $f(x)$, but another polynomial (called "master" of f), differing only in the constant coefficient by a small – but making the difference – multiple.

This permits a new approach to cube computation, proving in practice

Theorem 1. *The cube computation formula* $u^3 = u^2 \cdot u$ *is not (always) optimal.*

3   Split and Cube: Long Integer Case

To explain the main idea, we'll use both long integer polynomial representation, as explained above, and a schematic one as well. If u is a long integer, we may highlight its high and low half – coefficients a_1 and a_0 in equation (1).

$$u = a_1 x + a_0 \Big|_{x=\mathcal{B}^n} \equiv \begin{array}{|c|c|} \hline a_1 & a_0 \\ \hline t & n \end{array}$$

The idea is that a multiplication (with nonlinear complexity) will be avoided at the lower price of some scalar (linear complexity) operations instead.

To focus on higher ("main") terms in complexity expressions, we consider only nonlinear operations for multiplications, squarings and cubings. Let $M_{n,m}$ (M_n if $n = m$) be the nonlinear complexity of the multiplication of two numbers with $n \geqslant m$ digits in base $\mathcal{B}$, respectively; with S_n the squaring complexity of one of them and with C_n the cubing complexity. We must compute (in the schematic representation, piling means adding)

$$u^3 = f(\mathcal{B}^n) = (a_1\mathcal{B}^n + a_0)^3 = \sum_{i=0}^{3} c_i\mathcal{B}^{ni} = a_1^3 x^3 + 3a_1^2 a_0 x^2 + 3a_1 a_0^2 x + a_0^3 \Big|_{x=\mathcal{B}^n}$$

With the classical "algorithm" $u^3 = u^2 \cdot u$ the complexity is $C_{2n} = S_{2n} + M_{4n,2n}$. Computing u^2 with Karatsuba method (in a range where it's more effective than schoolbook method), we have that $C_{2n} = 3S_n + M_{4n,2n}$.

Consider now the following easily factorizable polynomial $g'(x)$, master of $f(x)$:

$$g'(x) = a_1^3 x^3 + 3a_1^2 a_0 x^2 + 3a_1 a_0^2 x + 9a_0^3 = (a_1^2 x^2 + 3a_0^2)(a_1 x + 3a_0) = g_1(x) \cdot g_2(x)$$

From the polynomial point of view, $g'(x)$ coefficients computation is similar to $f(x)$'s, as they differ only by a 9 in the constant term: there is nothing particularly appealing in it.

On the contrary, from the long integer point of view, things are not as easy, but more interesting. Compute $A = a_1^2$ and $B = a_0^2$ (nonlinear complexity: $2S_n$) and focus on the product $g'(\mathcal{B}^n) = (a_1^2 x^2 + 3a_0^2)(a_1 x + 3a_0)|_{x=\mathcal{B}^n}$ (nonlinear complexity: $M_{4n,2n}$).

$$g'(\mathcal{B}^n) \equiv \begin{array}{|c|c|} \hline A = a_1^2 & 3B = 3a_0^2 \\ \hline \end{array} \cdot \begin{array}{|c|c|} \hline a_1 & 3a_0 \\ \hline \end{array}$$

Even if this is not really what we want to compute, note that the total nonlinear complexity is now *smaller* than before: one squaring less gives $C'_{2n} = 2S_n + M_{4n,2n}$. Anyway, there are some points to consider to recover u^3 from $g'(\mathcal{B}^n)$:

(i) Multiplication by 3 can generate carries – and therefore memory management issues – to be taken care of.

(ii) Recomposition (set $x = \mathcal{B}^n$) unfortunately mixes things up the wrong way if unbalanced Toom-3 is used.

Although (i) is not complicated to manage, as regards (ii), it happens that unbalanced Toom-3 method computing $g'(\mathcal{B}^n)$ splits the longest factor in four parts (a number possibly differing by 1 or 2 from n is below indicated with $\sim n$).

$$\begin{array}{|c|c|c|c|}\hline A_1 & A_0 & 3B_1 & 3B_0 \\ \hline \end{array} \cdot \begin{array}{|c|c|}\hline a_1 & 3a_0 \\ \hline \end{array}$$

$$\underset{\sim n \quad\ n \quad\ \sim n \quad\ n}{} \quad \underset{n \quad \sim n}{}$$

so that the final division by 9, needed to have $u^3 = f(\mathcal{B}^n)$, is incompatible with the recomposition. In fact, as $g'(\mathcal{B}^n)$ is

$$g'(\mathcal{B}^n) = \sum_{i=0}^{4} c_i' x^i \Bigg|_{x=\mathcal{B}^n} = (A_1a_1)x^4 + (A_0a_1 + 3A_1a_0)x^3 + 3(A_0a_0 + B_1a_1)x^2 +$$
$$(3B_0a_1 + 9B_1a_0)x + 9B_0a_0 \big|_{x=\mathcal{B}^n}$$

we have a problem: $9a_0^3 = 9B_1a_0x + 9B_0a_0|_{x=\mathcal{B}^n}$ cannot be obtained by a linear combination of c_i' coefficients. It "appears" only *after* recomposition, but to obtain a_0^3 we should divide by 9 both $c_0' = 9B_0a_0$ (this is not a problem, it is explicitly computed) and only the second addend $9B_1a_0$ (that was *not* explicitly computed) summing to c_1'. One must therefore proceed in a slightly different way, considering instead the following $g(x)$-product.

$$g(x) = (A_1x^3 + A_0x^2 + 3B_1x + 27B_0)(a_1x + 3a_0) \equiv \begin{array}{|c|c|c|c|}\hline A_1 & A_0 & 3B_1 & 27B_0 \\ \hline \end{array} \cdot \begin{array}{|c|c|}\hline a_1 & 3a_0 \\ \hline \end{array}$$

This time we obtain a more appropriate result.

$$g(\mathcal{B}^n) = \sum_{i=0}^{4} c_i'' x^i \Bigg|_{x=\mathcal{B}^n} = (A_1a_1)x^4 + (A_0a_1 + 3A_1a_0)x^3 + 3(A_0a_0 + B_1a_1)x^2 +$$
$$(27B_0a_1 + 9B_1a_0)x + 81B_0a_0 \big|_{x=\mathcal{B}^n}$$

Note that $c_i'' = c_i$ for $i = 2, 3, 4$. Now we can appropriately divide c_1'' by 9 and c_0'' by 81 ($O(n)$ operations), correctly obtaining c_1 and c_0, and therefore $u^3 = f(\mathcal{B}^n)$.

$$g(\mathcal{B}^n) \equiv$$

A_1a_1		$3B_1a_1$		$81B_0a_0$
	A_0a_1		$27B_0a_1$	$\uparrow$
	$3A_1a_0$		$9B_1a_0$	c_0''
		$3A_0a_0$		c_1''

$$\Downarrow$$

$$u^3 = f(\mathcal{B}^n) \equiv$$

A_1a_1		$3B_1a_1$		B_0a_0
	A_0a_1		$3B_0a_1$	$\uparrow$
	$3A_1a_0$		B_1a_0	c_0
		$3A_0a_0$		c_1

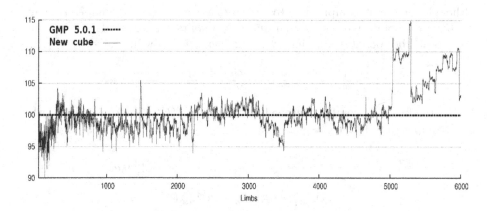

Fig. 1. New cube algorithm versus GMP-5.0.1 relative timings

We point out that it is possible in practice to avoid the final explicit division by 81, by slightly modifying Toom-3 unbalanced method: compute $B_0 b$ instead of $81 B_0 b$, and in the matrix inversion phase – we consider the inversion sequence presented in [4] – just substitute a subtraction with a multiply-by-81-and-subtract operation. A contains a high-level implementation of the algorithm in PARI/GP [8], which the reader is invited to refer to.

As (unbalanced) Toom-3 computes 5 products, the new nonlinear complexity is $C'_{2n} = 2S_n + 5M_n < 3S_n + 5M_n = C_{2n}$. A very rough analysis, counting only the number of nonlinear operations and forgetting about their nature and linear operations as well, tells that just 7 instead of 8 operations are needed, so that the relative gain cannot exceed $1/8 = 12.5\%$. In Fig. 1 a relative percentage timing comparison of our C language implementation using GMP library operations versus GMP library itself (release 5.0.1) is shown[1]. The software was compiled with gcc 4.3.2 on an Intel Core 2 Duo (3 GHz) processor machine. As the graph shows, the more consistent gain is obtained when u is from 30 to 230 limbs long, while smaller gains are possible elsewhere, till around 5000 limbs.

4   Split and Cube: Polynomial Case

For what concern the generic polynomial case ($d > 1$), with the classical cube algorithm the coefficients of $p(x)^2$ must be computed first: this can be done with a Toom approach requiring $2d + 1$ squarings of linear combinations of a_i coefficients (we suppose that the computation of linear combinations has a negligible complexity with respect to coefficients squaring/multiplication).

Whatever splitting of $f(x)$ for the new algorithm works: if $d_0 + d_1 = d - 1$ we may write

---

[1] The C code is freely available on request.

$$p(x) = \left(\sum_{i=d_0+1}^{d} a_i x^{i-1} \right) x + \left(\sum_{i=0}^{d_0} a_i x^i \right) = p_1(x)x + p_0(x)$$

so that the master polynomial to be computed is the following one. Note that if $\deg(p_0) > 0$, more than one coefficient must be divided by 9.

$$p'(x) = \left(p_1^2(x)x^2 + 3p_0(x)^2 \right) \left(p_1(x)x + 3p_0(x) \right)$$

To compute its first factor with Toom approach as well, $(2d_1 + 1) + (2d_0 + 1) = 2(d_0+d_1+1) = 2(d-1+1) = 2d$ squarings are now needed, one less than before. This was expected, of course, and shows that this gain is actually constant; it depends neither on the degree of $p(x)$ nor on its splitting.

It is reasonable to suppose the existence of a threshold for the degree of $p(x)$ beyond which the new cube algorithm is no more worth while, depending on the nature of elements and the operations complexity in $\mathbb{A}$, on the method used for squaring (for example, if $\mathbb{A} = \mathbb{Z}$ – but also in other cases – FFT could be used instead of Toom, and comparisons become more tricky), on implementation details, etc. In practice, the candidate polynomial case with the best relative gain (if any) is the quadratic $(d = 2)$ one.

5   Generic Long Integer Exponentiation

For $1 < e \in \mathbb{N}$, generic exponentiation $U = u^e$ is usually performed by a binary algorithm – in the following box, consider the representation $(1, e_k, \ldots, e_0)$ of the exponent e in base 2 (with $e_i \in \{0,1\}$) and let $e_{(2)} = (2, e_k, 2, e_{k-1}, \ldots, 2, e_0)$ its (redundant) expansion, codifying power computation.

Binary exponentiation algorithm
| ```
let U = u
for i = k to 0 step −1 do
        U ← U²
            if eᵢ = 1 then U ← U · u
return U
``` |

With the help of the new cube algorithm, it is possible to conceive other exponentiation methods. Anyway, always keep in mind that the new cube algorithm is effective only in a certain interval, and therefore thresholds should always be taken care of.

We present different possibilities mixing cube and square exponents expansion, obtaining in some case a nice speedup, despite the new cubing algorithm.

5.1   Exponentiation: Ternary Expansion

The first considered possibility was ternary exponent expansion and the corresponding power algorithm obtained by $e_{(3)}$, defined similarly as $e_{(2)}$, but the
```

comparison with GMP – see [9] – showed that it is effective only for some very specific exponent. We therefore tried other ways to exploit binary and ternary expansions mixing.

5.2 Exponentiation: Mixed Binary and Ternary Expansion

The use of a mixed *binary-ternary* exponent expansion seems to pay more. Let $e > 3$ and $n_1(e) = {}^{\#}\{i \mid e_i = 1\}$. The idea is to interlace the cubing with binary expansion at the "more promising" points. To localize these points we reason on $n_1(\varepsilon_\ell)$ and $n_1(\lfloor \varepsilon_\ell/3 \rfloor)$ for exponents ε_ℓ defined as follows.

Let $0 \leqslant \ell \leqslant k$: consider all binaries sub-expansions $(1, e_k, \ldots, e_\ell)$ of $(1, e_k, \ldots, e_0)$, corresponding to exponent ε_ℓ. Consider now the index $i = \max\limits_{\ell=0,\ldots,k} \{\ell \mid 0 \leqslant n_1(\varepsilon_\ell) - n_1(\lfloor \varepsilon_\ell/3 \rfloor) - (\varepsilon_\ell \bmod 3)$ is maximum$\}$ or $i = -1$ if the maximum is negative. Once i is defined, we consider the algorithm synthesized by the sequence produced by the below recursive E_1 exponentiation function, whose behavior on basic cases ($e = 1, 2, 3$) is trivially defined; in all other cases it is

$$E_1(e) = \begin{cases} (E_1(\lfloor e/2^i \rfloor), 3, e_{i-1}, 2, \ldots, 2, e_0) & \text{if } i \geqslant 0 \\ e_{(2)} & \text{if } i = -1 \end{cases}$$

Another scenario is obtained if $i = \min\limits_{\ell=0,\ldots,k} \{\ell \mid 0 < n_1(\varepsilon_\ell) - n_1(\lfloor \varepsilon_\ell/3 \rfloor) - (\varepsilon_\ell \bmod 3)$ is maximum$\}$, giving the E_2 function, defined similarly as above.

Example 1. $E_1(42) = (3, 0, 2, 1, 3, 0, 2, 0)$. The corresponding power raising computation sequence is $u \to u^3 \to u^6 \to u^7 \to u^{21} \to u^{42}$. $E_2(42)$ coincides instead with binary decomposition.

As a comparison, we show in table 5.3 the different decompositions of exponents up to 100 given by the $E_1(\cdot)$ and $E_2(\cdot)$ functions, only when they differ. For brevity, we omit all zero entries. We observe that E_1 tends to use cubes more often and possibly earlier (with smaller operands), while E_2 uses fewer cubes, and later (larger operands).

5.3 Exponentiation: First Ternary and Then Binary Expansion

Always trying to apply the new cubing algorithm as soon as possible, the *ternary-binary* expansion considers the most 4 significant bits of the binary expansion of $e > 7$, looking for one/two possible uses of the new cube algorithm, as the following function suggests, continuing the exponentiation with the binary algorithm.

$$E_3(e) = \begin{cases} (3, 0, 3, 0, 2, e_{k-3}, \ldots) & \text{if } e_k e_{k-1} e_{k-2} = 001 \\ (2, 0, 2, 1, 3, 0, 2, e_{k-3}, \ldots) & \text{if } e_k e_{k-1} e_{k-2} = 111 \\ (3, 0, 2, e_{k-1}, \ldots) & \text{if } e_k = 1 \text{ and } e_{k-1} e_{k-2} \neq 01, 11 \end{cases}$$

Table 1. Comparison: E_1 and E_2 exponents expansions

e	$E_1(e)$	$E_2(e)$	e	$E_1(e)$	$E_2(e)$
6	[3,2]	[2,3]	54	[3,3,3,2]	[2,2,2,1,2,3]
7	[3,2,1]	[2,3,1]	55	[3,3,3,2,1]	[2,2,2,1,2,3,1]
9	[3,3]	[2,2,2,1]	56	[3,2,1,2,2,2]	[2,3,1,2,2,2]
12	[3,2,2]	[2,2,3]	57	[3,2,1,2,2,2,1]	[2,2,2,1,2,1,3]
13	[3,2,2,1]	[2,2,3,1]	58	[3,2,1,2,2,1,2]	[2,3,1,2,2,1,2]
14	[3,2,1,2]	[2,3,1,2]	59	[3,2,1,2,2,1,2,1]	[2,3,1,2,2,1,2,1]
18	[3,3,2]	[2,2,2,1,2]	60	[2,2,1,3,2,2]	[2,2,1,2,2,3]
19	[3,3,2,1]	[2,2,2,1,2,1]	61	[2,2,1,3,2,2,1]	[2,2,1,2,2,3,1]
21	[3,2,1,3]	[2,2,1,2,2,1]	62	[2,2,1,3,2,1,2]	[2,2,1,2,3,1,2]
24	[3,2,2,2]	[2,2,2,3]	63	[3,2,1,3,3]	[2,2,1,2,2,1,3]
25	[3,2,2,2,1]	[2,2,2,3,1]	72	[3,3,2,2,2]	[2,2,2,1,2,2,2]
26	[3,2,2,1,2]	[2,2,3,1,2]	73	[3,3,2,2,2,1]	[2,2,2,1,2,2,2,1]
27	[3,3,3]	[2,2,2,1,3]	74	[3,3,2,2,1,2]	[2,2,2,1,2,2,1,2]
28	[3,2,1,2,2]	[2,3,1,2,2]	75	[3,2,2,2,1,3]	[2,2,2,3,1,3]
29	[3,2,1,2,2,1]	[2,3,1,2,2,1]	76	[3,3,2,1,2,2]	[2,2,2,1,2,1,2,2]
30	[2,2,1,3,2]	[2,2,1,2,3]	77	[3,3,2,1,2,2,1]	[2,2,2,1,2,1,2,2,1]
31	[2,2,1,3,2,1]	[2,2,1,2,3,1]	78	[3,2,2,1,3,2]	[2,2,3,1,2,3]
36	[3,3,2,2]	[2,2,2,1,2,2]	79	[3,2,2,1,3,2,1]	[2,2,3,1,2,3,1]
37	[3,3,2,2,1]	[2,2,2,1,2,2,1]	84	[3,2,1,3,2,2]	[2,2,1,2,2,1,2,2]
38	[3,3,2,1,2]	[2,2,2,1,2,1,2]	85	[3,2,1,3,2,2,1]	[2,2,1,2,2,1,2,2,1]
39	[3,2,2,1,3]	[2,2,3,1,3]	86	[3,2,1,3,2,1,2]	[2,2,1,2,2,1,2,1,2]
42	[3,2,1,3,2]	[2,2,1,2,2,1,2]	87	[3,2,1,2,2,1,3]	[2,3,1,2,2,1,3]
43	[3,2,1,3,2,1]	[2,2,1,2,2,1,2,1]	90	[2,2,1,3,3,2]	[2,2,1,2,1,2,2,1,2]
45	[2,2,1,3,3]	[2,2,1,2,1,2,2,1]	91	[2,2,1,3,3,2,1]	[2,2,1,2,1,2,2,1,2,1]
48	[3,2,2,2,2]	[2,2,2,2,3]	93	[2,2,1,3,2,1,3]	[2,2,1,2,1,2,1,2,2,1]
49	[3,2,2,2,2,1]	[2,2,2,2,3,1]	96	[3,2,2,2,2,2]	[2,2,2,2,2,3]
50	[3,2,2,2,1,2]	[2,2,2,3,1,2]	97	[3,2,2,2,2,2,1]	[2,2,2,2,2,3,1]
52	[3,2,2,1,2,2]	[2,2,3,1,2,2]	98	[3,2,2,2,2,1,2]	[2,2,2,2,3,1,2]
53	[3,2,2,1,2,2,1]	[2,2,3,1,2,2,1]	100	[3,2,2,2,1,2,2]	[2,2,2,3,1,2,2]

5.4   Results

We observed that the cube operation is, anyway, very important per se. In fact, by mixing binary and ternary exponentiation we obtained quite large savings, even outside the interval in which the new cube algorithm is effective.

Some examples are shown in Fig. 2 and 3, while Fig. 4 shows our results for limbs from 1 to 1200 (abscissas) and exponents from 3 to 63 (ordinates), showing the fastest exponentiation algorithm for each case. Red points correspond to GMP winning, black ones to E_1, blue to E_2, green to E_3, while white lines refer to exponents for which all methods fall back to binary exponentiation, so that they're all essentially equivalent.

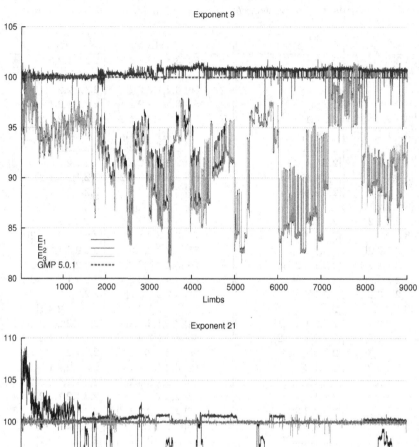

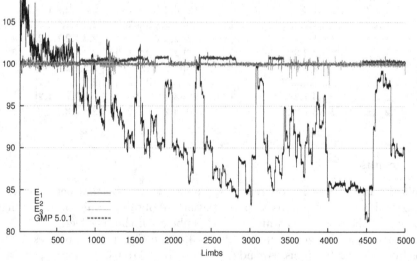

Fig. 2. E_1, E_2 and E_3 versus GMP-5.0.1 exponentiation relative timings for exponents 9, 21.

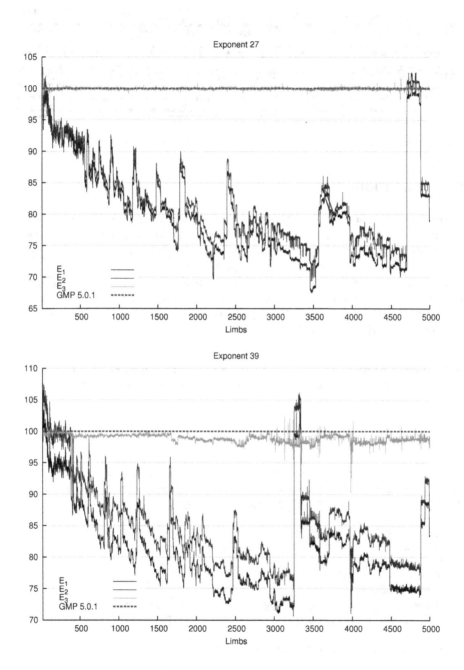

Fig. 3. E_1, E_2 and E_3 versus GMP-5.0.1 exponentiation relative timings for exponents 27, 39.

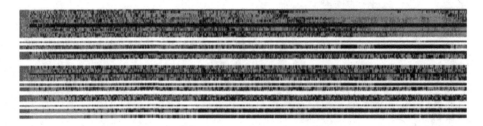

Fig. 4. Best exponentiation method: x-axis : limbs (1-1200), y-axis : exponents (3-63)

6   Conclusions

In this work the classical algorithm to compute the third power of a long integer was proved not to be (always) optimal. A new algorithm using a combination of Karatsuba and unbalanced Toom-3 methods paying some linear operations more for a nonlinear operation less was proposed.

Used as a new basic tool for generic long integer power raising computation, it contributed to obtain new, faster generic exponentiation procedures. Cubing proved to be effective in practice: comparison with GMP library showed that the obtained saving can reach 25 %.

Acknowledgments. The second author was partially funded by project "Robustezza e tolleranza ai guasti in reti e grafi", Sapienza University of Rome, Italy.

A   Code Implementation

We report here a PARI/GP high-level implementation of the new algorithm for cube computation, valid for both polynomials (balanced splitting is chosen: $d_0 = \lceil (d-1)/2 \rceil, d_1 = \lfloor (d-1)/2 \rfloor$) and long integers.

```
cube_long(a = 2) = {
 local(d,a0,a1,A0,A1,A00,A01,A10,A11,A00orig,H,tmp);

 \\ Compute a0 and a1.
 d = poldegree(a);
 if (d==0, H = 2^(ceil(log(a)/log(2)+1)>>1); \\ Long integer case
        , H = x^ceil((d+1)/2);                \\ Polynomial case
    );
 tmp = divrem(a,H); a1 = tmp[1]; a0 = tmp[2];

 \\ Identify high and low parts of a1^2 and a0^2.
 A1  = a1^2;       A0  = a0^2;   \\ 2S(n)
 tmp = divrem(A0,H); A01 = tmp[1]; A00orig = tmp[2];
 tmp = divrem(A1,H); A11 = tmp[1]; A10     = tmp[2];
```

```
A01 *= 3; A00 = 27*A00orig;

\\ Unbalanced Toom-3 (toom42) for AA = A11*x^3 + A10*x^2 + A01*x + A00;
\\ Evaluation                       BB = a1*x + 3*a0;
\\ in (oo,2,1,-1,0)
\\      AA                      BB
W2   = A10 + A00;
Wm1  = A11 + A01;
W0   = W2 + Wm1;
W2   = W2 - Wm1;        Wm1  = 3*a0;
                       Winf = Wm1 + a1;
W1   = W0*Winf;        \\ Evaluation in 1     M(n)
                       Winf = Winf + a1;
                       W0   = Wm1  - a1;
Wm1  = W0*W2;          \\ Evaluation in -1    M(n)
W0   = A11<<1 + A10;
W0   = W0<<1  + A01;
W0   = W0<<1  + A00;
W2   = W0*Winf;        \\ Evaluation in 2     M(n)
Winf = A11*a1;         \\ Evaluation in oo    M(n)
W0   = A00orig*a0;     \\ Evaluation in 0 (/81) M(n)


\\ Interpolation:
W2  = W2 - Wm1;        \\ [15 9 3 3 0 ]
W2 /= 3;               \\ [ 5 3 1 1 0 ]
Wm1 = (W1 - Wm1)>>1;   \\ [ 0 1 0 1 0 ]
W1  = W1  - 81*W0;     \\ [ 1 1 1 1 0 ]
W2  = (W2 - W1)>>1;    \\ [ 2 1 0 0 0 ]
W1  = W1  - Wm1;       \\ [ 1 0 1 0 0 ]
W2  = W2  - Winf<<1;   \\ [ 0 1 0 0 0 ]
W1  = W1  - Winf;      \\ [ 0 0 1 0 0 ]
Wm1 = Wm1 - W2;        \\ [ 0 0 0 1 0 ]
Wm1 /= 9;              \\ Extra division.


\\ Recomposition:
return ((((Winf*H + W2)*H + W1)*H + Wm1)*H + W0);
}
```

References

1. Karatsuba, A.A., Ofman, Y.: Multiplication of multidigit numbers on automata. Soviet Physics Doklady 7(7), 595–596 (1963)
2. Toom, A.L.: The complexity of a scheme of functional elements realizing the multiplication of integers. Soviet Mathematics Doklady 3, 714–716 (1963)
3. Cook, S.A.: On the minimum computation time of functions. PhD thesis, Department of Mathematics, Harvard University (1966)
4. Bodrato, M., Zanoni, A.: Integer and polynomial multiplication: towards optimal Toom-Cook matrices. In: Brown, C. (ed.) ISSAC 2007: Proceedings of the 2007 International Symposium on Symbolic and Algebraic Computation, pp. 17–24. ACM, New York (2007)

5. Fürer, M.: Faster integer multiplication. In: Johnson, D.S., Feige, U. (eds.) STOC, pp. 57–66. ACM (2007)
6. Schönhage, A., Strassen, V.: Schnelle Multiplikation großer Zahlen. Computing 7(3-4), 281–292 (1971)
7. Granlund, T., et al.: The GNU multiple precision (GMP) library (2010), http://gmplib.org/
8. PARI/GP: PARI/GP, version 2.5.0. The PARI Group, Bordeaux (2012), http://pari.math.u-bordeaux.fr/
9. Zanoni, A.: Another sugar cube, please! or sweetening third powers computation. Technical Report 632, Centro "Vito Volterra", Università di Roma "Tor Vergata" (January 2010)

Lightweight Abstraction
for Mathematical Computation in Java

Pavel Bourdykine and Stephen M. Watt

Department of Computer Science,
University of Western Ontario,
London, Canada
pbourdyk@csd.uwo.ca, Stephen.Watt@uwo.ca

Abstract. Many object-oriented programming languages provide type safety by allowing programmers to introduce distinct object types. In the case of Java, having objects as the sole abstraction mechanism also introduces a considerable or even prohibitive cost, especially when dealing with small objects over primitive types. Consequently, Java library implementations frequently avoid abstraction and are not type safe in practice. Many applications, including computer algebra, use values logically belonging to many different non-interchangable types. Languages such as Java are then either unsafe or inefficient to use in these applications. We present a solution allowing type safety in Java with little performance penalty. We do this by introducing a specialzed kind of object that provides distinct types for type checking, but which can always be removed entirely at compile time. In our implementation, programs are compiled twice, first with objects to verify type safety, and then with the objects removed for efficiency. This gives significant performance gains across a range of tests, including the generic SciGMark tests.

1 Introduction

A large part of the art of programming language design lies in how one assembles a multitude of ideas that are in principle distinct into a few simple constructs that work well together. When this is done well, it can be beautiful. When this is done badly, it can make programs inefficient and error-prone. This paper argues that this is what has happened in languages such as Java, where objects are the sole data abstraction mechanism, and we present a solution.

A simple example of separate considerations that can be nicely combined is given by the modern **return** statement. In principle, setting the value of a function and the transfer of control back to the caller are completely separate ideas. Indeed, in older programming languages these were done separately. One might easily want to perform some clean up actions, such as closing files, returning resources or updating global state, after the return value is determined. However, in these cases, using a temporary variable with modern **return** is neither costly nor dangerous. At first sight, using objects as the sole data abstraction mechanism would seem to be a similar happy combination. Abstract data types

V.P. Gerdt et al. (Eds.): CASC 2012, LNCS 7442, pp. 47–59, 2012.

typically have several fields in a hidden representation and provide operations on the abstract values, just as with classes in an object-oriented world. But there are several problems with this:

Abstraction is not just hiding record fields. Providing data abstraction only via objects forces all thinking about abstraction into the model of field visibility in composite structures. Quite often, one wishes to consider simple values as elements of a distinct type. For example, even though window IDs may be represented as integers, it would enhance program safety if they were treated as a distinct type. Likewise for values in different prime fields should be of different types from the integers and from each other. Additionally, composite data is often not represented as fields in an object. For example, it is not uncommon to represent colors as 32-bit integers with bit fields representing component values. Abstraction can help ensure that only integers intended to be color information are used as such.

Abstraction does not always need dynamic allocation, inheritance, synchronization, or other heavy-weight mechanisms. Sometimes we want abstraction only to ensure that programs do not depend on details that may later change and to enhance safety by ensuring values are not used inappropriately. It may be known from the outset that these values will not ever be used in any fancy ways. For example, we may know that there will never be any derived types from colors, there will be no subtle multiprocessing on single color values, *etc.* Requiring all of these features to be supported on abstract values first has a cost, and second reduces flexibility to have these abstract values treated in other interesting ways.

Abstraction is not used if it is too inefficient or onerous. When data abstraction carries a significant efficiency penalty *and* thought on the part of the programmer, then it is not used.

In languages such as C++, where objects and primitive types are on a similar footing, the extra cost need not be large. Nevertheless, even here, mechanisms have been proposed for opaque type definitions in C++ [1]. In languages such as Java, where there is a strong distinction between primitive types and object types, the cost to use objects is many times that of using primitive types. Programs are inefficient, programmers circumvent the type system or both. This has many obvious problems. If Java were not such a widely adopted language, we could reject it as being ill suited in these circumstances. As it is, some solution is needed.

The contribution of this paper is to show how light weight abstraction may be provided in Java. This provides type safety without introducing any significant inefficiencies. It is therefore suitable for creating light-weight abstract types for computationally intensive, efficiency-critical tasks such as computer algebra and scientific computing. Section 2 shows how this may be achieved by introducing object types with sufficient restrictions that they are guaranteed to be removed at compile time. While these ideas are presented for Java, the same ideas could equally well be applied in other settings. Section 3 then describes a tool that implements this mechanism that can be used in conjunction with standard Java compilers. Section 4 presents performance results, comparing the usual use of objects, the present light-weight abstraction mechanism and raw primitive types. These comparisons are made using the SciGMark test suite and details are shown for polynomial and matrix multiplication. Finally Section 5 concludes the paper.

2 Opaque Types in Java

To deal with the problems outlined, we introduce the notion of *opaque* types in the Java programming language. These types allow development of Java code that is reusable, elegant, and efficient. *Opaque* types are meant to be used as regular object types that can be represented internally by any other Java type with a focus on representation via primitive built-in Java types. The new types are required to behave and act like regular object types in the way they interact with the Java class hierarchy and the static type-checker. An example of this kind of application may be an object that has a small finite number of different states that can be intuitively represented by a set of bit patterns. Although this can be implemented similarly to something written in assembly language, by using *int* types, resulting in code that is quite efficient, the code's extensibility would suffer. Moreover, like assembly, this type of code is difficult to maintain, and debug[2,3]. This may lead to errors that could have been easily avoided if object types were used.

This approach encompasses a core notion of *opacity*. High level Java objects do not necessarily have to be represented or compiled as such. Objects simply serve as identification handles for static type-checking prior to compilation. The underlying type of these objects may be anything suitable for internally representing the construction. In this fashion, an alternative *String* object may be represented by a character array allowing for operations very similar to those on strings implemented in C or C++. In turn, a more complex object may be represented by such an alternative *String* thus creating an artificial class hierarchy that remains consistent and type-safe. In this work, however, we are concerned mostly with objects that may be represented by primitive types in order to boost performance.

Along with the optimized version of the *opaque* type the regular unchanged version of the class is kept for reference and debugging purposes. Leaving the user code unchanged after compilation allows for more straight forward top-level design where good Object Oriented Design practices may be followed. The user may also choose to compile the *opaque-typed* code and run it as is, without conversion, in order to ensure correctness. Keeping both versions of the class demonstrates the type safety of *opaque* Java types as either version of the project will produce identical execution results.

In order to implement Java *opaque* types, we introduce a set of type rules that have to be followed in order to use such objects safely and efficiently. These rules may be used by a preprocessor to transform the user's regular objects into those for which the generated code will use the underlying primitive types. We now give a more detailed description of these rules.

2.1 *Opaque* Type Rules

We use a Java code annotation (called *Opaque*) to identify classes as *opaque* types. Java annotations allow embedding of meta-data directly into Java source code. "Annotations do not directly affect program semantics, but they do affect the way programs are treated by tools and libraries, which can in turn affect the semantics of the running program."[4] The annotation has a single *String* type field that denotes the primitive representation type of the opaque object. For example, the annotation @Opaque("int") indicates that the object is *opaque* and that its primitive representation is of type *int*. Currently, the annotation field serves as a way to quickly identify the underlying type and speed up *opaque type* file analysis but could be left out in later versions of the solution. The single annotation dictates all the required information to the preprocessor. The next restrictions/rules must be followed in order to guarantee successful conversion consistent with the Java language standard:

- **Rule 1** object must have a single `protected` field of the underlying type unless it is a subclass of an *opaque type*
- **Rule 2** object constructor(s) must be declared `private`
- **Rule 3** all methods accessing or modifying the underlying type field representation must be declared `static` (or `final static` if no subclasses override the methods)

Rule 1 enforces *opaque* type representation and assures that it matches with the type specified by the annotation. The field (from here on in referred to as `rep`) takes place of the *opaque* object whenever it appears in translated user code. It is important that its uses are properly implemented and ensures there are no compilation issues post-conversion. The approach to having a single field for the representation is similar to that used in Aldor [5].

If the new *opaque* object `extends` an *opaque* type (the inheritance property detected by the preprocessing utility), the new object must *not* include a `rep` field in its declaration. The `rep` field is instead inherited from the superclass and bares the same primitive type. This ensures consistency in method inheritance and conversion.

Rule 2 follows the Java convention that only object types require a constructor. Since the new *opaque* object is to be converted to its underlying primitive type representation wherever it is used, its constructor must remain `private`. Creating new instances of the *opaque* object is still possible through the use of the `static` method "New". This method should be implemented by the user as a means of converting from the underlying primitive type to the object type primarily for testing purposes and initial implementation of code that uses *opaque* types. The typical implementation is outlined in Figure 1(a).

Rule 3 places a restriction on the other methods possibly acting on the object representation. Default visibility `static` methods allow inheritance and class access to regularly used operations within the new object. At first glance this may seem limiting for using the object; however, since object instances are all converted to the underlying primitive type, only class methods remain as valid

```
// a. Opaque object,
//    typical "New" implementation
@Opaque("short")
public class MyOpaqueObject {
    protected short rep;
    private MyOpaqueObject(short r) {
        rep = r;
    }
    ...
    public static MyOpaqueObject
    New(short r) {
        return new MyOpaqueObject(r);
    }
}


// b. Opaque object before conversion
@Opaque("int")
public class BaseClass {
    protected int rep;
    private BaseClass(short r) {
        rep = (int) r;
    }
    public static void
    operator(BaseClass bc, short modifier){
        ...
    }
    ...
}
```

```
// c. Opaque object after conversion
public class BaseClass {
    protected int rep;
    private BaseClass(short r){rep = (int) r;}

    public static void
    operator(int bc, short modifier) {
        ...
    }
    ...
}
// d. Regular main class
@Opaque("user")
public class TopLevel {
    public static void main(String[] args){
        OpaqueType var = OpaqueType.New(param);
        ...
    }
}
// e. Opaque array initialization
@Opaque("user")
public class TopLevel {
    public static void main(String[] args){
        OpaqueType[] ots =
                    new OpaqueType[DATA_SIZE];
        for(int i=0; i < DATA_SIZE; i++)
            ots[i] = OpaqueType.New(param);
        ...
    }
}
```

Fig. 1. Opaque object creation and use

operations that can act upon the object's actual implementation, i.e., its **rep** field. This method declaration simplifies the preprocessor task of handling extension quickly and efficiently and assures the *opaque* object is not inflated by non-static behaviors.

The approach takes advantage of the way Java class hierarchy works by allowing subclasses to preserve the "is a" relationship and properly inherit methods with default visibility. Properties of method overloading are also preserved due to use of default visibility and the requirement of using the class name whenever a method is called.

2.2 Converted Classes

The style for new object creation is modified slightly when attempting to make use of opaque types. It is useful to illustrate exactly what changes in the type declaration following invocation of the code conversion utility. Figures 1(b) and 1(c) show the original version of a simple class and its converted result respectively.

In Figure 1(b), object **BaseClass** is a Java opaque type represented by the built-in Java **int** type. The important details to notice regarding this class are its **protected int rep** field, **private** constructor and **static void operator** method. These three points are required by the semantic rules outlined in Section 2.1 and allow **BaseClass** to be converted and compiled as its underlying built-in type (**int**).

The converted `BaseClass` can be seen in Figure 1(c). In the new version of the class, the *@Opaque("int")* annotation has been removed, and the `static` method `operator` has been modified to use only arguments of the proper underlying types. The class retains its "high level" handle - `BaseClass`. Hence the user code that makes use of the class only needs minor typing modifications.

2.3   Opaque User Classes

Java classes that declare or make use of opaque objects are also annotated with the *@Opaque* annotation. Instead of an underlying `rep` type, the user classes contain the keyword "user" as the single parameter to the annotation. A user class may look as simple as in Figure 1(d).

Classes declaring objects of opaque type that are not opaque themselves (annotated as *@Opaque("user")*) undergo only minor changes during the conversion step. Opaque types are constructed in a way such that their declaration, initialization and usage do not require any object-exclusive syntax aside from declaring data structures whose elements are opaquely typed. The most common and primitive of these data structures is an array. Declaration of Java arrays containing opaque typed members is syntactically identical to any other array declaration (for any number of dimensions). Initialization, however, is dictated by the nature of the opaque members themselves - each opaque object is initialized via the `New` method as opposed to using the Java `new` keyword. This simple yet notable concept is summarized in Figure 1(e).

2.4   Annotation Processing Example

The structure that opaque annotations impose on Java source code is non-hierarchical despite playing a role in the hierarchical class structure of Java. The traversal through this construction of annotated source files is straight forward for the most part and the conversions applied to the code are often influenced directly by information contained in the same processing step. Figure 2 illustrates some subtleties during the conversion step that arise when analyzing a deep class hierarchy for opaque objects.

The analysis of such a hierarchy takes place as follows. The *Opaque*-annotated classes are identified among the source files and a list of them is stored along with their representation (taken from the `String`-type annotation argument). In Figure 2 this list would consist of pairs `BaseClass & int`, `ChildClassOne & int`, `ChildClassTwo & int`, `ChildClassThree & int`. The annotations make the suggested representation clear, but we still have to check that the class does not attempt to use a differently typed field. As you can tell, the underlying representation property is inherited, in this case all the way down the hierarchy from `BaseClass`. Method inheritance is taken care of by standard Java, for example, the `operator` method of `ChildClassThree` is overwritten for only that class and the overloaded method works as expected. The goal of our approach is to make the preserved object properties as intuitive as possible (i.e., make them work as the programmer would expect) during application of the *opaque* types mechanism.

```
@Opaque("int")                          @Opaque("int")
public class BaseClass {                public class ChildClassTwo
    protected int rep;                      extends BaseClass
    private BaseClass(short r){         {
        rep = (int) r;                      private ChildClassTwo(long r){
    }                                           rep = (int) r;
    public static void operator             }
        (BaseClass bc,                      ...
        short modifier)                 }
    { ... }
    ...                                 @Opaque("int")
}                                       public class ChildClassThree
                                            extends ChildClassTwo
@Opaque("int")                          {
public class ChildClassOne                  private ChildClassThree(int r){
    extends BaseClass                           rep = r;
{                                           }
    private ChildClassOne(short r){         public static ChildClassThree operator
        rep = (int) r;                          (ChildClassTwo modifier, short cc){
    }                                           ...
    ...                                     }
}                                           ...
                                        }
```

Fig. 2. Inheritance during conversion

3   Java Implementation

Implementing opaque Java types requires careful considerations in order to abide by the set restrictions and still make the new data declaration forms useful. For performance, ideally, the underlying representation of a particular type could be determined during the compilation process and the underlying type used for code generation. This kind of automated optimization would mean a seamless implementation of a significant performance gain. However, allowing the programmer to specify the underlying type of the opaque object allows for greater flexibility for accomplishing a certain task, even if at the cost of some efficiency.

A sophisticated mechanism to determine the underlying object representation on the fly could be an area of significant research, however, at this time we have elected to make the choice explicit. Hence the steps to build projects containing opaque types are quite straight forward.

Code conversion utility. In order to realize the potential of Java *opaque* types we need to develop a dual view of the annotated objects to the compiler. The first is the object view – necessary to take advantage of Java's inherent ability to handle a rich type hierarchy. The second is the underlying representation view, the one to be used during optimization and code generation phases of compilation. This dual representation is achieved using a code conversion utility written in Java itself making use of **Pattern** and **Matcher** classes from the `java.util.regex` package.

These classes provide a convenient way to identify where and how *opaque* types are used and apply conversions directly to Java source code. This allows the utility to finish its tasks in a timely manner without complicating the process of going from regular-looking objects (*opaque* types) to the immediate underlying representation.

The utility performs the following steps:

1. identify all recently modified Java source files in target project
2. sort source files into regular, opaque typed, and opaque user classes
3. build record of all opaque types and their underlying representations
4. convert all opaque sources

Automating the building process. Utilizing a pre-processor-like code conversion application prior to compilation complicates the building process by adding a necessary intermediate step to the routine mechanism. However, Java is a flexible language with a relatively long standing industry and research history. By this virtue a number of tools have been developed that augment various features of the language in particular when it comes to its compilation and building process. One of such tools is the Ant scripting language[6]. We use an Ant build script to perform the following tasks:

1. back up original source files
2. invoke converter on files modified since last invocation
3. compile newly converted files

Eclipse IDE. Integration into a main-stream development environment may seem like an extraneous task; however, this discussion follows naturally due to the Eclipse's ability to use Ant build scripts instead of the default compiler or build-chain. Implementing the build script directly into the Eclipse IDE allows the user to seamlessly develop code utilizing *opaque* types in the IDE.

Further details of the implementation are described in the first author's master thesis [7].

4   Performance Results

If regular Java objects performed as well as built-in types, there would be no need to invent a new mechanism for abstraction. This, however, is not the case. Primitive types in Java perform far better than objects.

We consider the overall application performance for synthetic tests by the time it takes the program to execute, and the memory consumed during its execution. Computational benchmark performance is compared using the number of floating operations per second performed by various implemented algorithms. We compare performance of Java code using regular objects, code which has been converted to use *opaque* objects, and code implemented with the use of primitive types only (dubbed "specialized"). For the purposes of measuring performance in such a way we have devised several synthetic tests that demonstrate *opaque* type advantages using brief implementations and included two modified benchmarks from the SciMark[8] and SciGMark[9] performance benchmark suites. The measurements for testing performance that could be adjusted to utilize *opaque* types most naturally have been included in this report. The particular benchmarks chosen from SciMark 2.0 and SciGMark 1.0 suites are dense polynomial multiplication with integer field coefficients originally developed for SciMark benchmark and modified by SciGMark and sparse matrix multiplication with real coefficient values.

The modified applications accomplish identical tasks and have minimal implementation differences aside from the use of *opaque* types and corresponding annotations. Along with the borrowed benchmarks, the synthetic tests that range from simple classes implementing only a few methods with shallow class hierarchy to classes with a large internal representation (e.g. a large integer array), several constructors, and a large number of methods are used to measure "bare bone" performance. All tests were executed 10 times in order to compensate for varying CPU and memory loads on different platforms. The averaged results were recorded and are shown next. The computationally intensive benchmarks were executed on large data sets in order to maximize the effect of data allocation and access on performance when dealing with objects versus more primitive structures in large quantities. This in turn increased result accuracy due to floating point operations being used as the measurement units. The simple tests were chosen to reflect varying uses and applications developers may encounter when writing Java code for a typical project.

Benchmark implementations were tested on several different platforms in order to demonstrate *opaque* types' independence of environment when increasing computational performance. The platforms used for testing were as follows:

- Intel C2Q Q6600 @ 2.4GHz, 4GB RAM, Windows 7 x86_64, JRE 1.7 (**lambda**)
- Intel I7-870@2.93GHz,16GB RAM, Ubuntu Server 10.04 x86_64, JRE 1.6(**tedium**)
- Intel Xeon E5620 @ 2.4GHz, 24GB RAM, Ubuntu 10.04 x86_64, JRE 1.6 (**z600**)
- Intel I5-660 @ 3.33GHz, 4GB RAM, Ubuntu 10.04 x86_64, JRE 1.6 (**PCA-45**)
- Intel C2D E4600@2.4GHz, 2GB RAM, Ubuntu 10.04 x86_64, JRE 1.6 (**orccapc02, orccapc03, orccapc04**)

Running of experiments on different platforms has also given us an opportunity to look at the variance in underlying software that affects the performance of Java applications using *opaque* types. The results were not significantly impacted by execution on different platforms, and even the JVM versions used did not incur a great deal of variance on the results.

Execution time and memory use were measured using built-in Java tools for determining system time (method `currentTimeMillis()` in `java.lang.System`), and tools for determining how memory is currently used by the Java Virtual Machine - `Runtime` methods called `totalMemory()` and `freeMemory()`. All tests measuring memory use were carefully designed to avoid involuntary garbage collection and execution time tests were averaged to account for varying CPU load during the experiments and were generally run at the highest CPU affinity.

Complex internal representations. Similarly to the *opaque* types used through this work, it is possible to represent object types by a single primitively typed array fields of fixed size. For example, an *opaque* type object may be represented by 256 bits, or an array of size 4 of type `long[]`. The next set of tests deals with objects represented by different sized arrays of primitively typed variables.

```
@Opaque("long[]")                              public class RegObject {
public class MyOpaqueObject {                      private long[] rep;
    protected long[] rep;                          public RegObject(long[] arg){
    private OpaqueObject(long[] arg){                  rep = new long[arg.length];
        rep = new long[arg.length];                    for (int i = 0; i < arg.length; i++)
        for (int i = 0; i < arg.length; i++)               rep[i] = arg[i];
            rep[i] = arg[i];                       }
    }                                              public void setBit(int i){
    public static OpaqueObject                         long mask = (long) (1 << (i % 64));
    New(long[] arg) {                                  rep[i / 64] |= mask;
        return new OpaqueObject(arg);              }
    }                                              ...
    public static OpaqueObject
    setBit(OpaqueObject o, int i) {             }
        long mask = (long) (1 << (i % 64));
        o.rep[i / 64] |= mask;
        return o;
    }
    ...
}
```

Fig. 3. Regular and *opaque* objects with array typed fields

The tests use the same, previously shown, metrics to measure execution speed and memory use. The implementation of the actual accomplished operation is kept as identical as possible to avoid performance differences due to algorithmic discrepancies. This assures that we compare directly the speed and size of regular objects versus *opaque* objects without introducing unnecessary bias.

Figure 3 illustrates an *OpaqueObject* represented by the `long[]` type and a *RegObject* that has a field of type `long[]`. Both objects have the similarly implemented method called *setBit*. Method *setBit* takes an argument of type `int` that corresponds to the bit number that must be turned on in the internal representation of the *OpaqueObject* or the field of the *RegObject* with 0 being the least significant bit. Imagine arranging either the internal `long[]` representation of *OpaqueObject* or the field of *RegObject* as sets of back-to-back 64 bit sets (each set represented by a `long` type value) where significance of the bits increases with the array index of the respective field. Thus operation *setBit* is potentially able to turn on a single bit in a bit set of size over 2,000,000,000. For the test, however, we limit the size of the `long` array to 4.

Table 1. Matrix Multiplication

PC	Code			Improvement	
	Generic (mflops)	Specialized (mflops)	Opaque (mflops)	Opaque vs. Generic	Opaque vs. Specialized
lambda	131.84	475.22	383.64	2.91	0.81
tedium	194.64	1199.08	968.4	4.98	0.81
z600	175.16	1044.22	833.06	4.76	0.80
PCA-45	158.82	1077.7	845.84	5.33	0.78
orccapc04	57.20	363.5	303.68	5.31	0.84
orccapc03	62.34	371.1	299.45	4.80	0.81
orccapc02	60.94	368.4	301.94	4.95	0.82
sodium	54.82	311.26	248.48	4.53	0.80
Overall average improvement:				4.70	0.81

Table 2. Polynomial Multiplication

PC	Code			Improvement	
	Generic (mflops)	Specialized (mflops)	Opaque (mflops)	Opaque vs. Generic	Opaque vs. Specialized
lambda	75.54	279.56	223.86	2.96	0.80
tedium	147.02	900.44	729.06	4.96	0.81
z600	131.32	800.68	639.42	4.87	0.80
PCA-45	136.54	910.64	723.38	5.30	0.79
orccapc04	56.29	355.15	285.91	5.08	0.81
orccapc03	54.98	350.90	288.3	5.24	0.82
orccapc02	57.84	355.32	287.92	4.98	0.81
sodium	38.90	223.68	179.52	4.61	0.80
Overall average improvement:				4.75	0.81

Putting it together. The next set of performance comparison tests consists of two standard benchmarks taken from the SciMark and SciGMark suites. In order to implement polynomial multiplication and sparse matrix multiplication benchmarks we build on the conventions established previously and reuse some implementations from the synthetic benchmarks.

The first test performed is sparse matrix multiplication with double precision floating point coefficients taken randomly from the complex number set. This is one of the most natural performance indicators for a language feature or a hardware benchmark. In this case, the test's aim is to demonstrate that it is possible to significantly increase the raw number of floating point calculators (measured here in millions of floating point operations per second) without losing correctness by reducing abstraction (or removing it altogether) in the implementation. Unfortunately, fully disposing of abstraction, as SciGMark implementation shows, highly obscures the code. Use of primitive types yields high performance and optimized execution, however, the code becomes more complex and is difficult to modify. The matrix sizes used for measurement are N by N matrices with $N = 10,000$ averaging $100,000$ non-zero coefficients per matrix.

The purpose of the implementation using *opaque* types is to preserve abstraction introduced by the generic object implementation utilized by SciGMark while pushing performance figures towards that of the specialized code. Figures 4(a), 4(b), 4(c) show snippets of the code implementing the underlying complex data and algorithms used in this benchmark. The included code shows implementation of the type creation, summing, and multiplication.

Performance results using these varying implementations of the complex data types are summarized in Table 1. Analyzing the data it's easy to conclude that without loss of much generality, the *opaque* implementation is on average 4.7 times faster than the general object implementation and is only about 20% slower than the specialized implementation from Figure 4(b).

The second benchmark used in our final set of tests is polynomial multiplication with dense polynomials of degree ≤ 40. The polynomial coefficients are once again taken from the complex set and are implemented in three different ways according to each multiplication algorithm (generic objects, specialized,

```
// a. Generic object implementation
public class Complex <R extends IRing<R>> {
    private R re;
    private R im;
    public Complex<R> create(R re, R im) { return new Complex<R>(re, im); }
    public Complex<R> s(Complex<R> o) { return new Complex<R>(re.s(o.re()),im.s(o.im())); }
    public Complex<R> m(Complex<R> o) {
        return new Complex<R>(re.m(o.re()).s(im.m(o.im())), re.m(o.im()).a(im.m(o.re())));
    }
}
// b. Specialized implementation
public class Complex {
    private double re;
    private double im;
    public Complex create(double re, double im) { return new Complex(re, im); }
    public Complex s(Complex o) { return new Complex(re + o.getRe(), im + o.getIm()); }
    public Complex m(Complex o) {
        return new Complex(re*o.getRe() + im*o.getIm(), re*o.getIm() + im*o.getRe());
    }
}
// c. Opaque object implementation
@Opaque("double[]")
public class Complex {
    protected double[] rep;
    public static Complex create(double re, double im) { return Complex.New(re, im); }
    public static Complex s(Complex o) {
        return Complex.New(s.rep[0]+o.rep[0], s.rep[1]+o.rep[1]); }
    public Complex m(Complex o) {
        return new Complex(s.rep[0]*o.rep[0]+s.rep[1]*o.rep[1],
                                    s.rep[0]*o.rep[1]+s.rep[1]*o.rep[0]);
    }
}
```

Fig. 4. Multiplication: generic, specialized and opaque object implementation

opaque). Table 2 summarizes obtained results measured in millions of floating point instructions per second with similar conclusions being drawn from this set of data as the sparse matrix multiplication.

Implementing dense polynomial multiplication using the proposed *opaque* typed method allows for an average of 4.75 times the number of operations per second while accomplishing the same task. The *opaque* implementation loses out to specialized code by an average of only 19%. This is an expected and impressive result considering how much abstraction is preserved through the use of *opaque* types.

5   Conclusions and Further Directions

We have observed that Java programmers and library designers have been forced to work around the language's abstraction mechanisms for performance-sensitive code. In practice, programs have used primitive types, such as int, when an abstraction should be used. The recent addition of Enumerations to the language help in some settings, but is of no help when the values are used in mathematical computations.

We have shown how type-safe, but very efficient programs may be obtained with the concept of an *opaque* type in Java. An *opaque* type is distinct and incompatible with its underlying representation type, which may be a primitive type or an object type. We have shown how *opaque* types may be provided via classes with only static methods, and annotated for handling with a software tool in a standard Java environment. *Opaque* types are type checked as though they were object types, but compiled as the actual representation values. This allows *opaque* values to benefit from all the optimizations on primitive types without relying on sophisticated data structure elimination optimizations.

At the moment, our software tool operates by compiling the code twice, but of course this could easily be integrated into any compiler. While we focus on Java for practical reasons, we expect the same observations and techniques to be directly applicable in other similar settings.

References

1. Brown, W.E.: Progress toward Opaque Typedefs for C++0X (2005)
2. Johnston, B.: Java programming today. Pearson Prentice Hall, Upper Saddle River (2004)
3. Koffman, E.B.: Objects, abstraction, data structures and design using Java. John Wiley and Sons (2005)
4. Sun Microsystems, Inc. Annotations (2004),
   `http://java.sun.com/j2se/1.5.0/docs/guide/language/annotations.html`
5. Watt, S.M.: Aldor. In: Grabmeier, J., Kaltofen, E., Weispfenning, V. (eds.) Handbook of Computer Algebra, pp. 265–270. Springer, Heidelberg (2003)
6. The Apache Ant Project (2010), `http://ant.apache.org`
7. Bourdykine, P.: Type Safety without Objects in Java, MSc. Thesis, U. Western Ontario (2009)
8. Miller, B., Pozo, R.: SciMark 2.0 Java Benchmark. National Institute of Standards and Technology (2004)
9. Dragan, L., Watt, S.M.: Performance Analysis of Generics for Scientific Computing. In: Proc. SYNASC 2005, pp. 93–100. IEEE Press (2005)

Calculation of Normal Forms
of the Euler–Poisson Equations[*]

Alexander D. Bruno[1] and Victor F. Edneral[2]

[1] Keldysh Institute for Applied Mathematics of RAS,
Miusskaya Sq. 4, Moscow, 125047, Russia
abruno@keldysh.ru
[2] Skobeltsyn Institute of Nuclear Physics,
Lomonosov Moscow State University,
Leninskie Gory 1, Moscow, 119991, Russia
edneral@theory.sinp.msu.ru

Abstract. In the paper [1], the special case of the Euler–Poisson equations describing movements of a heavy rigid body with a fixed point is considered. Among stationary points of the system, two of one-parameter families were chosen. These families correspond to the resonance of eigenvalues $(0, 0, \lambda, -\lambda, 2\lambda, -2\lambda)$ of the matrix of the linear part of the system, also in [1] it was conjectured the absence of the additional first integral (with respect to well-known 3 integrals (2)) near these families, except of classical cases of global integrability. In this paper, the supposition is proved by calculations of coefficients of the normal form.

Keywords: Euler–Poisson equations, resonant normal form, computer algebra.

1 Introduction

The Euler–Poisson system consists of six equations and describes the motion of a rigid body with a fixed point [2]

$$
\begin{aligned}
&A\dot{p} + (C - B)qr = Mg(z_0\gamma_2 - y_0\gamma_3) \ , \\
&B\dot{q} + (A - C)pr = Mg(x_0\gamma_3 - z_0\gamma_1) \ , \\
&C\dot{r} + (B - A)pq = Mg(y_0\gamma_1 - x_0\gamma_2) \ , \\
&\dot{\gamma}_1 = r\gamma_2 - q\gamma_3 \ , \\
&\dot{\gamma}_2 = p\gamma_3 - r\gamma_1 \ , \\
&\dot{\gamma}_3 = q\gamma_1 - p\gamma_2
\end{aligned}
\tag{1}
$$

where $A, B, C, M, x_0, y_0, z_0$ are real constants. A, B, C are the principal moments of inertia, and variables $\gamma_1, \gamma_2, \gamma_3$ are the Euler angles.

---

[*] The authors are supported by the Russian Foundation for Basic Research Grant No. 11-01-00023-a.

V.P. Gerdt et al. (Eds.): CASC 2012, LNCS 7442, pp. 60–71, 2012.

The system (1) has three first integrals

$$F_1 \stackrel{\text{def}}{=} Ap^2 + Bq^2 + Cr^2 + 2Mg(x_0\gamma_1 + y_0\gamma_2 + z_0\gamma_3) = \text{const} ,$$
$$F_2 \stackrel{\text{def}}{=} Ap\gamma_1 + Bq\gamma_2 + Cr\gamma_3 = \text{const} , \tag{2}$$
$$F_3 \stackrel{\text{def}}{=} \gamma_1^2 + \gamma_2^2 + \gamma_3^2 = \text{const} = 1 .$$

System (1) has also the linear automorphism

$$t, p, q, r, \gamma_1, \gamma_2, \gamma_3 \to -it, ip, iq, ir, -\gamma_1, -\gamma_2, -\gamma_3 . \tag{3}$$

For the special case

$$A = B, \ C/B = c, \ Mgx_0/B = -1, \ y_0 = z_0 = 0 ,$$

system (1) is

$$\begin{aligned}
\dot{p} &= (1-c)qr , \\
\dot{q} &= (c-1)pr - \gamma_3 , \\
\dot{r} &= \gamma_2/c , \\
\dot{\gamma}_1 &= r\gamma_2 - q\gamma_3 , \\
\dot{\gamma}_2 &= p\gamma_3 - r\gamma_1 , \\
\dot{\gamma}_3 &= q\gamma_1 - p\gamma_2
\end{aligned} \tag{4}$$

where c is a single parameter $c \in \mathbb{R}\,(0, 2]$. If $c = 1$ or $c = 1/2$ then system (4) has an additional (the fourth with respect to (20)) first integral. These are classical cases of integrability by Lagrange and Sofia Kovalevskaya.

The system (4) has a two-parameter (c, p_0) family of stationary points

$$\begin{aligned}
p &= p_0 = \text{const}, \ q = q_0 = 0, \ r = r_0 = 0 , \\
\gamma_1 &= \gamma_1^0 = 1, \ \gamma_2 = \gamma_2^0 = 0, \ \gamma_3 = \gamma_3^0 = 0 .
\end{aligned} \tag{5}$$

The resonance $(0, 0, \lambda, -\lambda, 2\lambda, -2\lambda)$ takes place at two one-parameter subfamilies which are defined by

$$p_0^2 = \frac{17 - 33c + 5\delta\sqrt{9 - 34c + 41c^2}}{8c} \stackrel{\text{def}}{=} p_0^2(\delta, c) , \tag{6}$$

where $\delta = \pm 1$. We denote families (6) as $\mathcal{F}_\pm$.

However, at the point

$$\begin{aligned}
\tilde{c} &= (-1 + \sqrt{5})/2 \approx 0.618034 \\
&\text{and} \\
p_0^2(\tilde{c}, \delta &= -1) = -(3 + \sqrt{5})/2 \approx -2.61803
\end{aligned} \tag{7}$$

all eigenvalues are zero, i.e., in (9) $\lambda_i = 0$, thus, the point (7) should be excluded from the family $\mathcal{F}_-$.

In this paper, the conjecture of the absence of an additional first integral near these families, except classical cases of global integrability, is proved by calculations of coefficients of the normal form. All calculations were performed using the MATHEMATICA package [4,5].

In the second Section of this paper, we study a common normal form structure at the resonance. In the third Section, we discuss the form of the formal first integrals of the normal form and prove Theorem 1 about necessary conditions of existence of an additional first integral, and in the fourth Section, we briefly describe the calculations which allow to apply that theorem in the studied case.

2   A Normal Form Structure at the Resonance

Here we will study the structure of the normal form of system (4) near stationary points of families $\mathcal{F}_\pm$. This description is similar to Section 10, Chapter V of the book [3]. Let its normal form be

$$\dot{z}_i = z_i g_i(Z), \quad i = 1, 2, 3, 4, 5, 6, \tag{8}$$

where the vector Λ of eigenvalues λ_i is

$$\Lambda = (0, 0, \lambda_3, -\lambda_3, \lambda_5, -\lambda_5), \quad \tilde{r}\lambda_5 = \tilde{s}\lambda_3, \quad 1 \le \tilde{r} < \tilde{s} . \tag{9}$$

We introduce resonant variables

$$\rho_1 = z_3 z_4, \quad \rho_2 = z_5 z_6, \quad w = z_3^{\tilde{s}} z_6^{\tilde{r}}, \quad \tilde{w} = z_4^{\tilde{s}} z_5^{\tilde{r}} . \tag{10}$$

Thus $w\tilde{w} = \rho_1^{\tilde{s}} \rho_2^{\tilde{r}}$.

Lemma 1. *At the resonance $\tilde{s}\lambda_3 = \tilde{r}\lambda_5$ the normal form (8) is*

$$\begin{aligned}
\dot{z}_i &= z_i \sum_{k=1}^{\infty} f_{ik}(w^k - \tilde{w}^k), \quad i = 1, 2, \\
\dot{\rho}_i &= \rho_i \sum_{k=1}^{\infty} F_{ik}(w^k - \tilde{w}^k), \quad i = 1, 2, \\
\dot{w} &= w \left[G_0 + \sum_{k=1}^{\infty} F_{3k} w^k + \sum_{k=1}^{\infty} H_{3k} \tilde{w}^k \right], \\
\dot{\tilde{w}} &= -\tilde{w} \left[G_0 + \sum_{k=1}^{\infty} F_{3k} \tilde{w}^k + \sum_{k=1}^{\infty} H_{3k} w^k \right] ,
\end{aligned} \tag{11}$$

where $f_{ik}, F_{ik}, G_0, H_{3k}$ are power series in z_1, z_2, ρ_1, ρ_2.

If $\tilde{r} + \tilde{s}$ is odd in decomposition (11) index k accepts only even values $2l$.

3   The First Integrals

According to [6] the first integral of normal form (8)

$$A = \sum a_Q Z^Q$$

contains only resonant terms for which

$$\langle Q, \Lambda \rangle = 0 .$$

Therefore, the first integral can be written down in the form of a power series in the resonant variables

$$A = a_0 + \sum_{m=1}^{\infty} a_m w^m + \sum_{m=1}^{\infty} b_m \tilde{w}^m ,$$

where a_0, a_m and b_m are power series in $\mathbf{z} \stackrel{\text{def}}{=} (z_1, z_2), \rho \stackrel{\text{def}}{=} (\rho_1, \rho_2)$. As a sequence of the automorphism (3)

$$t, z_1, z_2, \rho_1, \rho_2, w, \tilde{w} \to -t, z_1, z_2, \rho_1, \rho_2, \tilde{w}, w$$

we have $a_m = b_m$, i.e., the integral looks like

$$A = a_0 + \sum_{m=1}^{\infty} a_m (w^m + \tilde{w}^m) . \tag{12}$$

If $\tilde{r} + \tilde{s}$ is odd then from the automorphism (3)

$$t, z_1, z_2, \rho_1, \rho_2, w, \tilde{w} \to t, z_1, z_2, \rho_1, \rho_2, -w, -\tilde{w}$$

it is easy to see that the expansion contains only even indexes $m = 2n$, i.e.

$$A = a_0 + \sum_{n=1}^{\infty} a_{2n} (w^{2n} + \tilde{w}^{2n}) . \tag{13}$$

For the first integral (12), its derivative in view of (11) should be identically equal to zero. Hence,

$$
\begin{aligned}
0 \equiv & \frac{\partial A}{\partial z_1} z_1 g_1 + \frac{\partial A}{\partial z_2} z_2 g_2 + \frac{\partial A}{\partial \rho_1} \rho_1 (g_3 + g_4) + \frac{\partial A}{\partial \rho_2} \rho_2 (g_5 + g_6) \\
& + \frac{\partial A}{\partial w} b_5(\mathbf{z}, \rho, w, \tilde{w}) - \frac{\partial A}{\partial \tilde{w}} b_5(\mathbf{z}, \rho, \tilde{w}, w) \\
= & \sum_{k=1}^{\infty} \left(\frac{\partial a_0}{\partial z_1} z_1 f_{1k} + \frac{\partial a_0}{\partial z_2} z_2 f_{2k} + \frac{\partial a_0}{\partial \rho_1} \rho_1 F_{1k} + \frac{\partial a_0}{\partial \rho_2} \rho_2 F_{2k} \right) (w^k - \tilde{w}^k) \\
& + \sum_{m=1}^{\infty} \sum_{k=1}^{\infty} \left(\frac{\partial a_m}{\partial z_1} z_1 f_{1k} + \frac{\partial a_m}{\partial z_2} z_2 f_{2k} + \frac{\partial a_m}{\partial \rho_1} \rho_1 F_{1k} + \frac{\partial a_m}{\partial \rho_2} \rho_2 F_{2k} \right) \\
& \times (w^k - \tilde{w}^k)(w^m + \tilde{w}^m) + \sum_{m=1}^{\infty} m a_m \left[G_0 (w^m - \tilde{w}^m) \right. \\
& \left. + \sum_{k=1}^{\infty} F_{3k} (w^{k+m} - \tilde{w}^{k+m}) + \sum_{k=1}^{\infty} H_{3k} (w^m \tilde{w}^k - w^k \tilde{w}^m) \right] .
\end{aligned}
\tag{14}
$$

Let $\tilde{r} + \tilde{s}$ be odd. Then in the equation (14), indexes k and m are even: $k = 2l$ and $m = 2n$. Therefore, writing out this equation up to terms of the total order in z_3, z_4, z_5, z_6 smaller than $4(\tilde{s} + \tilde{r})$, we obtain

$$\frac{\partial a_0}{\partial z_1} z_1 f_{12} + \frac{\partial a_0}{\partial z_2} z_2 f_{22} + \frac{\partial a_0}{\partial \rho_1} \rho_1 F_{12} + \frac{\partial a_0}{\partial \rho_2} \rho_2 F_{22} + 2 a_2 G_0 = 0 , \tag{15}$$

where in each member there are terms with orders smaller than $2(\tilde{r} + \tilde{s})$ only. For odd $\tilde{r} + \tilde{s}$, its least possible value is equal to 3 ($\tilde{r} = 1$, $\tilde{s} = 2$). Equality (15) should be satisfied at least for the $\mathbf{z}, \rho$ free terms and for the terms linear in each of $\mathbf{z}, \rho$. Let

$$
\begin{aligned}
z_i f_{i2} &\overset{\text{def}}{=} \xi_i + \eta_{i1} z_1 + \eta_{i2} z_2 + \eta_{i3} \rho_1 + \eta_{i4} \rho_2 + \cdots \ , \\
\rho_j F_{j2} &\overset{\text{def}}{=} \xi_{j+2} + \eta_{j+2,1} z_1 + \eta_{j+2,2} z_2 + \eta_{j+2,3} \rho_1 + \eta_{j+2,4} \rho_2 + \cdots \ , \\
G_0 &\overset{\text{def}}{=} \zeta_1 z_1 + \zeta_2 z_2 + \zeta_3 \rho_1 + \zeta_4 \rho_2 + \cdots \ , \\
a_0 &\overset{\text{def}}{=} \text{const} + \alpha_1 z_1 + \alpha_2 z_2 + \alpha_3 \rho_1 + \alpha_4 \rho_2 + \cdots \ , \\
a_2 &\overset{\text{def}}{=} \beta + \ldots, \quad i, j = 1, 2 \ .
\end{aligned}
\tag{16}
$$

Equation (15) for the free term on the left-hand side gives the equality

$$
\alpha_1 \xi_1 + \alpha_2 \xi_2 + \alpha_3 \xi_3 + \alpha_4 \xi_4 = 0 \ .
\tag{17}
$$

If the vector $\Xi \overset{\text{def}}{=} (\xi_1, \xi_2, \xi_3, \xi_4) \neq 0$ then equation (17) has three-dimensional set of solutions $\alpha = (\alpha_1, \alpha_2, \alpha_3, \alpha_4)$. If $\Xi = 0$ then equality (15) for linear in the $\mathbf{z}, \rho$ terms implies four equalities

$$
\sum_{i=1}^{4} \eta_{ij} \alpha_i + 2\zeta_j \beta = 0, \quad j = 1, 2, 3, 4 \ .
\tag{18}
$$

The dimension of solutions (α, β) of the system (18) is equal $(5 - \text{rank } M)$, where M is a 4×5 matrix which consists of 4 vectors

$$
(\eta_{1j}, \eta_{2j}, \eta_{3j}, \eta_{4j}, \zeta_j), \quad j = 1, 2, 3, 4 \ .
\tag{19}
$$

As the initial system (4) has the three first integrals (2)

$$
\begin{aligned}
F_1 &= p^2 + q^2 + cr^2 - 2\gamma_1 = \text{const} \ , \\
F_2 &= p\gamma_1 + q\gamma_2 + cr\gamma_3 = \text{const} \ , \\
F_3 &= \gamma_1^2 + \gamma_2^2 + \gamma_3^2 = \text{const} = 1 \ ,
\end{aligned}
\tag{20}
$$

its normal form (8) also has three corresponding first integrals of form (12), (16) with vectors $\alpha^{(j)} = (\alpha_1^{(j)}, \alpha_2^{(j)}, \alpha_3^{(j)}, \alpha_4^{(j)})$ and constants $\beta^{(j)}$ ($j = 1, 2, 3$).

We form a 3×4 matrix $\bar{\alpha} = (\alpha_i^{(j)})$, where $i = 1, 2, 3, 4$; $j = 1, 2, 3$. The additional first integral is locally independent from known three. It is possible only in two cases:

1. the vector $\Xi \neq 0$ and $\text{rank } \bar{\alpha} < 3$;
2. the vector $\Xi = 0$ and $\text{rank } M = 1$.

Let the vector $V = (v_1, v_2, v_3, v_4)$ be the external product of vectors $\alpha^{(1)}, \alpha^{(2)}$, and $\alpha^{(3)}$. For its calculation it is necessary to write the vector $U = (u_1, u_2, u_3, u_4)$ over the matrix $\bar{\alpha}$ and to calculate a determinant of this resulting square matrix. Then we calculate $\det = \sum_{i=1}^{4} v_i u_i$, where v_i are the components of the external product. We will say that the formal integral (12) is *locally independent* from known integrals, if its linear approximation in z, ρ, w is linearly independent on the first approximations of the known integrals. Thus, we have proved the following theorem:

Theorem 1. *For existence of the additional formal integral at a point of families $\mathcal{F}_{\pm}$ it is necessary the satisfaction at this point of one of two conditions:*

1. $\Xi \neq 0$ *and* $V = 0$;
2. $\Xi = 0$ *and* $\operatorname{rank} M < 2$.

4   The Case of $\tilde{r} = 1$, $\tilde{s} = 2$

In this case, $\tilde{r} + \tilde{s} = 3$, i.e., it is odd, resonant variables (10) are

$$w = z_3^2 z_6, \quad \tilde{w} = z_4^2 z_5 .$$

The first integrals look like (13). Let the normal form be calculated

$$\dot{z}_j = z_j g_j \overset{\text{def}}{=} z_j \sum g_{jQ} Z^Q, \quad j = 1, \ldots, 6 . \tag{21}$$

We specify a connection of coefficients in the equation (17) and in the system (18) with coefficients in (21). Let

$$\begin{aligned}
G_1 &\overset{\text{def}}{=} g_3 + g_4 \overset{\text{def}}{=} \sum G_{1Q} Z^Q , \\
G_2 &\overset{\text{def}}{=} g_5 + g_6 \overset{\text{def}}{=} \sum G_{2Q} Z^Q , \\
G_3 &\overset{\text{def}}{=} 2g_3 + g_6 \overset{\text{def}}{=} \sum G_{3Q} Z^Q .
\end{aligned} \tag{22}$$

Then

$$\begin{aligned}
\xi_1 &= g_{1(-1,0,4,0,0,2)}, & \xi_2 &= g_{2(0,-1,4,0,0,2)} , \\
\xi_3 &= G_{1(0,0,3,-1,0,2)}, & \xi_4 &= G_{2(0,0,4,0,-1,1)} , \\
\eta_{11} &= g_{1(0,0,4,0,0,2)}, & \eta_{12} &= g_{1(-1,1,4,0,0,2)} , \\
\eta_{13} &= g_{1(-1,0,5,1,0,2)}, & \eta_{14} &= g_{1(-1,0,4,0,1,3)} , \\
\eta_{21} &= g_{2(1,-1,4,0,0,2)}, & \eta_{22} &= g_{2(0,0,4,0,0,2)} , \\
\eta_{23} &= g_{2(0,-1,5,1,0,2)}, & \eta_{24} &= g_{2(0,-1,4,0,1,3)} , \\
\eta_{31} &= G_{1(1,0,3,-1,0,2)}, & \eta_{32} &= G_{1(0,1,3,-1,0,2)} , \\
\eta_{33} &= G_{1(0,0,4,0,0,2)}, & \eta_{34} &= G_{1(0,0,3,-1,1,3)} , \\
\eta_{41} &= G_{2(1,0,4,0,-1,1)}, & \eta_{42} &= G_{2(0,1,4,0,-1,1)} , \\
\eta_{43} &= G_{2(0,0,5,1,-1,1)}, & \eta_{44} &= G_{2(0,0,4,0,0,2)} , \\
\zeta_1 &= G_{3(1,0,0,0,0,0)}, & \zeta_2 &= G_{3(0,1,0,0,0,0)} , \\
\zeta_3 &= G_{3(0,0,1,1,0,0)}, & \zeta_4 &= G_{3(0,0,0,0,1,1)} .
\end{aligned} \tag{23}$$

For the integral

$$A = \sum a_Q Z^Q \tag{24}$$

in the notations (13) and (16) we get

$$\begin{aligned}
\alpha_1 &= a_{(1,0,0,0,0,0)}, & \alpha_2 &= a_{(0,1,0,0,0,0)} , \\
\alpha_3 &= a_{(0,0,1,1,0,0)}, & \alpha_4 &= a_{(0,0,0,0,1,1)} , \\
\beta &= a_{(0,0,4,0,0,2)} .
\end{aligned} \tag{25}$$

Let us notice that for a calculation of the vector V, i.e., $\alpha_i^{(j)}$, it is necessary to calculate normal form and integrals (20) up to the 2nd order, for the vector Ξ — up to the 4th, and the matrix M, i.e. η_{ij}, — up to the 7th order.

4.1   Calculation of Known Integrals at the Resonance 1:2

Along curves $\mathcal{F}_+$ and $\mathcal{F}_-$, the normal forms of system (4) up to terms of the first order (i.e., up to the terms which are square free in variables of the system) were analytically calculated. Thus, we calculated also the three first integrals (20) in coordinates of the normal form. In particular, coefficients $\alpha_1, \alpha_2, \alpha_3, \alpha_4$ for each of these three integrals in (16) were obtained as functions in $\delta = \pm 1$ and $c \in \mathbb{R}(0, 2]$.

$\alpha^{(j)} = (\alpha_1^{(j)}, \alpha_2^{(j)}, \alpha_3^{(j)}, \alpha_4^{(j)})$ are the vectors of coefficients of the integrals F_j, $j = 1, 2, 3$ in (20). According to the text before the statement of Theorem 1 we form from the vectors $\alpha^{(1)}, \alpha^{(2)}$, and $\alpha^{(3)}$ their external product $V = (v_1, v_2, v_3, v_4)$. It has appeared that the vector V along the curves $\mathcal{F}_+$ and $\mathcal{F}_-$ can be calculated analytically as follows. Firstly we make the uniformization

$$c = \frac{18h}{80 + 34h - h^2} , \tag{26}$$

i.e., we replace the parameter c by the parameter h. This uniformization has two branches. We choose the single-valued branch where $c \in \mathbb{R}$ unambiguously corresponds to $h \in \mathbb{R}$ and which is defined by the interval

$$- 2.20937 \approx 17 - 3\sqrt{41} < h < 17 + 3\sqrt{41} \approx 36.2094 , \tag{27}$$

which ends are roots of a denominator in (26). At $\delta = 1$, i.e., on $\mathcal{F}_+$, it turns out

$$p_0 = \frac{1}{6}\sqrt{-4 + \frac{640}{h} - \frac{h}{2}} . \tag{28}$$

Components v_i of the external product V are

$$v_1 = \frac{-(320+(40-h)h)^2(80+h^2)}{1296h^3} ,$$
$$v_2 = \frac{-(320+(40-h)h)^2}{7776h^4\sqrt{2(32-h)(40+h)/h}} \times$$
$$\times(-102400+h(-41600-h(5040+(40-h)h))) ,$$
$$v_3 = \frac{-(320+(40-h)h)^2}{18h(40+h)} ,$$
$$v_4 = \frac{(320+(40-h)h)^2}{9(32-h)h^2} .$$

The equation $V = 0$, i.e., the system of four equations

$$v_i = 0, \quad i = 1, 2, 3, 4 , \tag{29}$$

has only two solutions

$$h_1 = 4(5 - 3\sqrt{5}) \approx -6.83282 \ ,$$
$$h_2 = 4(5 + 3\sqrt{5}) \approx 46.83282 \ . \tag{30}$$

Both these solutions lay outside the interval (27). Hence, at $\delta = 1$, external product $V \neq 0$.

Similarly, at $\delta = -1$ we have

$$p_0 = \frac{1}{3}\sqrt{-1 + \frac{10}{h} - 2h} \ , \tag{31}$$

$$v_1 = \frac{-(20 + (10-h)h)^2(80+h^2)}{81h^3} \ ,$$

$$v_2 = \frac{\sqrt{-1 + (10/h) - 2h}(20 + (10-h)h)^2}{243h^3(h-2)(2h+5)} \times$$
$$\times (-800 + h(-400 + h(630 + h(-65 + 2h)))) \ ,$$

$$v_3 = \frac{16(20 + (10-h)h)^2}{9h^2(h-2)} \ ,$$

$$v_4 = \frac{(20 + (10-h)h)^2}{9h(2h+5)} \ .$$

The system of equations (29) has thus only two solutions

$$h_3 = 5 - 3\sqrt{5} \approx -1.7082 \ ,$$
$$h_4 = 5 + 3\sqrt{5} \approx 11.7082 \ . \tag{32}$$

Both of them lay in the interval (27). With respect to (26) that solutions correspond to values c

$$c(h_3) = -(\sqrt{5}+1)/2 \approx -1.618034 \ ,$$
$$c(h_4) = (\sqrt{5}-1)/2 \approx 0.618034 \ .$$

Only the last value of c lays in semi-interval $\mathbb{R}\,(0,2]$, and along the curve $\mathcal{F}_-$ it corresponds to the exclusive point (7). At the exclusive point, the matrix of the linear part has not simple elementary divisors and the developed theory does not work. Hence along the curve $\mathcal{F}_-$ the external product $V \neq 0$.

Thus, at the resonance 1:2 the external product V anywhere in the mechanical area does not equal to zero, i.e., the first condition of Theorem 1 is not satisfied. Below we examine the second series of the conditions of existence of the additional formal integral.

4.2   The Case $\varXi = 0$

According to (16) and (23), from coefficients of the normal form it is possible to calculate the vector $(\xi_1, \xi_2, \xi_3, \xi_4) \overset{\text{def}}{=} \varXi$ as a functions in $\delta = \pm 1$ and $c \in \mathbb{R}\,(0,2]$. In papers [7,8], the normal forms up to the terms of the 4th order have been calculated, and from these normal forms, the values ξ_3, ξ_4 were calculated as well. These calculations used the uniformization (34) allowing to get rid of double radicals at diagonalization of the linear part of system (4) and were made on some grid of rational values of h. It was not possible to calculate the normal form up to the 4th order completely in analytical form in view of the necessity to treat extremely complicated results containing expressions with square roots of polynomials in parameter h which do not allow any uniformization. The calculation, however, was performed completely in the rational arithmetic with deduction of all roots from rational numbers, i.e., without any rounding off.

It has appeared that the values ξ_3 and ξ_4 are equal to zero only simultaneously and only at

$$
\begin{array}{llll}
c_1 = 1, & c_2 = 1/2, & c_3 \approx 0.2527783, & \text{for} \quad \delta = 1 \ ; \\
c_1 = 1, & c_2 = 1/2, & c_4 \approx 0.0452287, & c_5 \approx 0.1893723 \ , \\
c_6 \approx 0.51292, & \text{for} \quad \delta = -1 \ .
\end{array}
\tag{33}
$$

The approximate values for points c_3, c_4, c_5, and c_6 here mean that intervals in which there is a change of a sign of ξ_3 and ξ_4 are known to us, and we bring values of the centers of these intervals. Below, in Subsection 4.3, boundaries of intervals are specified.

The additional calculations carried out after the publication of papers [7,8] have shown that at the points (33), values ξ_1 and ξ_2 are equal to zero also according to (23) and (16). Therefore, there $\Xi = 0$, and for checking Theorem 1 it is necessary to calculate at points (33) the rank of the matrix M with respect to (19), (16) and (23).

4.3 Calculation of the Rank of the Matrix M at Points (33)

To calculate the entries of the matrix M, the coefficients of the normal form were calculated up to the 7th order. As rational points $c_1 = 1$ and $c_2 = 1/2$ are known precisely, exact values of coefficients of the normal form are obtained there, i.e., we know exact expressions for the matrix M. At $c = 1$ for $\delta = \pm 1$ it has appeared that rank $M = 0$, i.e., all entries of this matrix are zero. At $c = 1/2$, the matrix M is

$$
M = \begin{pmatrix}
0 & -111i/(16\sqrt{2}) & 117i/28 & -7i/32 & -i/2 \\
0 & 111i/16 & -117i/(64\sqrt{2}) & 7i/(16\sqrt{2}) & i/\sqrt{2} \\
0 & 111i/(2\sqrt{2}) & -117i/16 & 7i/4 & 4i \\
0 & -333i/(32\sqrt{2}) & 351i/256 & -21i/64 & -3i/4
\end{pmatrix} \quad \text{for} \quad \delta = 1 \ ,
$$

$$
M = \begin{pmatrix}
0 & -24/\sqrt{7} & -85i/4 & 0 & -7i/2 \\
0 & -96i/7 & 85/\sqrt{7} & 0 & 2\sqrt{7} \\
0 & 576/(49\sqrt{7}) & 510i/49 & 0 & 12i/7 \\
0 & 219/(7\sqrt{7}) & 6205i/224 & 0 & 73i/16
\end{pmatrix} \quad \text{for} \quad \delta = -1 \ .
$$

In the both values of δ rank $M = 1$. It agrees with the theory above.

At other points (33) where $\Xi = 0$, the c values can be calculated only approximately, that complicates the calculation of the rank of the matrix M there. To overcome this difficulty, for each point c_i $(i = 3, 4, 5, 6)$ three numbers $c_i^{(1)} < c_i^{(2)} < c_i^{(3)}$ were chosen in such a way that $c_i^{(1)}$ and $c_i^{(3)}$ laid at borders of the interval which contained the point $c_i^{(2)}$. At these three points, the corresponding minors of the third and second order of the matrix M were calculated, and the matrixes $ResM3$ and $ResM2$ were formed from them accordingly. These matrixes were constructed so: a) a check was made for each of minors that all of its elements are pure real or pure imaginary. In the last case, the imaginary unit was eliminated. After that each minor should be pure real; b) monotony of the sequence of values of the minor at three specified above values of c_i of the parameter c was checked. If these three values behaved monotonously then: if signs of the first and the last values of the minor were opposite, the element of the matrix $ResM3$ or $ResM2$ was assigned to zero, otherwise the element was replaced with a string from these three values of the minor that allows to analyze, whether vanishing the minor inside of the interval $(c_i^{(1)}, c_i^{(3)})$ is possible. The value $c_i^{(2)}$ at an internal point of the interval allows to clear an opportunity of zeroing the minor in a parabolic way.

All minors were calculated in exact arithmetic, and results presented below are the approximated values obtained from exact calculations of corresponding analytical values of the minors by floating-point arithmetic with internal accuracy of 24 digits.

Here it was used different from (26) uniformization

$$c = \frac{18h}{h^2 + 34h - 80} \tag{34}$$

with

$$-17 + 3\sqrt{41} \approx 2.20937 < h < \infty , \tag{35}$$

that define the unique correspondence of $h \in \mathbb{R}$ with values of $c \in \mathbb{R}$.

Let us describe results for the point c_3. For it, $c_3^{(1)}$ is

$$c = 2759625/10917334 \approx 0.2527746 ,$$
$$h = 4906/125 \approx 39.248 > 2.20937 ,$$
$$p_0 = \sqrt{23684086/12265}/15 ,$$

here

$$\Xi = (-0.0030096i, -0.0173395i, 0.000254485i, -0.000350348i) ,$$

$c_3^{(2)}$ is

$$c = 78494000/310525001 \approx 0.2527784 ,$$
$$h = 39247/1000 \approx 39.247 ,$$
$$p_0 = (3/10)\sqrt{18712389/196235} ,$$

here

$$\Xi = (-0.0013087i, -0.00753993i, 0.000110666i, -0.00015235i) ,$$

$c_3^{(3)}$ is

$$c = 176607000/698653129 \approx 0.252782 \ ,$$
$$h = 19623/500 \approx 39.246 \ ,$$
$$p_0 = \sqrt{378906379/196230}/15 \ ,$$

here

$$\Xi = (0.000391896i, 0.00225786i, -0.0000331409i, 0.0000456228i) \ .$$

It is easy to see that a change of a sign of elements of the vector Ξ occurs in the interval $(c_3^{(2)}, c_3^{(3)})$.

It has appeared that the matrix $ResM3$ consists only of zeroes. The first string of the matrix $ResM2$ consists of zeroes also but all the others consist of elements — strings of the kind $(a(c_3^{(1)}), a(c_3^{(2)}), a(c_3^{(3)}))$, and they are numbers of the same sign closed among themselves, and some of them are large numbers which have an order near 10^4. Hence, at the point c_3 we have

$$\operatorname{rank} M = 2 \ . \tag{36}$$

The situations with the rank M at the points c_4, c_5, c_6 are the same, and values of $h_4, h_5, h_6 > 2.20937$ satisfy inequality (35). Therefore, we omit the detailed descriptions of them.

So, at points c_3, c_4, c_5, c_6 equality (36) takes place, i.e., the second of conditions of Theorem 1 is not satisfied. According to this theorem at the points above, the system (4) has no additional formal integral, i.e., it is not integrable.

5   Conclusion

The considered system (4) has no additional first integral which would be independent of classical ones (20). So the system is integrable at the parameter values $c = 1$ and $c = 1/2$ only.

References

1. Bruno, A.D.: Theory of Normal Forms of the Euler–Poisson Equations. Preprint No. 100, Keldysh Institute for Applied Mathematics of RAS, Moscow (2005), http://library.keldysh.ru/preprint.asp?lg=e&id=2005-100 (abstract), http://dl.dropbox.com/u/59058738/Preprint100.pdf (full text in Russian)
2. Golubev, V.V.: Lectures on the Integration of the Equation of Motion of a Rigid Body about of Fixed Point. Gostehizdat, Moscow (1953) (in Russian); NSF Israel Program for Scientific Translations, Washington (1960) (in English)
3. Bruno, A.D.: Power Geometry in Algebraic and Differential Equations. Nauka, Moscow (1998); Elsevier, Amsterdam (2000)
4. Edneral, V.F., Khanin, R.: Application of the resonant normal form to high order nonlinear ODEs using MATHEMATICA. Nuclear Instruments and Methods in Physics Research A 502(2-3), 643–645 (2003)

5. Edneral, V.F.: An Algorithm for Construction of Normal Forms. In: Ganzha, V.G., Mayr, E.W., Vorozhtsov, E.V. (eds.) CASC 2007. LNCS, vol. 4770, pp. 134–142. Springer, Heidelberg (2007)
6. Bruno, A.D., Sadov, S.Y.: Formal integral of a divergentless system. Matem. Zametki 57(6), 803–813 (1995); Mathematical Notes 57(6), 565–572 (1995)
7. Bruno, A.D., Edneral, V.F.: Normal Forms and Integrability of ODE Systems. In: Ganzha, V.G., Mayr, E.W., Vorozhtsov, E.V. (eds.) CASC 2005. LNCS, vol. 3718, pp. 65–74. Springer, Heidelberg (2005)
8. Bruno, A.D., Edneral, V.F.: Normal form and integrability of ODE systems. Programmirovanie 32(3), 22–29 (2006); Programming and Computer Software 32(3), 1–6 (2006)

Stability of Equilibrium Positions in the Spatial Circular Restricted Four-Body Problem

Dzmitry A. Budzko[1] and Alexander N. Prokopenya[2,3]

[1] Brest State University,
Kosmonavtov bul. 21, 224016, Brest, Belarus
master_booblik@tut.by
[2] Warsaw University of Live Sciences
Nowoursynowska str. 159, 02-787 Warsaw, Poland
[3] Collegium Mazovia Innovative University in Siedlce ,
ul. Sokolowska 161, 08-110, Siedlce, Poland
alexander_prokopenya@sggw.pl

Abstract. We study stability of equilibrium positions in the spatial circular restricted four-body problem formulated on the basis of Lagrange's triangular solution of the three-body problem. Using the computer algebra system Mathematica, we have constructed Birkhoff's type canonical transformation, reducing the Hamiltonian function to the normal form up to the fourth order in perturbations. Applying Arnold's and Markeev's theorems, we have proved stability of three equilibrium positions for the majority of initial conditions in case of mass parameters of the system belonging to the domain of the solutions linear stability, except for the points in the parameter plane for which the third and fourth order resonance conditions are fulfilled.

1 Introduction

The theory of stability of the Hamiltonian systems is developed quite well, and a number of problems of motion stability have been solved (see, for example, [1]). A classical example is a problem of libration points stability in the restricted three-body problem [2,3] that was introduced first by Euler in connection with his lunar theory. This problem is highly interesting for applications, and so it has been a major topic in celestial mechanics during the past two hundred years. Finally some general methods for studying the stability of Hamiltonian systems have been developed [3,4,5]. However, application of these methods involves very bulky symbolic calculations which can be reasonably done only with computer and modern software such as the computer algebra system *Mathematica* [6], for example. Besides, stability analysis of more complicated dynamical systems requires improvement of available computing technique and designing new efficient algorithms of calculation, and this stimulates further investigations in this field.

In our previous paper [7], we have considered the circular restricted four-body problem formulated on the basis of Lagrange's triangular solution of the three-body problem. Remind that within the framework of this problem, three point

V.P. Gerdt et al. (Eds.): CASC 2012, LNCS 7442, pp. 72–83, 2012.

particles P_0, P_1, P_2 having masses m_0, m_1, m_2, respectively, move uniformly on circular Keplerian orbits around their common center of mass and form an equilateral triangle at any instant of time. We are interested in the motion of the fourth particle P_3 of negligible mass that moves in the gravitational field of P_0, P_1, and P_2. It has been shown that for small values of the system parameters, the problem has eight equilibrium solutions of which only three may be stable. We have studied the stability of equilibrium solutions in the planar case (see [7]) when the particle P_3 is constrained to move in the xOy plane. The present work is a generalization of [7] and is devoted to the stability analysis of equilibrium solutions in the spatial circular restricted four-body problem when the system has three degrees of freedom.

In Section 2, we describe equilibrium solutions and analyze their linear stability. In Section 3, we discuss the algorithm for normalization of the third-order term in the Hamiltonian expansion and analyze the stability of the equilibrium solutions under the third-order resonance. In Section 4, we normalize the fourth-order term of the Hamiltonian and apply theorems of Arnold and Markeev. Finally, we conclude in Section 5.

2   Linear Stability of Equilibrium Solutions

In the rotating frame of reference, where the particles P_0, P_1, P_2 are fixed in the xOy plane at points $(0,0)$, $(1,0)$, $(1/2, \sqrt{3}/2)$, respectively, the Hamiltonian function of the system can be written in the form

$$\mathcal{H} = \frac{1}{2}\left(p_x^2 + p_y^2 + p_z^2\right) - xp_y + yp_x + \frac{1}{1+\mu_1+\mu_2}\left(\left(\mu_1 + \frac{\mu_2}{2}\right)x + \frac{\mu_2\sqrt{3}}{2}y - \right.$$

$$\left. -\frac{1}{\sqrt{x^2+y^2+z^2}} - \frac{\mu_1}{\sqrt{(x-1)^2+y^2+z^2}} - \frac{2\mu_2}{\sqrt{(2x-1)^2+(2y-\sqrt{3})^2+z^2}}\right) \quad (1)$$

where x, p_x, y, p_y, and z, p_z are three pairs of canonically conjugate coordinate and momentum. Two mass parameters are defined as

$$\mu_1 = m_1/m_0 , \quad \mu_2 = m_2/m_0 .$$

Using the Hamiltonian (1), one can easily write the equations of motion of the particle P_3 and show that its equilibrium positions lie only in the xOy plane ($z = 0$). The corresponding equilibrium coordinates are determined as the solutions of the following algebraic system

$$(y - x\sqrt{3})\left(\frac{1}{(x^2+y^2)^{3/2}} - 1\right) = \mu_1(y+\sqrt{3}(x-1))\left(\frac{1}{((x-1)^2+y^2)^{3/2}} - 1\right) ,$$

$$y\left(\frac{1}{(x^2+y^2)^{3/2}} - 1\right) = \mu_2(y+\sqrt{3}(x-1))\left(1 - \frac{1}{\left(\left(x-\frac{1}{2}\right)^2 + \left(y-\frac{\sqrt{3}}{2}\right)^2\right)^{\frac{3}{2}}}\right) \quad (2)$$

Each equation of the system (2) determines a curve in the xOy plane, which can be easily visualized with the *Mathematica* built-in function *ContourPlot*, for example. So geometrically any equilibrium position of the particle P_3 corresponds to an intersection point of two curves (Fig. 1).

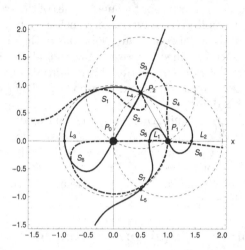

Fig. 1. Eight equilibrium positions $S_1, \ldots, S_8$ for $\mu_1 = 0.2$, $\mu_2 = 0.15$

Note that for any given $\mu_1 > 0$ the solid line in Fig. 1, determined by the first equation of system (2), is fixed, and it always passes through the points P_0, P_1, P_2, and $(1/2, -\sqrt{3}/2)$. In case of $\mu_2 = 0$, the thick dashed line, determined by the second equation of system (2), degenerates into the line $y = 0$ and the circle $x^2 + y^2 = 1$. Hence, system (2) has three roots at the Ox axis and two roots at points $(1/2, \pm\sqrt{3}/2)$, and these five roots correspond to the libration points L_1, L_2, L_3, and L_4, L_5 in the three-body problem (see [2], [3]). Increasing the value of μ_2, one can observe that the three equilibrium points located on the Ox axis when $\mu_2 = 0$, as well as the equilibrium point $(1/2, -\sqrt{3}/2)$, gradually move in the xOy plane along the solid line (the points S_5 – S_8 in Fig. 1). The point $x = 1/2$, $y = \sqrt{3}/2$ generates four new equilibrium positions (the points S_1 – S_4 in Fig. 1), one by each branch of the solid line outgoing the point P_2. Thus, graphical analysis indicates that there are eight roots of system (2) for small values of parameters μ_1, μ_2.

The problem of solving system (2) has been analyzed in detail in [8]. So we assume here that all equilibrium positions (x_0, y_0) of the particle P_3 can be found in the xOy plane for any values of parameters μ_1, μ_2. We can then expand the Hamiltonian (1) into Taylor series in the neighborhood of some equilibrium point and represent it in the form

$$H = H_2 + H_3 + H_4 + \ldots , \tag{3}$$

where H_k is the kth order homogeneous polynomial with respect to the canonical variables x, y, z, p_x, p_y, p_z. Note that zero-order term H_0 in (3) has been omitted as a constant, which doesn't influence the equations of motion, and the first-order term H_1 is equal to zero owing to equations determining equilibrium positions. Therefore, the first non-zero term in the expansion (3) is a quadratic one that is

$$H_2 = \frac{1}{2}\left(p_x^2 + p_y^2 + p_z^2\right) - p_y x + p_x y + h_{20}x^2 + h_{11}xy + h_{02}y^2 + h_{2z}z^2 , \quad (4)$$

where

$$h_{20} = \frac{1}{2(1 + \mu_1 + \mu_2)}\left(\frac{y_0^2 - 2x_0^2}{(x_0^2 + y_0^2)^{5/2}} + \mu_1 \frac{y_0^2 - 2(x_0 - 1)^2}{((x_0 - 1)^2 + y_0^2)^{5/2}} + \right.$$
$$\left. +\mu_2 \frac{(y_0 - \sqrt{3}/2)^2 - 2(x_0 - 1/2)^2}{((x_0 - 1/2)^2 + (y_0 - \sqrt{3}/2)^2)^{5/2}}\right) ,$$

$$h_{11} = -\frac{3}{1 + \mu_1 + \mu_2}\left(\frac{x_0 y_0}{(x_0^2 + y_0^2)^{5/2}} + \mu_1 \frac{(x_0 - 1)y_0}{((x_0 - 1)^2 + y_0^2)^{5/2}} + \right.$$
$$\left. +\mu_2 \frac{(x_0 - 1/2)(y_0 - \sqrt{3}/2)}{((x_0 - 1/2)^2 + (y_0 - \sqrt{3}/2)^2)^{5/2}}\right) ,$$

$$h_{02} = \frac{1}{2(1 + \mu_1 + \mu_2)}\left(\frac{x_0^2 - 2y_0^2}{(x_0^2 + y_0^2)^{5/2}} + \mu_1 \frac{(x_0 - 1)^2 - 2y_0^2}{((x_0 - 1)^2 + y_0^2)^{5/2}} + \right.$$
$$\left. +\mu_2 \frac{(x_0 - 1/2)^2 - 2(y_0 - \sqrt{3}/2)^2}{((x_0 - 1/2)^2 + (y_0 - \sqrt{3}/2)^2)^{5/2}}\right) ,$$

$$h_{2z} = \frac{1}{2(1 + \mu_1 + \mu_2)}\left(\frac{1}{(x_0^2 + y_0^2)^{3/2}} + \frac{\mu_1}{((x_0 - 1)^2 + y_0^2)^{3/2}} + \right.$$
$$\left. +\frac{\mu_2}{((x_0 - 1/2)^2 + (y_0 - \sqrt{3}/2)^2)^{3/2}}\right) .$$

One can readily check that the linearized equations of motion determined by the quadratic part H_2 of the Hamiltonian (3) form a sixth-order linear system of differential equations with constant coefficients. Characteristic exponents $\lambda_1, \ldots, \lambda_6$ for such a system can be easily found and are represented in the form

$$\lambda_{1,2} = \pm i\sigma_1 , \quad \lambda_{3,4} = \pm i\sigma_2 , \quad \lambda_{5,6} = \pm i\sigma_3 , \quad (5)$$

where i is the imaginary unit, and the frequencies $\sigma_1, \sigma_2, \sigma_3$ are given by

$$\sigma_{1,2} = \left(1 + h_{20} + h_{02} \pm \sqrt{h_{20}^2 + h_{02}^2 + h_{11}^2 - 2h_{20}h_{02} + 4h_{20} + 4h_{02}}\right)^{1/2} , \quad (6)$$

$$\sigma_3 = \sqrt{2h_{2z}} . \quad (7)$$

One can readily deduce from (4) and (7) that σ_3 is a real number for all possible values of the mass parameters μ_1, μ_2. Analysis of the frequencies (6) for all eight equilibrium positions (see Fig. 1) has shown [7] that for the points S_2, S_3, S_5, S_6, S_8 at least one frequency has an imaginary part for any values of parameters μ_1, μ_2.

Therefore, these five equilibrium positions are unstable. Equilibrium points S_1, S_4 and S_7 are stable in linear approximation if parameters μ_1, μ_2 are smaller than their values on the stability boundaries which are determined by the condition $\sigma_1 = \sigma_2$. The corresponding curve for the equilibrium point S_7 is shown in the $\mu_1 O \mu_2$ plane in Fig. 2 together with some resonance curves.

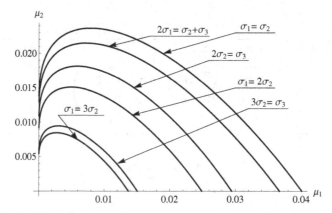

Fig. 2. Stability domain and resonance curves for equilibrium position S_7

3 Normalization of the Third-Order Term H_3

As in the planar case [7], the problem of equilibrium positions S_1, S_4, S_7 stability can be solved only in a strict nonlinear formulation based on Arnold's and Markeev's theorems. It becomes clear as soon as we normalize the quadratic part H_2 in the Hamiltonian expansion (3). An algorithm for constructing the corresponding canonical transformation is described in detail in [10]. Doing necessary symbolic calculations, we obtain the second order term H_2 in the form

$$H_2 = \frac{1}{2}\left(\sigma_1(p_1^2 + q_1^2) - \sigma_2(p_2^2 + q_2^2) + \sigma_3(p_3^2 + q_3^2)\right) , \qquad (8)$$

where p_1, q_1, p_2, q_2, and p_3, q_3 are three pairs of new canonically conjugated variables.

It is obvious that the quadratic form (8) is neither positive nor negative defined function and, hence, we cannot conclude on stability or instability of equilibrium solutions, using the principle of linearized stability [11]. Therefore, the stability problem can be solved only in a strict nonlinear formulation. As stability analysis of the equilibrium positions S_1, S_4, S_7 is done in a similar way we'll analyze only stability of the point S_7 in detail.

To normalize the third order term H_3 in the Hamiltonian (4) we use the method of constructing the Birkhoff's type real-valued canonical transformation described in [12]. It should be noted that, in contrast to [12], the system under consideration has two parameters μ_1 and μ_2 and three degrees of freedom, and due to this reason the calculations are much more bulky and difficult (see [7]).

After realization of the first canonical transformation, normalizing the quadratic part H_2, the third order term H_3 becomes

$$H_3 = \sum_{i+j+k+l+m+n=3} h^{(3)}_{ijklmn} q_1^i q_2^j q_2^k p_1^l p_2^m p_3^n \; . \tag{9}$$

Due to their bulk we do not provide here the corresponding expressions for $h^{(3)}_{ijklmn}$. And we'd like to find such canonical transformation that the third-order term H_3 in the expansion (3) was eliminated. Generating function for such transformation can be sought in the form of third-degree polynomial

$$S(\tilde{p}_1, \tilde{p}_2, \tilde{p}_3, q_1, q_2, q_3) = q_1 \tilde{p}_1 + q_2 \tilde{p}_3 + q_3 \tilde{p}_2 + \sum_{i+j+k+l+m+n=3} s^{(3)}_{ijklmn} q_1^i q_2^j q_3^k \tilde{p}_1^l \tilde{p}_2^m \tilde{p}_3^n \; , \tag{10}$$

where coefficients $s^{(3)}_{ijklmn}$ are to be found. Then new momenta $\tilde{p}_1, \tilde{p}_2, \tilde{p}_3$ and coordinates $\tilde{q}_1, \tilde{q}_2, \tilde{q}_3$ are determined by the following relationships

$$\tilde{q}_k = \frac{\partial S}{\partial \tilde{p}_k} \; , \quad p_k = \frac{\partial S}{\partial q_k} \; , k = 1, 2, 3 \; . \tag{11}$$

Note that these relationships are equations with respect to the former canonical variables q_1, q_2, q_3, p_1, p_2, p_3. On substituting (10) into (11) and solving these equations, we find q_1, q_2, q_3, p_1, p_2, p_3 in the form of second-degree polynomials in the new canonical variables $\tilde{q}_1$, $\tilde{q}_2$, $\tilde{q}_3$, $\tilde{p}_1$, $\tilde{p}_2$, $\tilde{p}_3$. Then we substitute the corresponding expressions into (8), (9) and expand the Hamiltonian $H = H_2 + H_3$ into Taylor series in powers of $\tilde{q}_1$, $\tilde{q}_2$, $\tilde{q}_3$, $\tilde{p}_1$, $\tilde{p}_2$, $\tilde{p}_3$. The expression obtained is again represented as a sum of homogeneous polynomials $\tilde{H}_k$ ($k = 2, 3, ...$) with respect to canonical variables $\tilde{q}_1$, $\tilde{q}_2$, $\tilde{q}_3$, $\tilde{p}_1$, $\tilde{p}_2$, $\tilde{p}_3$. One can readily check that the second-order term $\tilde{H}_2$ preserves the form (8), while the third-order term $\tilde{H}_3$ is a sum of 56 terms of the form

$$\tilde{h}^{(3)}_{ijklmn} \tilde{q}_1^i \tilde{q}_2^j \tilde{q}_3^k \tilde{p}_1^l \tilde{p}_2^m \tilde{p}_3^n \quad (i + j + k + l + m + n = 3) \tag{12}$$

with new coefficients $\tilde{h}^{(3)}_{ijklmn}$ which are expressed as linear functions of old coefficients $h^{(3)}_{ijklmn}$ and unknown coefficients $s^{(3)}_{ijklmn}$ of the generating function (10). Obviously, the third-order term $\tilde{H}_3$ would be eliminated if all the coefficients $\tilde{h}^{(3)}_{ijklmn}$ in (12) were equal to zero. Therefore, we can try to solve the system of fifty six equations of the form $\tilde{h}^{(3)}_{ijklmn} = 0$ and to find the coefficients $s^{(3)}_{ijklmn}$ of the corresponding canonical transformation (11).

Analysis of the coefficients $\tilde{h}^{(3)}_{ijklmn}$ shows that in fact we have several independent subsystems for determination of unknown coefficients $s^{(3)}_{ijklmn}$. As all these subsystems are solved similarly, we consider only two of them to demonstrate the most important peculiarities of their solving. The first subsystem is formed by three coefficients of $\tilde{p}_1 \tilde{p}_2^2$, $\tilde{p}_1 \tilde{q}_2^2$, $\tilde{q}_1 \tilde{q}_2 \tilde{p}_2$ in the expression for $\tilde{H}_3$. It determines the coefficients $s^{(3)}_{010110}$, $s^{(3)}_{100020}$, $s^{(3)}_{120000}$ and is given by

$$\tilde{h}^{(3)}_{000120} = h^{(3)}_{000120} + s^{(3)}_{100020} \sigma_1 - s^{(3)}_{010110} \sigma_2 \; ,$$

$$\tilde{h}^{(3)}_{020100} = h^{(3)}_{020100} + s^{(3)}_{120000}\sigma_1 + s^{(3)}_{010110}\sigma_2 \, ,$$

$$\tilde{h}^{(3)}_{110010} = h^{(3)}_{110010} - s^{(3)}_{010110}\sigma_1 + 2s^{(3)}_{100020}\sigma_2 - 2s^{(3)}_{120000}\sigma_2 \, . \tag{13}$$

Note that coefficients $s^{(3)}_{010110}$, $s^{(3)}_{100020}$, $s^{(3)}_{120000}$ appear only in the expressions for $\tilde{h}^{(3)}_{ijklmn}$ given in (13) and so they are completely determined by this system. It has a unique solution for any values of $\tilde{h}^{(3)}_{ijklmn}$ if its determinant being equal to $\sigma_1(4\sigma_2^2 - \sigma_1^2)$ is not zero. In such a case we can set $\tilde{h}^{(3)}_{000120} = \tilde{h}^{(3)}_{020100} = \tilde{h}^{(3)}_{110010} = 0$ and find the corresponding coefficients $s^{(3)}_{010110}$, $s^{(3)}_{100020}$, $s^{(3)}_{120000}$. Therefore, if $\sigma_1 \neq 0$ and the conditions

$$\sigma_1 \pm 2\sigma_2 \neq 0 \tag{14}$$

are fulfilled the three terms (12) with coefficients $\tilde{h}^{(3)}_{000120}$, $\tilde{h}^{(3)}_{020100}$, $\tilde{h}^{(3)}_{110010}$ are eliminated in $\tilde{H}_3$ by means of the canonical transformation (11).

The second subsystem determines the coefficients $s^{(3)}_{001020}$, $s^{(3)}_{010011}$, $s^{(3)}_{021000}$ and is given by

$$\tilde{h}^{(3)}_{000021} = -s^{(3)}_{010011}\sigma_2 + s^{(3)}_{001020}\sigma_3 \, ,$$

$$\tilde{h}^{(3)}_{011010} = 2s^{(3)}_{001020}\sigma_2 - 2s^{(3)}_{021000}\sigma_2 - s^{(3)}_{010011}\sigma_3 \, ,$$

$$\tilde{h}^{(3)}_{020001} = s^{(3)}_{010011}\sigma_2 + s^{(3)}_{021000}\sigma_3 \, . \tag{15}$$

Determinant of its matrix is equal to $\sigma_3(4\sigma_2^2 - \sigma_3^2)$ and, hence, it has a unique solution for any values of $\tilde{h}^{(3)}_{ijklmn}$ if $\sigma_3 \neq 0$ and $2\sigma_2 \pm \sigma_3 \neq 0$. However, one can readily see that in case of $\tilde{h}^{(3)}_{000021} = \tilde{h}^{(3)}_{011010} = \tilde{h}^{(3)}_{020001} = 0$ subsystem (15) has a trivial solution $s^{(3)}_{001020} = s^{(3)}_{010011} = s^{(3)}_{021000} = 0$ even if $\sigma_3 = 0$ or $2\sigma_2 \pm \sigma_3 = 0$. Therefore, the corresponding three terms (12) are eliminated in $\tilde{H}_3$ by means of the canonical transformation (11) and third-order resonance of the form $2\sigma_2 \pm \sigma_3 = 0$ has no influence on stability of equilibrium solutions.

Inspection of the remaining 50 coefficients $\tilde{h}^{(3)}_{ijklmn}$ shows that if $\sigma_1 \neq 0$, $\sigma_2 \neq 0$, $\sigma_3 \neq 0$ and the conditions

$$2\sigma_1 \pm \sigma_2 \neq 0 \, , \quad \sigma_1 \pm 2\sigma_3 \neq 0 \, , \quad \sigma_2 \pm 2\sigma_3 \neq 0 \tag{16}$$

are fulfilled, in addition to (14), all the coefficients $s^{(3)}_{ijklmn}$ of the canonical transformation (11) are found in a unique way. In this case, we can set $\tilde{h}^{(3)}_{ijklmn} = 0$ and find the canonical transformation such that the third-order term $\tilde{H}_3$ in the Hamiltonian (3) vanishes. Note that conditions (14) and (16) imply an absence of third-order resonances in the system (see [1]).

Analyzing frequencies (6) and (7), we obtain that for the linearly stable equilibrium point S_7 there exist values of parameters μ_1, μ_2, for which the condition of third-order resonance $\sigma_1 - 2\sigma_2 = 0$ is fulfilled (see Fig. 2). Thus, for the points (μ_1, μ_2) in the $\mu_1 O \mu_2$ plane located on the corresponding resonance curve, condition (14) is not fulfilled and, hence, the system (13) does not have a solution in case of $\tilde{h}^{(3)}_{ijklmn} = 0$. For the same reason the coefficients $\tilde{h}^{(3)}_{100020}$, $\tilde{h}^{(3)}_{120000}$, $\tilde{h}^{(3)}_{010110}$

can not be eliminated under the third-order resonance, as well. It means that the corresponding six resonance terms in $\tilde{H}_3$ cannot be eliminated.

Nevertheless, we can require the following conditions to be fulfilled

$$\tilde{h}^{(3)}_{000120} = \frac{B_1}{2\sqrt{2}}, \quad \tilde{h}^{(3)}_{020100} = -\frac{B_1}{2\sqrt{2}}, \quad \tilde{h}^{(3)}_{110010} = -\frac{B_1}{\sqrt{2}}, \qquad (17)$$

$$\tilde{h}^{(3)}_{010110} = \frac{B_2}{\sqrt{2}}, \quad \tilde{h}^{(3)}_{100020} = \frac{B_2}{2\sqrt{2}}, \quad \tilde{h}^{(3)}_{120000} = -\frac{B_2}{2\sqrt{2}}, \qquad (18)$$

where B_1, B_2 are constants. Solving the systems of equations (17), (18), we obtain the corresponding coefficients $s^{(3)}_{ijklmn}$ of the canonical transformation (11) and find the constants B_1, B_2 as

$$B_1 = \frac{1}{\sqrt{2}}(h^{(3)}_{000120} - h^{(3)}_{020100} - h^{(3)}_{110010}) \,,$$

$$B_2 = \frac{1}{\sqrt{2}}(h^{(3)}_{010110} + h^{(3)}_{100020} - h^{(3)}_{120000}) \,. \qquad (19)$$

Then the Hamiltonian (3) takes the form

$$\tilde{H} = \frac{1}{2}\sigma_1 \left(\tilde{q}_1^2 + \tilde{p}_1^2\right) - \frac{1}{2}\sigma_2 \left(\tilde{q}_2^2 + \tilde{p}_2^2\right) + \frac{B_1}{2\sqrt{2}} \left(\tilde{p}_1\tilde{p}_2^2 - \tilde{p}_1\tilde{q}_2^2 - 2\tilde{q}_1\tilde{q}_2\tilde{p}_2\right) +$$

$$+ \frac{B_2}{2\sqrt{2}} \left(\tilde{q}_1\tilde{p}_2^2 - \tilde{q}_1\tilde{q}_2^2 + 2\tilde{q}_2\tilde{p}_1\tilde{p}_2\right) + \tilde{H}_4 + \dots . \qquad (20)$$

Using the standard canonical transformation

$$\tilde{q}_1 = \sqrt{2\tau_1} \sin(\varphi_1 + \alpha) \,, \quad \tilde{p}_1 = \sqrt{2\tau_1} \cos(\varphi_1 + \alpha) \,,$$

$$\tilde{q}_2 = \sqrt{2\tau_2} \sin\varphi_2 \,, \quad \tilde{p}_2 = \sqrt{2\tau_2} \cos\varphi_2 \,, \qquad (21)$$

where parameter α is determined by the relationships

$$\cos\alpha = \frac{B_1}{B}, \quad \sin\alpha = \frac{B_2}{B}, \quad B = \sqrt{B_1^2 + B_2^2},$$

we rewrite the Hamiltonian (20) as

$$\tilde{H} = \sigma_1\tau_1 - \sigma_2\tau_2 + B\tau_2\sqrt{\tau_1}\cos(\varphi_1 + 2\varphi_2) + \tilde{H}_4(\varphi_1, \varphi_2, \tau_1, \tau_2) + \dots . \qquad (22)$$

We have done numerical analysis of parameter B for the equilibrium point S_7 under third-order resonance and shown that it is not equal to zero for all points (μ_1, μ_2) belonging to the resonance curve (see Fig. 3). Therefore, applying Markeev's theorem [1], we can conclude that equilibrium point S_7 in the circular restricted four-body problem, formulated on the basis of Lagrange's triangular solutions, is unstable under third-order resonance of the form $\sigma_1 = 2\sigma_2$.

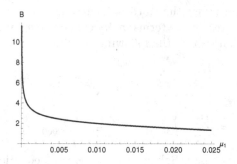

Fig. 3. Parameter B as function of μ_1 in the case of resonance $\sigma_1 = 2\sigma_2$ for S_7

4   Normalization of the Fourth-Order Term H_4

Let us assume that the condition $\sigma_1 \neq 2\sigma_2$ is fulfilled and that there is no resonance in the system up to the third order inclusively. Then after normalization of the second and third order terms we obtain the Hamiltonian (3) in the form

$$\tilde{H} = \tilde{H}_2 + \tilde{H}_4 + \dots \, , \tag{23}$$

where the second order term $\tilde{H}_2$ preserves the normal form (8). The third-order term $\tilde{H}_3$ is absent, and the fourth-order term $\tilde{H}_4$ may be written as

$$\tilde{H}_4 = \sum_{i+j+k+l+m+n=4} \tilde{h}^{(4)}_{ijklmn} \tilde{q}_1^i \tilde{q}_2^j \tilde{q}_3^k \tilde{p}_1^l \tilde{p}_2^m \tilde{p}_3^n \, . \tag{24}$$

The sum (24) contains 86 terms but coefficients $\tilde{h}^{(4)}_{ijklmn}$ are very cumbersome, and we do not write them here. Again we look for the function

$$S(p_1^*, p_2^*, p_3^*, \tilde{q}_1, \tilde{q}_2, \tilde{q}_3) = \tilde{q}_1 p_1^* + \tilde{q}_2 p_2^* + \tilde{q}_3 p_3^* + \sum_{i+j+k+l+m+n=4} s^{(4)}_{ijklmn} \tilde{q}_1^i \tilde{q}_2^j \tilde{q}_3^k p_1^{*l} p_2^{*m} p_3^{*n},$$

$$\tag{25}$$

generating the canonical transformation reducing the fourth-order term $\tilde{H}_4$ to the simplest form. New momenta p_1^*, p_2^*, p_3^* and coordinates q_1^*, q_2^*, q_3^* are determined by the relationships

$$q_k^* = \frac{\partial S}{\partial p_k^*} \, , \tilde{p}_k = \frac{\partial S}{\partial \tilde{q}_k} \, , \ (k = 1, 2, 3) \, . \tag{26}$$

Resolving (26) with respect to the old canonical variables in the neighborhood of the point $q_1^* = q_2^* = q_3^* = p_1^* = p_2^* = p_3^* = 0$ and substituting the solution into (23), we expand the Hamiltonian $\tilde{H}$ into the Taylor series in $q_1^*, q_2^*, q_3^*, p_1^*, p_2^*, p_3^*$. Obviously, the second order term H_2^* in this expansion again has the normal form (8). The third-order term H_3^* is absent, and the fourth-order term H_4^* is a sum of 126 terms of the form

$$h^{*(4)}_{ijklmn} q_1^{*i} q_2^{*j} q_3^{*j} p_1^{*k} p_2^{*l} p_3^{*l} \ (i + j + k + l + m + n = 4) \, ,$$

where new coefficients $h^{*(4)}_{ijklmn}$ are linear functions of unknown coefficients $s^{(4)}_{ijklmn}$ determining the generating function (25).

Analysis of the coefficients $h^{*(4)}_{ijklmn}$ shows that they are again divided into several independent groups and each group forms a system of equations determining some coefficients $s^{(4)}_{ijklmn}$. If the following conditions

$$\sigma_1 \neq 0, \quad \sigma_2 \neq 0, \quad \sigma_3 \neq 0, \quad \sigma_1 \pm \sigma_2 \neq 0, \quad \sigma_1 \pm \sigma_3 \neq 0, \quad \sigma_2 \pm \sigma_3 \neq 0 ,$$

$$\sigma_1 \pm 3\sigma_2 \neq 0, \quad \sigma_1 \pm 3\sigma_3 \neq 0, \quad 3\sigma_1 \pm \sigma_2 \neq 0 ,$$

$$\sigma_2 \pm 3\sigma_3 \neq 0, \quad \sigma_1 \pm \sigma_2 \pm 2\sigma_3 \neq 0 , \tag{27}$$

are fulfilled we can solve the equations $h^{*(4)}_{ijklmn} = 0$ and find the coefficients $s^{(4)}_{ijklmn}$ of the canonical transformation (26) that eliminates the corresponding terms in (24). However, there are 21 terms in the expansion (24) which can not be eliminated. They can be only simplified in such a way that the fourth-order term $\tilde{H}_4$ takes the form

$$H^*_4 = \frac{1}{4} \left(c_{11}(p_1^{*2} + q_1^{*2})^2 + c_{22}(p_2^{*2} + q_2^{*2})^2 + c_{33}(p_3^{*2} + q_3^{*2})^2 + \right.$$

$$+ c_{12}(p_1^{*2} + q_1^{*2})(p_2^{*2} + q_2^{*2}) + c_{13}(p_1^{*2} + q_1^{*2})(p_3^{*2} + q_3^{*2}) +$$

$$\left. + c_{23}(p_2^{*2} + q_2^{*2})(p_3^{*2} + q_3^{*2}) \right) .$$

Then, using the standard canonical transformation

$$q_k^* = \sqrt{2\tau_k} \sin \varphi_k , \quad p_k^* = \sqrt{2\tau_k} \cos \varphi_k , \quad (k = 1, 2, 3) , \tag{28}$$

we rewrite the Hamiltonian (23) as

$$H^* = H^{(0)} + H_5^*(\varphi_1, \tau_1, \varphi_2, \tau_2, \varphi_3, \tau_3) + \dots , \tag{29}$$

where

$$H^{(0)} = \sigma_1\tau_1 - \sigma_2\tau_2 + \sigma_3\tau_3 + c_{11}\tau_1^2 + c_{22}\tau_2^2 + c_{33}\tau_3^2 + c_{12}\tau_1\tau_2 + c_{13}\tau_1\tau_3 + c_{23}\tau_2\tau_3 , \tag{30}$$

and $H_5^*(\varphi_1, \tau_1, \varphi_2, \tau_2, \varphi_3, \tau_3)$ is the fifth-order term in the expansion (23).

Now we can apply the classical Arnold theorem [4] which states that in the case of absence of resonances up to the fourth order (included) an equilibrium solution of the Hamiltonian system is stable for the majority of initial conditions if the following condition is fulfilled

$$D_3 = \det \left(\frac{\partial^2 H^{(0)}}{\partial \tau_i \partial \tau_j} \right) \neq 0 \tag{31}$$

for $\tau_1 = \tau_2 = \tau_3 = 0$.

Numerical analysis of the determinant (31) shows that $D_3 \neq 0$ in the domain of linear stability of equilibrium position S_7 which is bounded by the curve $\sigma_1 = \sigma_2$ in the $\mu_1 O \mu_2$ plane (see Fig. 2). Cross section of the surface $D_3 = D_3(\mu_1, \mu_2)$ by the plane $\mu_1 = 0.0160642$ shown in Fig. 4 demonstrates dependence of D_3 on parameters μ_1, μ_2. Note that the point of singularity of D_3 corresponds to the case of third-order resonance $\sigma_1 = 2\sigma_2$.

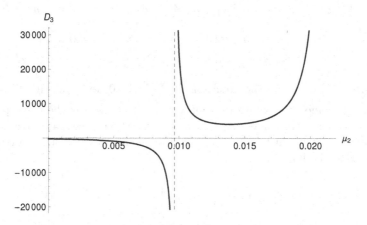

Fig. 4. Determinant D_3 as function of μ_2 for the point S_7, $\mu_1 = 0.0160642$

Analysis of the frequencies (6) and (7) shows that there exist such points in the domain of linear stability of equilibrium position S_7 in the $\mu_1 O \mu_2$ plane (see Fig. 2) for which fourth-order resonance conditions of the form $\sigma_1 = 3\sigma_2$, $3\sigma_2 = \sigma_3$, $2\sigma_1 = \sigma_2 + \sigma_3$ are fulfilled. It should be noted that the last two resonances do not influence on stability of equilibrium position S_7 because the corresponding resonance terms in the expansion of the Hamiltonian (24) are eliminated by the normalizing canonical transformation. The case of resonance $\sigma_1 = 3\sigma_2$ was studied in detail in [7] where it was shown that this resonance produces instability of the equilibrium solution S_1. In the spatial case considered, under fourth-order resonance $\sigma_1 = 3\sigma_2$, the equilibrium point S_7 is unstable too.

5 Conclusion

We have studied the stability of the equilibrium point S_7 in the spatial circular restricted four-body problem formulated on the basis of the Lagrange triangular solution of the three-body problem. We proved that this equilibrium point is stable almost for all initial conditions for any values of parameters μ_1, μ_2 from the domain of its linear stability in the $\mu_1 O \mu_2$ plane but for the resonance curves $\sigma_1 = 2\sigma_2$ and $\sigma_1 = 3\sigma_2$ shown in Fig. 2. The equilibrium point S_7 is unstable under these two resonances while the resonances $2\sigma_2 = \sigma_3$, $3\sigma_2 = \sigma_3$ and $2\sigma_1 = \sigma_2 + \sigma_3$ do not influence its stability.

Note that all numerical and symbolic computations and visualization of the obtained results have been done with the computer algebra system *Mathematica*. And all the calculations can be easily repeated for other equilibrium points.

References

1. Markeev, A.P.: Stability of the Hamiltonian systems. In: Matrosov, V.M., Rumyantsev, V.V., Karapetyan, A.V. (eds.) Nonlinear Mechanics, pp. 114–130. Fizmatlit, Moscow (2001) (in Russian)
2. Szebehely, V.: Theory of Orbits. The Restricted Problem of Three Bodies. Academic Press, New York (1967)
3. Markeev, A.P.: Libration Points in Celestial Mechanics and Cosmodynamics. Nauka, Moscow (1978) (in Russian)
4. Arnold, V.I.: Small denominators and problems of stability of motion in classical and celestial mechanics. Uspekhi Math. Nauk 18(6), 91–192 (1963) (in Russian)
5. Moser, J.: Lectures on the Hamiltonian Systems. Mir, Moscow (1973) (in Russian)
6. Wolfram, S.: The Mathematica Book, 4th edn. Wolfram Media/Cambridge University Press (1999)
7. Budzko, D.A., Prokopenya, A.N.: On the Stability of Equilibrium Positions in the Circular Restricted Four-Body Problem. In: Gerdt, V.P., Koepf, W., Mayr, E.W., Vorozhtsov, E.V. (eds.) CASC 2011. LNCS, vol. 6885, pp. 88–100. Springer, Heidelberg (2011)
8. Budzko, D.A., Prokopenya, A.N.: Symbolic-numerical analysis of equilibrium solutions in a restricted four-body problem. Programming and Computer Software 36(2), 68–74 (2010)
9. Birkhoff, G.D.: Dynamical Systems. GITTL, Moscow (1941) (in Russian)
10. Budzko, D.A., Prokopenya, A.N., Weil, J.A.: Quadratic normalization of the Hamiltonian in restricted four-body problem. Vestnik BrSTU. Physics, Mathematics, Informatics (5), 82–85 (2009) (in Russian)
11. Liapunov, A.M.: General Problem about the Stability of Motion. Gostekhizdat, Moscow (1950) (in Russian)
12. Gadomski, L., Grebenikov, E.A., Prokopenya, A.N.: Studying the stability of equilibrium solutions in the planar circular restricted four-body problem. Nonlinear Oscillations 10(1), 66–82 (2007)

Computing Hopf Bifurcations in Chemical Reaction Networks Using Reaction Coordinates

Hassan Errami[1], Werner M. Seiler[1], Markus Eiswirth[2], and Andreas Weber[3]

[1] Institut für Mathematik, Universität Kassel, Kassel, Germany
errami@uni-kassel.de, seiler@mathematik.uni-kassel.de
[2] Fritz-Haber Institut der Max-Planck-Gesellschaft, Berlin, Germany and Ertl Center for Electrochemisty and Catalysis, Gwangju Institute of Science and Technology (GIST), South Korea
eiswirth@fhi-berlin.mpg.de
[3] Institut für Informatik II, Universität Bonn, Bonn, Germany
weber@cs.uni-bonn.de

Abstract. The analysis of dynamic of chemical reaction networks by computing Hopf bifurcation is a method to understand the qualitative behavior of the network due to its relation to the existence of oscillations. For low dimensional reaction systems without additional constraints Hopf bifurcation can be computed by reducing the question of its occurrence to quantifier elimination problems on real closed fields. However deciding its occurrence in high dimensional system has proven to be difficult in practice. In this paper we present a fully algorithmic technique to compute Hopf bifurcation fixed point for reaction systems with linear conservation laws using reaction coordinates instead of concentration coordinates, a technique that extends the range of networks, which can be analyzed in practice, considerably.

1 Introduction

In chemical and biochemical systems, reactions networks can be represented as a set of reactions. If it is assumed they follow mass action kinetics then the dynamics of these reactions can be represented by ordinary differential equations (ODE) for systems without additional constraints or differential algebraic equations (DAE) for systems with constraints. Particularly, in complex systems it is sometimes difficult to estimate the values of the parameters of these equations, hence the simulation studies involving the kinetics is a daunting task. Nevertheless, quite a few things about the dynamics can be concluded from the structure of the reaction network itself. In this context there has been a surge of algebraic methods, which are based on the structure of network and the associated stoichiometry of the chemical species. These methods provide a way to understand the qualitative behaviour (e.g. steady states, stability, bifurcations, oscillations, etc) of the network. The analysis of chemical reaction networks by detecting of the occurrence of Hopf bifurcation attracts especially more and more interests in chemical and biological field due to its linkage to oscillatory behaviour.

V.P. Gerdt et al. (Eds.): CASC 2012, LNCS 7442, pp. 84–97, 2012.

A fully algebraic method for the computation of Hopf bifurcation fixed points for systems with polynomial vector field has already been introduced by El Kahoui and Weber [1] using the powerful technique of quantifier elimination on real closed fields [2]. This technique has already been applied to mass action kinetics of small dimension [3]. Although the method is complete in theory it fails for systems of higher dimensions in practice.

However the detection of Hopf bifurcation in high dimensional systems and in systems with constraints as is the case in chemical and biochemical systems has proven to be difficult. A central method to overcome this difficulty is called stoichiometric network analysis (SNA). This method has been introduced by Clark in 1980 [4] and based on the analysis of the system dynamic in the flux space instead in the concentration space and expand the steady states into a combination of subnetworks using convex geometry. For the steady state loci new coordinates that are called *reaction coordinates* can be introduced. These methods have been used in several "hand computations" in a semi-algorithmic way for parametric systems, the most elaborate being described in [5].

Our algorithmic method presented in this paper uses and combines the ideas of these methods and extends them to a new approach for computing Hopf bifurcation in complex systems using reaction coordinates also allowing systems with linear constraints.

2   Chemical Reaction Networks

A chemical reaction occurs when two or more chemical species react to become new chemical species. This process is usually presented by an equation where the *reactants* are given on the left hand side of an arrow and the *products* on the right hand side, the numbers next to the species called *stoichiometric coefficients* present the amount to which a chemical species participates in a reaction and the parameter on the arrow called *rate constant* stands for an experimental constant influencing the reaction velocity. A chemical reaction is called *irreversible*, if it proceeds only in one direction, and is called *reversible*, if it proceeds in either directions. An example of a chemical reaction, as it usually appears in the literature, is the following:

$$A + B \xrightarrow{k} 3A + C$$

In this reaction, one unit of chemical *species* A and one of B react (at reaction rate k) to form three units of A and one of C. The concentrations of these three species, denoted by x_a, x_b and x_c, will change in time as the reaction occurs. Under the assumption of *mass-action kinetics*, species A and B react at a rate proportional to the product of their concentrations, where the proportionality constant is the rate constant k. Noting that the reaction yields a net change of two units in the amount of A [6,7,5], we obtain the following corresponding differential equations:

$$\dot{x}_a = 2kx_ax_b$$
$$\dot{x}_b = -kx_ax_b$$
$$\dot{x}_c = kx_ax_b \tag{1}$$

A *chemical reaction network* is a finite set of chemical reactions. It can be presented as a finite directed graph whose vertices are labeled by complexes and whose edges are labeled by parameters(reaction rate constants). Specifically, the digraph is denoted $G = (V, E)$, with vertex set $V = \{1, 2, ..., m\}$ and edge set $E \subseteq \{(i, j) \in V \times V : i \neq j\}$. A network is reversible if the graph G is undirected, in which case each undirected edge has two labels k_{ij} and k_{ji} [7,6].

2.1 Flux Cone and Reaction Coordinates

Clarke [4] has introduced a method called stoichiometric network analysis (SNA) to analyze the stability of chemical reaction networks. The idea of SNA is to observe the dynamics of the system in the reaction space instead of concentration space. This leads to expand the steady state into a combination of subnetworks that form a convex cone in the flux-space called *flux cone* [8]. In section 6.2 we discuss the computation of the *flux cone* in and for detailed description of the concepts of SNA we refer to the seminal work of Clarke [4].

2.2 Constraints in Chemical Reaction Networks

The differential equations in chemical reaction networks usually are constrained reflecting various physical conservation laws. The situation found in chemical reaction networks can easily be generalized, and we will provide an analysis for the situation for the case of *pseudolinear ordinary differential equations* in general, which will contain all cases of constraints for chemical reaction systems discussed in this paper, as an instance.

3 Pseudolinear Ordinary Differential Equations

The following material represents a slight generalisation of results already well-known for systems appearing in reaction kinetics (see e.g. [22] and references therein). The basic underlying property of the considered differential equations is captured by the following definition.

Definition 1. *We call an autonomous system of ordinary differential equations* $\dot{x} = \phi(x)$ *for an unknown function* $x : \mathbb{R} \to \mathbb{R}^n$ *pseudolinear, if its right hand side can be written in the form* $\phi(x) = N\psi(x)$ *with a constant matrix* $N \in \mathbb{R}^{n \times m}$ *and some vector valued function* $\psi : \mathbb{R}^n \to \mathbb{R}^m$.

Obviously, any *polynomially* nonlinear system can be written in such a form, if we take for $\psi(x)$ the vector of all terms appearing on the right hand side of the system. As one can see from the following two lemmata, the pseudolinear structure is of interest only in the case that the matrix N does not possess full row rank and hence the range of N is not the full space $\mathbb{R}^n$. In the sequel, we will always assume that the function ψ satisfies $m \geq n$, as this is usually the case in applications like reaction kinetics.

Lemma 1. *For a pseudolinear system* $\dot{\mathbf{x}} = N\psi(\mathbf{x})$ *any affine subspace of the form* $\mathcal{A}_{\mathbf{y}} = \mathbf{y} + \operatorname{im} N \subseteq \mathbb{R}^n$ *for an arbitrary constant vector* $\mathbf{y} \in \mathbb{R}^n$ *defines an invariant manifold.*

Proof. Obviously, we have $\dot{\mathbf{x}}(t) \in \operatorname{im} N$ for all times t and $T_{\mathbf{x}}\mathcal{A}_{\mathbf{y}} = \operatorname{im} N$ for all points $\mathbf{x} \in \mathcal{A}_{\mathbf{y}}$ by definition of an affine space. Thus, if $\mathbf{x}(0) \in \mathcal{A}_{\mathbf{y}}$, then the whole trajectory will stay in $\mathcal{A}_{\mathbf{y}}$. □

Remark 1. For the application in reaction kinetics, the following minor strengthening of Lemma 1 is of interest. Assume that the function ψ satisfies additionally $\psi(\mathbf{x}) \in \mathbb{R}_{\geq 0}^m$ for all $\mathbf{x} \in \mathbb{R}_{\geq 0}^n$ which is for example trivially the case when each component of ψ is a polynomial with positive coefficients. If we solve our differential equation for non-negative initial data $\mathbf{x}(0) = \mathbf{x}_0 \in \mathbb{R}_{\geq 0}^n$, then the solution always stays in the convex polyhedral cone $\mathbf{x}_0 + \left\{ \sum_{i=1}^m \lambda_i \mathbf{n}_i \mid \forall i : \lambda_i \geq 0 \right\}$ where the vectors $\mathbf{n}_i$ are the columns of the matrix N. Indeed, in this case the tangent vector $\dot{\mathbf{x}}(t)$ along the trajectory is trivially always a non-negative linear combination of the columns of N.

Lemma 2. *Let* $\mathbf{v}^T \cdot \mathbf{x} = Const$ *for some vector* $\mathbf{v} \in \mathbb{R}^n$ *be a* linear *conservation law of a pseudolinear system* $\dot{\mathbf{x}} = N\psi(\mathbf{x})$ *such that* $\operatorname{im} \psi$ *is not contained in a hyperplane. Then* $\mathbf{v} \in \ker N^T$. *Conversely, any vector* $\mathbf{v} \in \ker N^T$ *induces a linear conservation law.*

Proof. Let us first assume that $\mathbf{v} \in \ker N^T$. Then

$$\frac{d}{dt}\left(\mathbf{v}^T \cdot \mathbf{x}\right) = \mathbf{v}^T N\psi(\mathbf{x}) = \left(N^T\mathbf{v}\right)^T \psi(\mathbf{x}) = 0 \,.$$

If $\mathbf{v}^T \cdot \mathbf{x} = Const$ is a conservation law, then differentiation with respect to time yields $\left(N^T\mathbf{v}\right)^T \psi(\mathbf{x}) = 0$. Because of our assumption on the function ψ, this implies that $N^T\mathbf{v} = 0$. □

By a classical result in linear algebra (the four "fundamental spaces" of a matrix), we have the direct sum decomposition $\mathbb{R}^n = \operatorname{im} N \oplus \ker N^T$ which is even an orthogonal decomposition with respect to the standard scalar product. Hence we may consider Lemma 1 as a corollary to Lemma 2, as the above described invariant manifolds are simply defined by all the linear conservation laws produced by Lemma 2.[1]

Remark 2. Gatermann and Huber [22] speak of a conservation law only in the case that $v_i \geq 0$ for all components v_i of the vector $\mathbf{v}$. In mathematics, we are not aware of such a restriction and cannot see any physical reasons to impose it.

---

[1] Note that in the special case most relevant for us, namely that each component of ψ is a different monomial, the assumption made in Lemma 2 is always satisfied.

4   Reduction to Invariant Manifolds

If a dynamical system admits invariant manifolds, we may consider a system of lower dimension by reducing to such a manifold. However, in general it may not be possible to derive explicitly the reduced system. Nevertheless, for many purposes like stability or bifurcation analysis one can easily reduce to smaller matrices. The following result describes such a reduction process in the linear case. It represents an elementary exercise in basic linear algebra. In order to avoid the inversion of matrices, we consider here $\mathbb{R}^n$ as a Euclidean space with respect to the standard scalar product.

Lemma 3. *Let A be the matrix of a linear mapping $\mathbb{R}^n \to \mathbb{R}^n$ for the standard basis and $\mathcal{U} \subseteq \mathbb{R}^n$ a k-dimensional A-invariant subspace. If the columns of the matrix $W \in \mathbb{R}^{n \times k}$ define an orthonormal basis of $\mathcal{U}$, then the restriction of the mapping to the subspace $\mathcal{U}$ with respect to the basis defined by W is given by the matrix $W^T A W \in \mathbb{R}^{k \times k}$.*

Proof. Considered as a linear map $\mathbb{R}^k \to \mathcal{U} \subseteq \mathbb{R}^n$, the matrix W defines a parametrisation of $\mathcal{U}$ with inverse $W^T : \mathcal{U} \to \mathbb{R}^k$. Indeed, $W^T W = \mathbb{1}_k$, since the columns of W are orthonormal. If $\mathbf{v} \in \mathcal{U}$, then $\mathbf{v} = W\mathbf{w}$ for some vector $\mathbf{w} \in \mathbb{R}^k$ and thus $W^T \mathbf{v} = (W^T W)\mathbf{w} = \mathbf{w}$ implying that $(WW^T)\mathbf{v} = W\mathbf{w} = \mathbf{v}$, i.e. the matrix $WW^T \in \mathbb{R}^{n \times n}$ describes $\mathrm{id}_{\mathcal{U}}$. By standard linear algebra, the matrix $W^T A W$ describes therefore the restriction of A to $\mathcal{U}$.   □

As a simple application, we note that in the case of a pseudolinear system $\dot{\mathbf{x}} = N\boldsymbol{\psi}(\mathbf{x})$ the stability properties of an equilibrium $\mathbf{x}_e$ of the pseudolinear system $\dot{\mathbf{x}} = N\boldsymbol{\psi}(\mathbf{x})$ are determined by the eigenstructure of the reduced Jacobian

$$J = W^T N \mathrm{Jac}\big(\boldsymbol{\psi}(\mathbf{x}_e)\big) W \in \mathbb{R}^{k \times k}$$

where the columns of W form an orthonormal basis of $\mathrm{im}\, N$. If parameters are present, then also for a bifurcation analysis the eigenstructure of this matrix and not of the full Jacobian (which is an n-dimensional matrix) are relevant.

5   Stability and Bifurcations for Semi-Explicit DAEs

The considerations indicated in the last section can be easily extended to more general situations, as they appear in the theory of DAEs. For simplicity (and as it suffices for our purposes), we assume that we are dealing with an autonomous system in the semi-explicit form

$$\dot{\mathbf{x}} = \mathbf{f}(\mathbf{x}), \qquad 0 = \mathbf{g}(\mathbf{x}) \tag{2}$$

where $\mathbf{f} : \mathbb{R}^n \to \mathbb{R}^n$ and $\mathbf{g} : \mathbb{R}^n \to \mathbb{R}^{n-k}$. Furthermore, we assume that the above system of ordinary differential equations is involutive,[2] i.e. that it contains already all its integrability conditions. This assumption is equivalent to the existence of a matrix valued function $M(\mathbf{x})$ such that

$$\mathrm{Jac}\big(\mathbf{g}(\mathbf{x})\big) \cdot \mathbf{f}(\mathbf{x}) = M(\mathbf{x}) \cdot \mathbf{g}(\mathbf{x}). \tag{3}$$

---

[2] See [23] for an introduction into the theory of involutive systems.

Thus one may say that the components of $\mathbf{g}$ are *weak* conservation laws, as their time derivatives vanish modulo the constraint equations $\mathbf{g}(\mathbf{x}) = 0$.

Let $\mathbf{x}_e$ be an equilibrium of (2), i. e. we have $\mathbf{f}(\mathbf{x}_e) = 0$ and $\mathbf{g}(\mathbf{x}_e) = 0$. We introduce the real matrices

$$A = \mathrm{Jac}\big(\mathbf{f}(\mathbf{x}_e)\big) \in \mathbb{R}^{n \times n}, \quad B = \mathrm{Jac}\big(\mathbf{g}(\mathbf{x}_e)\big) \in \mathbb{R}^{(n-k) \times n}.$$

For simplicity, we assume in the sequel that the matrix B has full rank (or, in other words, that our algebraic constraints are independent) and thus that $\ker B$ is a k-dimensional subspace. The proof of the next result demonstrates clearly why the assumption that the system (2) is involutive is important, as the relation (3) is crucial for it.

Lemma 4. *The subspace* $\ker B$ *is A-invariant.*

Proof. Set $\bar{M} = M(\mathbf{x}_e)$. Differentiating (3) and evaluating the result at $\mathbf{x} = \mathbf{x}_e$ yields the relation $BA = \bar{M}B$. Hence, if $\mathbf{v} \in \ker B$, then also $A\mathbf{v} \in \ker B$ since $B(A\mathbf{v}) = \bar{M}(B\mathbf{v}) = 0$. $\qquad\square$

Remark 3. In the case that (2) is a linear system, i. e. by assuming that $\mathbf{x}_e = 0$ we may write $\mathbf{f}(\mathbf{x}) = A\mathbf{x}$ and $\mathbf{g}(\mathbf{x}) = B\mathbf{x}$, we can easily revert the argument in the proof of Lemma 4 and thus conclude that now (2) is involutive, if and only if $\ker B$ is A-invariant.

Proposition 1. *Let the columns of the matrix $W \in \mathbb{R}^{n \times k}$ define an orthonormal basis of $\ker B$. The linear stability of the equilibrium $\mathbf{x}_e$ is then decided by the eigenstructure of the matrix $W^T A W$.*

Proof. Linearisation around the equilibrium $\mathbf{x}_e$ yields the associated variational system $\dot{\mathbf{z}} = A\mathbf{z}$, $B\mathbf{z} = 0$. We complete W to an orthogonal matrix $\widehat{W}$ by adding some further columns and perform the coordinate transformation $\mathbf{z} = \widehat{W}\mathbf{y}$. This yields the system $\dot{\mathbf{y}} = \widehat{W}^T A \widehat{W}\mathbf{y}$, $B\widehat{W}\mathbf{y} = 0$. Since by construction the columns of W span $\ker B$, the second equation implies that only the upper k components of $\mathbf{y}$ may be different of zero. Furthermore, Lemma 4 implies that the matrix $\widehat{W}^T A \widehat{W}\mathbf{y}$ is in block triangular form with the left upper $k \times k$ block given by $W^T A W$. If we denote the upper part of $\mathbf{y}$ by $\tilde{\mathbf{y}}$, we obtain thus the equivalent reduced system $\dot{\tilde{\mathbf{y}}} = W^T A W \tilde{\mathbf{y}}$ which implies our claim. $\qquad\square$

Remark 4. Let $\mathbf{v} \in \mathbb{R}^k$ be a (generalised) eigenvector of the reduced matrix $W^T A W$, i. e. we have $(W^T A W - \lambda \mathbb{1}_k)^\ell \mathbf{v} = 0$ for some $\ell > 0$ and $\lambda \in \mathbb{R}$. Since $W^T W = \mathbb{1}_k$ and WW^T defines the identity map on $\ker B$ (see the proof of Lemma 3), we obtain $W^T(A - \lambda \mathbb{1}_n)^\ell W\mathbf{v} = 0$ implying that $W\mathbf{v} \in \mathbb{R}^n$ is a (generalised) eigenvector of A for the same eigenvalue λ, since the matrix W^T defines an injective map. Thus every eigenvalue of the reduced matrix $W^T A W$ is also an eigenvalue of A.

Remark 5. It is also not difficult to interpret the remaining (generalised) eigenvectors of A. By construction, they are transversal to the constraint manifold defined by $\mathbf{g}(\mathbf{x}) = 0$ and they describe whether this manifold is attractive or repellent for the flow of the unconstrained system $\dot{\mathbf{x}} = \mathbf{f}(\mathbf{x})$. While this is for example of considerable importance for the numerical integration of (2), as it describes the drift off the constraint manifold due to rounding and discretisation errors, it has no influence on the stability of the exact flow of (2).

The irrelevance of the remaining (generalised) eigenvectors of A becomes also apparent from the following argument. Recall that the differential part of (2) defines what is often called an *underlying differential equation* for the DAE, i. e. an unconstrained differential equation which possesses for initial data satisfying the constraints the same solution as the DAE. Consider now the modified system obtained by adding to the right hand side of the differential part an arbitrary linear combination of the algebraic part. It is easy to see that the arising DAE (which simply has a different underlying equation)

$$\dot{\mathbf{x}} = \mathbf{f}(\mathbf{x}) + L(\mathbf{x})\mathbf{g}(\mathbf{x}), \qquad 0 = \mathbf{g}(\mathbf{x}),$$

where $L(\mathbf{x})$ is a matrix valued function of appropriate dimensions, possesses exactly the same solutions as (2); in particular $\mathbf{x}_e$ is still an equilibrium. If we proceed as above with the linear stability analysis of $\mathbf{x}_e$, the matrix B remains unchanged, whereas A is transformed into the modified matrix $\tilde{A} = A + \bar{L}B$ with $\bar{L} = L(\mathbf{x}_e)$. Obviously, $\ker B$ is also $\tilde{A}$-invariant and furthermore $W^T\tilde{A}W = W^TAW$, if the columns of W form a basis of $\ker B$ as in Proposition 1.

Thus all (generalised) eigenvectors lying in $\ker B$ are equal for A and $\tilde{A}$ and thus the stability of $\mathbf{x}_e$ is not affected by this transformation. However, the remaining (generalised) eigenvectors may change arbitrarily. One can for example show that by a suitable choice of the matrix L one may always achieve that the constraint manifold becomes attractive.

6   Algorithms for Computing Hopf Bifurcations in Chemical Reaction Networks Using Reaction Coordinates

In this section we present an algorithmic approach for computing the Hopf bifurcation in chemical systems. Our approach is mainly based on three methods already presented in this paper: stoichiometric network analysis, method for reduction of manifold for systems with conservation laws, and techniques of quantifier elimination on real closed field. The pseudo code given in Fig. 1 and outline the main steps of our algorithm, which are detailed in the following subsections.

6.1   Pre-processing: Step 1

For starting the analysis of a chemical network we need two significant pieces of information to describe all reaction laws. The first information describes the occurrence of the species in each reaction. This can be presented by a stoichiometric matrix $\mathcal{S}$, where the species build the rows and the reactions build the

Input: a chemical reaction network $\mathcal{N}$ with $\dim(\mathcal{N}) = n$.
Output: statement about the existence of Hopf-bifurcation.

1: Generate the stoichiometric matrix $\mathcal{S}$ and kinetic matrix $\mathcal{K}$ from the reaction network.
2: Compute the minimal set $\mathcal{E}$ of the vectors generating the flux cone.
3: For $d = 1 \ldots n$: Compute all d-faces and of the flux cone (subsystems).
   For each subsystem $\mathcal{N}_i$ **do**
       4: Compute the transformed Jacobian Jac_i of $\mathcal{N}_i$ using $\mathcal{K}$, $\mathcal{S}$ and flux cone coordinates j_i´s
       5: **If** Jac_i is singular compute the reduced manifold of Jac_i calling the result also Jac_i
       6: Compute the characteristic polynomial of Jac_i
       7: Compute the Hurwitz determinant of Jac_i
       8: Compute the Hopf-existence condition for $\mathcal{N}_i$
       9: Generate the first-order existentially quantified formula $\mathcal{F}_i$ expressing Hopf-existence condition, the constraints on concentrations and the constraints on the cone coordinates
       10: Reduce and simplify the generated formulae
Output: The disjunction of $\mathcal{F}_i$ yields a criterion for the existence of a Hopf bifurcation fixed point, It can be computed lazily for increasing d and the subsystems.

Fig. 1. Algorithms for Computing Hopf Bifurcations in Chemical Reaction Networks Using Flux Coordinates

columns. Each entry of the matrix presents the difference of the number of produced and consumed molecules of the corresponding species in the corresponding reaction. The second information describes the velocities of the reactions. This can be presented by flux vector $v(x, k)$ or by kinetic matrix $\mathcal{K}$. The entries of this matrix present the information whether species is a reactant(entry = stoichiometric coefficient of species) and affects consequently the velocity of the reaction or not (entry = 0). To enable the computational analysis of a chemical networks the reactions should be presented in a format that enables its accurate representation and allows the computational extraction of needed data. For our computations we use the XML based and in biological research widely used format SBML [20]. As pre-processing step we parse the SBML file presenting the chemical network using Java library JSBML [19] to generate the stoichiometric matrix and kinetic matrix.

6.2   Geometrical Computations: Step 2 and 3

To analyse a chemical system one is interested in the stationary reaction behaviour, which is observable in experiments, i.e one investigates the solution set of

$$\mathcal{S}v(x, k) = 0. \tag{4}$$

The set of stationary solutions is usually considered in the concentration space $\mathbb{R}^n_+$, i.e in the variables x. Instead of the variables $x \in \mathbb{R}^n_+$ we will consider the variables z representing $v(x, k)$ which are called reaction coordinates or reaction rate coordinates and thus we consider the set of stationary solutions in the space of reaction rates $\mathbb{R}^l_+$. A first advantage is that the Jacobian in the space of reaction rates is of the following form, cf. [5]:

$$\mathrm{Jac}(x) = \widehat{\mathrm{Jac}}(z)\mathrm{diag}(1/x_1, ..., 1/x_m). \tag{5}$$

As long as we split each reversible reaction into two irreversible reactions (forward and backward directions) the flux through this reactions must be greater than or equal to zero, i.e

$$v(x, k) \geq 0 \tag{6}$$

The set of all possible stationary solutions over the network $\mathcal{N}$ that fulfil the equation (4) and the constraint (6) defines the convex polyhedral cone *flux cone* [4,18] and determine a minimal set of generating vectors $\mathcal{E}$, which are called *extreme rays* or *extreme currents*. Each vector z can then as linear combination of the vector set $\mathcal{E}$ with nonnegative coefficients j_i's called *convex parameters*.

To compute the extreme currents we need to integrate algorithms that allow to deal with polyhedral computations. In our current implementation we use POLYMAKE in the step 2 of our algorithm to compute the extreme currents $\mathcal{E}$ for a generating stoichiometric matrix $\mathcal{S}$. POLYMAKE is an open source software tool written in Perl and C++ and designed for the algorithmic treatment of polytopes and polyhedra [9].

Computing extreme currents $\mathcal{E}$ is the basis for simplifying the analysis of chemical networks by its decomposing into minimal steady-state generating subnetworks. The influence of a subnetwork on the full network dynamics (i.e., how much the given subnetwork plays a part in creating a certain steady state) depends on the convex parameters j_i [4,21]. From a chemical perspective the Hopf bifurcation occurrs mostly in the spaces formed by two or three adjacent extreme currents, i.e detecting the Hopf bifurcation in subsystems can be restricted on subsystems combined by two faces or three faces of the flux cone. As step 3 of our algorithms we compute all subsystems generated by the 2- and 3-faces using also POLYMAKE. Our algorithm can also handle d-faces for $d > 3$ yielding a complete method in theory, but the restriction to $d = 2, 3$ will be of greatest practical interest.

6.3   Transformation of the Jacobian: Step 4

Gatermann et al. [5] proved that the Jacobian of reaction coordinates z can be transformed into the follwing form:

$$\widehat{\mathrm{Jac}}(z) = \mathcal{S}\mathrm{diag}(z)\mathcal{K}^t \tag{7}$$

If x is a steady state we transform into convex coordinate j_i with $z = \sum_i^d j_i \mathcal{E}_i$ with d being the dimensionality of the face. When we replace $\widehat{\mathrm{Jac}}(z)$ in the equation 7 we obtain the new Jacobian Jac_d in reactions space:

$$\mathrm{Jac}_d(x) = \mathcal{S}\mathrm{diag}(\sum_i^d j_i \mathcal{E}_i)\mathcal{K}^t \mathrm{diag}(1/x_1, ..., 1/x_m) \tag{8}$$

6.4 Jacobian of Reduced Manifold: Step 5

Chemical reaction networks with conservation laws give rise to singularity of
the Jacobian of the entire polynomial system presenting the network and also
of some Jacobian matrices of the computed subsystems. To compute the Hopf
condition the Jacobian matrices should be transformed to nonsingular matrices.
Therefore we reduce them in step 5 of our algorithm by computing the Jacobian
Jac_i of reduced manifolds using the method presented in sect. 3.

6.5 Generating and Reducing Quantified Formulae: Steps 6–10

Our aim in the last steps is obtaining a simple formula that gives a clear statement if a Hopf bifurcation occur in the system. We firstly give a semi-algebraic
description of Hopf bifurcation by use of the Hurwitz determinants, and produces a first-order formula which is transformed into a quantifier-free formula.
Using the positivity conditions on all parameters we can use *positive quantifier
elimination* [10,3] implemented in REDLOG [11,12], which had been originally
driven by the efficient implementation of quantifier elimination based on virtual substitution methods [13,14,15]. For formula simplification and as "fallback
method" we use QEPCAD B [16].

7 Computation Examples

7.1 Example1: Phosphofructokinase Reaction

As a first example we consider the main example used in the hand computation
in [5]—the phosphofructokinase reaction.

It yields the following system of ordinary differential equations:

$$\dot{x}_1 = k_{21}x_1^2 x_2 + k_{46} - k_{64}x_1 - k_{34}x_1 + k_{43}x_3$$
$$\dot{x}_2 = -k_{21}x_1^2 x_2 + k_{56} - k_{65}x_2$$
$$\dot{x}_3 = k_{34}x_1 - k_{43}x_3 \tag{9}$$

This problem has already been investigated using its formulation in reaction
coordinates in [3]. Using currently available quantifier elimination packages the
problem could not be solved in its parametric form. Only when using the existential closure on the parameters it could be shown by successful quantifier elimina-
tions performed in REDLOG that there exist positive parameters for which there
exists a Hopf bifurcations fixed point in the positive orthant. When redoing the
experiments we found that the situation described in [3] still applies.

The results on the subsystems involving 2-faces and 3-faces are summarized in Table 1. A Hopf bifurcation can be found using the two-face involving two extreme currents {E3,E4} in less than one second of computation time. The 3-faces {E1,E3,E4} and {E2,E3,E4} extending this two-face require some seconds of computation time to find a Hopf bifurcation fixed point. All other faces do not contain a Hopf bifurcation fixed point.

Table 1. Computation of Hopf bifurcation in the phosphofructokinase reaction using reaction coordinates

Subsystem	Result	Time (ms)
{E1}	false	12
{E2}	false	12
{E3}	false	12
{E4}	false	10
{E1,E2}	false	12
{E1,E3}	false	10
{E1,E4}	false	14
{E2,E3}	false	11
{E2,E4}	false	11
{E3,E4}	true	207
{E1,E2,E3}	false	9
{E1,E2,E4}	false	10
{E1,E3,E4}	true	8146
{E2,E3,E4}	true	1621

7.2   Example 2: Enzymatic Transfer of Calcium Ions

Our second example is also investigated in [5]—the enzymatic transfer of calcium ions, Ca^{++}, across the cellmembranes.

It yields the following system:

$$\dot{x}_1 = -k_{12}x_1 + k_{21} + k_{43}x_1x_2 + k_{56}x_4 - k_{65}x_1x_3$$
$$\dot{x}_2 = -k_{43}x_1x_2 + k_{76}x_4$$
$$\dot{x}_3 = k_{56}x_4 - k_{65}x_1x_3 + k_{76}x_4$$
$$\dot{x}_4 = -k_{56}x_4 + k_{65}x_1x_3 - k_{76}x_4 \tag{10}$$

For this system the Jacobian matrix is singular—hence in the classical sense there are no Hopf bifurcations. But in the in reduced system we find that there are Hopf bifurcations—and we can compute them in concentration space as well as using reaction coordinates. The results and computation times are summarized in Table 2.

7.3   Example 3: Model of Calcium Oscillations

The following model of calcium oscillations contains a fractional exponent ε. It is discussed in [17].

Table 2. Enzymatic transfer of calcium ions: Computation of Hopf bifurcation in reaction space and concentration space after reduction of manifold

System	result	time(ms)
{E1}	false	9
{E2}	false	8
{E3}	false	10
{E1,E2}	true	111
{E1,E3}	false	8
{E2,E3}	false	7
{E1,E2,E3}	true	13972
Polynomial system in CS	true	94

$$\dot{x} = k_1 - k_5 xz$$
$$\dot{y} = k_2 x - 4k_3 y^2 + 4k_4 z - k_6 y^\varepsilon$$
$$\dot{z} = k_3 y^2 - k_4 z \tag{11}$$

$$S = \begin{pmatrix} 1 & 0 & 0 & 0 & -1 & 0 \\ 0 & 1 & -4 & 4 & 0 & -1 \\ 0 & 0 & 1 & -1 & 0 & 0 \end{pmatrix}$$

$$K = \begin{pmatrix} 0 & 1 & 0 & 0 & 1 & 0 \\ 0 & 0 & 2 & 0 & 0 & \varepsilon \\ 0 & 0 & 0 & 1 & 1 & 0 \end{pmatrix}$$

In concentration space the solution of a quantifier elimination problem works only for integer values of the parameter ε—as it occurs in the exponent, and the techniques of quantifier elimination over the ordered field of the reals is restricted to polynomials (or rational functions).

However, in the formulation in reaction coordinates the parameter ε occurs as a variable with values in the real closed field used in the computations.

Hence for a given subsystem we cannot only ask the decision question whether there exists a Hopf bifurcation fixed point, but we can ask the question with a free parameter ε.

The answer—a quantifier free formula involving ε—gives the condition for ε, for which a Hopf bifurcation occurs for the subsystem. When using subsystems resulting from 2-faces we did not find Hopf bifurcations, but for the parameteric question on 3-faces we obtained the following answer in less than 10sec of computation time using the combination of REDLOG and QEPCAD B:

$$\varepsilon + 2 > 0 \wedge 4\varepsilon - 1 < 0$$

Hence for $\varepsilon \in (-2, 0.25)$ we have shown that Hopf bifurcation fixed points exist (for suitable reaction constants). Using numerical simulations for this model Reidl et al. [17] could not find Hopf bifurcations for values of the parameter ε bigger than about 0.05.

Acknowledgement. This research was supported in part by *Deutsche Forschungsgemeinschaft* within SPP 1489.

References

1. El Kahoui, M., Weber, A.: Deciding Hopf bifurcations by quantifier elimination in a software-component architecture. Journal of Symbolic Computation 30(2), 161–179 (2000)
2. Tarski, A.: A Decision Method for Elementary Algebra and Geometry, 2nd edn. University of California Press, Berkeley (1951)
3. Sturm, T., Weber, A., Abdel-Rahman, E., El Kahoui, M.: Investigating algebraic and logical algorithms to solve Hopf bifurcation problems in algebraic biology. Mathematics in Computer Science 2(3) (2009), Special Issue on Symbolic Computation in Biology
4. Clarke, B.L.: Stability of Complex Reaction Networks. Advances in Chemical Physics, vol. XLIII. Wiley Online Library (1980)
5. Gatermann, K., Eiswirth, M., Sensse, A.: Toric ideals and graph theory to analyze Hopf bifurcations in mass action systems. Journal of Symbolic Computation 40(6), 1361–1382 (2005)
6. Shiu, A.J.: Algebraic methods for biochemical reaction network theory. Phd thesis, University of California, Berkeley (2010)
7. Pérez Millán, M., Dickenstein, A., Shiu, A., Conradi, C.: Chemical reaction systems with toric steady states. Bulletin of Mathematical Biology, 1–29 (October 2011)
8. Wagner, C., Urbanczik, R.: The geometry of the flux cone of a metabolic network. Biophysical Journal 89(6), 3837–3845 (2005)
9. Gawrilow, E., Joswig, M.: Polymake: a framework for analyzing convex polytopes. In: Kalai, G., Ziegler, G.M. (eds.) Polytopes—Combinatorics and Computation. Oberwolfach Seminars, vol. 29, pp. 43–73. Birkhäuser, Basel (2000), 10.1007/978-3-0348-8438-9_2
10. Sturm, T.F., Weber, A.: Investigating Generic Methods to Solve Hopf Bifurcation Problems in Algebraic Biology. In: Horimoto, K., Regensburger, G., Rosenkranz, M., Yoshida, H. (eds.) AB 2008. LNCS, vol. 5147, pp. 200–215. Springer, Heidelberg (2008)
11. Dolzmann, A., Sturm, T.: REDLOG: Computer algebra meets computer logic. ACM SIGSAM Bulletin 31(2), 2–9 (1997)
12. Sturm, T.: Redlog online resources for applied quantifier elimination. Acta Academiae Aboensis, Ser. B 67(2), 177–191 (2007)
13. Weispfenning, V.: The complexity of linear problems in fields. Journal of Symbolic Computation 5(1&2), 3–27 (1988)
14. Weispfenning, V.: Quantifier elimination for real algebra—the quadratic case and beyond. Applicable Algebra in Engineering Communication and Computing 8(2), 85–101 (1997)
15. Dolzmann, A., Sturm, T.: Simplification of quantifier-free formulae over ordered fields. Journal of Symbolic Computation 24(2), 209–231 (1997)
16. Brown, C.W.: QEPCAD B: A system for computing with semi-algebraic sets via cylindrical algebraic decomposition. ACM SIGSAM Bulletin 38(1), 23–24 (2004)
17. Reidl, J., Borowski, P., Sensse, A., Starke, J., Zapotocky, M., Eiswirth, M.: Model of calcium oscillations due to negative feedback in olfactory cilia. Biophysical Journal 90(4), 1147–1155 (2006)

18. Larhlimi, A.: New Concepts and Tools in Constraint-based Analysis of Metabolic Networks. Dissertation, University Berlin, Germany
19. Dräger, A., Rodriguez, N., Dumousseau, M., Dörr, A., Wrzodek, C., Keller, R., Fröhlich, S., Novère, N.L., Zell, A., Hucka, M.: JSBML: a flexible and entirely Java-based library for working with SBML. Bioinformatics 4 (2011)
20. Hucka, M., Smith, L., Wilkinson, D., Bergmann, F., Hoops, S., Keating, S., Sahle, S., Schaff, J.: The Systems Biology Markup Language (SBML): Language Specification for Level 3 Version 1 Core. In: Nature Precedings (October 2010)
21. Domijan, A., Kirkilionis, M.: Bistability and oscillations in chemical reaction networks. Journal of Mathematical Biology 59(4), 467–501 (2009)
22. Gatermann, K., Huber, B.: A family of sparse polynomial systems arising in chemical reaction systems. J. Symb. Comp. 33, 275–305 (2002)
23. Seiler, W.: Involution — The Formal Theory of Differential Equations and its Applications in Computer Algebra. Algorithms and Computation in Mathematics, vol. 24. Springer, Heidelberg (2009)

Comprehensive Involutive Systems

Vladimir Gerdt[1] and Amir Hashemi[2]

[1] Laboratory of Information Technologies, Joint Institute for Nuclear Research
141980 Dubna, Russia
gerdt@jinr.ru
[2] Department of Mathematical Sciences, Isfahan University of Technology
Isfahan, 84156-83111, Iran
Amir.Hashemi@cc.iut.ac.ir

Abstract. In this paper we consider parametric ideals and introduce a notion of *comprehensive involutive system*. This notion plays the same role in theory of involutive bases as the notion of comprehensive Gröbner system in theory of Gröbner bases. Given a parametric ideal, the space of parameters is decomposed into a finite set of cells. Each cell yields the corresponding involutive basis of the ideal for the values of parameters in that cell. Using the Gerdt–Blinkov algorithm described in [6] for computing involutive bases and also the Montes DisPGB algorithm for computing comprehensive Gröbner systems [13], we present an algorithm for construction of comprehensive involutive systems. The proposed algorithm has been implemented in `Maple`, and we provide an illustrative example showing the step-by-step construction of comprehensive involutive system by our algorithm.

1 Introduction

One of the most important algorithmic objects in computational algebraic geometry is *Gröbner basis*. The notion of Gröbner basis was introduced and an algorithm for its construction was designed in 1965 by Buchberger in his Ph.D. thesis [3]. Later on, he discovered [4] two criteria for detecting some useless reductions that made the Gröbner bases method a practical tool to solve a wide class of problems in polynomial ideal theory and in many other research areas of science and engineering [5]. We refer to the monograph [2] for details on the theory of Gröbner bases.

The concept of comprehensive Gröbner bases can be considered as an extension of these bases for polynomials over fields to polynomials with parametric coefficients. This extension plays an important role in application to constructive algebraic geometry, robotics, electrical network, automatic theorem proving and so on (see, for example, [11,12,13,14]). *Comprehensive Gröbner bases* and equivalent to them *comprehensive Gröbner systems* were introduced in 1992 by Weispfenning [22]. He proved that any parametric polynomial ideal has a comprehensive Gröbner basis and described an algorithm to compute it. In 2002, Montes [13] proposed a more efficient algorithm (DisPGB) for computing comprehensive Gröbner systems. A year later Weispfenning in [21] proved the existence of a canonical comprehensive Gröbner basis. In 2003, Sato and Suzuki

V.P. Gerdt et al. (Eds.): CASC 2012, LNCS 7442, pp. 98–116, 2012.

[17] introduced the concept of alternative comprehensive Gröbner bases. Then in 2006, Manubens and Montes in [11] by using discriminant ideal improved DISPGB, and in [12] they introduced an algorithm for computing minimal canonical Gröbner systems. Also in 2006, Sato and Suzuki [18] (see also [19]) suggested an important computational improvement for comprehensive Gröbner bases by constructing the reduced Gröbner bases in polynomial rings over ground fields. In 2010, Kapur, Sun and Wang [10], by combining Weispfenning's algorithm [22] with Suzuki and Sato's algorithm [18], proposed a new algorithm for computing comprehensive Gröbner systems. More recently, in 2010, Montes and Wibmer in [15] presented the GRÖBNERCOVER algorithm (its implementation in Singular is available at http://www-ma2.upc.edu/~montes/) which computes a finite partition of the parameter space into locally closed subsets together with polynomial data and such that the reduced Gröbner basis for given values of parameters can immediately be determined from the partition.

Involutive bases form an important class of Gröbner bases. The theory of involutive bases goes back to the seminal works of French mathematician Janet. In the 20s of the last century, he developed [9] a constructive approach to analysis of certain systems of partial differential equations based on their completion to involution (cf. [20]). Inspired by the involution methods described in the book by Pommaret [16], Zharkov and Blinkov [23] introduced the concept of *involutive polynomial bases* in commutative algebra in the full analogy with the concept of involutive systems of homogeneous linear partial differential equations with constant coefficients and in one dependent variable. Besides, Zharkov and Blinkov designed the first algorithm for construction of involutive polynomial bases. The particular form of an involutive basis they used is nowadays called *Pommaret basis* [20].

Gerdt and Blinkov [7] proposed a more general concept of involutive bases for polynomial ideals and designed efficient algorithmic methods to construct such bases. The underlying idea of the involutive approach is to translate the methods originating from Janet's approach into the polynomial ideals theory in order to provide a method for construction of involutive bases by combining algorithmic ideas in the theory of Gröbner bases with constructive ideas in the theory of involutive differential systems. In doing so, Gerdt and Blinkov [7] introduced the concept of *involutive division*. Moreover, they derived the involutive form of Buchberger's criteria. This led to a strong computational tool which is a serious alternative to the conventional Buchberger algorithm. We refer to Seiler's book [20] for a comprehensive study and application of involution to commutative algebra and geometric theory of partial differential equations.

In this paper, we introduce a notion of *comprehensive involutive systems*. For a parametric ideal, we decompose the space of parameters into a finite set of cells, and for each cell we yield the corresponding involutive basis of the ideal. Thereby, for each values of parameters, we find first a cell containing these values. Then, by substituting these values into the corresponding basis, we get the involutive basis of the given ideal. Based on the Gerdt–Blinkov involutive (abbreviated below by GBI) algorithm as described in [6] and also the Montes

DisPGB algorithm [13], we present an algorithm for constructing comprehensive involutive systems. The proposed algorithm has been implemented in `Maple`, and we provide an illustrative example showing the step-by-step results of the algorithm.

The paper is structured as follows. Section 2 contains the basic definitions and notations related to comprehensive Gröbner systems, and a short description of the DisPGB algorithm. The basic definitions and notations from the theory of involutive bases are given in Section 3. In Section 4, the notion of comprehensive involutive system is introduced, and an algorithm for construction of such systems is described. In Section 5, we give an example illustrating in detail the performance of the algorithm of Section 4.

2   Comprehensive Gröbner Systems

In this section, we recall the basic definitions and notations in theory of comprehensive Gröbner systems and briefly describe the DisPGB algorithm.

Let $R = K[\mathbf{x}]$ be a polynomial ring, where $\mathbf{x} = x_1, \ldots, x_n$ is a sequence of variables and K is an arbitrary field. Below, we denote a monomial $x_1^{\alpha_1} \cdots x_n^{\alpha_n} \in R$ by $\mathbf{x}^{\alpha}$ where $\alpha = (\alpha_1, \ldots, \alpha_n) \in \mathbb{N}^n$ is a sequence of non-negative integers. We shall use the notations $\deg_i(\mathbf{x}^{\alpha}) := \alpha_i$, $\deg(\mathbf{x}^{\alpha}) := \sum_{i=1}^{n} \alpha_i$. An *admissible* monomial ordering on R is a total order $\prec$ on the set of all monomials such that for any $\alpha, \beta, \gamma \in \mathbb{N}^n$ the following holds:

$$\mathbf{x}^{\alpha} \succ \mathbf{x}^{\beta} \Longrightarrow \mathbf{x}^{\alpha+\gamma} \succ \mathbf{x}^{\beta+\gamma}, \qquad \mathbf{x}^{\alpha} \neq 1 \Longrightarrow \mathbf{x}^{\alpha} \succ 1.$$

A typical example of admissible monomial ordering is the *lexicographical ordering*, denoted by $\prec_{\text{lex}}$. If $\mathbf{x}^{\alpha}, \mathbf{x}^{\beta} \in R$ are two monomials, then $\mathbf{x}^{\alpha} \prec_{\text{lex}} \mathbf{x}^{\beta}$ if the leftmost nonzero entry of $\beta - \alpha$ is positive. Another typical example is the *degree-reverse-lexicographical ordering* denoted by $\prec_{\text{degrevlex}}$ and defined as $\mathbf{x}^{\alpha} \prec_{\text{degrevlex}} \mathbf{x}^{\beta}$ if $\deg(\mathbf{x}^{\alpha}) > \deg(\mathbf{x}^{\beta})$ or $\deg(\mathbf{x}^{\alpha}) = \deg(\mathbf{x}^{\beta})$ and the rightmost nonzero entry of $\beta - \alpha$ is negative.

We shall write $I = \langle f_1, \ldots, f_k \rangle$ for the ideal I in R generated by the polynomials $f_1, \ldots, f_k \in R$. Let $f \in R$ and $\prec$ be a monomial ordering on R. The *leading monomial* of f is the largest monomial (with respect to $\prec$) occurring in f, and we denote it by $\text{LM}(f)$. If $F \subset R$ is a set of polynomials, then we denote by $\text{LM}(F)$ the set $\{\text{LM}(f) \mid f \in F\}$ of its leading monomials. The *leading coefficient* of f, denoted by $\text{LC}(f)$, is the coefficient of $\text{LM}(f)$. The *leading term* of f is $\text{LT}(f) = \text{LC}(f)\text{LM}(f)$. The *leading term ideal* of I is defined as $\text{LT}(I) = \langle \text{LT}(f) \mid f \in I \rangle$.

A finite set $G = \{g_1, \ldots, g_k\} \subset I$ is called a *Gröbner basis* of I if $\text{LT}(I) = \langle \text{LT}(g_1), \ldots, \text{LT}(g_k) \rangle$. For more details and definitions related to Gröbner bases we refer to [2].

Now consider $F = \{f_1, \ldots, f_k\} \subset S := K[\mathbf{a}, \mathbf{x}]$ where $\mathbf{a} = a_1, \ldots, a_m$ is a sequence of parameters. Let $\prec_{\mathbf{x}}$ (resp. $\prec_{\mathbf{a}}$) be a monomial ordering for the power products of x_i's (resp. a_i's). We also need a compatible *elimination product ordering* $\prec_{\mathbf{x},\mathbf{a}}$. It is defined as follows: For all $\alpha, \gamma \in \mathbb{Z}_{\geq 0}^n$ and $\beta, \delta \in \mathbb{Z}_{\geq 0}^m$

$$\mathbf{x}^\gamma \mathbf{a}^\delta \prec_{\mathbf{x},\mathbf{a}} \mathbf{x}^\alpha \mathbf{a}^\beta \iff \mathbf{x}^\gamma \prec_{\mathbf{x}} \mathbf{x}^\alpha \text{ or } \mathbf{x}^\gamma = \mathbf{x}^\alpha \text{ and } \mathbf{a}^\delta \prec_{\mathbf{a}} \mathbf{a}^\beta .$$

Now, we recall the definition of a comprehensive Gröbner system for a parametric ideal.

Definition 1. ([22]) *A triple set* $\{(G_i, N_i, W_i)\}_{i=1}^\ell$ *is called a* comprehensive Gröbner system *for* $\langle F \rangle$ *w.r.t* $\prec_{\mathbf{x},\mathbf{a}}$ *if for any* i *and any homomorphism* σ : $K[\mathbf{a}] \to K'$ *(where* K' *is a field extension of* K*) satisfying*

$$(i) \ (\forall p \in N_i \subset K[\mathbf{a}]) \ [\sigma(p) = 0], \qquad (ii) \ (\forall q \in W_i \subset K[\mathbf{a}]) \ [\sigma(q) \neq 0]$$

we have $\sigma(G_i)$ *is a Gröbner basis for* $\sigma(\langle F \rangle) \subset K'[\mathbf{x}]$ *w.r.t.* $\prec_{\mathbf{x}}$.

For simplification, we shall use the abbreviation CGS to refer to a comprehensive Gröbner system, and CGSs in the plural case. For each i, the set N_i (resp. W_i) is called a (resp. non-) null conditions set. The pair (N_i, W_i) is called a *specification* of the homomorphism σ if both conditions in the above definition are satisfied.

Example 1. Let $F = \{ax^2y - y^3, bx + y^2\} \subset K[a, b, x, y]$ where $\mathbf{a} = a, b$ and $\mathbf{x} = x, y$. Let us consider the lexicographical monomial ordering $b \prec_{\mathrm{lex}} a$ on the parameters and on the variables $y \prec_{\mathrm{lex}} x$ as well. Using the DISPGB algorithm we can compute a CGS for $\langle F \rangle$ which is equal to

$$
\begin{array}{lll}
\{-b^2y^3 + ay^5, bx + y^2\} & \{\ \} & \{a, b\} \\
\{x^2y, y^2\} & \{b\} & \{a\} \\
\{y^3, bx + y^2\} & \{a\} & \{b\} \\
\{y^2\} & \{a, b\} & \{\ \}.
\end{array}
$$

For instance, if $a = 0, b = 2$, then the third element of this system corresponds to this specialization. Therefore, $\{y^3, 2x + y^2\}$ is a Gröbner basis for the ideal $\langle F \rangle|_{a=0, b=2} = \langle -y^3, 2x + y^2 \rangle$.

Remark that, by the above definition, a CGS is not unique for a given parametric ideal, and one can find other partitions for the space of parameters, and, therefore, other CGSs for the parametric ideal.

Now, we briefly describe the Montes DISPGB algorithm to compute CGSs for parametric ideals (see [13,11]). The main idea of DISPGB is based on discussing the nullity or not w.r.t. a given specification (N, W) for the leading coefficients of the polynomials appearing at each step (this process is performed by the NEWCOND subalgorithm). Let us consider a set $F \subset S$ of parametric polynomials. Given a polynomial $f \in F$ and a specification (N, W), NEWCOND is called. Three cases are possible: If $\mathrm{LC}(f)$ specializes to zero w.r.t. (N, W), we replace f by $f - \mathrm{LT}(f)$, and then start again. If $\mathrm{LC}(f)$ specializes to a nonzero element we continue with the next polynomial in F. Otherwise (if $\mathrm{LC}(f)$ is not decidable, i.e. we can't decide whether or not it is null w.r.t. (N, W)), the subalgorithm BRANCH is called to create two complementary cases by assuming $\mathrm{LC}(f) = 0$ and $\mathrm{LC}(f) \neq 0$. Therefore, two new disjoint branches with the specifications $(N \cup \{\mathrm{LC}(f)\}, W)$ and $(N, W \cup \{\mathrm{LC}(f)\})$ are made. This procedure is continued until every polynomial in F has a nonnull leading coefficient w.r.t.

the current specification. Then, we proceed with CONDPGB: This algorithm receives, as an input, a set of parametric polynomials and a specification (N, W) and, by applying Buchberger's algorithm, creates new polynomials. When a new polynomial is generated, NEWCOND verifies whether its leading coefficient leads to a new condition or not. If a new condition is found, then the subalgorithm stops, and BRANCH is called to make two new disjoint branches. Otherwise, the process is continued and computes a Gröbner basis for $\langle F \rangle$, according to the current specification. The collection of these bases, together with the corresponding specifications yields a CGS for $\langle F \rangle$.

3   Involutive Bases

Now we recall the basic definitions and notations concerning involutive bases and present below the general definition of involutive bases. First of all, we describe the cornerstone notion of *involutive division* [7] as a restricted monomial division [6] which, together with a monomial ordering, determines properties of an involutive basis. This makes the main difference between involutive bases and Gröbner bases. The idea behind involutive division is to partition the variables into two subsets of multiplicative and nonmultiplicative variables, and only the multiplicative variables can be used in the divisibility relation.

Definition 2. [7,6] *An involutive division $\mathcal{L}$ on the set of monomials of R is given, if for any finite set U of monomials and any $u \in U$, the set of variables is partitioned into subsets $M_{\mathcal{L}}(u, U)$ of multiplicative and $NM_{\mathcal{L}}(u, U)$ of nonmultiplicative variables such that*

1. $u, v \in U$, $u\mathcal{L}(u, U) \cap v\mathcal{L}(v, U) \neq \emptyset \Longrightarrow u \in v\mathcal{L}(v, U)$ or $v \in u\mathcal{L}(u, U)$,
2. $v \in U$, $v \in u\mathcal{L}(u, U) \Longrightarrow \mathcal{L}(v, U) \subset \mathcal{L}(u, U)$,
3. $u \in V$ and $V \subset U \Longrightarrow \mathcal{L}(u, U) \subset \mathcal{L}(u, V)$,

where $\mathcal{L}(u, U)$ denotes the set of all monomials in the variables in $M_{\mathcal{L}}(u, U)$. If $v \in u\mathcal{L}(u, U)$, then we call u an $\mathcal{L}-$(involutive) divisor of v, and we write $u|_{\mathcal{L}}v$. If v has no involutive divisor in a set U, then it is $\mathcal{L}-$irreducible modulo U.

In this paper, we are concerned with the wide class [8] of involutive divisions determined by a permutation ρ on the indices of variables and by a total monomial ordering $\sqsupset$ which is either admissible or the inverse of an admissible ordering. This class is defined by

$$(\forall u \in U) \ [\ NM_{\sqsupset}(u, U) = \bigcup_{v \in U \setminus \{u\}} NM_{\sqsupset}(u, \{u, v\}) \ ] \tag{1}$$

where

$$NM_{\sqsupset}(u, \{u, v\}) := \begin{cases} \text{if } u \sqsupset v \text{ or } (u \sqsubset v \wedge v \mid u) \text{ then } \emptyset \\ \text{else } \{x_{\rho(i)}\}, \ i = \min\{j \mid \deg_{\rho(j)}(u) < \deg_{\rho(j)}(v)\}. \end{cases} \tag{2}$$

Remark 1. The involutive Janet division introduced and studied in [7] is generated by formulae (1)–(2) if $\sqsupset$ is the lexicographic monomial ordering $\succ_{\mathrm{lex}}$ and ρ is the identical permutation. The partition of variables used by Janet himself [9] (see also [20]) is generated by $\succ_{\mathrm{lex}}$ as well with the permutation which is inverse to the identical one:

$$\rho = \begin{pmatrix} 1 & 2 & \dots n \\ n & n-1 & \dots 1 \end{pmatrix}.$$

Throughout this paper $\mathcal{L}$ is assumed to be a division of the class (1)–(2). Now, we define an involutive basis.

Definition 3. *Let $I \subset R$ be an ideal, $\prec$ be a monomial ordering on R and $\mathcal{L}$ be an involutive division. A finite set $G \subset I$ is an* involutive basis *of I if for all $f \in I$ there exists $g \in G$ such that $\mathrm{LM}(g)|_{\mathcal{L}}\mathrm{LM}(f)$. An involutive basis G is* minimal *if for any other involutive basis $\tilde{G}$ the inclusion $\mathrm{LM}(G) \subseteq \mathrm{LM}(\tilde{G})$ holds.*

From this definition and from that for Gröbner basis [3,2] it follows that an involutive basis of an ideal is its Gröbner basis, but the converse is not always true.

Remark 2. By using an involutive division in the division algorithm for polynomial rings, we obtain an involutive division algorithm. If G is an involutive basis for an involutive division $\mathcal{L}$, we use $\mathrm{NF}_{\mathcal{L}}(f, G)$ to denote $\mathcal{L}-normal\ form$ of f modulo G, i.e. the remainder of f on the involutive division by G. A polynomial set F is $\mathcal{L}-autoreduced$ if $f = \mathrm{NF}_{\mathcal{L}}(f, F \setminus \{f\})$ for every $f \in F$.

The following theorem provides an algorithmic characterization of involutive bases which is an involutive analogue of the Buchberger characterization of Gröbner bases.

Theorem 1. *([7,8]) Given an ideal $I \subset R$, an admissible monomial ordering $\prec$ on R and an involutive division $\mathcal{L}$, a finite subset $G \subset I$ is an involutive basis of I if for each $f \in G$ and each $x \in NM_{\mathcal{L}}(\mathrm{LM}(f), \mathrm{LM}(G))$ the equality $\mathrm{NF}_{\mathcal{L}}(xf, G) = 0$ holds. An involutive basis exists for any I, $\mathcal{L}$ and $\prec$. A monic and $\mathcal{L}$-autoreduced involutive basis is uniquely defined by I and $\prec$.*

4   Comprehensive Involutive Systems

In this section, like the concept of comprehensive Gröbner systems, we define the new notion of comprehensive involutive system for a parametric ideal. Then, based on the GBI algorithm [6] and the Montes DisPGB algorithm [13], we propose an algorithm for computing comprehensive involutive systems.

Definition 4. *Consider a finite set of parametric polynomials $F \subset S = K[\mathbf{a}, \mathbf{x}]$ where K is a field, $\mathbf{x} = x_1, \dots, x_n$ is a sequence of variables and $\mathbf{a} = a_1, \dots, a_m$ is a sequence of parameters, $\prec_{\mathbf{x}}$ (resp. $\prec_{\mathbf{a}}$) is a monomial ordering involving the x_i's (resp. a_i's), and $\mathcal{L}$ is an involutive division on $K[\mathbf{x}]$. Let $M =$*

$\{(G_i, N_i, W_i)\}_{i=1}^{\ell}$ be a finite triple set where sets $N_i, W_i \subset K[\mathbf{a}]$ and $G_i \subset S$ are finite. The set M is called an $(\mathcal{L}-)$comprehensive involutive system for $\langle F \rangle$ w.r.t $\prec_{\mathbf{x,a}}$ if for each i and for each homomorphism $\sigma : K[a] \rightarrow K'$ (where K' is a field extension of K) satisfying

$$(i) \; (\forall p \in N_i) \, [\, \sigma(p) = 0 \,], \qquad (ii) \; (\forall q \in W_i) \, [\, \sigma(q) \neq 0 \,]$$

$\sigma(G_i)$ is an $(\mathcal{L}-)$involutive basis for $\sigma(\langle F \rangle) \subset K'[\mathbf{x}]$. We use the abbreviation CIS (resp. CISs) to stand for comprehensive involutive system (resp. systems). M is called minimal, if for each i, the set $\sigma(G_i)$ is a minimal involutive basis.

Given a CGS, one can straightforwardly compute a CIS by using the following proposition.

Proposition 1. *Let* $G = \{g_1, \ldots, g_k\}$ *be a minimal Gröbner basis of an ideal* $I \subset K[x_1, \ldots, x_n]$ *for a monomial ordering* $\prec$. *Let* $h_i = \max_{g \in G} \{\deg_i(\mathrm{LM}(g))\}$. *Then the set of products*

$$\{mg \mid g \in G, \; m \text{ is a monomial s.t. } (\forall i) \, [\, \deg_i(m) \leq h_i - \deg_i(\mathrm{LM}(g)) \,]\} \quad (3)$$

is an $\mathcal{L}$-*involutive basis of* I.

Proof. Denote $\mathrm{LM}(G)$ by U. From (1)–(2) it follows

$$(\forall u \in U) \; (\forall x_i \in NM_{\mathcal{L}}(u, U))) \, [\, \deg_i(u) < h_i \,]. \quad (4)$$

It is also clear that if we enlarge G with a (not necessarily nonmultiplicative) prolongation gx_j of its element $g \in G$ such that $\deg_j(\mathrm{LM}(g)) < h_j$, then (4) holds for the enlarged leading monomial set $U := U \cup \{\mathrm{LM}(g)x_j\}$ as well. Consider completion $\bar{G}$ of the polynomial set G with all possible prolongations of its elements satisfying (3) and denote the monomial set $\mathrm{LM}(\bar{G})$ by $\bar{U}$. Then

$$(\forall u \in \bar{U}) \; (\forall x \in NM_{\mathcal{L}}(u, U)) \; (\exists v \in \bar{U}) \, [\, v \mid_{\mathcal{L}} ux \,].$$

This means, by Theorem 1, that the monomial set $\bar{U}$ is an involutive basis of $\langle \mathrm{LM}(G) \rangle$. Now, since G is a Gröbner basis of I we have $\mathrm{LT}(I) = \langle \mathrm{LM}(G) \rangle$, and hence $\mathrm{LT}(I) = \langle \mathrm{LM}(\bar{G}) \rangle$. Therefore, $\bar{G}$ is an involutive basis of I by Definition 3. □

Example 2. Let $F = \{ax^2, by^2\} \subset \mathbb{K}[\mathbf{a}, \mathbf{x}]$ where $\mathbf{a} = a, b$ and $\mathbf{x} = x, y$. Let also $b \prec_{lex} a$ and $y \prec_{lex} x$. Then, F is a CGS for any sets of null and nonnull conditions. Using Proposition 1, we can construct the following Janet basis of $\langle F \rangle$ which is a GIS for any sets of null and nonnull conditions:

$$\{ax^2, by^2, ayx^2, ay^2x^2, bxy^2, bx^2y^2\}.$$

On the other hand, the algorithm that we present below computes the following minimal Janet CIS for $\langle F \rangle$:

$\{ax^2, by^2, bxy^2\}$	$\{\,\}$	$\{a, b\}$
$\{ax^2\}$	$\{b\}$	$\{a\}$
$\{by^2\}$	$\{a\}$	$\{b\}$
$\{0\}$	$\{a, b\}$	$\{\,\}$.

Remark 3. Using Proposition 1, we cannot directly compute a minimal CIS from a given CGS. Indeed, to do this, we must examine the leading coefficients of each Gröbner basis in the CGS, and this may lead to further partitions of the space of parameters. Moreover, the CIS computed by this way may be too large, since many prolongations constructed by means of (3) may be useless. That is why, based on the GBI algorithm [6] and on the Montes DISPGB algorithm [13], we propose a more efficient algorithm for computing minimal CISs.

Now we describe the structure of polynomials that is used in our new algorithm. To avoid unnecessary reductions (during the computation of involutive bases) by applying the involutive form of Buchberger's criteria (see [6]), we need to supply polynomials with additional structural information.

Definition 5. *[6] An* ancestor *of a polynomial* $f \in F \subset R \setminus \{0\}$, *denoted by* $\mathrm{anc}(f)$, *is a polynomial* $g \in F$ *of the smallest* $\deg(\mathrm{LM}(g))$ *among those satisfying* $\mathrm{LM}(f) = u\mathrm{LM}(g)$ *where* u *is either the unit monomial or a power product of nonmultiplicative variables for* $\mathrm{LM}(g)$ *and such that* $\mathrm{NF}_{\mathcal{L}}(f - ug, F \setminus \{f\}) = 0$ *if* $f \neq ug$.

Algorithm COMINVSYS

Input: F, a set of polynomials; $\mathcal{L}$, an involutive division; $\prec_{\mathbf{x}}$, a monomial ordering on the variables; $\prec_{\mathbf{a}}$, a monomial ordering on the parameters
Output: a minimal CIS for $\langle F \rangle$
1: global: List, ind;
2: List:=Null;
3: ind:=1;
4: $B := \{[F[i], F[i], \emptyset] \mid i = 1, \ldots, |F|\}$;
5: $G := \{\mathrm{BRANCH}([F[1], F[1], \emptyset], B, \{\ \}, \{\ \}, \{\ \})\}$;
6: **for** i **from** 2 **to** $|F|$ **do**
7:    ind:=ind+1;
8:    $G := \{\mathrm{BRANCH}([F[i], F[i], \emptyset], A[2], A[3], A[4], A[5]) \mid A \in G\}$;
9: **od**
10: **Return** (List)

Below we show how to use the concept in Definition 5 to apply the involutive form of Buchberger's criteria. In what follows, we store each polynomial f as the $p = [f, g, V]$ where $f = \mathrm{poly}(p)$ is the polynomial part of p, $g = \mathrm{anc}(p)$ is the ancestor of f and $V = NM(p)$ is the list of nonmultiplicative variables of f have been already used to construct prolongations of f (see the **for**-loop 20-23 in the subalgorithm GBI). If P is a set of triples, we denote by $\mathrm{poly}(P)$ the set $\{\mathrm{poly}(p) \mid p \in P\}$. If no confusion arises, we may refer to a triple p instead of $\mathrm{poly}(p)$, and vice versa.

We consider now the main algorithm COMINVSYS which computes a minimal CIS for a given ideal. It should be noted that we use the subalgorithms NEW-COND and CANSPEC (resp. TAILNORMALFORM) as they have (resp. it has) been presented in [13] (resp. [6]), and recall them for the sake of completeness. Also, we use the subalgorithm BRANCH (resp. GBI , HEADREDUCE and HEADNOR-MALFORM) from [13] (resp. [6]) with some appropriate modifications.

Subalgorithm BRANCH

Input: p, a triple; B, a specializing basis; N, a set of null conditions; W, a set of nonnull conditions; P, a set of non-examined triples

Output: It stores the refined (B', N', W', P'), and creates two new vertices when necessary or marks the vertex as terminal

1: $p = [f, g, V]$;
2: $(test, N, W) := \text{CANSPEC}(N, W)$;
3: **if** $test$=false **then**
4:     **Return** STOP (incompatible specification has been detected)
5: **fi**
6: $(cd, f', N', W') := \text{NEWCOND}(f, N, W)$;
7: $p := [f', \overline{g}^{N'}, V]$ ($\overline{g}^{N'}$ denotes the remainder of the division of g by N');
8: **if** $\text{ind} < |F|$ and $cd \neq \{ \ \}$ **then**
9:     $\text{BRANCH}(p, B, N', W' \cup cd, P)$; $\text{BRANCH}(p, B, N' \cup cd, W', P)$;
10: **fi**
11: **if** $\text{ind} < |F|$ and $cd = \{ \ \}$ **then**
12:     **Return** $(p, \overline{B}^{N'}, N', W', P)$
13: **fi**
14: **if** $cd = \{ \ \}$ **then**
15:     $(test, p', B', N', W', P') := \text{GBI} \ (B, N', W', P)$;
16:     **if** $test$ **then**
17:         $\text{List} := \text{List}, (B', N', W')$;
18:     **else**
19:         $\text{BRANCH}(p', B', N', W', P')$;
20:     **fi**
21: **else**
22:     $\text{BRANCH}(p, B, N', W' \cup cd, P)$; $\text{BRANCH}(p, B, N' \cup cd, W', P)$;
23: **fi**

In the main algorithm, List is a global variable to which we add any computed involutive basis together with its corresponding specification to form the final CIS. That is why, at the beginning of computation we must set it to the empty list (see BRANCH). Note that here and in BRANCH, we use $|F|$ to denote the number of polynomials in the input set F. The variable ind is also a global variable, and we use it to examine all the leading coefficients of the elements in F (see BRANCH). Once we are sure about the non-nullity of these coefficients, then we start the involutive basis computation. Indeed, BRANCH inputs a triple $p = [f, g, V]$, a set B of examined and processed polynomials, a set N of null conditions, a set W of nonnull conditions and a set P of non-processed polynomials. Then, it analyses the leading coefficient of f w.r.t. N and W. Now, two cases are possible:

- ind$< |F|$: If $LC(f)$ is not decidable by N and W then we create two complementary cases by assuming $LC(f) = 0$ and $LC(f) \neq 0$. Then we pass to the next polynomial in F.
- ind$= |F|$: We are now sure that we have examined all the leading coefficients of the elements in F (except possibly the very last one which is to be f). If $LC(f)$ is not decidable by N and W then we again create two complementary cases with $LC(f) = 0$ and $LC(f) \neq 0$. Otherwise, we continue to process the polynomials in P by using the GBI algorithm. If $P = \emptyset$ this means that B is an involutive basis consistent with the conditions in N and W, and we add (B, N, W) to List.

Subalgorithm CanSpec

Input: N, a set of null conditions; W, a set of nonnull conditions
Output: true if N and W are compatible and false otherwise; (N', W'), a
   quasi-canonical representation of (N, W)

1: $W' :=$FacVar$(\{\overline{q}^N : q \in W\})$; $test:=$true; $N' := N$; $h := \prod_{q \in W} q$;
2: **if** $h \in \sqrt{\langle N' \rangle}$ **then**
3:    $test:=$false; $N' := \{1\}$;
4:    **Return** $(test, N', W')$;
5: **fi**
6: $flag:=$true;
7: **while** $flag$ **do**
8:    $flag:=$false;
9:    $N'':=$ Remove any factor of a polynomial in N' that belongs to W';
10:   **if** $N'' \neq N'$ **then**
11:      $flag:=$true;
12:      $N':=$ a Gröbner basis of $\langle N'' \rangle$ w.r.t. $\prec_{\mathbf{a}}$;
13:      $W' :=$FacVar$(\{\overline{q}^{N'} : q \in W'\})$;
14:   **fi**
15: **od**
16: **Return** $(test, N', W')$

It is worth noting that if the input specification of BRANCH is incompatible, then it stops the process only for the corresponding branch, and continues the construction of other branches. Moreover, using the above notations, if ind$< |F|$ and no new condition is detected, then BRANCH returns an element of the form $(p, \overline{B}^{N'}, N', W', P)$ where p is a triple, N', W' are two sets of conditions, $\overline{B}^{N'}$ is the normal form of a specializing basis B and P is a set of non-examined triples. Otherwise, it calls itself to create the new branches. Finally, if ind$= |F|$, then the algorithm does not return anything and completes the global variable List.

The subalgorithm CanSpec produces a quasi-canonical representation for a given specification. Its subalgorithm FacVar invoked in lines 1 and 13 returns the set of factors of its input polynomial.

Definition 6. ([13]) *A specification* (N, W) *is called quasi-canonical if*

- *N is the reduced Gröbner basis w.r.t. $\prec_a$ of the ideal containing all polynomials that specialize to zero in $K[\mathbf{a}]$.*
- *The polynomials in W specializing to non-zero are reduced modulo N and irreducible over $K[\mathbf{a}]$*
- *$\prod_{q \in W} q \notin \sqrt{\langle N \rangle}$.*
- *The polynomials in N are square-free over $K[\mathbf{a}]$.*
- *If some $p \in N$ is factorized, then no factor of p belongs to W.*

Subalgorithm NewCond

Input: f, a parametric polynomial; N, a set of null conditions; W, a set of nonnull conditions

Output: cd, a new condition; f', a parametric polynomial; N', a set of null conditions; W', a set of nonnull conditions

1: $f' := f$; $test$:=true; $N' := N$; cd:={ };
2: **while** $test$ **do**
3:     **if** $\mathrm{LC}(f') \in \sqrt{\langle N' \rangle}$ **then**
4:         $N' :=$ a Gröbner basis for $\langle N', \mathrm{LC}(f') \rangle$ w.r.t. $\prec_a$;
5:         $f' := f' - \mathrm{LT}(f)$;
6:     **else**
7:         $test$:=false;
8:     **fi**
9: **od**
10: $f' := \overline{f'}^{N'}$;
11: $W' := \{\overline{w}^{N'} \mid w \in W\}$;
12: $cd := cd \cup \mathrm{FacVar}(\mathrm{LC}(f')) \setminus W'$;
13: **Return**(cd, f', N', W')

We describe now the NewCond subalgorithm. When it is invoked in line 6 of Branch with the input data (f, N, W), one of the two following cases may occur:

1. If $\mathrm{LC}(f)$ is decidable w.r.t. the specification (N, W), then the subalgorithm returns:
   (i) NewCond$(f - \mathrm{LT}(f), N, W)$ in the case when $\mathrm{LC}(f)$ specializes to zero w.r.t. (N, W).
   (ii) $(\emptyset, f, N, W)$ in the case when $\mathrm{LC}(f)$ does not specialize to zero w.r.t. (N, W).
2. If $\mathrm{LC}(f)$ is not decidable w.r.t (N, W), then NewCond returns (cd, f, N, W) where set cd contains one of the non-decidable factors (w.r.t (N, W)) of $\mathrm{LC}(f)$.

It should be emphasized that $\text{FACVAR}(\text{LC}(f')) \setminus W'$ in line 12 returns only one factor of $\text{LC}(f')$.

Subalgorithm GBI

Input: B, a specializing basis; N, a set of null conditions; W, set of nonnull conditions; P, set of non-examined triples

Output: If $test$=true, a minimal involutive basis for $\langle B \rangle$ w.r.t. $\mathcal{L}$ and $\prec_{\mathbf{x},\mathbf{a}}$; otherwise, it returns a triple so that we must discuss the leading coefficient of its polynomial part

```
 1: if P = ∅ then
 2:    Select p ∈ B with no proper divisor of LM(poly(p)) in LM(poly(B))
 3:    T := {p};   Q := B \ {p};
 4: else
 5:    T := B;   Q := P;
 6: fi
 7: while Q ≠ ∅ do
 8:    (test, p, T, N, W, Q') := HEADREDUCE(T, N, W, Q);
 9:    if test =false then
10:       Return (false, p, T, N, W, Q')
11:    fi
12:    Q := Q';
13:    Select and remove p ∈ Q with no proper divisor of LM(poly(p)) in
         LM(poly(Q));
14:    if poly(p) = anc(p) then
15:       for q ∈ T whose LM(poly(q)) is a proper multiple of LM(poly(p)) do
16:          Q := Q ∪ {q};   T := T \ {q};
17:       od
18:    fi
19:    h := TAILNORMALFORM(p, T);   T := T ∪ {{h, anc(p), NM(p)}};
20:    for q ∈ T and x ∈ NM_L(LM(poly(q)), LM(poly(T))) \ NM(q) do
21:       Q := Q ∪ {{x poly(q), anc(q), ∅}};
22:       NM(q) := NM(q) ∩ NM_L(LM(poly(q)), LM(poly(T))) ∪ {x};
23:    od
24: od
25: Return (true, 0, T, N, W, { })
```

The subalgorithm GBI, is an extension of the algorithm INVOLUTIVEBASIS II described in [6]. The latter algorithm computes involutive bases and applies the involutive form of Buchberger's criteria to avoid some unnecessary reductions [7] (see also [1,6]). The criteria are applied in the subalgorithm HEADNORMALFORM (see line 7) that is invoked in line 5 of GBI.

Proposition 2. ([7,6]) *Let $I \subset R$ be an ideal and $G \subset I$ be a finite set. Let also $\prec$ be a monomial ordering on R and $\mathcal{L}$ be an involutive division. Then G is an $\mathcal{L}$−involutive basis of I if for all $f \in G$ and for all $x \in NM_{\mathcal{L}}(\text{LM}(f), \text{LM}(G))$ one of the two conditions holds:*

1. $\mathrm{NF}_{\mathcal{L}}(xf, G) = 0$.
2. *There exists* $g \in G$ *with* $\mathrm{LM}(g)|_{\mathcal{L}}\mathrm{LM}(xf)$ *satisfying one of the following conditions:*
   (C_1) $\mathrm{LM}(\mathrm{anc}(f))\mathrm{LM}(\mathrm{anc}(g)) = \mathrm{LM}(xf)$,
   (C_2) $\mathrm{lcm}(\mathrm{LM}(\mathrm{anc}(f)), \mathrm{LM}(\mathrm{anc}(g)))$ *is a proper divisor of* $\mathrm{LM}(xf)$.

The subalgorithm GBI invokes three its own subalgorithms HEADREDUCE, TAILNORMALFORM and HEADNORMALFORM. The subalgorithm HEADREDUCE performs the involutive head reduction of polynomials in the input set of triples modulo the input specializing basis. The subalgorithm TAILNORMALFORM (resp. HEADNORMALFORM) invoked in line 19 of GBI (resp. in line 4 of HEADREDUCE) computes the involutive tail normal form (resp. the involutive head normal form) of the polynomial in the input triple modulo the input specializing basis.

Subalgorithm HEADREDUCE

Input: B, a specializing basis; N, a set of null conditions; W, a set of nonnull conditions; P a set of non-examined triples
Output: If *test*=true, the $\mathcal{L}$-head reduced form of P modulo B; otherwise, it returns a triple such that we must examine the leading coefficient of its polynomial part
1: $S := P$; $Q := \emptyset$;
2: **while** $S \neq \emptyset$ **do**
3:     Select and remove $p \in S$;
4:     $(test, h, B, N, W) :=$HEADNORMALFORM(p, B, N, W);
5:     **if** *test*=false **then**
6:         **Return** (false, $p, B, N, W, S \cup Q$)
7:     **fi**
8:     **if** $h \neq 0$ **then**
9:         **if** $\mathrm{LM}(\mathrm{poly}(p)) \neq \mathrm{LM}(h)$ **then**
10:             $Q := Q \cup \{\{h, h, \emptyset\}\}$;
11:         **else**
12:             $Q := Q \cup \{p\}$;
13:         **fi**
14:     **else**
15:         **if** $\mathrm{LM}(\mathrm{poly}(p)) = \mathrm{LM}(\mathrm{anc}(p))$ **then**
16:             $S := S \setminus \{q \in S \mid \mathrm{anc}(q) = \mathrm{poly}(p)\}$;
17:         **fi**
18:     **fi**
19: **od**
20: **Return** (true, $0, B, N, W, Q$)

In HEADNORMALFORM, the Boolean expression Criteria(p, g) is true if at leat one of the conditions (C_1) or (C_2) in Proposition 2 are satisfied for p and g, false otherwise. We refer to [6] for more details on GBI and on its subalgorithms.

Subalgorithm TAILNORMALFORM

Input: p, a triple; B, a set of triples
Output: $\mathcal{L}$-normal form of poly(p) modulo poly(B)
1: $h := \text{poly}(p)$;
2: $G := \text{poly}(B)$;
3: **while** h has a term t which is $\mathcal{L}$−reducible modulo G **do**
4:    Select $g \in G$ with $\text{LM}(g)|_{\mathcal{L}} t$;
5:    $h := h - g \frac{t}{\text{LT}(g)}$;
6: **od**
7: **Return** (h)

Subalgorithm HEADNORMALFORM

Input: p, a triple; B, a specializing basis; N, a set of null conditions; W, set of nonnull conditions
Output: If *test*=true, the $\mathcal{L}$-head normal form of poly(p) modulo B; otherwise, a polynomial whose leading coefficient must be examined
1: $h := \text{poly}(p)$; $G := \text{poly}(B)$;
2: **if** $\text{LM}(h)$ is $\mathcal{L}$-irreducible modulo G **then**
3:    **Return** (true, h, B, N, W)
4: **else**
5:    Select $g \in G$ with $\text{LM}(\text{poly}(g))|_{\mathcal{L}} \text{LM}(h)$;
6:    **if** $\text{LM}(h) \neq \text{LM}(\text{anc}(p))$ **then**
7:      **if** Criteria(p, g) **then**
8:        **Return** (true, $0, B, N, W$)
9:      **fi**
10:   **else**
11:      **while** $h \neq 0$ and $\text{LM}(h)$ is $\mathcal{L}$-reducible modulo G **do**
12:        Select $g \in G$ with $\text{LM}(g)|_{\mathcal{L}} \text{LM}(h)$;
13:        $h := h - g \frac{\text{LT}(h)}{\text{LT}(g)}$;
14:        $(cd, h', N', W') :=\text{NEWCOND}(h, N, W)$;
15:        **if** $cd \neq \emptyset$ **then**
16:          **Return** (false, h', B, N', W')
17:        **fi**
18:      **od**
19:   **fi**
20: **fi**
21: **Return** (true, h, B, N, W)

Theorem 2. *Algorithm* COMINVSYS *terminates in finitely many steps, and computes a minimal* CIS *for its input ideal.*

Proof. Let $I = \langle F \rangle$ where $F = \{f_1, \ldots, f_k\} \subset K[\mathbf{a}, \mathbf{x}]$ is a parametric set, $\mathbf{x} = x_1, \ldots, x_n$ (resp. $\mathbf{a} = a_1, \ldots, a_m$) is a sequence of variables (resp. parameters).

Let $\prec_{\mathbf{x}}$ (resp. $\prec_{\mathbf{a}}$) be a monomial ordering involving the x_i's (resp. a_i's), and $\mathcal{L}$ be an involutive division on $K[\mathbf{x}]$.

Suppose that COMINVSYS receives F as an input. To prove the *termination*, we use the fact that $K[\mathbf{a}]$ is a Noetherian ring. When BRANCH is called, the leading coefficient of some polynomial $f \in I$ is analyzed. For this purpose, the subalgorithm NEWCOND determines whether $\mathrm{LC}(f)$ is decidable or not w.r.t. the given specification (N, W). Two alternative cases can take place:

- $\mathrm{LC}(f)$ is decidable and we check the global variable ind. Now if ind$< k$, then we study the next polynomial in F. Otherwise, GBI is called. If all the leading coefficients of the examined polynomials (to compute a minimal involutive basis) are decidable, then the output, say G, is a minimal involutive basis of I w.r.t. (N, W), and we add (G, N, W) to List. Otherwise, two new branches are created by calling BRANCH (cf. the second case given below). In doing so, the minimality of G and the termination of its computation is provided by the structure of GBI algorithm (see [6]).

- $\mathrm{LC}(f)$ is not decidable and we create two branches with $(N, W \cup cd)$ and $(N \cup cd, W)$, where cd is the one-element set containing the new condition derived from $\mathrm{LC}(f)$.

Thus, in the second case, the branch for which N (resp. W) is assumed, increases the ideal $\langle N \rangle \subset K[\mathbf{a}]$ (resp. $\langle W \rangle \subset K[\mathbf{a}]$). Note that we replace N by a Gröbner basis of its ideal (see line 4 in NEWCOND). Since the ascending chains of ideals stabilize, the algorithm terminates. This argument was inspired by the proof in [13], Theorem 16.

To prove the *correctness*, assume that $M = \{(G_i, N_i, W_i)\}_{i=1}^{\ell}$ is the output of COMINVSYS for the input is F (note that we have used the fact the this algorithm terminates in finitely many steps). Consider integer $1 \le i \le \ell$ homomorphism $\sigma : K[\mathbf{a}] \to K'$ where (N_i, W_i) is a specification of σ and K' is a field extension of K.

We have to show that for each $f \in G_i$ and $x \in NM_{\mathcal{L}}(\mathrm{LM}(\sigma(f)), \mathrm{LM}(\sigma(G_i)))$, in accordance with Theorem 1, the equality $\mathrm{NF}_{\mathcal{L}}(\sigma(xf), \sigma(G_i)) = 0$ holds. By using 'reductio ad absurdum', suppose $g = \mathrm{NF}_{\mathcal{L}}(\sigma(xf), \sigma(G_i))$ and $g \ne 0$. Since (G_i, N_i, W_i) has been added to List in BRANCH, the leading coefficients of the polynomials in the subalgorithm GBI, examined at computation of a minimal involutive basis for F, are decidable w.r.t. (N_i, W_i). Furthermore, $f \in G_i$ implies that in the course of GBI xf is added to Q, the set of all nonmultiplicative prolongations that must be examined (see the notations used in GBI). Then, HEADREDUCE is called to perform the $\mathcal{L}$-head reduction of the elements of Q modulo the last computed basis $T \subset G_i$. The computed $\mathcal{L}$-head normal form of xf is further reduced by invoking TAILNORMALFORM which performs the $\mathcal{L}$-tail reduction. By the above notations, g is the result of this step. Thus, g should be added to $T \subset G_i$. It follows that $\mathrm{NF}_{\mathcal{L}}(\sigma(xf), \sigma(G_i)) = 0$, a contradiction, and this completes the proof. $\qquad\square$

5 Example

Now we give an example to illustrate the step by step construction of a minimal CIS by the algorithm CoMINVSYS proposed and described in the previous section[1].

For the input $F = \{ax^2, by^2\} \subset \mathbb{K}[a, b, x, y]$ from Example 2, Janet division and the lexicographic monomial ordering with $b \prec_{\text{lex}} a$ and $y \prec_{\text{lex}} x$ the algorithm performs as follows:

$\rightarrow$CoMINVSYS$(F, \mathcal{L}, \prec_{lex}, \prec_{lex})$

   List $:= Null$; ind $:= 1$; $k := 2$;

   $B := \{[ax^2, ax^2, \emptyset], [by^2, by^2, \emptyset]\}$

   $\rightarrow$BRANCH$([ax^2, ax^2, \emptyset], B, \{\,\}, \{\,\}, \{\,\})$

     $\rightarrow$NEWCOND$(ax^2, \{\,\}, \{\,\}) = (\{a\}, \{\,\}, \{\,\})$

     $\rightarrow$BRANCH$([ax^2, ax^2, \emptyset], B, \{\,\}, \{a\}, \{\,\})$

       $\rightarrow$NEWCOND$(ax^2, \{\,\}, \{a\}) = (\{\,\}, \{\,\}, \{a\})$

   $G := \{([ax^2, ax^2, \emptyset], B, \{\,\}, \{a\}, \{\,\})\}$

     $\rightarrow$BRANCH$([ax^2, ax^2, \emptyset], B, \{a\}, \{\,\}, \{\,\})$

       $\rightarrow$NEWCOND$(ax^2, \{a\}, \{\,\}) = (\{\,\}, \{a\}, \{\,\})$

   $G := \big\{([ax^2, ax^2, \emptyset], B, \{\,\}, \{a\}, \{\,\}), ([ax^2, ax^2, \emptyset], \{[0, 0, \emptyset], [by^2, by^2, \emptyset]\}, \{a\}, \{\,\}, \{\,\})\big\}$

   ind $:= 2$;

   $A = ([ax^2, ax^2, \emptyset], B, \{\,\}, \{a\}, \{\,\})$

   $\rightarrow$BRANCH$([by^2, by^2, \emptyset], B, \{\,\}, \{a\}, \{\,\})$

     $\rightarrow$NEWCOND$(by^2, \{\,\}, \{a\}) = (\{b\}, \{\,\}, \{\,\})$

     $\rightarrow$BRANCH$([by^2, by^2, \emptyset], B, \{\,\}, \{a, b\}, \{\,\})$

     (* further BRANCH$([by^2, by^2, \emptyset], B, \{b\}, \{a\}, \{\,\})$ is executed*)

     $\rightarrow$NEWCOND$(by^2, \{\,\}, \{a, b\}) = (\{\,\}, \{\,\}, \{a, b\})$

     ind $\geq k = 2$

     $cd = \{\,\}$

     $\rightarrow$GBI $(B, \{\,\}, \{a, b\}, \{\,\})$

       $T := \{[by^2, by^2, \emptyset]\}$

       $Q := \{[ax^2, ax^2, \emptyset]\}$

       $\rightarrow$HEADREDUCE$(T, \{\,\}, \{a, b\}, Q)$

         $\rightarrow$HEADNORMALFORM$([ax^2, ax^2, \emptyset], T, \{\}, \{a, b\}) = (true, ax^2, T, \{\}, \{a, b\})$

       HEADREDUCE returns $(true, 0, T, \{\,\}, \{a, b\}, Q)$

       $p := [ax^2, ax^2, \emptyset]$

       $Q = \{\,\}$

         $\rightarrow$TAILNORMALFORM$(p, T) = ax^2$

       $T := \{[by^2, by^2, \emptyset], [ax^2, ax^2, \emptyset]\}$

       $Q := \{[bxy^2, by^2, \emptyset]\}$

[1] The Maple code of our implementation of the algorithm CoMINVSYS for the Janet division is available at the Web pages http://invo.jinr.ru and http://amirhashemi.iut.ac.ir/software.html

$\rightarrow$HEADREDUCE$(T, \{\,\}, \{a, b\}, Q) = (true, 0, T, \{\,\}, \{a, b\}, Q)$

$p := [bxy^2, by^2, \emptyset]$

$Q = \{\,\}$

$\quad\rightarrow$TAILNORMALFORM$(p, T) = bxy^2$

$T := \{[by^2, by^2, \emptyset], [ax^2, ax^2, \emptyset], [bxy^2, by^2, \emptyset]\}$

$Q := \{[bx^2y^2, by^2, \emptyset]\}$

$\quad\rightarrow$HEADREDUCE$(T, \{\,\}, \{a, b\}, Q) = (true, 0, T, \{\,\}, \{a, b\}, \{\,\})$

$Q := \{\,\}$

$\rightarrow$GBI returns $(true, 0, \{by^2, ax^2, bxy^2\}, \{\,\}, \{a, b\})$

List $:= (\{by^2, ax^2, bxy^2\}, \{\,\}, \{a, b\})$

$B = \{[ax^2, ax^2, \emptyset], [0, 0, \emptyset]\}$

$\rightarrow$BRANCH$([by^2, by^2, \emptyset], B, \{b\}, \{a\}, \{\,\})$

$\quad\rightarrow$NEWCOND$(by^2, \{b\}, \{a\}) = (\{\,\}, \{b\}, \{a\})$

ind $\geq k = 2$

$cd = \{\,\}$

$\rightarrow$GBI $(B, \{b\}, \{a\}, \{\,\}) = (true, 0, \{ax^2\}, \{b\}, \{a\})$

List $:= (\{by^2, ax^2, bxy^2\}, \{\,\}, \{a, b\}), (\{ax^2\}, \{b\}, \{a\})$

(* Return back to COMINVSYS *)

$A = ([ax^2, ax^2, \emptyset], \{[0, 0, \emptyset], [by^2, by^2, \emptyset]\}, \{a\}, \{\,\}, \{\,\})$

$B = \{[0, 0, \emptyset], [by^2, by^2, \emptyset]\}$

$\rightarrow$BRANCH$([by^2, by^2, \emptyset], B, \{\,\}, \{a\}, \{\,\})$

$\quad\rightarrow$NEWCOND$(by^2, \{a\}, \{\,\}) = (\{b\}, \{\,\}, \{\,\})$

$\quad\rightarrow$BRANCH$([by^2, by^2, \emptyset], B, \{a\}, \{b\}, \{\,\})$

(* further BRANCH$([by^2, by^2, \emptyset], B, \{a, b\}, \{\,\}, \{\,\})$ is executed *)

$\quad\rightarrow$NEWCOND$(by^2, \{a\}, \{b\}) = (\{\,\}, \{a\}, \{b\})$

ind $\geq k = 2$

$cd = \{\,\}$

$\rightarrow$GBI $(B, \{a\}, \{b\}, \{\,\}) = (true, 0, \{by^2\}, \{a\}, \{b\})$

List $:= (\{by^2, ax^2, bxy^2\}, \{\,\}, \{a, b\}), (\{ax^2\}, \{b\}, \{a\}), (\{by^2\}, \{a\}, \{b\})$

$B = \{[0, 0, \emptyset], [0, 0, \emptyset]\}$

$\rightarrow$BRANCH$([by^2, by^2, \emptyset], B, \{a, b\}, \{\,\}, \{\,\})$

$\quad\rightarrow$NEWCOND$(by^2, \{a, b\}, \{\,\}) = (\{\,\}, \{a, b\}, \{\,\})$

ind $\geq k = 2$

$cd = \{\,\}$

$\rightarrow$GBI $(B, \{a, b\}, \{\,\}, \{\,\}) = (true, 0, \{0\}, \{a, b\}, \{\,\})$

List $:= (\{by^2, ax^2, bxy^2\}, \{\,\}, \{a, b\}), (\{ax^2\}, \{b\}, \{a\}), (\{by^2\}, \{a\}, \{b\}), (\{0\}, \{a, b\}, \{\,\})$

Acknowledgements. The main part of research presented in the paper was done during the stay of the second author (A.H.) at the Joint Institute for Nuclear Research in Dubna, Russia. He would like to thank the first author (V.G.) for the invitation, hospitality, and support. The contribution of the first author was partially supported by grants 01-01-00200, 12-07-00294 from the Russian Foundation for Basic Research and by grant 3802.2012.2 from the Ministry of Education and Science of the Russian Federation.

References

1. Apel, J., Hemmecke, R.: Detecting unnecessary reductions in an involutive basis computation. J. Symbolic Computation 40, 1131–1149 (2005)
2. Becker, T., Weispfenning, T.: Gröbner Bases: a Computational Approach to Commutative Algebra. Graduate Texts in Mathematics, vol. 141. Springer, New York (1993)
3. Buchberger, B.: Ein Algorithms zum Auffinden der Basiselemente des Restklassenrings nach einem nuildimensionalen Polynomideal. PhD thesis, Universität Innsbruck (1965)
4. Buchberger, B.: A Criterion for Detecting Unnecessary Reductions in the Cconstruction of Gröbner Bases. In: Ng, K.W. (ed.) EUROSAM 1979. LNCS, vol. 72, pp. 3–21. Springer, Heidelberg (1979)
5. Buchberger, B., Winkler, F. (eds.): Gröbner Bases and Applications. London Mathematical Society Lecture Note Series, vol. 251. Cambridge University Press, Cambridge (1998)
6. Gerdt, V.P.: Involutive algorithms for computing Gröbner bases. In: Cojocaru, S., Pfister, G., Ufnarovski, V. (eds.) Computational Commutative and Non-Commutative Algebraic Geometry, pp. 199–225. IOS Press, Amstrerdam (2005) (arXiv:math/0501111)
7. Gerdt, V.P., Blinkov, Y.A.: Involutive bases of polynomial ideals. Mathematics and Computers in Simulation 45, 519–542 (1998)
8. Gerdt, V.P., Blinkov, Y.A.: Involutive Division Generated by an Antigraded Monomial Ordering. In: Gerdt, V.P., Koepf, W., Mayr, E.W., Vorozhtsov, E.V. (eds.) CASC 2011. LNCS, vol. 6885, pp. 158–174. Springer, Heidelberg (2011)
9. Janet, M.: Les Systèmes d'Équations aux Dérivées Partielles. Journal de Mathématique 3, 65–151 (1920)
10. Kapur, D., Sun, Y., Wand, D.: A new algorithm for computing comprehensive Gröbner systems. In: Watt, S.M. (ed.) Proc. ISSAC 2010, pp. 29–36. ACM Press, New York (2010)
11. Manubens, M., Montes, A.: Improving DISPGB algorithm using the discriminant ideal. J. Symbolic Computation 41, 1245–1263 (2006)
12. Manubens, M., Montes, A.: Minimal canonical comprehensive Gröbner systems. J. Symbolic Computation 44, 463–478 (2009)
13. Montes, A.: A new algorithm for discussing Gröbner bases with parameters. J. Symbolic Computation 33, 183–208 (2002)
14. Montes, A.: Solving the load flow problem using Gröbner bases. SIGSAM Bulletin 29, 1–13 (1995)
15. Montes, A., Wibmer, M.: Gröbner bases for polynomial systems with parameters. J. Symbolic Computation 45, 1391–1425 (2010)
16. Pommaret, J.-F.: Systems of Partial Differential Equations and Lie Pseudogroups. Mathematics and its Applications, vol. 14. Gordon & Breach Science Publishers, New York (1978)
17. Sato, Y., Suzuki, A.: An alternative approach to comprehensive Gröbner bases. J. Symbolic Computation 36, 649–667 (2003)

18. Sato, Y., Suzuki, A.: A simple algorithm to compute comprehensive Gröbner bases using Gröbner bases. In: Trager, B.M. (ed.) Proc. ISSAC 2006, pp. 326–331. ACM Press, New York (2006)
19. Suzuki, A.: Computation of Full Comprehensive Gröbner Bases. In: Ganzha, V.G., Mayr, E.W., Vorozhtsov, E.V. (eds.) CASC 2005. LNCS, vol. 3718, pp. 431–444. Springer, Heidelberg (2005)
20. Seiler, W.M.: Involution - The Formal Theory of Differential Equations and its Applications in Computer Algebra. Algorithms and Computation in Mathematics, vol. 24. Springer, Berlin (2010)
21. Weispfenning, V.: Cannonical comprehensive Gröbner bases. J. Symbolic Computation 36, 669–683 (2003)
22. Weispfenning, V.: Comprehensive Gröbner bases. J. Symbolic Computation 14, 1–29 (1992)
23. Zharkov, A.Y., Blinkov, Y.A.: Involutive approach to investigating polynomial systems. Mathematics and Computers in Simulation 42, 323–332 (1996)

A Polynomial-Time Algorithm for the Jacobson Form of a Matrix of Ore Polynomials

Mark Giesbrecht[1] and Albert Heinle[2]

[1] Cheriton School of Computer Science, University of Waterloo, Canada
[2] Lehrstuhl D für Mathematik, RWTH Aachen University, Aachen, Germany

Abstract. We present a new algorithm to compute the Jacobson form of a matrix A of polynomials over the Ore domain $\mathsf{F}(z)[x; \sigma, \delta]^{n \times n}$, for a field F. The algorithm produces unimodular U, V and the diagonal Jacobson form J such that $UAV = J$. It requires time polynomial in $\deg_x(A)$, $\deg_z(A)$ and n. We also present tight bounds on the degrees of entries in U, V and J. The algorithm is probabilistic of the Las Vegas type: we assume we are able to generate random elements of F at unit cost, and will always produces correct output within the expected time. The main idea is that a randomized, unimodular, preconditioning of A will have a Hermite form whose diagonal is equal to that of the Jacobson form. From this the reduction to the Jacobson form is easy. Polynomial-time algorithms for the Hermite form have already been established.

1 Introduction

The Jacobson normal form is a fundamental invariant of matrices over a ring of Ore polynomials. Much like the Smith normal form over a commutative principal ideal domain, it captures important information about the structure of the solution space of a matrix over the ring, and many important geometric properties of its system of shift or differential equations.

In this paper we consider the problem of computing canonical forms of matrices of Ore polynomials over a function field $\mathsf{F}(z)$. Let $\sigma : \mathsf{F}(z) \to \mathsf{F}(z)$ be an automorphism of $\mathsf{F}(z)$ and $\delta : \mathsf{F}(z) \to \mathsf{F}(z)$ be a σ-derivation. That is, for any $a, b \in \mathsf{F}(z)$, $\delta(a + b) = \delta(a) + \delta(b)$ and $\delta(ab) = \sigma(a)\delta(b) + \delta(a)b$. We then define $\mathsf{F}(z)[x; \sigma, \delta]$ as the set of polynomials in $\mathsf{F}(z)[x]$ under the usual addition, but with multiplication defined by $xa = \sigma(a)x + \delta(a)$, for any $a \in \mathsf{F}(z)$. This is well-known to be a left (and right) principal ideal domain, with a straightforward euclidean algorithm (see Ore (1933)).

Cohn (1985), Proposition 8.3.1 shows that we may assume that we are in either the pure differential case (with $\sigma(a) = a$), or the pure difference case, with $\delta(a) = 0$. In this paper, we will constrain ourselves still further to the shift polynomials and the differential polynomials over $\mathsf{F}(z)$, where F is a field of characteristic 0.

(1) $\sigma(z) = \mathcal{S}(z) = z + 1$ is the so-called *shift* automorphism of $\mathsf{F}(z)$, and δ identically zero on F. Then $\mathsf{F}(z)[x; \mathcal{S}, 0]$ is generally referred to as the ring of *shift polynomials*. We write $\mathsf{F}(z)[\partial; \mathcal{S}]$ for this ring.

V.P. Gerdt et al. (Eds.): CASC 2012, LNCS 7442, pp. 117–128, 2012.

(2) $\delta(z) = 1$ and $\sigma(z) = z$, so $\delta(h(z)) = h'(z)$ for any $h \in \mathsf{F}(z)$ with h' its usual derivative. Then $\mathsf{F}(z)[x; \sigma, \delta]$ is called the ring of *differential polynomials*. We write $\mathsf{F}(z)[x; \prime]$ for this ring.

More general Ore polynomials (in particular, in fields of finite characteristic) will be treated in the journal version of this paper.

Let $a, b \in \mathsf{F}(z)[x; \sigma, \delta]$. Following Jacobson (1943), Chapter 3, we say that a is a *total divisor* of $b \neq 0$ if there exists a two-sided ideal $\mathcal{I}$ such that $a\mathsf{F}(z)[x; \sigma, \delta] \supseteq \mathcal{I} \supseteq b\mathsf{F}(z)[x; \sigma, \delta]$. We say that two elements $a, b \in \mathsf{F}(z)[x; \sigma, \delta]$ are *similar* if there exists a $u \in \mathsf{F}(z)[x; \sigma, \delta] \setminus \{0\}$ such that $b = \mathrm{lclm}(a, u)u^{-1}$. A matrix $U \in \mathsf{F}[\partial; \sigma, \delta]^{n \times n}$ is said to be *unimodular* if there exists a matrix $V \in \mathsf{F}[\partial; \sigma, \delta]^{n \times n}$ such that $UV = I$ (i.e., the inverse is also a matrix over $\mathsf{F}[\partial; \sigma, \delta]$).

Let $A \in \mathsf{F}[\partial; \sigma, \delta]^{n \times n}$. Jacobson (1943), Theorem 3.16, shows that there exist unimodular matrices $U, V \in \mathsf{F}[\partial; \sigma, \delta]^{n \times n}$ such that

$$J = UAV = \mathrm{diag}(s_1, \ldots, s_r, 0, \ldots, 0),$$

where s_i is a total divisor of s_{i+1} for $1 \leq i < r$. We call J the *Jacobson* form of A, and the diagonal entries of J are unique up to the notion of similarity given above. For the rings $\mathsf{F}(z)[\partial; \mathcal{S}]$ and $\mathsf{F}(z)[\partial; \prime]$ we establish stronger statements about the shape of the Jacobson form. In particular, we show that if for shift polynomials $\mathsf{R} = \mathsf{F}(z)[x; \mathcal{S}]$, there exist unimodular $U, V \in \mathsf{R}^{n \times n}$ such that

$$J = UAV = \mathrm{diag}(1, \ldots, 1, x, \ldots, x, x^2, \ldots, x^2, \ldots, x^k, \ldots, x^k, \varphi x^k, 0, \ldots, 0),$$

where $\varphi \in \mathsf{R}$ is monic. For differential polynomials $\mathsf{R} = \mathsf{F}(z)[x; \prime]$, it is well-known that there exists unimodular $U, V \in \mathsf{R}^{n \times n}$ such that

$$J = UAV = \mathrm{diag}(1, \ldots, 1, \varphi, 0 \ldots, 0),$$

where $\varphi \in \mathsf{R}$ is monic.

Finding normal forms of matrices is as old as the term matrix itself in mathematics. A primary aim is to obtain a diagonal matrix after a finite number of reversible matrix operations. For matrices with entries in a commutative ring there has been impressive progress in computing the Smith normal form, and the improvements in complexity have resulted directly in the best implementations. The Jacobson form is the natural generalization of the Smith form in a noncommutative (left) principal ideal domain. Commutative techniques do not directly generalize (for one thing there is no straightforward determinant), but our goal is to transfer some of this algorithmic technology to the non-commutative case.

Over the past few years, a number of algorithms and implementations have been developed for computing the Jacobson form. The initial definition of the Jacobson form (Jacobson, 1943) was essentially algorithmic, reducing the problem to computing diagonalizations of 2×2 matrices, which can be done directly using GCRDs and LCLMs. Unfortunately, this approach lacks not only efficiency in terms of ring operations, but also results in extreme coefficient growth.

Recent methods of Levandovskyy and Schindelar (2012) have developed a algorithm based on Gröbner basis theory. An implementation of it is available in the computer algebra system SINGULAR. A second approach by Robertz et al. implementing the algorithm described in Cohn (1985) can be found in the Janet library for MAPLE.. Another approach is proposed by Middeke (2008) for differential polynomials, making use of a cyclic vector computation. This algorithm requires time polynomial in the system dimension and order, but coefficient growth is not accounted for. Finally, the dissertation of Middeke (2011) considers an FGLM-like approach to converting a matrix of differential polynomials from the Popov to Jacobson form.

Our goal in this paper is to establish rigorous polynomial-time bounds on the cost of computing the Jacobson form, in terms of the dimension, degree and coefficient bound on the input. We tried to avoid Gröbner bases and cyclic vectors, because we do not have sufficiently strong statements about their size or complexity. Our primary tool in this work is the polynomial-time algorithm for computing the *Hermite form* of a matrix of Ore polynomials, introduced at the CASC 2009 conference by Giesbrecht and Kim (2009, 2012).

Definition 1.1. *Let* $R = F(z)[x; \sigma, \delta]$ *be an Ore polynomial ring and* $A \in R^{n \times n}$ *with full row rank. There exists a unimodular matrix* $Q \in R^{n \times n}$, *such that* $H = QA$ *is an upper triangular matrix with the property that*

- *The diagonal entries* H_{ii} *of are monic;*
- *Each superdiagonal entry is of degree (in* x) *lower than the diagonal in its column (i.e.,* $\deg_x H_{ji} < \deg_x H_{ii}$ *for* $1 \leq j < i \leq n$)

The Hermite form (with monic diagonals) is unique.

Giesbrecht and Kim (2009, 2012) establishes the following (polynomial-time) cost and degree bounds for computing the Hermite form:

Fact 1.2. *Let* $A \in F[z][x; \sigma, \delta]$ *have full row rank with entries of degree at most* d *in* x, *and of degree at most* e *in* z. *Let* $H \in F(z)[x; \sigma, \delta]^{n \times n}$ *be the Hermite form of* A *and* $U \in F(z)[x; \sigma, \delta]^{n \times n}$ *such that* $UA = H$. *Then*

(a) *We can compute the Hermite form* $H \in F(z)[x; \sigma, \delta]^{n \times n}$ *of* A, *and* $U \in F(z)[x; \sigma, \delta]^{n \times n}$ *such that* $UA = H$ *with a deterministic algorithm that requires* $O(n^9 d^3 e)$ *operations in* F;
(b) $\deg_x H_{ij} \leq nd$, $\deg_z H_{ij} \in O(n^2 de)$ *and* $\deg_z U_{ij} \in O(n^2 de)$ *for* $1 \leq i, j \leq n$.

Our approach to computing the Jacobson form follows the method of Kaltofen et al. (1990) for computing the Smith normal form of a polynomial matrix. This algorithm randomly preconditions the input matrix by multiplying by random unimodular matrices on the left and the right, and then computes a left and right echelon/Hermite form. The resulting matrix is shown to be in diagonal Smith form with high probability.

Our algorithm follows a similar path, but the unimodular preconditioner must be somewhat more powerful to attain the desired Jacobson form. In this current paper we will only pursue our algorithm for differential polynomials, though the method should work well for shift polynomials as well.

The remainder of this paper is as follows. In Section 2 we establish stronger versions of the Jacobson form for differential and shift polynomials. In Section 3 we show the reduction from computing the Jacobson form to computing the Hermite form, while in Section 4 we demonstrate degree bounds and complexity for our algorithms. Finally, we offer some conclusions and future directions in Section 5.

2   Strong Jacobson Form

In this section we establish the existence of the strong Jacobson form for polynomials over the shift and differential rings.

Theorem 2.1. *Let* $R = F(z)[x; S]$ *be the ring of shift polynomials, and* $A \in R^{n \times n}$. *Then there exist unimodular matrices* $U, V \in R^{n \times n}$ *such that*

$$J = UAV = \mathrm{diag}(1, \ldots, 1, x, \ldots, x, \ldots, x^k, \ldots, x^k, \varphi x^k, 0, \ldots, 0), \qquad (2.1)$$

where $\varphi \in R$.

Proof. We may assume that A is in Jacobson form,

$$A = \mathrm{diag}(f_1, \ldots, f_r, 0, \ldots, 0) \in R^{n \times n},$$

with f_i a total divisor of f_{i+1} for $1 \leq i < r$, though perhaps not with the nice shape of (2.1). We work though the 2×2 diagonal submatrices of A in sequence. Let $\mathrm{diag}(f, g)$ be such a submatrix, where f divides g from both sides. This means there exists $w, \tilde{w} \in R$, such that $g = wf = f\tilde{w}$. If f is 1 or a power of x already, we can continue with the next submatrix. Without loss of generality f has the form $(x^i + a_{i-1}x^{i-1} + \ldots + a_0)x^\mu$, where $a_j \in F(z)$ for $0 \leq j < i$, $\mu \in \mathbb{Z}_{\geq 0}$ and at least one a_j is not equal to zero. Now perform the following unimodular transformation on the given submatrix:

$$\begin{bmatrix} f & 0 \\ 0 & wf \end{bmatrix} \rightsquigarrow \begin{bmatrix} 1 & 0 \\ -S^{i+\mu+\deg(w)}(z)w & 1 \end{bmatrix} \begin{bmatrix} f & 0 \\ 0 & wf \end{bmatrix} \begin{bmatrix} 1 & 0 \\ z & 1 \end{bmatrix} = \begin{bmatrix} f & 0 \\ wfz - S^{i+\mu+\deg(w)}(z)wf & wf \end{bmatrix}.$$

The term $wfz - S^{i+\mu+\deg(w)}(z)wf$ is constructed such that it has degree in x strictly lower than that of wf.

We claim that f is not a right divisor of $wfz - S^{i+\mu+\deg(w)}(z)wf$. Suppose conversely that it is still a right divisor. Then

$$wfz - S^{i+\mu+\deg(w)}(z)wf = hf$$

for some $h \in R$. Since clearly $S^{i+\mu+\deg(w)}(z)wf$ is divisible by f from the right, wfz must also be divisible by f from the right. But this means that fz is equal to $(az + b)f$ for some $a, b \in F$. This is only possible if f is either 1 or a power of x. This is a contradiction to our choice of f. Thus, if we perform a GCRD computation on these two polynomials, we will get a polynomial of a strictly smaller degree. This action can be performed, until just x^μ is left. Continuing similarly with the next 2×2 submatrix, the shape (2.1) is established. □

The following characterization of the matrix of differential polynomials is well-known. It follows immediately from the fact that $F(z)[x; \prime]$ is a simple ring.

Theorem 2.2. *Let* $R = F(z)[x; \prime]$ *be the ring of differential polynomials, and* $A \in R^{n \times n}$. *Then there exist unimodular matrices* $U, V \in R^{n \times n}$ *such that*

$$J = UAV = \mathrm{diag}(1, \dots, 1, \varphi, 0 \dots, 0),$$

for some $\varphi \in R$.

3   Reducing Computing Jacobson Form to Hermite Form

In this section we present our technique for computing the Jacobson form of a matrix of Ore polynomials. Ultimately, it is a simple reduction to computing the Hermite form of a preconditioned matrix. We present it only for the ring $R = F(z)[x; \prime]$. An analogous method should work for the ring of shift polynomials, and will be developed in a later paper. We begin with some preparatory work.

3.1   On Divisibility

We first demonstrate that right multiplication by an element of $F[z]$, i.e., by a unit in R, transforms a polynomial to be relatively prime to the original.

Lemma 3.1. *Given* $h \in R$, *nontrivial in* x, *there exists a* $w \in F[z]$ *with* $\deg_z(w) \leq deg_x(h)$, *such that* $GCRD(h, hw) = 1$.

Proof. Without loss of generality assume h is normalized to be monic and has the form $x^n + h_{n-1}x^{n-1} + \dots + h_1 x + h_0$.
Case 1: h is irreducible.
    The only monic right divisor of h of positive degree is h itself. Thus, brought into normal form (i.e., with leading coefficient one), h and hw should be the same polynomial. We have $lc(h) = 1$, $lc(hw) = w$, $tc(h) = h_0$, and $tc(hw) = h_0 w + h_1 \delta(w) + \dots + h_n \delta^n(w)$, where $lc : R \to F(z)$ and $tc : R \to F(z)$ extract the leading and tailing coefficients respectively. The choice of w, such that the tail coefficients are different, is always possible. If you normalize both polynomials from the left and subtract them, then you get a polynomial of strict lower degree in x and not equal 0. This is due to the fact that the tail coefficient of hw after normalizing has the form

$$h_0 + \frac{h_1 \delta(w) + \dots + h_n \delta^n(w)}{w}, \tag{3.1}$$

and you can choose w such that the fraction above does not equal 0. Since h was assumed to be irreducible, we can reduce these polynomial further to 1 with a linear combination of h and hw (otherwise we would get a nontrivial GCRD of two irreducible polynomials).

Case 2: $h = h_1 \cdots h_m$, with h_i irreducible for $1 \le i \le m$.

In this case the proof is complicated by non-commutativity. Multiplication with w will affect the rightmost factor. If there is just one factorization we can again use the argument from case 1, and we are done.

If we have more than one factorization, things become interesting. We first show that a multiplication by w for the rightmost factor h_m in one factorization $h_1 \cdots h_m$ cannot be equal to $\tilde{h}_m$ for the rightmost factor $\tilde{h}_m$ of an arbitrary other factorization $\tilde{h}_1 \cdots \tilde{h}_m$. Suppose this equality holds. Then

$$h_1 \cdots h_{m-1} h_m w = \tilde{h}_1 \cdots \tilde{h}_{m-1} \tilde{h}_m w,$$

where we can directly see, that then

$$h_1 \cdots h_{m-1} = \tilde{h}_1 \cdots \tilde{h}_{m-1},$$

which means, that we already dealt with the same factorization, a contradiction. Thus, we cannot get the same rightmost factor via multiplication by a unit from the right. Now we can use the same argument as in case 1 and see that the GCRD of the rightmost factors will be 1. □

Remark 3.2. The condition on the tailing coefficient (3.1) in the proof shows us, that we can also always find for $f \ne g \in R$ a $w \in F[z]$ such that $GCRD(f, fw) = 1$ $GCRD(g, gw) = 1$.

In the second case of the proof above it was not necessary that we were just looking at h, because we can look at any left multiple of h and get the same result. Thus, we can guarantee that we will obtain, with high probability, coprime elements by premultiplication by a suitable random element.

Corollary 3.3. *For any* $f, g \in R$, *there exists a* $w \in F[z]$ *of degree at most* $\max\{\deg(f), \deg(g)\}$ *such that* $GCRD(fw, g) = 1$.

Lemma 3.4. *Let* $f, g \in R$ *have* $\deg_x f = n$ *and* $\deg_x g = m$, *and assume* f *and* g *have degree at most* e *in* z. *Let* $w \in F[z]$ *be chosen randomly of degree* $d = \max\{m, n\}$, *with coefficients chosen from a subset of* F *of size at least* $n(n + m)(n + e)$. *Then*

$$\text{Prob}\left\{GCRD(f, gw) = 1\right\} \ge 1 - \frac{1}{n}.$$

Proof. Assume the coefficients of w are independent indeterminates commuting with x. Consider the condition that $GCRD(f, gw) = 1$. We can reformulate this as a skew-Sylvester resultant condition in the coefficients of f and gw over $F(z)$. That is, there exists a matrix $\text{Syl}(f, gw) \in F(z)^{(n+m) \times (n+m)}$ such that $D = \det \text{Syl}(f, gw) \in F(z)$ is nonzero if and only if $GCRD(f, gw) = 1$. By Corollary 3.3 we know D is not identically zero. It is easily derived from the Leibniz formula for the determinant that $\deg_z D \le (n+m)(n+e)$. The probability stated then follows immediately from the Schwarz-Zippel Lemma (Schwartz, 1980).   □

We now use these basic results to construct a generic preconditioning matrix for A. First consider the case of a 2×2 matrix $A \in \mathsf{R}^{2\times 2}$, with Hermite form

$$H = \begin{pmatrix} f & g \\ 0 & h \end{pmatrix} = UA$$

for some unimodular $U \in \mathsf{R}^{2\times 2}$. We then precondition A by multiplying it by

$$Q = \begin{pmatrix} 1 & 0 \\ w & 1 \end{pmatrix},$$

where $w \in \mathsf{F}[z]$ is chosen randomly of degree $\max\{\deg(f), \deg(g), \deg(h)\}$, so

$$UAQ = \begin{pmatrix} f+gw & g \\ hw & h \end{pmatrix}.$$

Our goal is to have the Hermite form of AQ have a 1 in the $(1,1)$ position. This is achieved exactly when $GCRD(f+gw, hw) = 1$. The following lemma will thus be useful.

Lemma 3.5. *Given $f, g, h \in \mathsf{R}$. Then there exists a $w \in \mathsf{F}[z]$ with $\deg(w) \le \max\{\deg_x(f), \deg_x(g), \deg_x(h)\}$ such that $GCRD(f+gw, hw) = 1$.*

Proof. We consider two different cases.
Case 1: $GCRD(g, h) = 1$. This implies $GCRD(gw, hw) = 1$ for all possible w. Then there exist $e, l \in \mathsf{R}$ such that $egw + lhw = 1$. Therefore – because we are aiming to obtain 1 as the GCRD – we would proceed by computing the GCRD of $ef + 1$ and hw. Lemma 3.1 shows the existence of appropriate w, such that $GCRD(ef + 1, hw) = 1$.

Case 2: $GCRD(g, h) \ne 1$. Without loss of generality, let g be the GCRD of h and g (using the euclidean algorithm we can transform $GCRD(f+gw, hw)$ into such a system, and f will just get an additional left factor). Since we can choose w, such that $GCRD(f, hw) = 1$, we have $e, l \in \mathsf{R}$, such that $ef + lhw = 1$. This means that we just have to compute the GCRD of hw and $1 + egw$. Let $\tilde{h}$ be such that $\tilde{h}g = h$. If we choose the left factors e_2, l_2, such that $e_2 egw + l_2 \tilde{h}gw = gw$, we know that h and e_2 have no common right divisor. Our GCRD problem is equivalent to $GCRD(e_2 + gw, \tilde{h}gw)$, which can be further transformed to $GCRD(\tilde{h}e_2, \tilde{h}gw)$ (since we have $\tilde{h}(e_2 + gw) - \tilde{h}gw = \tilde{h}e_2$). As we have seen in Remark 3.2, We can adjust our choice of w to fulfill the conditions $GCRD(f, hw) = 1$ and $GCRD(\tilde{h}e_2, \tilde{h}gw) = 1$.   □

A similar resultant argument to Lemma 3.4 now demonstrates that for a random choice of w we obtain our co-primality condition. We leave the proof to the reader.

Lemma 3.6. *Given $f, g, h \in R$, with $d = \max\{\deg_x(f), \deg_x(g), \deg_x(h)\}$, and $e = \max\{\deg_z(f), \deg_z(g), \deg_z(h)\}$. Let $w \in R$ have degree d, and suppose its coefficients are chosen from a subset of F of size at least $n(n+d)(n+e)$. Then*

$$\mathrm{Prob}\left\{GCRD(f + gw, hw) = 1\right\} \geq 1 - \frac{1}{n}.$$

This implies that for *any* matrix $A \in \mathsf{R}^{2 \times 2}$ and a randomly selected $w \in \mathsf{F}[z]$ of appropriate degree we obtain with high probability

$$A \begin{bmatrix} 1 & 0 \\ w & 1 \end{bmatrix} = U \begin{bmatrix} 1 & * \\ 0 & h \end{bmatrix} = U \begin{bmatrix} 1 & 0 \\ 0 & h \end{bmatrix} V,$$

where $h \in R$ and $U, V \in \mathsf{R}^{2 \times 2}$ are unimodular matrices. Hence A has the Jacobson form $\mathrm{diag}(1, h)$. This is accomplished with one Hermite form computation on a matrix of the same degree in x, and not too much higher degree in z, than that of A.

Remark 3.7. With that we obtain an extra property for our resulting Hermite form: Since we can find such a $w \in \mathsf{F}[z]$, such that $GCRD(f + gw, hw) = 1$, there exist e, l, k, m, such that

$$\begin{bmatrix} e & l \\ k & m \end{bmatrix} \begin{bmatrix} f + gw & g \\ hw & h \end{bmatrix} = \begin{bmatrix} 1 & eg + lh \\ 0 & kg + mh \end{bmatrix}. \tag{3.2}$$

Now, we know, that the following equalities do hold:

$$ef + egw + lhw = 1 \iff egw + lhw = 1 - ef \iff eg + lh = w^{-1} - efw^{-1},$$

and similarly we get

$$kf + kgw + mhw = 0 \iff kgw + mhw = -kf \iff kg + mh = -kfw^{-1}.$$

This means that, on the right hand side of our equation (3.2), we have

$$\begin{bmatrix} 1 & w^{-1} - efw^{-1} \\ 0 & -kfw^{-1} \end{bmatrix}.$$

Therefore, for our next computation (i.e., if we just considered the 2×2 submatrix with this and computed the new Hermite form), we would deal with that same f as right factor multiplied by a unit from the right in the upper left corner of the next 2×2 submatrix and will be able to perform our computations there.

We now generalize this technique to $n \times n$ matrices over R.

Theorem 3.8. *Let $A \in \mathsf{R}^{n \times n}$ have full row rank. Let Q be a lower triangular, banded, unimodular matrix of the form*

$$\begin{bmatrix} 1 & 0 & 0 & \dots & 0 \\ w_1 & 1 & 0 & \dots & 0 \\ 0 & \ddots & \ddots & \ddots & \vdots \\ \vdots & \ddots & \ddots & \ddots & 0 \\ 0 & \dots & 0 & w_{n-1} & 1 \end{bmatrix} \in \mathsf{R}^{n \times n},$$

where $w_i \in \mathsf{F}[z]$ for $i \in \{1, \ldots, n-1\}$, $\deg(w_i) = i \cdot n \cdot d$ and d is the maximum degree of the entries in A. Then with high probability the diagonal of the Hermite form of $B = AQ$ is $\mathrm{diag}(1, 1, \ldots, 1, m)$, where $m \in \mathsf{F}(z)[x;']$.

Proof. Let H be the Hermite form of A and have the form

$$
\begin{bmatrix}
f_1 & h_1 & * & \cdots & * \\
0 & f_2 & h_2 & \cdots & * \\
0 & \ddots & \ddots & \ddots & \vdots \\
\vdots & \ddots & \ddots & \ddots & h_{n-1} \\
0 & \cdots & 0 & 0 & f_n
\end{bmatrix},
$$

By Giesbrecht and Kim (2012), Theorem 3.6, we know that the sum of the degrees of the diagonal entries of the Hermite form of A equals $n \cdot d$. Thus we can regard nd as an upper bound for the degrees of the f_i. If we now multiply the matrix

$$
\begin{bmatrix}
1 & 0 & 0 & \cdots & 0 \\
w_1 & 1 & 0 & \cdots & 0 \\
0_{n-2 \times 1} & 0_{n-2 \times 1} & I_{n-2} &
\end{bmatrix}
$$

from the right, we obtain the following in the upper left 2×2 submatrix:

$$
\begin{bmatrix}
f_1 + h_1 w_1 & h_1 \\
f_2 w_1 & f_2
\end{bmatrix}.
$$

As we have seen in the remark above, after calculation of the Hermite form of this resulting matrix, we get with high probability

$$
\begin{bmatrix}
1 & * & * & \cdots & * \\
0 & k f_1 w_1^{-1} & * & \cdots & * \\
0 & \ddots & \ddots & \ddots & \vdots \\
\vdots & \ddots & \ddots & \ddots & h_{n-1} \\
0 & \cdots & & 0 & 0 & f_n
\end{bmatrix}.
$$

The entry $k f_1 w_1^{-1}$ has degree at most $2 \cdot n \cdot d$, where we see, why we have chosen the degree $2 \cdot n \cdot d$ for w_2. After $n-1$ such steps we obtain a Hermite form with 1s on the diagonal, and an entry in $\mathsf{F}(z)[x; \delta]$ ☐

This leads us to the following simple algorithm to compute the Jacobson form by just calculating the Hermite Form after preconditioning.

Algorithm 1. JacobsonViaHermite: Compute the Jacobson normal form of a matrix over the differential polynomials

Input: $A \in \mathsf{F}(z)[x;']^{n \times n}$, $n \in \mathbb{N}$,

Output: The Jacobson normal form of A

Preconditions:

- Existence of an algorithm HERMITE to calculate the Hermite normal form of a given matrix over $\mathsf{F}(z)[x;']$
- Existence of an algorithm RANDPOLY which computes a random polynomial of specified degree with coefficients chosen from a specified set.

1: $d \leftarrow \max\{\deg(A_{i,j}) \mid i,j \in \{1,\ldots,n\}\}$
2: **for** i from 1 to $n-1$ **do**
3:     $w_i \leftarrow$ RANDPOLY(degree $= i \cdot n \cdot d$)
4: **end for**
5: Construct a matrix W, such that
$$W_{ij} \leftarrow \begin{cases} 1 & \text{if } i = j \\ w_i & \text{if } i = j + 1 \\ 0 & \text{otherwise} \end{cases}$$
6: result $\leftarrow$ HERMITE($A \cdot W$)
7: **if** result$_{ii} \neq 1$ for any $i \in \{1,\ldots,n-1\}$ **then**
8:     FAIL {With low probability this happens}
9: **end if**
10: Eliminate the off diagonal entries in result by simple column operations
11: **return** result

3.2 Experimental Implementation and Results

We have written an experimental implementation in MAPLE as a proof of concept of our algorithm.

Since there are no other implementations of the calculation of the Hermite form available for Ore rings, we used the standard way of calculating the Hermite form, i.e. by repeated GCRD computations. Since the Hermite form of a matrix is unique, the choice of algorithm is just a matter of calculation speed.

One problem with the preconditioning approach is that the diagonal entries become "ugly" (recall that they are only unique up to the equivalence described in the introduction). We illustrate this with an example as follows.

Example 3.9. Consider matrix A:

$$\begin{bmatrix} 1 + zx & z^2 + zx \\ z + (z+1)x & 5 + 10x \end{bmatrix}.$$

Its Jacobson form, calculated by SINGULAR, has as its nontrivial entry:

$$(45z - 10 - 11z^2 - z^4 + 2z^5) + (2z^5 + 3z^4 - 12z^3 + 10z + 2z^2)x + (2z^4 - 19z^3 + 9z^2)x^2.$$

Calculating the Jacobson form with the approach of calculating a lot of GCRDs or GCLDs respectively results in the polynomial:

$$(-3z^3+z^5-4z^2+3z+10)+(-8z^3+z^2+z^5+z^4+13z+19)x+(-10z^3+8z^2+z^4+9z)x^2.$$

If we precondition the matrix in the described way, the output of SINGULAR stays the same, but the output of the straightforward approach is the polynomial:

$$88360z^9 - 384554z^8 + 243285z^7 + 1104036z^6 - 4428356z^5 + 2474570z^4 + 3533537z^3$$
$$- 3915039z^2 + 1431017z - 150930$$
$$+ (88360z^9 - 31114z^8 - 948071z^7 + 5093247z^6 - 7538458z^5 + 5740077z^4 - 1935190z^3$$
$$- 20353z^2 + 154797z + 10621)x$$
$$+ (-739659z^3 + 137249z^2 + 5031z + 1769774z^4 - 2553232 + z^5 + 2133343z^6$$
$$- 1003074z^7 + 88360z^8)x^2.$$

The calculation time was as expected similar to just calculating a Hermite form. Both answers are "correct", but the Groebner-based approach has the effect of reducing coefficient size and degree. An important future task could be to find a normal form for a polynomial in this notion of weak similarity. This normal form should have as simple coefficients as possible.

The demonstration here is simply that the algorithm works, not that we would beat previous heuristic algorithms in practice. The primary goal of this work is to demonstrate a polynomial-time algorithm, which we hope will ultimately lead to faster methods for computing and a better understanding of the Jacobson form.

4   Degree Bounds and Complexity

The cost of the algorithm described for the Jacobson normal form is just the cost of a single preconditioning step (a matrix multiplication), plus the cost of computing a Hermite form (for which we use the algorithm of Giesbrecht and Kim (2009)). The growth in the degree of the input matrix after the precondition is an additive factor of $O(n^2d)$, which is largely dominated by the cost of computing the Hermite form. We thus obtain the following theorem.

Theorem 4.1. Let $A \in \mathsf{F}(z)[x;']^{n \times n}$ have full row rank, with $deg_x(A_{ij}) \leq d$ for $1 \leq i, j \leq n$, and $\deg_z(A_{ij}) \leq e$.

(a) We can compute the Jacobson form J of A, and unimodular matrices U, V such that $J = UAV$, with an expected number of $O(n^9d^3e)$ operations in F. The algorithm is probabilistic of the Las Vegas type, and always returns the correct solution.

(b) If $J = \mathrm{diag}(1, \ldots, 1, s_n)$, then $\deg_x(s_n) \leq nd$, and $\deg_x U_{ij}, \deg_x V_{ij} \leq nd$.

(c) $deg_z H_{ij} \in O(n^2de)$ and $deg_z U_{ij} \in O(n^2de)$ for $1 \leq i, j \leq n$.

Proof. Part (a) follows directly from the algorithm and the preceding analysis. Part (b) and (c) follow from the degree bounds over on the Hermite form over Ore polynomial rings in Giesbrecht and Kim (2009, 2012).

Of course a faster algorithm for computing the Hermite form would directly yield a faster algorithm for computing the Jacobson form of an input matrix.

5 Conclusion and Future Work

In this paper, we have developed a probabilistic algorithm for computing the Jacobson form of a square matrix with entries in the ring of differential polynomials which can also easily be generalized to the non-sqare case. The complexity of our algorithm depends on the complexity of calculating the Hermite form of a matrix with entries in $F(z)[x; ']$. Using the algorithm of Giesbrecht and Kim (2009) we establish a polynomial-time algorithm for the Jacobson form of a matrix of differential polynomials. We also establish polynomial bounds on the entries in the Jacobson form and on the transformation matrices. While we do not necessarily anticipate that this will ultimately be the most practical method to compute the Jacobson form, we hope that the techniques presented will be helpful in developing effective implementations. Future work will involve a generalization to more general Ore polynomial rings (in particular shift-polynomials), as well as asymptotically faster algorithms.

Acknowledgements. The authors thank Viktor Levandovskyy for his helpful ideas and encouragement, and the anonymous referees for their comments.

References

Cohn, P.: Free Rings and their Relations. Academic Press, London (1985)

Giesbrecht, M., Kim, M.S.: On Computing the Hermite Form of a Matrix of Differential Polynomials. In: Gerdt, V.P., Mayr, E.W., Vorozhtsov, E.V. (eds.) CASC 2009. LNCS, vol. 5743, pp. 118–129. Springer, Heidelberg (2009), doi: 10.1007/978-3-642-04103-7_12

Giesbrecht, M., Kim, M.: Computing the Hermite form of a matrix of Ore polynomials, (submitted for publication, 2012), ArXiv: 0906.4121

Jacobson, N.: The Theory of Rings. American Math. Soc., New York (1943)

Kaltofen, E., Krishnamoorthy, M.S., Saunders, B.D.: Parallel algorithms for matrix normal forms. Linear Algebra and its Applications 136, 189–208 (1990)

Levandovskyy, V., Schindelar, K.: Computing diagonal form and Jacobson normal form of a matrix using Gröbner bases. Journal of Symbolic Computation (in press, 2012)

Middeke, J.: A polynomial-time algorithm for the Jacobson form for matrices of differential operators. Technical Report 08-13, Research Institute for Symbolic Computation (RISC), Linz, Austria (2008)

Middeke, J.: A computational view on normal forms of matrices of Ore polynomials. PhD thesis, Research Institute for Symbolic Computation, Johannes Kepler University, Linz, Austria (2011)

Ore, O.: Theory of non-commutative polynomials. Annals of Math 34, 480–508 (1933)

Schwartz, J.T.: Fast probabilistic algorithms for verification of polynomial identities. J. Assoc. Computing Machinery 27, 701–717 (1980)

The Resonant Center Problem for a 2:-3 Resonant Cubic Lotka–Volterra System

Jaume Giné[1], Colin Christopher[2], Mateja Prešern[3],
Valery G. Romanovski[4,5], and Natalie L. Shcheglova[6]

[1] Departament de Matemàtica, Universitat de Lleida,
Av. Jaume II, 69, 25001 Lleida, Spain
`gine@matematica.udl.cat`
[2] School of Computing and Mathematics, Plymouth University,
Plymouth PL4 8AA, UK
`C.Christopher@plymouth.ac.uk`
[3] Department of Mathematics and Statistics, University of Strathclyde,
26 Richmond street, Glasgow G1 1XH, United Kingdom
`mateja.presern@strath.ac.uk`
[4] CAMTP - Center for Applied Mathematics and Theoretical Physics,
University of Maribor, Krekova 2, Maribor SI-2000, Slovenia
[5] Faculty of Natural Science and Mathematics, University of Maribor,
Koroška cesta 160, SI-2000 Maribor, Slovenia
`valery.romanovsky@uni-mb.si`
[6] Faculty of Mechanics and Mathematics, Belarusian State University,
4, Nezavisimosti avenue, 220030, Minsk, Belarus
`shcheglova@tut.by`

Abstract. Using tools of computer algebra we derive the conditions for the cubic Lotka–Volterra system $\dot{x} = x(2 - a_{20}x^2 - a_{11}xy - a_{02}y^2)$, $\dot{y} = y(-3 + b_{20}x^2 + b_{11}xy + b_{02}y^2)$ to be linearizable and to admit a first integral of the form $\Phi(x, y) = x^3 y^2 + \cdots$ in a neighborhood of the origin, in which case the origin is called a $2 : -3$ resonant center.

Keywords: resonant center problem, polynomial systems of differential equations, first integral.

1991 *Mathematics Subject classification*: Primary 34C14; Secondary 34A26, 37C27, 34C25.

1 Introduction

In this paper we consider a polynomial vector field in $\mathbb{C}^2$ with a $p : -q$ resonant elementary singular point, i.e.,

$$\dot{x} = p\,x + P(x, y), \quad \dot{y} = -q\,y + Q(x, y), \tag{1}$$

where $p, q \in \mathbb{Z}$ with $p, q > 0$ and P and Q are polynomials. The interest in these elementary singular points arises from the fact that there is a *resonant center* defined for this type of singular points. A resonant center is a generalization of the

V.P. Gerdt et al. (Eds.): CASC 2012, LNCS 7442, pp. 129–142, 2012.

concept of a real center to systems of ordinary differential equations in $\mathbb{C}^2$ of the form (1), see [8,22,24]. The classical real center problem goes back to Poincaré and Lyapunov, see [18,20], and has been studied extensively in hundreds of works, see for instance [6,12,13,22] and references therein. We have the following definition of a resonant center or focus, coming from Dulac [10] (see also [24]).

Definition 1. *A $p : -q$ resonant elementary singular point of an analytic system is a center if there exists a local meromorphic first integral $\Phi = x^p y^q + h.o.t.$*

Without loss of generality we can write system (1) in the form

$$
\begin{aligned}
\dot{x} &= px - \sum_{(i,j)\in S} a_{ij} x^{i+1} y^j, \\
\dot{y} &= -qy + \sum_{(i,j)\in S} b_{ji} x^j y^{i+1},
\end{aligned}
\tag{2}
$$

where $p, q \in \mathbb{N}$, $GCD(p, q) = 1$, and where S is the set

$$
S = \{(u_k, v_k) : u_k + v_k \geq 1, k = 1, \ldots, \ell\} \subset \mathbb{N}_{-1} \times \mathbb{N}_0,
$$

where $\mathbb{N}$ denotes the set of natural numbers and for a non-negative integer n, $\mathbb{N}_{-n} = \{-n, \ldots, -1, 0\} \cup \mathbb{N}$. The notation (2) simply emphasizes that we take into account only non-zero coefficients of the polynomials of interest. This will simplify formulas which occur later.

The condition that a function

$$
\Psi(x, y) = x^q y^p + \sum_{\substack{i+j>p+q \\ i,j\in\mathbb{N}_0}} v_{i-q,j-p} x^i y^j
\tag{3}
$$

be a first integral of (1) (the indexing has been chosen so as to be in agreement with algorithm of [21, Appendix]) is the identity

$$
D(\Psi) \stackrel{\text{def}}{=} \frac{\partial \Psi}{\partial x}(px + P(x, y)) + \frac{\partial \Psi}{\partial y}(-qy + Q(x, y)) \equiv 0,
\tag{4}
$$

which yields

$$
\begin{aligned}
&\left(q x^{q-1} y^p + \sum_{i+j>p+q} i v_{i-q,j-p} x^{i-1} y^j \right) \left(px - \sum_{(m,n)\in S} a_{mn} x^{m+1} y^n \right) \\
&+ \left(p x^q y^{p-1} + \sum_{i+j>p+q} j v_{i-q,j-p} x^i y^{j-1} \right) \left(-qy + \sum_{(m,n)\in S} b_{nm} x^n y^{m+1} \right) \equiv 0.
\end{aligned}
\tag{5}
$$

We augment the set of coefficients in (3) with the collection $J = \{v_{-q+s,q-s} : s = 0, \ldots, p+q\}$, where, in agreement with formula (3), we set $v_{00} = 1$ and $v_{mn} = 0$ for all other elements of J, so that elements of J are coefficients of terms of degree $p + q$ in $\Psi(x, y)$. We also set $a_{mn} = b_{nm} = 0$ for $(m, n) \notin S$. With these conventions, for $(k_1, k_2) \in \mathbb{N}_{-q} \times \mathbb{N}_{-p}$, the coefficient g_{k_1,k_2} of $x^{k_1+q}y^{k_2+p}$ in (5) is zero for $k_1 + k_2 \le 0$ and

$$g_{k_1,k_2} = (pk_1 - qk_2)v_{k_1,k_2}$$
$$- \sum_{\substack{s_1+s_2=0 \\ s_1 \ge -q, \, s_2 \ge -p}}^{k_1+k_2-1} [(s_1 + q)a_{k_1-s_1,k_2-s_2} - (s_2 + p)b_{k_1-s_1,k_2-s_2}]v_{s_1,s_2}. \quad (6)$$

for $k_1 + k_2 \ge 1$.

This formula can be used recursively to construct a formal first integral Ψ for system (2): at the first stage finding all v_{k_1,k_2} for which $k_1 + k_2 = 1$, at the second all v_{k_1,k_2} for which $k_1 + k_2 = 2$, and so on. For any pair k_1 and k_2, if $qk_1 \ne pk_2$ and if all coefficients v_{ℓ_1,ℓ_2} are already known for $\ell_1 + \ell_2 < k_1 + k_2$, then v_{k_1,k_2} is uniquely determined by (6) and the condition that g_{k_1,k_2} be zero. But at each of the stages $k_1 + k_2 = k(p + q)$, $k \in \mathbb{N}$ (and only at these stages, since $GCD(p,q) = 1$) there occurs the one "resonance" pair $(k_1, k_2) = (kq, kp)$ for which $qk_1 = pk_2$. Hence, for this pair, (6) becomes

$$g_{kq,kp} = - \sum_{\substack{s_1+s_2=0 \\ s_1 \ge -q, s_2 \ge -p}}^{kq+kp-1} [(s_1 + q)a_{k_1-s_1,k_2-s_2} - (s_2 + p)b_{k_1-s_1,k_2-s_2}]v_{s_1,s_2}, \quad (7)$$

so that the process of constructing a first integral Ψ only succeeds at this step if the expression on the right-hand side of (7) is zero. In this case, the value of $v_{k_1,k_2} = v_{kq,kp}$ is not determined by equation (6) and may be assigned arbitrarily.

It is evident from (6) that for all pairs of indices $(k_1, k_2) \in \mathbb{N}_{-q} \times \mathbb{N}_{-p}$, v_{k_1,k_2} is a polynomial function of the coefficients of (2), that is, it is an element of the polynomial ring $\mathbb{C}[a, b]$ (where (a, b) is the 2ℓ-tuple of the coefficients of system (2)), hence by (7) so are the expressions $g_{kq,kp}$ for all k. The polynomial $g_{kq,kp}$, which can be regarded as the k-th "obstruction" to the existence of the integral (3), is called the k-th focus quantity of system (2). Thus, the set of all systems inside the family (2), which admit a first integral of the form (3), is the zero set (the variety) of the ideal $\mathcal{B} = \langle g_{q,p}, g_{2q,2p}, \ldots \rangle$, called the Bautin ideal of (2). To find this variety in practice, one can compute focus quantities until the step k_0 at which the chain of ideals $\sqrt{\mathcal{B}_1} \subset \sqrt{\mathcal{B}_2} \subset \sqrt{\mathcal{B}_3} \subset \ldots$ stabilizes, which can be easily verified using the radical membership test[1] (here $\mathcal{B}_k = \langle g_{q,p}, \ldots, g_{kq,kp} \rangle$ and $\sqrt{\mathcal{B}_k}$ denotes the radical of the ideal $\mathcal{B}_k$). Then, using an appropriate computer algebra system (routines of SINGULAR [16] are usually very efficient to perform this task), one computes the irreducible decomposition of the variety of $\mathcal{B}_{k_0}$ and then for each component of the decomposition proves that corresponding systems indeed admit first integrals of the form (3).

For the $1 : -2$ resonant singular point and when P and Q in (1) are quadratic polynomials, the integrability problem is completely solved in [11,24] where necessary and sufficient conditions (20 cases) are given. Moreover, in [8], necessary

---

[1] The test says that given a polynomial $f \in \mathbb{C}[x_1, \ldots, x_n]$ and an ideal $J = \langle f_1, \ldots, f_s \rangle \subset \mathbb{C}[x_1, \ldots, x_n]$, f vanishes on the variety of the ideal J if and only if the reduced Groebner basis of $\langle f_1, \ldots, f_s, 1 - wf \rangle \subset \mathbb{C}[w, x_1, \ldots, x_n]$ is $\{1\}$.

and sufficient conditions (15 cases) for linearizability of the system are given. In [24], some sufficient center conditions for the $p : -q$ resonant singular point of a quadratic system are given. The most studied case is the quadratic Lotka–Volterra system, i.e.,

$$\dot{x} = x + ax^2 + bxy, \quad \dot{y} = -\lambda y + cxy + dy^2, \qquad (\lambda > 0). \qquad (8)$$

The necessary and sufficient conditions for integrability and linearizability of system (8) are already known for $\lambda \in \mathbb{N}$ [8,24]; that is, the $1 : -n$ resonant cases. In [15], some sufficient conditions are given for a general choice of λ. In particular, when $\lambda = p/2$ or $\lambda = 2/p$ with $p \in \mathbb{N}^+$, the necessary and sufficient conditions for integrability and linearizability are given. In [19], authors continue the study of the quadratic Lotka–Volterra system (8) and present sufficient conditions for integrability of Lotka–Volterra systems with $3 : -q$ resonance. In particular cases of $3 : -5$ and $3 : -4$ resonances, necessary and sufficient conditions for integrability of the systems are also given.

The $1 : -3$ resonant centers on $\mathbb{C}^2$ with homogeneous cubic nonlinearities were studied in [17]. This case corresponds to system (1) with $p = 1$, $q = 3$, where P and Q are homogeneous cubic polynomials. In [5], the necessary conditions and distinct sufficient conditions are derived for $1 : -q$ resonant centers of the homogeneous cubic Lotka–Volterra system.

Note that all studied cases involve rather laborious calculations related to decompositions of affine varieties defined by focus quantities. The complexity of calculations for different pairs p and q is difficult to estimate in advance, however it appears it depends on the structure of focus quantities as described in Section 3 of [21].

In this paper we focus our attention on the homogeneous cubic $2 : -3$ resonant Lotka–Volterra system, i.e., systems of the form

$$\begin{aligned} \dot{x} &= x(2 - a_{20}x^2 - a_{11}xy - a_{02}y^2), \\ \dot{y} &= y(-3 + b_{20}x^2 + b_{11}xy + b_{02}y^2). \end{aligned} \qquad (9)$$

The main aim is to give conditions for integrability and linearizability of system (9).

2   Conditions for Integrability and Linearizability

In order to obtain conditions for linearizability and integrability of system (9) we first recall a result obtained in [5] for $q \in \mathbb{N}$, which can be generalized directly for $q \in \mathbb{Q}^+$ with $q > 1$. The result follows.

Theorem 1. *The system*

$$\begin{aligned} \dot{x} &= x(1 - a_{20}x^2 - a_{11}xy - a_{02}y^2), \\ \dot{y} &= y(-q + b_{20}x^2 + b_{11}xy + b_{02}y^2) \end{aligned} \qquad (10)$$

has a resonant center at the origin if one of the following conditions holds:

(1) $a_{11} = b_{20} = b_{11} = 0$;

(2) $a_{11} = (q-2)a_{20} - b_{20} = 0$;

(3) $qa_{20}a_{11} + a_{11}b_{20} + (q-2)a_{20}b_{11} - b_{20}b_{11} = 0$,

$qa_{20}a_{02} + (q-1)a_{20}b_{02} - b_{20}b_{02} = 0$,

$qa_{11}a_{02} - qa_{02}b_{11} + (2q-1)a_{11}b_{02} - b_{11}b_{02} = 0$.

Indeed, from the proofs presented in [5], case (2) of Theorem 1 is valid for any $q \in \mathbb{R}$ with $q > 1$ and cases (1) and (3) are valid for any $q \in \mathbb{R}$ with $q > 0$.

The following theorem gives some sufficient conditions for linearizability of system (9).

Theorem 2. *System (9) has a linearizable resonant center at the origin if one of the following conditions holds:*

(α) $a_{11} = b_{11} = 1$,

(1) $a_{20} - b_{20} = 27a_{02}^2b_{20}^2 - 9a_{02}b_{20}^2b_{02} + 144a_{02}b_{20} - 28b_{20}b_{02} + 48 = 0$;

(2) $b_{20} = a_{20} = 0$;

(β) $a_{11} = 1, b_{11} = 0$,

(1) $a_{20} - b_{20} = 27a_{02}^2b_{20}^2 - 9a_{02}b_{20}^2b_{02} + 396a_{02}b_{20} - 52b_{20}b_{02} + 360 = 0$;

(2) $b_{20} = a_{20} = 0$;

(γ) $a_{11} = 0, b_{11} = 1$,

(1) $3a_{02}b_{20} - b_{20}b_{02} + 6 = a_{20} - b_{20} = 0$;

(2) $a_{02} = a_{20} - b_{20} = 0$;

Proof. We are going to prove that under conditions described in Theorem 2, system (9) is linearizable. We did not find the complete set of linearizability conditions, but those presented are obtained using the method developed in [8]. To this end, we make substitutions $v = xy$ and $w = y^2$. In these new coordinates, system (9) takes the form

$$\dot{v} = v\left(-w + (b_{20} - a_{20})v^2 + (b_{11} - a_{11})vw + (b_{02} - a_{02})w^2\right),$$
$$\dot{w} = 2w\left(-3w + b_{20}v^2 + b_{11}vw + b_{02}w^2\right). \tag{11}$$

We now set $a_{20} = b_{20}$, and (11) becomes the quadratic system

$$\dot{v} = -v + (b_{11} - a_{11})v^2 + (b_{02} - a_{02})vw,$$
$$\dot{w} = -6w + 2b_{20}v^2 + 2b_{11}vw + 2b_{02}w^2, \tag{12}$$

which has a resonant node at the origin. By the Poincaré–Lyapunov normal form theory (see e.g. [1,2]), an analytic system

$$\dot{u} = -u + \sum_{j+k=2}^{\infty} U_{jk}u^jv^k, \qquad \dot{v} = -nv + \sum_{j+k=2}^{\infty} V_{jk}u^jv^k,$$

can by a convergent transformation

$$\xi = u + \sum_{j+k=2}^{\infty} \alpha_{jk}u^jv^k, \qquad \eta = v + \sum_{j+k=2}^{\infty} \beta_{jk}u^jv^k, \tag{13}$$

be brought to the normal form

$$\dot{\xi} = -\xi, \quad \dot{\eta} = -n\eta + a\xi^n. \tag{14}$$

Hence, system (12) is linearizable if and only if the resonant monomial ($a\xi^6$ in the normal form of the second equation) is zero. Alternatively, we seek an analytic separatrix $w = \sum_{i>0} a_i v^i$ (and the only obstruction is in the terms of order 6). Computing this separatrix for system (12) we find that the coefficient a_6 is

$$b_{20}(360a_{11}^4 - 516a_{11}^3 b_{11} + 240a_{11}^2 b_{11}^2 - 36a_{11}b_{11}^3 + 396a_{02}a_{11}^2 b_{20} - 52a_{11}^2 b_{02}b_{20}$$
$$-306a_{02}a_{11}b_{11}b_{20} + 24a_{11}b_{02}b_{11}b_{20} + 54a_{02}b_{11}^2 b_{20} + 27a_{02}^2 b_{20}^2 - 9a_{02}b_{02}b_{20}^2)/36. \tag{15}$$

Setting $a_{11} = b_{11} = 1$ in condition (15) then gives

$$b_{20}(48 + 144a_{02}b_{20} - 28b_{02}b_{20} + 27a_{02}^2 b_{20}^2 - 9a_{02}b_{02}b_{20}^2)/36,$$

which yields subcases (1) and (2) of (α).
Setting $a_{11} = 1, b_{11} = 0$ in condition (15) gives

$$b_{20}(360 + 396a_{02}b_{20} - 52b_{02}b_{20} + 27a_{02}^2 b_{20}^2 - 9a_{02}b_{02}b_{20}^2)/36,$$

which yields subcases (1) and (2) of (β).
Finally, by setting $a_{11} = 0, b_{11} = 1$, condition (15) becomes $a_{02}b_{20}^2(6 + 3a_{02}b_{20} - b_{02}b_{20})/4$, which yields subcases (1) and (2) of (γ).

For all cases of Theorem 2 there exists an analytic change of coordinates

$$v_1 = v(1 + O(v, w)) = xy(1 + O(x, y)), \quad w_1 = w(1 + O(v, w)) = y^2(1 + O(x, y)),$$

which brings the node to the linear system. The linear system has a first integral v_1^6/w_1 which pulls back to a first integral of the form $y^4x^6(1+O(x, y))$. Extracting the root of this first integral we obtain a first integral of the form $y^2x^3(1+O(x, y))$ for all cases of Theorem 2.

We recall that the Darboux factor of system (1) is a polynomial $f(x, y)$, such that

$$\frac{\partial f}{\partial x}(px + P) + \frac{\partial f}{\partial y}(-qy + Q) = Kf,$$

where $K(x, y)$ is a polynomial called the *cofactor*. A simple computation shows that if there are Darboux factors $f_1, f_2, \ldots, f_k$ with the cofactors $K_1, K_2, \ldots, K_k$ satisfying $\sum_{i=1}^k \alpha_i K_i = 0$, then $H = f_1^{\alpha_1} \cdots f_k^{\alpha_k}$ is a first integral of (2), and if

$$\sum_{i=1}^k \alpha_i K_i + P_x' + Q_y' = 0,$$

then the equation admits the integrating factor $\mu = f_1^{\alpha_1} \cdots f_k^{\alpha_k}$, see for instance [3,4]. If system (2) has an integrating factor of this form then it usually admits also a first integral of the form (3) and, therefore, has center at the origin (see e.g. [8,22]).

The following theorem is the main result of this paper.

Theorem 3. *System* (9) *has a resonant center at the origin if one of the following conditions holds:*

(α) $a_{11} = b_{11} = 1$,

    (1) $a_{20} - b_{20} = 27a_{02}^2 b_{20}^2 - 9a_{02}b_{20}^2 b_{02} + 144a_{02}b_{20} - 28b_{20}b_{02} + 48 = 0;$

    (2) $b_{20} = a_{20} = 0;$

    (3) $b_{02} = a_{20} = 0;$

(β) $a_{11} = 1$, $b_{11} = 0$,

    (1) $a_{20} - b_{20} = 27a_{02}^2 b_{20}^2 - 9a_{02}b_{20}^2 b_{02} + 396a_{02}b_{20} - 52b_{20}b_{02} + 360 = 0;$

    (2) $b_{20} = a_{20} = 0;$

    (3) $a_{20}b_{02} + 6 = b_{20} = a_{02} - b_{02} = 0;$

    (4) $a_{20}b_{02} - 6 = b_{20} = 3a_{02} + 4b_{02} = 0;$

    (5) $b_{20}b_{02} + 18 = 3a_{02} + 4b_{02} = a_{20} + 3b_{20} = 0;$

    (6) $3a_{02} + 4b_{02} = 3a_{20} + 2b_{20} = 0;$

(γ) $a_{11} = 0$, $b_{11} = 1$,

    (1) $b_{02} = a_{02} = 0;$

    (2) $3a_{02}b_{20} - b_{20}b_{02} + 6 = a_{20} - b_{20} = 0;$

    (3) $a_{02} = a_{20} - b_{20} = 0;$

    (4) $a_{20} + 2b_{20} = 0;$

(δ) $a_{11} = b_{11} = 0$,

    (1) $3a_{20}a_{02} + a_{20}b_{02} - 2b_{20}b_{02} = 0;$

    (2) $a_{02} = 0;$

    (3) $b_{20} = 0;$

    (4) $a_{20} + 2b_{20} = 0.$

Proof. Computing the conditions. To compute focus quantities of system (9) we use the algorithm in [21], which is derived from formulae (6) and (7).

Following the algorithm and using a straightforward modification of the computer code in [22, Figure 6.1], we compute 12 focus quantities $g_{3,2}, \ldots, g_{36,24}$, where $g_{q(2k+1),p(2k+1)} = 0$ for $k = 0, \ldots, 5$, $g_{6,4} = (1512a_{11}^4 a_{20} + 216a_{02}a_{11}^2 a_{20}^2 + 36a_{11}^2 a_{20}^2 b_{02} - 1764a_{11}^3 a_{20}b_{11} - 288a_{02}a_{11}a_{20}^2 b_{11} - 72a_{11}a_{20}^2 b_{02}b_{11} + 672a_{11}^2 a_{20}b_{11}^2 + 72a_{02}a_{20}^2 b_{11}^2 + 20a_{20}^2 b_{02}b_{11}^2 - 84a_{11}a_{20}b_{11}^3 + 1008a_{11}^4 b_{20} + 2196a_{02}a_{11}^2 a_{20}b_{20} + 63\,a_{02}^2\,a_{20}^2\,b_{20} + 576\,a_{11}^2\,a_{20}\,b_{02}\,b_{20} + 21\,a_{02}\,a_{20}^2\,b_{02}\,b_{20} - 1848\,a_{11}^3\,b_{11}\,b_{20} - 1386a_{02}a_{11}a_{20}b_{11}b_{20} - 272a_{11}a_{20}b_{02}b_{11}b_{20} + 1008a_{11}^2 b_{11}^2 b_{20} + 198a_{02}a_{20}b_{11}^2 b_{20} + 20a_{20}b_{02}b_{11}^2 b_{20} - 168a_{11}b_{11}^3 b_{20} + 360a_{02}a_{11}^2 b_{20}^2 + 126a_{02}^2 a_{20}b_{20}^2 - 976a_{11}^2 b_{02}b_{20}^2 - 468a_{02}a_{11}b_{11}b_{20}^2 + 512a_{11}b_{02}b_{11}b_{20}^2 + 108a_{02}b_{11}^2 b_{20}^2 - 40b_{02}b_{11}^2 b_{20}^2 - 84a_{02}b_{02}b_{20}^3)/504$, and the rest of the polynomials are too long to be presented here. Next, we need to find the decomposition of the variety of the ideal $\mathcal{B}_{12} = \langle g_{6,4}, g_{12,8}, \ldots, g_{36,24} \rangle$ (the ideal is defined by 6 nonzero polynomials). We expected it could be done using the routine *minAssGTZ* of SINGULAR (in [9]) which computes minimal associate primes of the polynomial ideal using the method of [14]. However, this turned out to be a very difficult computational task, and we were unable to complete the computation – neither working over the field of rational numbers nor in the field of characteristic 32003. Note that, using a rescaling $x \to \alpha x$,

$y \to \beta y$, one can set any nonzero pair of coefficients (a_{kj}, b_{mn}) to $(1, 1)$, except the pair (a_{11}, b_{11}). In the pair (a_{11}, b_{11}) by a rescaling we can set only one of coefficients a_{11}, b_{11} equal to one, but the other one remains arbitrary. We tried to find the decomposition of the variety of the ideal $\mathcal{B}_{12} = \langle g_{6,4}, g_{12,8}, \ldots, g_{36,24} \rangle$ with a_{11} set to one (and the other coefficients being arbitrary) and with b_{11} set to one (and the other coefficients being arbitrary). In both cases, we were unable to complete computations at our facilities working in the field of characteristic 32003.

We, therefore, limit our consideration to the cases when either one or both coefficients a_{11}, b_{11} in system (9) are equal to zero, or both coefficients are equal to 1. That is we consider the following 4 cases: (α) $a_{11} = b_{11} = 1$, (β) $a_{11} = 1$, $b_{11} = 0$, (γ) $a_{11} = 0$, $b_{11} = 1$ and (δ) $a_{11} = b_{11} = 0$.

In case (δ), computing in the field of characteristic zero, we obtain conditions (δ) of the theorem.

In case (β), computations with $minAssGTZ$ in the field of characteristic 32003 yield the list L presented in line 2 of Figure 1. Using the code from Figure 1 (which is based on the rational reconstruction algorithm of [23]) we obtain conditions (β) of the theorem. A simple check shows that each of conditions (β) yields vanishing of all polynomials of the ideal $\mathcal{B}_{12}$. However, since modular computations are enforced, some components of the irreducible decomposition of the variety $\mathbf{V}(\mathcal{B}_{12})$ of the ideal $\mathcal{B}_{12}$ can be lost. To check that the decomposition is correct, we use the function $intersect$ of SINGULAR to compute $J = \cap_{k=1}^{6} J_k$, where J_k are ideals defined by conditions (1)–(6) of case (β) in Theorem 3. Then, using the radical membership test, we verify that each polynomial of J vanishes at $\mathbf{V}(\mathcal{B}_{12})$. This means that (1)–(6) of (β) give the correct decomposition of the variety of the ideal $\mathcal{B}_{12}$ with $a_{11} = 1$, $b_{11} = 0$. Thus, we have obtained necessary conditions of integrability for this case.

Similarly, we obtain necessary conditions for cases $a_{11} = b_{11} = 1$ and $a_{11} = 0$, $b_{11} = 1$ (conditions (α) and (γ), respectively, of the theorem).

Proof of sufficiency. Cases (3) of (α), (6) of (β) and (1) of (γ) satisfy condition (3) of Theorem 1; cases (4) of (γ) and (4) of (δ) satisfy condition (2) of Theorem 1; case (3) of (δ) satisfies condition (1) of Theorem 1. Thus, they are sufficient conditions for system (9) to have a resonant center at the origin.

Furthermore, cases (1) and (2) of (α) in Theorem 3 correspond to cases (1) and (2) of (α) of Theorem 2. However, we present another proof for case (2) of (α). The corresponding system is

$$\dot{x} = 2x - x^2 y - a_{02} xy^2, \quad \dot{y} = -3y + xy^2 + b_{02} y^2, \tag{16}$$

and it has two algebraic invariant curves: $l_1 = x$ and $l_2 = y$. In this case, we do not find enough invariant curves to construct a Darboux first integral or an integrating factor. Although we are not able to find a closed form for a first integral of system (16), we are going to prove that such analytic first integral exists. We look for a formal first integral of the form

$$\phi(x, y) = \sum_{k=2}^{\infty} h_k(x) y^k. \tag{17}$$

For each $k = 0, 1, 2, \ldots$, functions $h_k(x)$ should satisfy the first-order linear differential equation

$$(k-2)b_{02}h_{k-2}(x) + (k-1)xh_{k-1}(x) - a_{02}xh'_{k-2}(x)$$
$$-x^2h'_{k-1}(x) - 3kh_k(x) + 2xh'_k(x) = 0. \qquad (18)$$

Solving this equation, we obtain $h_2(x) = x^3$, $h_3 = -x^4$, $h_4 = -x^3(3a_{02} - 2b_{02} - 3x^2)/6$, taking the integration constants equal to zero. Using induction on k we wish to show that

$$h_k(x) = p_k(x) \quad \text{for} \quad k \geq 2,$$

where $p_k(x)$ are polynomials of degree $k + 1$. Hence, we assume that for $k = 2, \ldots, m-1$, there exist polynomials h_k satisfying (18), such that $\deg(h_k) = k+1$. We then solve the linear differential equation (18) for $k = m$ and obtain

$$h_m(x) = Cx^{3m/2} + \frac{1}{2}x^{3m/2} \int x^{-1-(3m)/2}g_m(x)dx, \qquad (19)$$

where $g_m(x) = (2-m)b_{02}h_{m-2}(x)+(1-m)xh_{m-1}(x)+x^2h'_{m-1}(x)+a_{02}xh'_{m-2}(x)$ and C is a constant of integration. Taking into account that, by hypothesis, $\deg(h_{m-1}(x)) = m$ and $\deg(h_{m-2}(x)) = m - 1$, we find that the degree of $g_m(x)$ is at most $m + 1$. Now, we must study whether the integral can give any logarithmic terms. Therefore, we must prove that terms involving x^{-1} do not appear in the integrand in (19). The exponents that can appear in the integrand are of the form

$$-1 - (3m)/2 + m + 1 - s, \quad \text{where} \quad s = 0, 1, \ldots, m + 1.$$

We want to know if this exponent, which, when simplified, is equal to $-m/2 - s$, can equal -1. This would imply that $m = 2 - 2s$ as $m \in \mathbb{N}$, thus, $s = 0, 1$ and $m = 0, 1$. However, since $m \geq 2$, there can be no logarithmic terms. Moreover, we can see that taking the constant of integration C equal to zero, $h_m(x)$ has degree at most $m + 1$. Hence, we proved that system (16) admits a formal first integral of the form (17). Consequently it has an analytic first integral around the origin.

Case (1) of (β) corresponds to the case (1) of (β) of Theorem 2.

Case (2) of (β) corresponds to the case (2) of (β) of Theorem 2. However, in this case there also exists a Darboux integrating factor. Indeed, the corresponding system is

$$\dot{x} = 2x - x^2y - a_{02}xy^2, \quad \dot{y} = -3y + b_{02}y^2.$$

It has four algebraic invariant curves: $l_1 = x$, $l_2 = y$ and $l_{3,4} = 1 \pm \dfrac{\sqrt{b_{02}}y}{\sqrt{3}}$, and it is possible to compute an integrating factor of the form

$$\mu = x^{-2}y^{-\frac{5}{3}}\left(1 - \frac{b_{02}y^2}{3}\right)^{-\frac{2}{3} - \frac{a_{02}}{2b_{02}}}.$$

By [8, Theorem 4.13], this means that the system also admits an analytic first integral (3).

Case (3) of (β). In this case, system (9) takes the form

$$\dot{x} = 2x + \frac{6}{b_{02}}x^3 - x^2y - b_{02}xy^2, \quad \dot{y} = -3y + b_{02}y^2.$$

This system has five algebraic invariant curves: $l_1 = x$, $l_2 = y$, $l_{3,4} = 1 \pm \frac{\sqrt{b_{02}}y}{\sqrt{3}}$ and $l_5 = 1 + \frac{3x^2}{b_{02}} - 2xy$, which allow to construct a Darboux integrating factor of the form $\mu = l_1^{-\frac{5}{2}}l_2^{-2}l_3^{\frac{3}{4}}l_4^{\frac{3}{4}}l_5^{-\frac{1}{4}}$. Again, by [8, Theorem 4.13], there exists an analytic integral (3).

Case (4) of (β). Here, the corresponding system is

$$\dot{x} = 2x - \frac{6}{b_{02}}x^3 - x^2y + \frac{4b_{02}}{3}xy^2, \quad \dot{y} = -3y + b_{02}y^2,$$

and it has six algebraic invariant curves: $l_1 = x$, $l_2 = y$, $l_{3,4} = 1 \pm \frac{\sqrt{b_{02}}y}{\sqrt{3}}$ and $l_{5,6} = 1 \pm \left(\frac{\sqrt{3}x}{\sqrt{b_{02}}} + \frac{\sqrt{b_{02}}y}{\sqrt{3}} \right)$, yielding the integrating factor

$$\mu = l_1^{-\frac{5}{2}}l_2^{-2}(l_3l_4)^{-\frac{5}{4}}(l_5l_6)^{-\frac{1}{4}}.$$

and, therefore, a first integral (3).

Case (5) of (β). The system of this case is written as

$$\dot{x} = 2x - \frac{54}{b_{02}}x^3 - x^2y + \frac{4}{3}b_{02}xy^2, \quad \dot{y} = -3y - \frac{18}{b_{02}}x^2y + b_{02}y^2.$$

If we take $b_{02} = 3$ by a scaling we find that the substitution

$$X = \frac{4x^2}{(-1+3xy+y^2)^2}, \quad Y = \frac{8x^3y(3x+y-1)(1+3x+y)}{(-1+3xy+y^2)^4}$$

blows down the origin of the system

$$\dot{x} = 2x - 18x^3 - x^2y + 4xy^2, \quad \dot{y} = -3y - 6x^2y + 3y^3$$

to a node (after scaling time by a factor $1 - 3xy - y^2$)

$$\dot{X} = 4X - 2Y - 9X^2, \quad \dot{Y} = 3Y(1 - 8X)$$

The first integral of the node is in the form $\tilde{\Phi} = Y^4/(X + kY + O(2))^3$, which pulls back to a first integral of the form

$$\Phi = x^6y^4(k + O(1)), \quad (k > 0 \text{ is a constant}).$$

We can then take square roots to obtain the first integral of our original system.

If b_{02} is negative (so choose $b_{02} = -3$), then we get a similar transformation with x replaced by ix and y by y/i, this transformation is still real if we multiply out the brackets corresponding to $(3x + y - 1)(1 + 3x + y)$ in the transformation above.

Case (1) of (γ) follows from case (2) of (β) by swapping a_{11} with b_{11}, and a_{20}, b_{20} with a_{02}, b_{02}, respectively.

Case (2) of (γ) corresponds to the case (1) of (γ) of Theorem 2.

Case (3) of (γ) corresponds to the case (2) of (γ) of Theorem 2. Another proof of integrability for this case can be obtained in a similar manner as the proof given above for case (2) of (α).

Case (2) of (δ). The corresponding system is

$$\dot{x} = 2x - a_{20}x^3, \quad \dot{y} = -3y + b_{20}x^2y + b_{02}y^2.$$

It has four algebraic invariant curves: $l_1 = x$, $l_2 = y$ and $l_{3,4} = 1 \pm \dfrac{\sqrt{a_{20}}x}{\sqrt{2}}$, and it is possible to compute an integrating factor of the form

$$\mu = x^{-4}y^{-3}\left(1 - \frac{a_{20}x^2}{2}\right)^{-\frac{1}{2} - \frac{b_{20}}{a_{20}}}.$$

Integration yields the first integral

$$\Psi = \frac{3x^3y^2}{3\left(1 - \frac{a_{20}x^2}{2}\right)^{\frac{3}{2} - \frac{b_{20}}{a_{20}}} - b_{02}y^2\, {}_2F_1\left(-\frac{3}{2}, \frac{b_{20}}{a_{20}} - \frac{1}{2}; -\frac{1}{2}; \frac{a_{20}x^2}{2}\right)}.$$

3   Concluding Remarks

We have obtained the necessary and sufficient conditions for integrability of system (9) with $a_{11} = b_{11}$ equal to 0 or 1, or one of the coefficients a_{11}, b_{11} equal to zero.

To obtain the necessary and sufficient conditions of integrability of general system (9) we need to compute the irreducible decomposition of the variety of the ideal $\mathcal{B}_{12}$ where only one of coefficients a_{11}, b_{11} is set to 1, but the calculations cannot be completed at our computational facilities because of computational complexity of the problem.

Acknowledgments.   The first author is partially supported by a MICINN/FEDER grant number MTM2011-22877 and by a Generalitat de Catalunya grant number 2009SGR 381. The fourth author acknowledges the support of the Slovenian Research Agency.

Appendix

In Figure 1 we present a code in MATHEMATICA to perform the rational reconstruction based on the algorithm in [23]. For the input, the code takes the ideal returned by $minAssGTZ$ of SINGULAR in the case $a_{11} = 1, b_{11} = 0$. The output are the conditions (β) of the Theorem 3.

```
In[1]:= $PreRead = ReplaceAll[#, {"(" :> "{", ")" :> "}"}] &;

In[2]:= L[1] = ideal
             (a20 - b20, a02^2*b20^2 - 10668*a02*b20^2*b02 - 10653*a02*b20 - 8299*b20*b02 + 10681);
        L[2] = ideal (b20, a20);
        L[3] = ideal (a02 + 10669*b02, a20 - 10667*b20);
        L[4] = ideal (a02 + 10669*b02, a20 + 3*b20, b02 + 18);
        L[5] = ideal (b20, a02 + 10669*b02, a20*b02 - 6);
        L[6] = ideal (b20, a02 - b02, a20*b02 + 6);

In[8]:= $PreRead =.

In[9]:= Sing2Math[L_Symbol] := Table[DownValues[L][[i, 2]], {i, 1, Length[DownValues[L]]}] /. {ideal -> 1}

In[10]:= RATCONVERT[c_, m_] := Block[{u = {1, 0, m}, v = {0, 1, c}, r},

         While[Sqrt[m/2] <= v[[3]], r = u - Quotient[u[[3]], v[[3]]] v; u = v; v = r]; If[Abs[v[[2]]] >= Sqrt[m/2], err, v[[3]]/v[[2]]]]

In[11]:= CenterCond[L_] := Table[ Factor[
             Replace[Sing2Math[L], (n_Integer | Times[n_Integer, x__]) :> If[n >= 0, RATCONVERT[n, 32003] x,
                 -RATCONVERT[-n, 32003] x], {3}][[i]]], {i, 1, Length[DownValues[L]]}]

In[12]:= CenterCond[L] // MatrixForm
```

Out[12]//MatrixForm=

$$\begin{pmatrix}
\{a20 - b20, \ \frac{1}{27}\left(360 + 396\, a02\, b20 - 52\, b02\, b20 + 27\, a02^2\, b20^2 - 9\, a02\, b02\, b20^2\right)\} \\
\{b20, a20\} \\
\{\frac{1}{3}\,(3\, a02 + 4\, b02), \ \frac{1}{3}\,(3\, a20 + 2\, b20)\} \\
\{\frac{1}{3}\,(3\, a02 + 4\, b02), \ a20 + 3\, b20, \ 18 + b02\, b20\} \\
\{b20, \ \frac{1}{3}\,(3\, a02 + 4\, b02), \ -6 + a20\, b02\} \\
\{b20, \ a02 - b02, \ 6 + a20\, b02\}
\end{pmatrix}$$

Fig. 1. MATHEMATICA code for rational reconstruction

In the code *PreRead*, there is a global variable the value of which, if set, is applied to the text or box form of every input expression before it is fed to MATHEMATICA. The function *Sing2Math* transforms the output L of the type *list* of SINGULAR to a list of MATHEMATICA, the *RATCONVERT* performs the rational reconstruction of a given number using the algorithm of [23], and applying *CenterCond* we obtain the rational reconstruction for the whole list L.

References

1. Bibikov, Y.N.: Local theory of nonlinear analytic ordinary differential equations. Lecture Notes in Mathematics, vol. 702. Springer, Heidelberg (1979)
2. Bruno, A.D.: A Local Method of Nonlinear Analysis for Differential Equations. Nauka, Moscow (1979) (in Russian); Local Methods in Nonlinear Differential Equations. Springer, Berlin (1989) (translated from Russian)
3. Chavarriga, J., Giacomini, H., Giné, J., Llibre, J.: On the integrability of two-dimensional flows. J. Differential Equations 157, 163–182 (1999)
4. Chavarriga, J., Giacomini, H., Giné, J., Llibre, J.: Darboux integrability and the inverse integrating factor. J. Differential Equations 194, 116–139 (2003)
5. Chen, X., Giné, J., Romanovski, V.G., Shafer, D.S.: The 1: − q resonant center problem for certain cubic Lotka–Volterra systems. Appl. Math. Comput. (to appear)
6. Christopher, C., Li, C.: Limit Cycles of Differential Equations. Birkhäuser, Basel (2007)
7. Christopher, C., Rousseau, C.: Nondegenerate linearizable centres of complex planar quadratic and symmetric cubic systems in $\mathbb{C}^2$. Publ. Mat. 45, 95–123 (2001)
8. Christopher, C., Mardešic, P., Rousseau, C.: Normalizable, integrable and linearizable saddle points for complex quadratic systems in $\mathbb{C}^2$. J. Dyn. Control Syst. 9, 311–363 (2003)
9. Decker, W., Pfister, G., Schönemann, H.A.: Singular 2.0 library for computing the primary decomposition and radical of ideals primdec.lib (2001)
10. Dulac, H.: Détermination et intégration d'une certaine classe d'équations différentielles ayant pour point singulier un centre. Bull. Sci. Math. 32, 230–252 (1908)
11. Fronville, A., Sadovski, A.P., Żołądek, H.: Solution of the 1 : −2 resonant center problem in the quadratic case. Fund. Math. 157, 191–207 (1998)
12. Giné, J.: On the number of algebraically independent Poincaré-Liapunov constants. Appl. Math. Comput. 188, 1870–1877 (2007)
13. Giné, J., Mallol, J.: Minimum number of ideal generators for a linear center perturbed by homogeneous polynomials. Nonlinear Anal. 71, e132–e137 (2009)
14. Gianni, P., Trager, B., Zacharias, G.: Gröbner bases and primary decomposition of polynomials. J. Symbolic Comput. 6, 146–167 (1988)
15. Gravel, S., Thibault, P.: Integrability and linearizability of the Lotka–Volterra System with a saddle point with rational hyperbolicity ratio. J. Differential Equations 184, 20–47 (2002)
16. Greuel, G.M., Pfister, G., Schönemann, H.: Singular 3.0. A Computer Algebra System for Polynomial Computations. Centre for Computer Algebra, University of Kaiserslautern (2005), http://www.singular.uni-kl.de
17. Hu, Z., Romanovski, V.G., Shafer, D.S.: 1 : −3 resonant centers on $\mathbb{C}^2$ with homogeneous cubic nonlinearities. Comput. Math. Appl. 56, 1927–1940 (2008)
18. Liapunov, M.A.: Problème général de la stabilité du mouvement. Ann. of Math. Stud. 17. Pricenton University Press, Princeton (1947)
19. Liu, C., Chen, G., Li, C.: Integrability and linearizability of the Lotka–Volterra systems. J. Differential Equations 198, 301–320 (2004)
20. Poincaré, H.: Mémoire sur les courbes définies par les équations différentielles. Journal de Mathématiques 37, 375–422 (1881); 8, 251–296 (1882), Oeuvres de Henri Poincaré, vol. I, pp. 3–84. Gauthier–Villars, Paris (1951)

21. Romanovski, V.G., Shafer, D.S.: On the center problem for $p : -q$ resonant polynomial vector fields, Bull. Belg. Math. Soc. Simon Stevin 15, 871–887 (2008)
22. Romanovski, V.G., Shafer, D.S.: The Center and Cyclicity Problems: A Computational Algebra Approach. Birkhäuser, Boston (2009)
23. Wang, P.S., Guy, M.J.T., Davenport, J.H.: P-adic reconstruction of rational numbers. SIGSAM Bull. 16, 2–3 (1982)
24. Żołądek, H.: The problem of center for resonant singular points of polynomial vector fields. J. Differential Equations 137, 94–118 (1997)

Complexity of Solving Systems with Few Independent Monomials and Applications to Mass-Action Kinetics

Dima Grigoriev[1] and Andreas Weber[2]

[1] CNRS, Mathématiques, Université de Lille, Villeneuve d'Ascq, 59655, France,
Dmitry.Grigoryev@math.univ-lille1.fr
[2] Institut für Informatik II, Universität Bonn, Friedrich-Ebert-Allee 144,
53113 Bonn, Germany,
weber@cs.uni-bonn.de

Abstract. We design an algorithm for finding solutions with nonzero coordinates of systems of polynomial equations which has a better complexity bound than for known algorithms when a system contains a few linearly independent monomials. For parametric binomial systems we construct an algorithm of polynomial complexity. We discuss the applications of these algorithms in the context of chemical reaction systems.

Keywords: Complexity of solving systems of polynomial equations, Smith form, toric systems, mass-action kinetics, chemical reaction networks.

1 Introduction

We study systems of polynomial equations with a few linearly independent monomials. To find solutions with nonzero coordinates of such systems we design in Sect. 2 an algorithm which makes use of a combination of the multiplicative structure on the monomials with the additive structure emerging from the linear equations on monomials called Gale duality and which was used in [1] for improving Khovanskii's bound on the number of real solutions of systems of fewnomials. This combination allows one to diminish the number of variables, being crucial since the latter brings the greatest contribution into the complexity of solving systems of polynomial equations. Moreover, the designed algorithm allows one to look for positive real solutions that is important in the applications to mass-action kinetics [2–7].

Note that the designed algorithm has a better complexity bound than the one just employing the known general methods for solving systems of polynomial equations [8, 9] or inequalities [10]. So more, it has better complexity bounds than the methods relying on Gröbner bases [11] or involutive divisions [12] which have double-exponential complexity upper and lower bound [13].

In Sect. 3, we expose an algorithm finding solutions of *parametrical* binomial systems with nonzero coordinates and parameters within polynomial complexity

V.P. Gerdt et al. (Eds.): CASC 2012, LNCS 7442, pp. 143–154, 2012.

which invokes computing the Smith canonical form of an integer matrix. Such systems also emerge in mass-action kinetics. Similar to Sect. 2 the algorithm allows one to look for positive real solutions. The polynomial complexity cannot be achieved using the general methods for solving systems of polynomial equations or, respectively, inequalities (as well as the Gröbner or involutive bases because the example of generators of an ideal from [13] with double-exponential complexity consists just of binomials).

In [14] a polynomial complexity algorithm is designed to test whether a binomial system has a finite number of affine solutions (including ones with zero coordinates). On the other hand, it is proved in [14] that the problem of counting the number of affine solutions of a binomial system is #P-complete. We observe also that the problem of testing whether a system of binomial equations extended by linear equations (being customary in biochemical reactions networks) has a positive solution, is NP-hard. Indeed, adding to a system of binomials $x_i \cdot y_i = 1$, $1 \le i \le n$ in $2 \cdot n$ variables linear equations $x_i + y_i = 5/2$, $1 \le i \le n$ and a single linear equation in the variables $x_1, \ldots, x_n$, we arrive to the knapsack problem.

Potential applications of these algorithms in the context of chemical reaction networks are discussed in Sect. 4. We also expose a computational example there.

As a related work we mention also [15] where an algorithm for solving systems of *quadratic* inequalities is designed with the complexity bound being good when the number of inequalities is rather small.

2   Polynomial Systems with a Few Linearly Independent Monomials

Any system of polynomial equations can be represented in a form

$$A \cdot Y = 0 \tag{1}$$

where $A = (a_{k,j})$, $1 \le k \le l$, $1 \le j \le m$ is a matrix, and $Y = (Y_j)$, $1 \le j \le m$ is a vector of monomials $Y_j = X_1^{y_{j,1}} \cdots X_n^{y_{j,n}}$ in the variables $X_1, \ldots, X_n$. An algorithm designed in this Section searches for solutions of (1) with non-vanishing coordinates $x_1, \ldots, x_n \in (\overline{\mathbb{Q}})^* := \overline{\mathbb{Q}} \setminus \{0\}$. The condition of non-vanishing coordinates is not too restrictive for the purposes of mass-action kinetics since in the latter one looks usually for solutions with positive real coordinates. Assume that $y_{j,i} \le d$, $1 \le j \le m$, $1 \le i \le n$, and that the entries $a_{k,j} \in \mathbb{Z}$ are integers, therein $|a_{k,j}| \le M$. The assumption on $a_{k,j}$ to be integers is adopted just for the sake of simplifying complexity bounds, one could consider by the same token algebraic entries $a_{k,j} \in \overline{\mathbb{Q}}$.

The considered form of systems of polynomial equations appears, in particular, in the study of stationary solutions of the dynamical equations of the mass-action kinetics [2–4, 6, 7].

In general, the algorithm solving systems (1) (with or without imposing the condition of non-vanishing coordinates of solutions) has complexity bound polynomial in l, d^{n^2}, $\log M$ [8, 9]. In this paper, we suggest an algorithm for solving systems with the complexity being better than in general when the difference $r := m - \text{rk}(A)$ is small enough.

The solutions of system (1) depend on r parameters $Z_1, \ldots, Z_r$. One can thus express monomials $Y_j = \sum_{1 \leq k \leq r} u_{j,k} \cdot Z_k$, $1 \leq j \leq m$ with suitable rationals $u_{j,k} \in \mathbb{Q}$.

One can bring the matrix $y := (y_{j,i})$ to the Smith canonical form. Namely, one can find integer square matrices $B = (b_{\alpha,\beta})$ of size $m \times m$ and $C = (c_{\gamma,\delta})$ of the size $n \times n$ such that $\det(B) = \det(C) = 1$, and the matrix $V = (v_{j,i}) := ByC$, $1 \leq j \leq m$, $1 \leq i \leq n$ has the following form. The only non-vanishing entries $v_{j,i}$ are on the diagonal $v_{j,j} \neq 0$, $1 \leq j \leq p$ where $p := \text{rk}(y)$. Moreover, $v_{1,1}|v_{2,2}| \cdots |v_{p,p}$, although we will not make use of this extra property on divisibility. The complexity of constructing matrices B, C is polynomial in n, m, $\log d$ [16]; moreover, one can make its parallel complexity poly-logarithmic [17]. In particular, $|b_{\alpha,\beta}|$, $|c_{\gamma,\delta}| \leq (d \cdot \min\{n, m\})^{O(\min\{n,m\})}$.

Consider polynomials $f_s = \prod_{1 \leq j \leq m} Y_j^{b_{s,j}} \in \mathbb{Q}[Z_1, \ldots, Z_r]$, $1 \leq s \leq m$. Then $\deg(f_s) \leq m \cdot (d \cdot \min\{n, m\})^{O(\min\{n,m\})}$. The input system (1) has a solution over $(\overline{\mathbb{Q}})^*$ iff the system of equations $f_{p+1} = \cdots = f_m = 1$ and inequation $f_1 \cdots f_p \neq 0$ has a solution in $Z_1, \ldots, Z_r$ over $\overline{\mathbb{Q}}$. In particular, among $f_{p+1}, \ldots, f_m$ the polynomial $f_q = Y_q$, $p < q \leq m$ occurs, when the monomial Y_q equals 1 identically (provided that the monomial 1 is among the monomials $Y_1, \ldots, Y_m$). The latter yields an equation $(f_q =) \sum_{1 \leq k \leq r} u_{q,k} \cdot Z_k = 1$. One can find the irreducible components of the constructible set of solutions of the latter system using [8, 9]. Any solution $(z_1, \ldots, z_r)$ of the latter system provides a solution of the input system as follows.

Denote the monomials $W_t := \prod_{1 \leq i \leq n} X_i^{c_{t,i}}$, $1 \leq t \leq n$. Then the equations $W_t^{v_{t,t}} = f_t$, $1 \leq t \leq p$ impose the conditions on W_t, $1 \leq t \leq p$, while $W_{p+1}, \ldots, W_n$ can be chosen as arbitrary non-zeros. Finally, having $W_1, \ldots, W_n$, one can come back to $X_1, \ldots, X_n$ by means of the matrix C^{-1}.

Sometimes, in the applications to chemistry one looks for positive real solutions $X_1 > 0, \ldots, X_n > 0$ of the input system (1) [3, 4, 6, 7]. The latter is equivalent to $W_1 > 0, \ldots, W_n > 0$. This imposes the condition $f_t > 0$, $1 \leq t \leq p$ and one can solve the system of inequalities $f_t > 0$, $1 \leq t \leq p$, $f_{p+1} = \cdots = f_m = 1$ over the reals with the help of [10]. After that W_t, $1 \leq t \leq p$ are obtained uniquely from the equations $W_t^{v_{t,t}} = f_t$, $1 \leq t \leq p$, while $W_{p+1} > 0, \ldots, W_n > 0$ can be chosen in an arbitrary way. Finally, we can summarize the results.

Proposition 1. *One can design an algorithm which finds the irreducible components of the constructible set of solutions with non-vanishing coordinates $x_1, \ldots, x_n$ of a system of polynomial equations (1) within complexity polynomial in l, n, m, $(d \cdot \min\{n, m\})^{O(\min\{n,m\}) \cdot r^2}$, $\log M$. Moreover, the algorithm can find positive real solutions of (1) also within the same complexity bound.*

Note that this complexity bound is better than the bound polynomial in l, d^{n^2}, $\log M$ from [8–10] when r is significantly smaller than n. As usually, the practical complexity bounds are apparently better than the established a priori bounds, especially when the complexity of bringing to the Smith form being small.

Remark 1. Using indeterminates $Z_1, \ldots, Z_r$ in a similar way to our proposal has been done by several authors, see e.g. [2, 3] and references therein. Also using the Smith normal form has been proposed in [2] as well as [3] (in addition to using logarithms or the Hermite normal form), but for computations the Hermite normal form or Gröbner basis methods have been used in these papers. Hence although several parts of our proposed algorithms have been around for the special case of chemical reaction networks for several years, but nevertheless in addition to the complexity analysis also our proposed algorithm seems to be new in its full form.

3 Parametric Binomial Systems

Now suppose that a matrix A at each of its rows contains at most two non-vanishing entries, and moreover every entry is a monomial of the form $\beta \cdot K^E :=$ $\beta \cdot K_1^{e_1} \cdots K_q^{e_q}$. Herein $\beta \in \mathbb{Q}$ and $K_1, \ldots, K_q$ play the role of parameters. Such parametric systems appear in the applications to mass-action kinetics [3, 4, 6, 7, 18]. In other words, each equation of (1) can be viewed as a binomial in the variables $X_1, \ldots, X_n, K_1, \ldots, K_q$. We pose a question, for which non-zero values of $K_1, \ldots, K_q$ the system (1) has a solution in non-vanishing $x_1, \ldots, x_n$? Alternatively, for which positive real values of $K_1, \ldots, K_q$ the system (1) has a positive real solution?

Rewrite now the system (1) of l binomials in the form

$$X^{G_j} = \beta_j \cdot K^{H_j}, \, 1 \leq j \leq l \tag{2}$$

where $X^{G_j} := X_1^{g_{j,1}} \cdots X_n^{g_{j,n}}$, $K^{H_j} := K_1^{h_{j,1}} \cdots K_q^{h_{j,q}}$. The algorithm brings the matrix $G := (g_{j,i})$, $1 \leq j \leq l$, $1 \leq i \leq n$ to the Smith canonical form. Thus, the algorithm yields integer unimodular matrices B', C' such that $B' \cdot G \cdot C'$ is in the Smith canonical form. Let $s := \mathrm{rk}(G)$ and the only non-vanishing entries of $B' \cdot G \cdot C'$ be its first s diagonal entries $g'_{1,1}, \ldots, g'_{s,s}$. Denote $B' =: (b'_{j,\alpha})$, $1 \leq j, \alpha \leq l$ and $\gamma_j \cdot K^{H'_j} := \prod_{1 \leq \alpha \leq l}(\beta_\alpha \cdot K^{H_\alpha})^{b'_{j,\alpha}}$.

The system (2) for given non-zero $K_1, \ldots, K_q$ has a solution in non-zero $X_1, \ldots, X_n$ iff

$$\gamma_j \cdot K^{H'_j} = 1, \, s + 1 \leq j \leq l. \tag{3}$$

In its turn, solvability of (2) in positive real solutions $X_1, \ldots, X_n$ for positive real $K_1, \ldots, K_q$ imposes extra conditions $\beta_1 > 0, \ldots, \beta_l > 0$.

For non-zero values of parameters $K_1, \ldots, K_q$ satisfying (3) one can find monomials $\prod_{1 \leq i \leq n} X_i^{c'_{\mu,i}}$, $1 \leq \mu \leq s$, where the matrix $C' =: (c'_{\mu,i})$, $1 \leq \mu, i \leq n$, from the equations $(\prod_{1 \leq i \leq n} X_i^{c'_{\mu,i}})^{g'_{\mu,\mu}} = \gamma_\mu \cdot K^{H'_\mu}$, $1 \leq \mu \leq s$, while the non-zero values of the monomials $\prod_{1 \leq i \leq n} X_i^{c'_{\mu,i}}$, $s + 1 \leq \mu \leq n$ are chosen in an arbitrary way. Then the algorithm uniquely finds $X_1, \ldots, X_n$ from the monomials

$\prod_{1\leq i\leq n} X_i^{c'_{\mu,i}}$, $1 \leq \mu \leq n$ with the help of the matrix $(C')^{-1}$. Respectively, for positive real $K_1,\ldots,K_q$ to get positive real $X_1,\ldots,X_n$ one chooses the positive values of the monomials $\prod_{1\leq i\leq n} X_i^{c'_{\mu,i}}$, $s+1 \leq \mu \leq n$ in an arbitrary way.

To describe the conditions on non-zero $K_1,\ldots,K_q$ satisfying (3), the algorithm brings $(l-s) \times q$ matrix $H' := (h'_{j,\alpha})$, $s+1 \leq j \leq l$, $1 \leq \alpha \leq q$, where the vector $H'_j =: (h'_{j,\alpha})$, $1 \leq \alpha \leq q$, to the Smith canonical form. Thus, the algorithm yields integer unimodular matrices $B'' = (b''_{j,\delta})$, $s+1 \leq j,\delta \leq l$, $C'' = (c''_{\mu,\alpha})$, $1 \leq \mu,\alpha \leq q$ such that the only non-vanishing entries of the matrix $B'' \cdot H' \cdot C''$ are its first t diagonal entries $h''_{1,1},\ldots,h''_{t,t}$, where $t = \mathrm{rk}(H')$.

Denote $\epsilon_j := \prod_{s+1\leq\delta\leq l} \gamma_\delta^{-b''_{j,\delta}}$, $s+1 \leq j \leq l$. Then (3) has a solution in non-zero $k_1,\ldots,k_q$ iff

$$\epsilon_j = 1,\ s+t+1 \leq j \leq l. \tag{4}$$

If (4) holds one can find the values of the monomials $\prod_{1\leq\alpha\leq q} K_\alpha^{c''_{\mu,\alpha}}$, $1 \leq \mu \leq t$ from the equalities $(\prod_{1\leq\alpha\leq q} K_\alpha^{c''_{\mu,\alpha}})^{h''_{\mu,\mu}} = \epsilon_{s+\mu}$, $1 \leq \mu \leq t$, while the non-zero values of the monomials $\prod_{1\leq\alpha\leq q} K_\alpha^{c''_{\mu,\alpha}}$, $t+1 \leq \mu \leq q$ are chosen in an arbitrary way. Respectively, the latter values are taken as arbitrary positive reals when one is looking for positive reals $K_1,\ldots,K_q$. After that, the algorithm finds uniquely $K_1,\ldots,K_q$ from the values of the monomials $\prod_{1\leq\alpha\leq q} K_\alpha^{c''_{\mu,\alpha}}$, $1 \leq \mu \leq q$ with the help of the matrix $(C'')^{-1}$.

Thus, the described algorithm applies twice the subroutine for constructing the Smith canonical form (and does not need to involve algorithms for solving systems of polynomial equations). Observe that solvability of (2) for non-zero $x_1,\ldots,x_n$, $k_1,\ldots,k_q$ is equivalent to solvability of (4). Each $\epsilon_j = \prod_{1\leq\alpha\leq l} \beta_\alpha^{\lambda_{j,\alpha}}$, $s+1 \leq j \leq l$ for appropriate integers $\lambda_{j,\alpha} \in \mathbb{Z}$ such that

$$|\lambda_{j,\alpha}| \leq (d \cdot \min\{l,n\})^{O(\min\{l,n\})} \cdot (d \cdot \min\{l,q\})^{O(\min\{l,q\})}$$

assuming that all the exponents in (2) satisfy inequalities $|g_{j,i}|, |h_{j,l}| \leq d$ (due to [16, 17]).

To verify (4) the algorithm constructs a *relative factorization* of $\beta_1,\ldots,\beta_l$ (for the sake of simplifying notations assume that all $\beta_1,\ldots,\beta_l$ are positive integers; for rational numbers one has to consider the absolute values of their numerators and denominators). Namely, the algorithm constructs by recursion nonnegative integers $\eta_1,\ldots,\eta_r$ pairwise relatively prime such that $\beta_\mu = \eta_1^{\kappa_{\mu,1}} \cdots \eta_r^{\kappa_{\mu,r}}$, $1 \leq \mu \leq l$ for suitable nonnegative integers $\kappa_{\mu,i}$. As a base of recursion the algorithm starts with $\beta_1,\ldots,\beta_l$. Assume that at some step the algorithm has constructed $\beta'_1,\ldots,\beta'_{l'}$ such that $(\beta'_1 \cdots \beta'_{l'})|(\beta_1 \cdots \beta_l)$. Take any pair β'_i, β'_j, $1 \leq i \neq j \leq l'$ for which $\theta := \mathrm{GCD}(\beta'_i,\beta'_j) \neq 1$ and replace the pair β'_i, β'_j by the triple $\theta, \beta'_i/\theta, \beta'_j/\theta$. If there is no such a pair the algorithm halts.

The product of the modified $(l'+1)$-tuple is a strict divisor of the product $(\beta'_1 \cdots \beta'_{l'})$ at the previous step of the algorithm. Hence after at most of $\log_2(\beta_1 \cdots \beta_l) \leq l \cdot \log_2 M$ steps the algorithm constructs the relative factorization $\eta_1,\ldots,\eta_r$. One can easily show that the latter is unique, although we don't make use of its uniqueness. The complexity of constructing $\eta_1, \ldots, \eta_r$ is bounded by a polynomial in l, $\log M$. In particular, $\sum_{1 \leq \mu \leq l, 1 \leq i \leq r} \kappa_{\mu,i}$ is also bounded by a polynomial in l, $\log M$.

Now the algorithm is able to verify equalities (4) representing each $\epsilon_j = \prod_{1 \leq i \leq r} \eta_i^{\nu_{j,i}}$, $s+t+1 \leq j \leq l$ as a product of powers of $\eta_1, \ldots, \eta_r$ for appropriate integers $\nu_{j,i}$ (perhaps, nonnegative). Then $\epsilon_j = 1$ iff $\nu_{j,i} = 0$, $1 \leq i \leq r$. The complexity of computing all $\nu_{j,i}$, $s + t + 1 \leq j \leq l$, $1 \leq i \leq r$ does not exceed a polynomial in n, l, q, $\log(d \cdot M)$. Finally, we can summarize the results obtained in this section.

Proposition 2. *One can solve a parametric binomial system (2) with non-zero values of both parameters $k_1, \ldots, k_q$ and variables $x_1, \ldots, x_n$ within polynomial complexity, i.e., within a polynomial in the size n, l, q, $\log(d \cdot M)$ of the input. Within the same complexity bound one can find positive real solutions of (2).*

Remark 2. In the proof of [19, Theorem 4.1] a similar application of the Smith normal form is used for the special case of binomial systems arising for so called "deficiency zero systems" of chemical reaction networks (see [2] for a definition or Sect. 4 below; please notice that [2] as the final journal version of [19] unfortunately no longer contains the cited algorithmic application of the Smith normal form). However, for general parametric binomial systems our algorithm applying twice the subroutine for constructing the Smith canonical form seems to be new—in addition to providing a complexity analysis.

4   Applications to Chemical Reaction Networks

There is a vast literature for chemical reaction networks with mass action kinetics. We refer to [2] and the cited literature therein for definitions relevant in our context.

In these systems the matrix A in (1) can be factored as

$$A = \tilde{Y} \cdot I_a \cdot I_k, \tag{5}$$

where $I_a = (i_{k,j})$, $1 \leq k \leq h$, $1 \leq j \leq m$ is an integer matrix with entries $0, 1, -1$, $\tilde{Y}$ is an $l \times h$-integer matrix with non-negative entries, and I_k is a matrix $k_{u,v}$ of reaction rates, which in general are seen as parameters for the system. The occurring dimensions can be interpreted as follows: n is the number of participating molecular species, l is the number of reactions, and m is the number of complexes.

Following [2] the *deficiency* of a chemical reaction network with an associated polynomial system of the form

$$\tilde{Y} \cdot I_a \cdot I_k \cdot Y$$

can be defined as

$$\mathrm{rk}\, I_a - \mathrm{rk}\, \tilde{Y} \cdot I_a.$$

Hence it is a non-negative integer.

4.1   Chemical Reaction Networks with Toric Steady States

Remarkably, many chemical reaction systems have the property that the steady state ideal of the corresponding polynomial system is a binomial ideal [18]. Using the terminology of [18] these systems are ones having toric steady states.

For a given chemical reaction network the property of having toric steady states is dependent on the parameters in general. A simple instance is given in [18, Example 2.3].

For chemical reaction networks with toric steady states for all admissible parameters Péres Millán et al. [18] establish criteria for the existence of positive equilibria, and also for so called multi-stationarity, which are basically linear algebra criteria.

However, in cases for which multi-stationarity is established, the criteria in [18] give no detailed information about the structure of the equilibria of the system, whereas our algorithm computes in polynomial time all equilibria hence allowing a detailed analysis of them.

On the other hand, in the algorithm presented in Sect. 3 we presume that already *the input system* is in the form of a parametric binomial system, whereas in [18] it is not necessary that the input system is of this form, but the main result in [18] gives sufficient conditions for a chemical reaction system to have toric steady states.

For a given chemical reaction network, which potentially has toric steady states, there are several possibilities to come up with a parametric binomial system that in turn can be solved by the algorithm presented in Sect. 3:

- Use the construction for a binomial system given in [6]. As this construction uses an enumeration of spanning trees of underlying graphs, its worst time complexity is exponential.
- As a sufficient condition one can check [18, Condition 3.1]. Then [18, Theorem 3.3] gives an easy construction for a binomial system generating the steady state ideal. A check of [18, Condition 3.1] for a given basis of $\ker A$ is easily doable. However, enumerating all possible bases of $\ker A$ yields exponential complexity. So one has to hope for suitable heuristics to come up with good test candidates among all bases of $\ker A$.
- Compute Gröbner bases (any monomial ordering would be sufficient). As already mentioned the worst time complexity is doubly-exponential, but the practical complexity could be much better for many relevant examples (as also there is freedom to use a suitable monomial ordering).

Although for all of these constructions the worst case complexity is (at least) exponential, it might nevertheless be interesting to explore their behavior for actual chemical reaction systems.

Moreover, the factorization $A = \tilde{Y} \cdot I_a \cdot I_k$—or other factorizations of the matrix A—might yield much simpler problems. For instance, for deficiency zero systems the fact that $\operatorname{rk} I_a = \operatorname{rk} \tilde{Y} \cdot I_a$ implies that only $I_a \cdot I_k$ has to be considered instead of $\tilde{Y} \cdot I_a \cdot I_k$.

Although the worst-case complexity of these methods all are worse than the one of the general algorithm given in Sect. 3, one can employ all of them and the latter algorithm using simple *coarse grained competitive parallelism*, which can be realized in many software infrastructure—e.g. the one already used a decade ago and described in [20].

Remark 3. Of course solving systems gives significantly more information than counting the number of solutions only. Also other forms of solution testing can be applied. One of these criteria is whether the projection onto one coordinate of all positive steady states of a system is unique. This property directly corresponds to the "absolute concentration robustness" [21], for which a special criterion for systems having deficiency of 1 is proven in [22].

4.2   Examples from the BioModels Database

We use examples stored in the BioModels database [23] in the following to discuss the practical relevance of the assumptions made above. Of course, for other example classes the situation might be different.

For most of the examples for which r being significantly smaller than n, the deficiency of the network is 0. Hence the Deficiency-zero-theorem already gives significant information about the uniqueness [2, 24] of equilibria for these cases and also the algorithm given in the proof of [19, Theorem 4.1] could compute these unique equilibria for a fixed set of parameter, i.e., a unique solution to the polynomial system. However, we are not aware of an implementation of this method, and our algorithm given in Sect. 3 is as efficient as the more restricted method of [19, Theorem 4.1].

However, there are also several example of networks having deficiency 1 or even higher deficiencies – for which no such general theorems are known, for which r being significantly smaller than n.

Example BIOMOD188. As an example we consider the model with the number 188 in the BioModels database, which was originally described in [25]. The network induces a stoichiometric matrix of size 20×20 and has a deficiency of 4. The dimension of the nullspace of the stoichiometric matrix A is 6. The rank of the exponent matrix Y is 11.

The polynomial system is as follows (with two instances of the 0-polynomial due to the automated construction of the system from the SBML description):

$$k_2 \cdot x_4 - k_6 \cdot x_1 + k_8 \cdot x_3 - k_9 \cdot x_2 \cdot x_1 + k_{10} \cdot x_3 - k_{16} \cdot x_1 \cdot x_6 + k_{17} \cdot x_9,$$
$$k_7 \cdot x_5 - k_9 \cdot x_2 \cdot x_1 + k_{10} \cdot x_3 - k_{14} \cdot x_2 \cdot x_6 + k_{15} \cdot x_8,$$
$$-k_8 \cdot x_3 + k_9 \cdot x_2 \cdot x_1 - k_{10} \cdot x_3, k_3 \cdot x_2 + k_4 \cdot x_8 - k_5 \cdot x_4, k_0 \cdot x_{12} - k_1 \cdot x_5,$$
$$k_{13} \cdot x_{10} \cdot x_7 - k_{19} \cdot x_6, -k_{13} \cdot x_{10} \cdot x_7 + k_{19} \cdot x_6,$$
$$k_{14} \cdot x_2 \cdot x_6 - k_{15} \cdot x_8, k_{16} \cdot x_1 \cdot x_6 - k_{17} \cdot x_9 - k_{18} \cdot x_9,$$
$$k_{11} - k_{12} \cdot x_{10}, k_1 \cdot x_5 + k_5 \cdot x_4 + k_6 \cdot x_1 + k_{12} \cdot x_{10} + k_{18} \cdot x_9,$$
$$-k_0 \cdot x_{12}, k_8 \cdot x_3, k_7 \cdot x_5, k_6 \cdot x_1 + k_{18} \cdot x_9, k_2 \cdot x_4, k_5 \cdot x_4, k_3 \cdot x_2 + k_4 \cdot x_8, 0, 0$$

Notice that the polynomial system no longer contains all of the variables in the set $\{x_1, x_2, \ldots, x_{20}\}$, but only the subset of cardinality 11 consisting of $\{x_1, x_2, x_3, x_4, x_5, x_6, x_7, x_8, x_9, x_{10}, x_{12}\}$.

Now if we consider the factorization of the polynomial system into

$$\tilde{Y} \cdot I_a \cdot I_k \cdot Y$$

with a diagonal matrix I_k one can consider the groupings according to the law of associativity:

$$(\tilde{Y} \cdot I_a \cdot I_k) \cdot Y \tag{6}$$
$$(\tilde{Y} \cdot I_a) \cdot (I_k \cdot Y). \tag{7}$$

If we consider the view expressed in (6) one has to perform the linear algebra over the field $\mathbb{Q}(k_1, \ldots, k_m)$ but having only the $x_1, \ldots, x_n$ as variables in subsequent steps. When considering the $k_1, \ldots, k_m$ as part of the monomials, i.e., taking the view expressed in (7) the linear algebra is over $\mathbb{Q}$ but the $k_1, \ldots, k_m$ have to be counted as variables in addition to the $x_1, \ldots, x_n$. Notice that the latter view is used by Clarke [26].

Using both approaches we find that some of the F_i ($1 \leq i \leq 20$) are zero, and hence the system does not have a solution with all non-zero entries.

When inspecting more closely which entries are zero, we find that—taking $k_1, \ldots, k_{20}$ as parameters— the $Y_3, Y_4, Y_5, Y_{10}, Y_{11}, Y_{12}, Y_{13}, Y_{16}$ are zero when expressed as linear combinations of the Z_j. When viewing the reaction constants as part of the monomials we obtain that $Y_3, Y_4, Y_5, Y_{10}, Y_{11}, Y_{13}, Y_{19}$ are zero. When resolving these conditions in terms of the x_i (and k_i) we obtain in the k-as-parameter-case the logical condition

$$x_3 = 0 \wedge x_4 = 0 \wedge x_5 = 0 \wedge x_{10} = 0 \wedge x_{12} = 0 \wedge 1 = 0 \tag{8}$$

and in the k-in-monomial-case

$$k_8 \cdot x_3 = 0 \wedge k_5 \cdot x_4 = 0 \wedge k_1 \cdot x_5 = 0 \wedge k_{12} \cdot x_{10} = 0 \wedge k_0 \cdot x_{12} = 0 \wedge$$
$$k_7 \cdot x_5 = 0 \wedge k_2 \cdot x_4 = 0 \wedge k_{11} = 0. \tag{9}$$

Hence there are no solutions unless $k_{11} = 0$, which is a condition leading to the inconsistency $1 = 0$ in (8)—without any further information in the k-as-parameter-case. When going back to the original description using the software infrastructure described in [27] we obtain the information that associated with constant k_{11} there is a creation of $\longrightarrow$ damDNA from "the environment"; moreover, associated with k_{12} there is a reaction denoted damDNA $\longrightarrow$ Sink. So there are some quasi-steady states involved in the SBML representation of the reaction system. Dealing rigorously with quasi-steady state approximations is an important line of research in algebraic biology (see e.g. [28]).

When considering solutions of polynomial systems we can apply the substitutions $k_{11} = 0, x_3 = 0, x_4 = 0, x_5 = 0, x_{10} = 0, x_{12} = 0$ and consider the resulting system.

Example BIOMOD053. As another example we take the model #53 from the BIOMOD database.

The chemical reaction network involves 6 species and has deficiency 2. The resulting polynomial system is as follows:

$$-k_1x_1x_2 + k_2x_3 - k_5x_1, -k_1x_1x_2 + k_2x_3 - k_7x_2x_5 + k_{10},$$
$$k_1x_1x_2 - k_2x_3 - k_3x_3 + k_4x_4 - k_9x_3 - k_{12}x_3, k_3x_3 - k_4x_4 - k_6x_4,$$
$$k_5x_1 - k_7x_2x_5 - k_8x_5 + k_{11}, k_6x_4 + k_7x_2x_5 + k_9x_3.$$

Our simple prototype implementation of our algorithms using the computational infrastructure of Maple can easily determine that the system has a solution with all-non-zero entries. Our algorithm can come up with a explicit representation of the solution after some minutes of computation time. The string representation of the output is big (about 1 MB). However, the big output size is mainly due to rather lengthy polynomial expressions in the parameters occurring in the solutions. The structure of the solutions in the symbols representing a "can be chosen arbitrarily" in the methods presented above (cf. Sect. 2) is much simpler.

5   Conclusion and Future Work

Although several related ideas have been around in the literature on algebraic methods for chemical reaction systems the full algorithmic development given above seems to be new—in addition to providing the complexity analysis.

In contrast to the known theorems developed in the context of chemical reaction network theory—which only work in special cases but give results entirely independent of the parameters—our algorithms are universally applicable.

It will be the topic of future research to systematically apply careful implementations of the algorithms given in this paper to the networks given in databases such as BioModels database and others. For this purpose we will integrate the implementation of the algorithms described in this paper into the general infrastructure described in [27]. By these tests, we will not only explore the practical limits of the methods but we also might get insight into the question whether some of the properties that hold for deficiency-zero and deficiency-one systems (such as unique positive steady states for a chemical compatibility class, or the absolute concentration robustness property for a certain subclass) also hold for systems of deficiency bigger than one—at least parametrically for relevant ranges of parameters.

Acknowledgements. The first author is grateful to the Max-Planck Institut für Mathematik, Bonn for its hospitality during writing this paper. We are grateful to M. Eiswirth for useful discussions initiating this research. The help of Satya Swarup Samal and Hassan Errami in generating the algebraic input for our computational examples out of the SBML descriptions is greatly appreciated by us. Also the authors are grateful to the anonymous referees for valuable remarks.

References

1. Bihan, F., Sottile, F.: Fewnomial bounds for completely mixed polynomial systems. Advances in Geometry 11(3), 541–556 (2011)
2. Gatermann, K., Huber, B.: A family of sparse polynomial systems arising in chemical reaction systems. Journal of Symbolic Computation 33(3), 275–305 (2002)
3. Gatermann, K., Wolfrum, M.: Bernstein's second theorem and Viro's method for sparse polynomial systems in chemistry. Advances in Applied Mathematics 34(2), 252–294 (2005)
4. Gatermann, K., Eiswirth, M., Sensse, A.: Toric ideals and graph theory to analyze Hopf bifurcations in mass action systems. Journal of Symbolic Computation 40(6), 1361–1382 (2005)
5. Domijan, M., Kirkilionis, M.: Bistability and oscillations in chemical reaction networks. Journal of Mathematical Biology 59(4), 467–501 (2009)
6. Craciun, G., Dickenstein, A., Shiu, A., Sturmfels, B.: Toric dynamical systems. Journal of Symbolic Computation 44(11), 1551–1565 (2009)
7. Sturm, T., Weber, A., Abdel-Rahman, E.O., El Kahoui, M.: Investigating algebraic and logical algorithms to solve Hopf bifurcation problems in algebraic biology. Mathematics in Computer Science 2(3), 493–515 (2009)
8. Chistov, A.L.: Algorithm of polynomial complexity for factoring polynomials and finding the components of varieties in subexponential time. Journal of Soviet Mathematics 34(4), 1838–1882 (1986)
9. Grigoriev, D.: Factorization of polynomials over a finite field and the solution of systems of algebraic equations. Journal of Soviet Mathematics 34(4), 1762–1803 (1986)
10. Grigoriev, D., Vorobjov, N.N.: Solving systems of polynomial inequalities in subexponential time. Journal of Symbolic Computation 5(1-2), 37–64 (1988)
11. Sturmfels, B.: Grobner bases and convex polytopes. University Lecture Series, vol. 8. American Mathematical Society, Providence (1996)
12. Gerdt, V.: Involutive methods applied to algebraic and differential systems. Constructive algebra and systems theory. Verh. Afd. Natuurkd. 1. Reeks. K. Ned. Akad. Wet., R. Neth. Acad. Arts Sci. 53, 245–250 (2006)
13. Mayr, E., Meyer, A.: The complexity of the word problems for commutative semigroups and polynomial ideals. Adv. in Math 46, 305–329 (1982)
14. Cattani, E., Dickenstein, A.: Counting solutions to binomial complete intersections. Journal of Complexity, 1–25 (2007)
15. Grigoriev, D., Pasechnik, D.V.: Polynomial-time computing over quadratic maps. I. sampling in real algebraic sets. Comput. Complexity 14, 20–52 (2005)
16. Frumkin, M.: An application of modular arithmetic to the construction of algorithms for solving systems of linear equations. Soviet Math. Dokl 229, 1067–1070 (1976)
17. Dumas, J.G., Saunders, B.D.: On efficient sparse integer matrix Smith normal form computations. Journal of Symbolic Computation 32, 71–99 (2001)
18. Pérez Millán, M., Dickenstein, A., Shiu, A., Conradi, C.: Chemical reaction systems with toric steady states. Bulletin of Mathematical Biology, 1–29 (October 2011)
19. Gatermann, K., Huber, B.: A family of sparse polynomial systems arising in chemical reaction systems. Technical Report Preprint SC 99-27, Konrad-Zuse-Zentrum für Informationstechnik Berlin (1999)
20. El Kahoui, M., Weber, A.: Deciding Hopf bifurcations by quantifier elimination in a software-component architecture. Journal of Symbolic Computation 30(2), 161–179 (2000)

21. Shinar, G., Feinberg, M.: Structural sources of robustness in biochemical reaction networks. Science 327(5971), 1389–1391 (2010)
22. Shinar, G., Feinberg, M.: Structural sources of robustness in biochemical reaction networks. supporting online material. Science 327(5971), 1389–1391 (2010)
23. Li, C., Donizelli, M., Rodriguez, N., Dharuri, H., Endler, L., Chelliah, V., Li, L., He, E., Henry, A., Stefan, M.I., Snoep, J.L., Hucka, M., Le Novère, N., Laibe, C.: BioModels database: An enhanced, curated and annotated resource for published quantitative kinetic models. BMC Systems Biology 4, 92 (2010)
24. Feinberg, M.: Stability of complex isothermal reactors–I. the deficiency zero and deficiency one theorems. Chemical Engineering Science 42(10), 2229–2268 (1987)
25. Proctor, C.: Explaining oscillations and variability in the p53-Mdm2 system. BMC Systems Biology 2, 75 (2008)
26. Clarke, B.L.: Complete set of steady states for the general stoichiometric dynamical system. The Journal of Chemical Physics 75(10), 4970–4979 (1981)
27. Samal, S.S., Errami, H., Weber, A.: A Software Infrastructure to Explore Algebraic Methods for Bio-Chemical Reaction Networks. In: Gerdt, V.P., et al. (eds.) CASC 2012, LNCS, vol. 7442, pp. 294–307. Springer, Heidelberg (2012)
28. Boulier, F., Lefranc, M., Lemaire, F., Morant, P.E.: Applying a Rigorous Quasi-Steady State Approximation Method for Proving the Absence of Oscillations in Models of Genetic Circuits. In: Horimoto, K., Regensburger, G., Rosenkranz, M., Yoshida, H. (eds.) AB 2008. LNCS, vol. 5147, pp. 56–64. Springer, Heidelberg (2008)

Symbolic-Numerical Calculations of High-$|m|$ Rydberg States and Decay Rates in Strong Magnetic Fields[*]

Alexander Gusev, Sergue Vinitsky, Ochbadrakh Chuluunbaatar,
Vladimir Gerdt, Luong Le Hai, and Vitaly Rostovtsev [**]

Joint Institute for Nuclear Research, Dubna, Moscow Region, Russia
gooseff@jinr.ru, vinitsky@theor.jinr.ru

Abstract. Symbolic-numeric solving of the boundary value problem for the Schrödinger equation in cylindrical coordinates is given. This problem describes the impurity states of a quantum wire or a hydrogen-like atom in a strong homogeneous magnetic field. It is solved by applying the Kantorovich method that reduces the problem to the boundary-value problem for a set of ordinary differential equations with respect to the longitudinal variables. The effective potentials of these equations are given by integrals over the transverse variable. The integrands are products of the transverse basis functions depending on the longitudinal variable as a parameter and their first derivatives. To solve the problem at high magnetic quantum numbers $|m|$ and study its solutions we present an algorithm implemented in Maple that allows to obtain *analytic expressions* for the effective potentials and for the transverse dipole moment matrix elements. The efficiency and accuracy of the derived algorithm and that of Kantorovich numerical scheme are confirmed by calculating eigenenergies and eigenfunctions, dipole moments and decay rates of low-excited Rydberg states at high $|m| \sim 200$ of a hydrogen atom in the laboratory homogeneous magnetic field $\gamma \sim 2.35 \times 10^{-5} (B \sim 6T)$.

1 Introduction

In earlier papers, we considered the application of the Kantorovich method for solving the discrete- and continuous-spectrum boundary-value problems (BVP) [1] for hydrogen-like atoms in magnetic field and the ion axial channelling problem in a crystal. The approach implies the use of a parametric basis of oblate spheroidal angular functions in spherical coordinates where the radial variable runs a semi-axis [2,3,4,5]. The method has been further developed in connection with calculations of spectral and optical characteristics of model semiconductor nanostructures, namely, quantum dots(QD), quantum wells(QW) and quantum

---

[*] This work was partially supported by the RFBR Grants Nos. 10-02-00200 and 11-01-00523.
[**] The coauthors (AG, SV, OC, VG, and LH) congratulate Vitaly Rostovtsev on turning 20 for the fourth time.

wires(QWr) [6,7,8,9]. For this purpose we used different parametric basis functions in appropriate coordinate systems. The functions were calculated by solving parametric eigenvalue problems by means of the program ODPEVP [10].

Taking into account the growing interest in problems possessing axial symmetry, like impurity states of QWr's or high-angular-momentum Rydberg states and quasi-stationary states imbedded in continuum of a hydrogen atom in magneto-optical traps [11,12,13], it is imperative to implement the Kantorovich scheme for solving the BVP for the longitudinal variable running the whole axis of a cylindrical coordinate system[8,9]. This would allow direct calculation of the main characteristics of a multichannel scattering problem, such as reflection and transmission coefficients matrices, recombination rates and ionization cross-sections for Rydberg states, and decay rates of the lowest bound states of manifolds with high values of the magnetic quantum number $|m|$ [11,12,13].

For the Schrödinger equation describing a hydrogen-like atom in a strong homogeneous magnetic field, the boundary-value problem (BVP) in cylindrical coordinates is reduced to solving a set of the longitudinal equations in the framework of the Kantorovich method. The effective potentials of these equations are given by integrals over the transverse variable, the integrands being products of transverse basis functions, depending on the longitudinal variable as a parameter, and their first derivatives with respect to the parameter. One can say that at high $|m|$, the discrete-spectrum problem is described by a system of two coupled 2D- and 1D-oscillators corresponding to the transverse ρ and longitudinal z variables, with the frequencies ω_ρ and ω_z, respectively. To analyze the low-excited Rydberg states of such system it is useful to have the solution *in an analytic form*. Indeed, for high $|m|$ we can consider the Coulomb potential as a perturbation with respect to the transversal centrifugal potential and the oscillator potential with the frequency $\omega_\rho = \gamma/2$. For the laboratory magnetic field $B = B_0 \gamma \sim 6T$, i.e., $\gamma \sim 2.35 \times 10^{-5}$, this is true at the *adiabatic parameter* values $\tilde{m} \sim 5.89$, where $\tilde{m}$ is defined as $\tilde{m} = (\omega_\rho/\omega_z)^{4/3} = |m|\gamma^{1/3}$. Under the condition $|m| \geq 6\gamma^{-1/3}$ we can approximate the Coulomb potential by a Taylor expansion in powers of the auxiliary transverse variable with respect to a specially chosen point with given accuracy in the region of its convergence. Then we can find the approximate transversal eigenvalues and eigenfunctions depending parametrically on the longitudinal variable, in the framework of a perturbation scheme and by using the eigenvalues and eigenfunctions of the 2D oscillator as unperturbed ones. To express *analytically* the transverse basis functions and eigenvalues, the corresponding effective potentials, and the transverse dipole moment matrix elements as well as perturbation solution of the BVP, we elaborate a symbolic-numerical algorithm (SNA) implemented in Maple. The efficiency and accuracy of the algorithm and that of the derived Kantorovich numerical scheme are confirmed by computation of eigenenergies and eigenfunctions, dipole moments and decay rates for the manifolds of high-$|m|$ low-excited Rydberg states of a hydrogen atom in the laboratory homogeneous magnetic field, and by comparison with the results obtained by other methods.

The paper is organized as follows. In Section 2, we briefly describe the reduction by the KM of the 3D eigenvalue problem at fixed values $|m|$ of magnetic quantum number to the 1D eigenvalue problem for a set of close-coupled longitudinal equations. In Sections 3 and 4, the algorithm for calculating the effective potentials and the transverse dipole moment matrix elements *in the analytic form* at large values of $|m|$ is presented. The algorithm has been implemented in Maple. To find the validity range of the method, in Section 5 we compare our results with the known ones obtained in the cylindrical coordinates. Decay rates of the lowest bound states of manifolds with high magnetic quantum number $|m|$ are also presented here. In Section 6, we conclude and discuss possible future applications of the described method.

2   Problem Statement in Cylindrical Coordinates

The component $\Psi(\rho, z)$ of the wave function $\Psi(\rho, z, \varphi) = \Psi(\rho, z) \exp(\imath m\varphi)/\sqrt{2\pi}$ of a hydrogen atom in an axially symmetric magnetic field $\boldsymbol{B} = (0, 0, B)$ in the cylindrical coordinates (ρ, z, φ) satisfies the 2D Schrödinger equation in the region $\Omega_c = \{0 < \rho < \infty \text{ and } -\infty < z < \infty\}$:

$$-\frac{\partial^2}{\partial z^2}\Psi(\rho, z) + A_c\Psi(\rho, z) = \epsilon\Psi(\rho, z), \quad A_c = -\frac{1}{\rho}\frac{\partial}{\partial \rho}\rho\frac{\partial}{\partial \rho} + m\gamma + U(\rho, z), (1)$$

$$U(\rho, z) = \frac{m^2}{\rho^2} + \frac{\gamma^2\rho^2}{4} + V_c(\rho, z), \quad V_c(\rho, z) = -\frac{2q}{\sqrt{\rho^2 + z^2}}. \tag{2}$$

Here $m = 0, \pm 1, \ldots$ is the magnetic quantum number, $\gamma = B/B_0 = \hbar\omega_c/(2Ry)$, $B_0 \cong 2.35 \times 10^5\,T$ is a dimensionless parameter which determines the field strength B, $\omega_c = eB/(m_e c) = eB_0\gamma/(m_e c)$ is the cyclotron frequency, and $U(\rho, z)$ is the potential energy (see Fig. 1a), q is Coulomb charge of nucleus. We use the atomic units $(a.u.)$ $\hbar = m_e = e = 1$ and assume the mass of the nucleus to be infinite. In these expressions, $\epsilon = 2E$, E is the energy (expressed in Rydbergs, $1\,Ry = (1/2)\,a.u.$) of the bound state $|m\sigma\rangle$ with fixed values of m and z-parity $\sigma = \pm 1$, and $\Psi(\rho, z) \equiv \Psi^{m\sigma}(\rho, z) = \sigma\Psi^{m\sigma}(\rho, -z)$ is the corresponding wave function. The boundary conditions in each $m\sigma$ subspace $L_2(\Omega)$ of the complete Hilbert space have the form

$$\lim_{\rho \to 0} \rho\frac{\partial\Psi(\rho, z)}{\partial\rho} = 0, \quad \text{for} \quad m = 0, \quad \text{and} \quad \Psi(0, z) = 0, \quad \text{for} \quad m \neq 0, \tag{3}$$

$$\lim_{\rho \to \infty} \Psi(\rho, z) = 0. \tag{4}$$

The eigenfunction $\Psi(\rho, z) \equiv \Psi_t(\rho, z) \in L_2(\Omega)$ of the discrete real-valued spectrum $\epsilon : \epsilon_1 < \epsilon_2 < \cdots \epsilon_t < \cdots < \gamma$ obeys the asymptotic boundary condition. Approximately this condition is replaced by the boundary condition of the second and/or first type at small and large $|z|$, but finite $|z| = z_{\max} \gg 1$,

$$\lim_{z \to 0} \frac{\partial\Psi(\rho, z)}{\partial z} = 0, \quad \sigma = +1, \quad \Psi(\rho, 0) = 0, \quad \sigma = -1, \tag{5}$$

$$\lim_{z \to \pm\infty} \Psi(\rho, z) = 0 \quad \to \quad \Psi(\rho, \pm|z_{\max}|) = 0. \tag{6}$$

In numerical calculation of the eigenvalues and eigenfunctions with given accuracy by programs KANTBP2 and ODPEVP realizing the finite element method, we used computational schemes derived from the Rayleigh–Ritz variational functional [1,10]

$$\mathcal{R}(\Psi_t, \epsilon_t) = \left(\int\limits_{-z_{\max}}^{z_{\max}} dz \int\limits_0^\infty \rho d\rho \frac{\partial \Psi_t(\rho, z)}{\partial z} \frac{\partial \Psi_t(\rho, z)}{\partial z} + \frac{\partial \Psi_t(\rho, z)}{\partial \rho} \frac{\partial \Psi_t(\rho, z)}{\partial \rho} \right. \tag{7}$$

$$\left. + \Psi_t(\rho, z)(m\gamma + U(\rho, z))\Psi_t(\rho, z) \right) / \int_{-z_{\max}}^{z_{\max}} dz \int_0^\infty \rho d\rho \Psi_t(\rho, z)\Psi_{t'}(\rho, z)$$

with the additional normalization and orthogonality conditions

$$\langle t|t' \rangle = \int_{-z_{\max}}^{z_{\max}} dz \int_0^\infty \rho d\rho \Psi_t(\rho, z)\Psi_{t'}(\rho, z) = 2 \int_0^{z_{\max}} dz \int_0^\infty \rho d\rho \Psi_t(\rho, z)\Psi_{t'}(\rho, z) = \delta_{tt'}. \tag{8}$$

For $m \neq 0$ eigenfunctions $\Psi_t(\rho, z) \sim \rho^{|m|/2}$ at small ρ. So, in numerical calculations, a reduced interval $[0 < \rho_{\min}, \rho_{\max} \gg 1]$ is conventionally used [8].

2.1   Kantorovich Reduction

Consider a formal expansion of the partial solution $\Psi_t^{m\sigma}(\rho, z)$ of Eqs. (1)–(4) corresponding to the eigenstate $|m\sigma t\rangle$ expanded in the finite set of one-dimensional basis functions $\{B_j^m(\rho; z)\}_{j=1}^{j_{\max}}$

$$\Psi_t^{m\sigma}(\rho, z) = \sum_{j=1}^{j_{\max}} B_j^m(\rho; z)\chi_j^{(m\sigma t)}(z). \tag{9}$$

In Eq. (9), the functions $\chi^{(t)}(z) \equiv \chi^{(m\sigma t)}(z)$, $(\chi^{(t)}(z))^T = (\chi_1^{(t)}(z),\dots,\chi_{j_{\max}}^{(t)}(z))$ are unknown, and the surface functions $B(\rho; z) = B^m(\rho; -z)$, $(B(\rho; z))^T = (B_1(\rho; z),\dots,B_{j_{\max}}(\rho; z))$ form an orthonormal basis for each value of the variable $z \in \mathcal{R}$ which is treated as a parameter.

In KM, the wave functions $B_j(\rho; z)$ (see Fig. 2) and the potential curves $E_j(z)$ (in Ry) are determined as solutions of the following eigenvalue problem

$$A_c B_j(\rho; z) = E_j(z)B_j(\rho; z), \tag{10}$$

with the operator A_c from (1)–(2) and the boundary conditions (3), (4) at each fixed $z \in \mathcal{R}$. Since the operator in the left-hand side of Eq. (10) is self-adjoint, its eigenfunctions are orthonormal

$$\left\langle B_i(\rho; z) \middle| B_j(\rho; z) \right\rangle_\rho = \int_0^\infty B_i(\rho; z)B_j(\rho; z)\rho d\rho = \delta_{ij}, \tag{11}$$

where δ_{ij} is the Kronecker symbol. Therefore, we transform the solution of the above problem into the solution of an eigenvalue problem for a set of $j_{\max}$

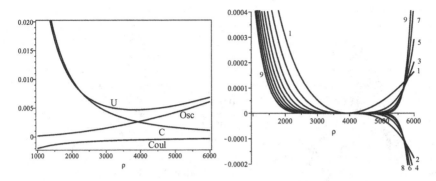

Fig. 1. Left panel: the profile of potential energy $U(\rho, z) = m^2/\rho^2 + \gamma^2\rho^2/4 + V_c(\rho, z)$ (U) in the plane $z = 0$ and its components, namely, the centrifugal (C), oscillator (Osc), and Coulomb (Coul) potentials. Right panel: the approximation errors $\delta U^{(j_{\max})}(\rho, z) \equiv \sum_{i=1}^{j_{\max}} U^{(i)}(\rho, z) - U(\rho, z)$ $(j_{\max} = 1, ..., 9)$ of the potential energy $U(\rho, z = 0)$. Here $q = -1$, $m = -200$, $\gamma = 2.553191 \cdot 10^{-5}$ $(B = 6\mathrm{T}, \tilde{m} \approx 5.89)$

ordinary second-order differential equations that determines the energy ϵ and the coefficients $\chi^{(i)}(z)$ of the expansion (9)

$$\left(-\mathbf{I}\frac{d^2}{dz^2} + \mathbf{U}(z) + \mathbf{Q}(z)\frac{d}{dz} + \frac{d\mathbf{Q}(z)}{dz}\right)\chi^{(t)}(z) = \epsilon_t \mathbf{I}\chi^{(t)}(z). \tag{12}$$

Here $\mathbf{I}$, $\mathbf{U}(z) = \mathbf{U}(-z)$, and $\mathbf{Q}(z) = -\mathbf{Q}(-z)$ are the $j_{\max} \times j_{\max}$ matrices whose elements are expressed as

$$U_{ij}(z) = E_i(z)\delta_{ij} + H_{ij}(z), \quad H_{ij}(z) = \int_0^\infty \frac{\partial B_i(\rho; z)}{\partial z}\frac{\partial B_j(\rho; z)}{\partial z}\rho d\rho, \tag{13}$$

$$I_{ij}(z) = \delta_{ij}, \quad Q_{ij}(z) = -Q_{ji}(z) = -\int_0^\infty B_i(\rho; z)\frac{\partial B_j(\rho; z)}{\partial z}\rho d\rho.$$

The discrete spectrum solutions $\epsilon : \epsilon_1 < \epsilon_2 < \cdots \epsilon_t < \cdots < \gamma$ at fixed m and parity $\sigma = \pm 1$ obey the asymptotic boundary condition and are orthonormal

$$\lim_{z \to 0}\left(\frac{d}{dz} - \mathbf{Q}(z)\right)\chi^{(t)}(z) = 0, \quad \sigma = +1, \quad \chi^{(t)}(0) = 0, \quad \sigma = -1, \tag{14}$$

$$\lim_{z \to \pm\infty}\chi^{(t)}(z) = 0 \quad \to \quad \chi^{(t)}(\pm z_{\max}) = 0, \tag{15}$$

$$\int_{-z_{\max}}^{z_{\max}}\left(\chi^{(t)}(z)\right)^T\chi^{(t')}(z)dz = 2\int_0^{z_{\max}}\left(\chi^{(t)}(z)\right)^T\chi^{(t')}(z)dz = \delta_{tt'}. \tag{16}$$

Remark 1. In diagonal adiabatic approximation

$$\left(-\frac{d^2}{dz^2} + U_{jj}(z)\right)\chi_j^{(v)}(z) = \epsilon_{jv}\chi_j^{(v)}(z) \tag{17}$$

discrete spectrum $\epsilon : \epsilon_{j1} < \epsilon_{j2} < \cdots \epsilon_{jv} < \cdots < \gamma$ numerated by number v that determines the number $v - 1$ of nodes of the solution $\chi_j^{(v)}(z)$ at fixed value j.

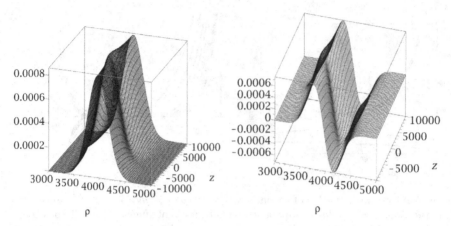

Fig. 2. The basis functions B_1 and B_2 for $m = -200$, $q = 1$, $\gamma = 2.553191 \cdot 10^{-5}$

3 Solving the Parametric Eigenvalue Problem at Large $|m|$

Step 1. In (10), (11) apply the transformation to a scaled variable x

$$x = \frac{\gamma \rho^2}{2}, \quad \rho = \frac{\sqrt{x}}{\sqrt{\gamma/2}}, \tag{18}$$

and put $\lambda_j(z) = E_j(z)/(2\gamma) = \lambda_j^{(0)} + m/2 + \delta\lambda_j(z)$, where $\lambda_j^{(0)} = n + (|m|+1)/2$. The eigenvalue problem reads

$$\left(-\frac{\partial}{\partial x} x \frac{\partial}{\partial x} + \frac{m^2}{4x} + \frac{x}{4} + \frac{m}{2} - \frac{q}{\gamma\sqrt{\frac{2x}{\gamma} + z^2}} - \lambda_j \right) B_j(x; z) = 0, \tag{19}$$

with a normalization condition

$$\frac{1}{\gamma} \int_0^\infty B_j(x; z)^2 dx = 1. \tag{20}$$

At $q = 0$, Eq. (19) without $m/2$ takes the form

$$L(n) B_j^{(0)}(x) = 0, \qquad L(n) = -\frac{\partial}{\partial x} x \frac{\partial}{\partial x} + \frac{m^2}{4x} + \frac{x}{4} - \lambda_j^{(0)}, \tag{21}$$

and has the regular and bounded solutions at

$$\lambda_j^{(0)} = n + (|m|+1)/2, \tag{22}$$

where the transverse quantum number $n \equiv N_\rho = j - 1 = 0, 1, \ldots$ determines the number of nodes of the solution $B_j^{(0)}(x) \equiv B_{nm}^{(0)}(x)$ with respect to the variable x. The normalized solutions of Eq. (21) take the form

$$B_j^{(0)}(x) = C_{n|m|}e^{-\frac{x}{2}}x^{\frac{|m|}{2}}L_n^{|m|}(x), \qquad C_{n|m|} = \left[\gamma\frac{n!}{(n+|m|)!}\right]^{\frac{1}{2}}, \qquad (23)$$

$$\frac{1}{\gamma}\int_0^\infty B_{nm}^{(0)}(x)B_{n'm}^{(0)}(x)dx = \delta_{nn'}, \tag{24}$$

where $L_n^{|m|}(x)$ are Laguerre polynomials [14].

Step 2. Substituting the notation $\delta\lambda_j(z) = \lambda_j(z) - \lambda_j^{(0)} - m/2 \equiv E_j(z)/(2\gamma) - (n + (m + |m| + 1)/2)$, and the Taylor expansion in the vicinity of the point $x_0 = x_s\gamma$:

$$V_c(x,z) = -\frac{q}{\gamma\sqrt{\frac{2x}{\gamma}+z^2}} = -\sum_{k=1}^{j_{\max}}V^{(k)}(x,z)\varepsilon^k = -\frac{\varepsilon q}{\gamma(z^2+2x_s)^{1/2}} \tag{25}$$

$$+\frac{\varepsilon q(x-x_s\gamma)}{\gamma^2(z^2+2x_s)^{3/2}} - \frac{3\varepsilon^2 q(x-x_s\gamma)^2}{2\gamma^3(z^2+2x_s)^{5/2}} + \frac{5\varepsilon^3 q(x-x_s\gamma)^3}{2\gamma^4(z^2+2x_s)^{7/2}} + O\left(\frac{\varepsilon^4}{(z^2+2x_s)^{9/2}}\right),$$

into Eq. (19) at $q \neq 0$, transform it to the following form

$$L(n)B_j(x;z) + \left(\sum_{k=1}^{j_{\max}}V^{(k)}(z)\varepsilon^k - \delta\lambda_j(z)\right)B_j(x;z) = 0. \tag{26}$$

Here ε is a formal parameter that will be put to be 1 in the final expression. The parameters $x_s = \rho_s^2/2$ and ρ_s approximately correspond to the minimum of the potential energy (2). In so doing, the Coulomb term is neglected. In the calculations we choose $\rho_s = \sqrt{2|m|/\gamma}$ under assumption that the condition $\gamma^2\rho^2/4 + m^2/\rho^2 \gg 2|q|/\rho$ is valid. The approximation errors $\delta U^{(j_{\max})}(\rho,z)$ at $j_{\max} = 1,...,9$ are illustrated in Fig. 1b. One can see that in the localization interval $\rho \in [3000, 5000]$ of the eigenfunction (19), the errors decrease with increasing order $j_{\max}$ (see Fig. 2). Performing Taylor expansion at $|z|/\rho_s \gg 1$, we arrive at the inverse power series that gives the same results as the perturbation theory in powers of $1/|z|$ [8].

Step 3. The solution of Eq. (26) is found in the form of perturbation expansion in powers of ε

$$\delta\lambda_j(z) = \sum_{k=1}^{k_{\max}}\varepsilon^k\lambda_n^{(k)}(z), \quad B_j(x;z) = B_n^{(0)}(x) + \sum_{k=0}^{k_{\max}}\varepsilon^k B_n^{(k)}(x,z). \tag{27}$$

Equating coefficients at the same powers of ε, we arrive at the system of inhomogeneous differential equations with respect to corrections $\lambda_n^{(k)}(z)$ and $B_n^{(k)}(x,z)$:

$$L(n)B_n^{(0)}(x) = 0 \equiv f_n^{(0)}(z), \tag{28}$$

$$L(n)B_n^{(k)}(x,z) = (\lambda_n^{(k)}(z) - V^{(k)}(z))B_n^{(0)}(x)$$

$$+ \sum_{p=1}^{k-1}(\lambda^{(k-p)}(z) - V^{(k-p)}(z))B_n^{(p)}(x,z) \equiv f_n^{(k)}(z), \quad k \geq 1.$$

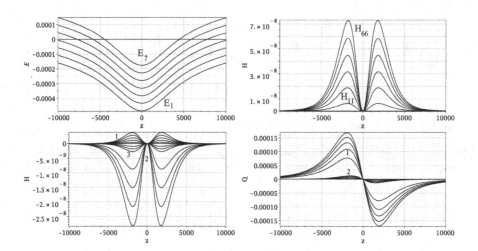

Fig. 3. The eigenvalues $E_j(z)$ and the effective potentials $H_{jj}(z)$, $H_{jj'}(z)$ (curves $H_{jj-1}(z)$, $j = 2, ..., 6$, are marked by number 1, curves $H_{jj-2}(z)$, $j = 3, ...6$, are marked by number 2 and curves $H_{jj-3}(z)$, $j = 4, ..., 6$, are marked by number 3) and $Q_{jj'}(z)$ (curves $Q_{jj-1}(z)$, $j = 2, ..., 6$, are marked by number 1, and curves $Q_{jj-2}(z)$, $j = 3, ..., 6$, are marked by number 2) for $m = -200$, $q = 1$, $\gamma = 2.553191 \cdot 10^{-5}$

To solve Eqs. (26) we used the *nonnormalized* orthogonal basis

$$B_{n+s}(x) = C_{n|m|}e^{-\frac{x}{2}}x^{\frac{|m|}{2}}L_{n+s}^{|m|}(x) = C_{n|m|}C_{n+s|m|}^{-1}B_{n+s,m}^{(0)}(x), \qquad (29)$$

$$\langle s|s'\rangle = \int_0^\infty B_{n+s}(x)B_{n+s'}(x)dx = \delta_{ss'}\gamma\frac{n!}{(n+|m|)!}\frac{(n+s+|m|)!}{(n+s)!}.$$

The action of the operators $L(n)$ and x on the functions $B_{n+s}(x)$ is defined by the relations

$$L(n)B_{n+s}(x) = sB_{n+s}(x), \qquad (30)$$
$$xB_{n+s}(x) = -(n+s+|m|)B_{n+s-1}(x) + (2(n+s)+|m|+1)B_{n+s}(x)$$
$$-(n+s+1)B_{n+s+1}(x)$$

that involve no fractional powers of quantum numbers n and m.

Step 4. Applying Eqs. (30), the right-hand side $f_n^{(k)}(z)$ and the solutions $B_n^{(k)}(x, z)$ of the system (28) are expanded over the *nonnormalized* basis states $B_{n+s}(x)$

$$B_n^{(k)}(x, z) = \sum_{s=-s_{\max}}^{s_{\max}} b_{n;s}^{(k)}(z)B_{n+s}(x), \quad f_n^{(k)}(z) = \sum_{s=-s_{\max}}^{s_{\max}} f_{n;s}^{(k)}(z)B_{n+s}(x). (31)$$

Then the recurrent set of linear algebraic equations for unknown *nonnormalized* coefficients $b_{n;s}^{(k)}(z)$ and corrections $\lambda_n^{(k)}(z)$ is obtained

$$sb_{n;s}^{(k)}(z) - f_{n;s}^{(k)}(z) = 0, \quad s = -s_{\max}, ..., s_{\max},$$

which is solved sequentially for $k = 1, 2, \ldots, k_{\max}$:

$$f_{n;0}^{(k)}(z) = 0 \quad \rightarrow \lambda_n^{(k)}(z); \quad b_{n;s}^{(k)}(z) = f_{n;s}^{(k)}(z)/s, \quad s = -s_{\max}, \ldots, s_{\max}, \quad s \neq 0.$$

The initial conditions (22) and $b_{n;s}^{(0)}(z) = \delta_{s0}$ follow from Eqs. (21) and (24).

Step 5. To obtain the normalized wave function $B_j(x; z)$ up to the kth order, the coefficient $b_0^{(k)}$ are defined by the following relation:

$$b_{n;0}^{(k)}(z) = -\frac{1}{2\gamma} \sum_{p=1}^{k-1} \sum_{s'=-s_{\max}}^{s_{\max}} \sum_{s=-s_{\max}}^{s_{\max}} b_{n;s}^{(k-p)}(z)\langle s|s'\rangle b_{n;s'}^{(p)}(z), \quad b_{n;0}^{(k=1)}(z) = 0.$$

As an example of the output file at steps 1–5, we display nonzero coefficients $\lambda_n^{(k)}(z)$, $b_{n;s}^{(k)}(z)$ of the expansions (27), (31) over the *nonnormalized* basis functions (29) up to $O(\varepsilon^2)$:

$$\lambda_n^{(0)} = n + (|m|+1)/2,$$

$$\lambda_n^{(1)}(z) = -\frac{q}{\gamma\sqrt{z^2+2x_s}} + \frac{q(2n+|m|+1)}{\gamma^2(z^2+2x_s)^{3/2}} - \frac{x_s q}{\gamma(z^2+2x_s)^{3/2}},$$

$$\lambda_n^{(2)}(z) = -q^2(2n+|m|+1)/(\gamma^4(z^2+2x_s)^3) - 3q[|m|^2+2+6n|m|$$
$$+6n^2+6n+3|m|-2\gamma(2n+|m|+1)x_s+x_s^2\gamma^2]/(2\gamma^3(z^2+2x_s)^{5/2}),$$

$$b_{n;0}^{(0)}(z) = 1, \tag{32}$$

$$b_{n;-1}^{(1)}(z) = -q(n+|m|)/(\gamma^2(z^2+2x_s)^{3/2}), \quad b_{n;1}^{(1)}(z) = q(n+1)/(\gamma^2(z^2+2x_s)^{3/2}),$$

$$b_{n;-2}^{(2)}(z) = q(n+|m|)(n+|m|-1)(2q-3\gamma\sqrt{(z^2+2x_s)})/(4\gamma^4(z^2+2x_s)^3),$$

$$b_{n;-1}^{(2)}(z) = q(n+|m|)(2q+3\gamma(2n+|m|-\gamma x_s)\sqrt{(z^2+2x_s)})/(\gamma^4(z^2+2x_s)^3),$$

$$b_{n;0}^{(2)}(z) = q^2(2n^2+2n+2n|m|+|m|+1)/(2\gamma^4(z^2+2x_s)^3),$$

$$b_{n;1}^{(2)}(z) = -q(n+1)(2q+3\gamma(2n+|m|+2-\gamma x_s)\sqrt{(z^2+2x_s)})/(\gamma^4(z^2+2x_s)^3),$$

$$b_{n;2}^{(2)}(z) = q(n+1)(n+2)(2q+3\gamma\sqrt{(z^2+2x_s)})/(4\gamma^4(z^2+2x_s)^3).$$

These expansions involve parameters $x_s = \rho_s^2/2$ and ρ_s that approximately corresponded to the minimum of the potential energy (2) and determined the point $x_0 = \gamma x_s$ of expansion of (25) of Coulomb potential $V_c(x, z)$.

Step 6. In terms of the scaled variable x, the expressions of the effective potentials $H_{ij}(z) = H_{ji}(z)$ and $Q_{ij}(z) = -Q_{ji}(z)$ take the form

$$H_{ij}(z) = \frac{1}{\gamma}\int_0^\infty dx \frac{\partial B_i(x;z)}{\partial z}\frac{\partial B_j(x;z)}{\partial z}, \quad Q_{ij}(z) = -\frac{1}{\gamma}\int_0^\infty dx B_i(x;z)\frac{\partial B_j(x;z)}{\partial z}. \tag{33}$$

To calculate them we expand the solution (26) over the *normalized* orthogonal basis $B_{n+s;m}^{(0)}(x)$ with the *normalized* coefficients $b_{n;n+s;m}^{(k)}(z)$,

$$B_j(x;z) \equiv B_j^m(x;z) = \sum_{k=0}^{k_{\max}} \varepsilon^k \sum_{s=-s_{\max}}^{s_{\max}} b_{n;n+s;m}^{(k)}(z) B_{n+s;m}^{(0)}(x). \tag{34}$$

The normalized coefficients $b^{(k)}_{n;n+s;m}(z)$ are expressed via $b^{(k)}_{n;s}(z)$,

$$b^{(k)}_{n;n+s;m}(z) = b^{(k)}_{n;s}(z)\sqrt{\frac{n!}{(n+|m|)!}\frac{(n+s+|m|)!}{(n+s)!}} \tag{35}$$

as follows from Eqs. (31), (34), and (29).

Step 7. As a result of substituting Eqs. (34) into Eq. (33), the matrix elements take the form

$$Q_{jj+t}(z) = -\sum_{k=0}^{k_{\max}} \varepsilon^k \sum_{k'=0}^{k} \sum_{s=\max(-s_{\max},-s_{\max}+t)}^{\min(s_{\max},s_{\max}+t)} b^{(k')}_{n;n+s;m}(z)\frac{db^{(k-k')}_{n+t;n+s;m}(z)}{dz},$$

$$H_{jj+t}(z) = \sum_{k=0}^{k_{\max}} \varepsilon^k \sum_{k'=0}^{k} \sum_{s=\max(-s_{\max},-s_{\max}+t)}^{\min(s_{\max},s_{\max}+t)} \frac{db^{(k')}_{n;n+s;m}(z)}{dz}\frac{db^{(k-k')}_{n+t;n+s;m}(z)}{dz}. \tag{36}$$

By collecting the coefficients at similar powers of ε in Eq. (36) the algorithm yields the final expansions of eigenvalues and effective potentials available in the output file

$$E_j(z) = \sum_{k=0}^{k_{\max}} E_j^{(k)}(z), \ H_{ij}(z) = \sum_{k=2}^{k_{\max}} H_{ij}^{(k)}(z), \ Q_{ij}(z) = \sum_{k=1}^{k_{\max}} Q_{ij}^{(k)}(z). \tag{37}$$

Successful runs of the Maple implementation of the algorithm were performed up to $k_{\max} = 6$ (the run time 30 s using Intel Core i5, 3.36 GHz, 4 GB). Below we present a few first nonzero coefficients derived *in the analytic form* ($j = n+1$):

$$E_j^{(0)} = 2\gamma(n + (m + |m| + 1)/2),$$

$$E_j^{(1)}(z) = -\frac{2q}{\sqrt{z^2+\rho_s^2}} + \frac{2q(2n+|m|+1)}{\gamma(z^2+\rho_s^2)^{3/2}} - \frac{\rho_s^2 q}{(z^2+\rho_s^2)^{3/2}},$$

$$E_j^{(2)}(z) = -\frac{2q^2(2n+|m|+1)}{\gamma^3(z^2+\rho_s^2)^3}$$

$$-\frac{3q[|m|^2+2+6n|m|+6n^2+6n+3|m|-\gamma(2n+|m|+1)\rho_s^2+\rho_s^4\gamma^2/4]}{\gamma^2(z^2+\rho_s^2)^{5/2}},$$

$$Q^{(1)}_{jj-1}(z) = -\sqrt{n}\sqrt{n+|m|}\frac{3zq}{\gamma^2(z^2+\rho_s^2)^{5/2}},$$

$$Q^{(2)}_{jj-1}(z) = -\sqrt{n}\sqrt{n+|m|}\left[\frac{15zq(2|m|+4n-\rho_s^2\gamma)}{2\gamma^3(z^2+\rho_s^2)^{7/2}} + \frac{12zq^2}{\gamma^4(z^2+\rho_s^2)^4}\right],$$

$$Q^{(2)}_{jj-2}(z) = -\sqrt{n}\sqrt{n-1}\sqrt{n+|m|}\sqrt{n+|m|-1}\frac{15qz}{4\gamma^3(z^2+\rho_s^2)^{7/2}},$$

$$H^{(2)}_{jj}(z) = 9q^2(2n^2+2n|m|+2n+|m|+1)\left[\frac{1}{\gamma^4(z^2+\rho_s^2)^4} - \frac{\rho_s^2}{\gamma^4(z^2+\rho_s^2)^5}\right],$$

$$H^{(2)}_{jj-2}(z) = -9q^2\sqrt{n}\sqrt{n-1}\sqrt{n+|m|}\sqrt{n+|m|-1}\left[\frac{1}{\gamma^4(z^2+\rho_s^2)^4} + \frac{\rho_s^2}{\gamma^4(z^2+\rho_s^2)^5}\right].$$

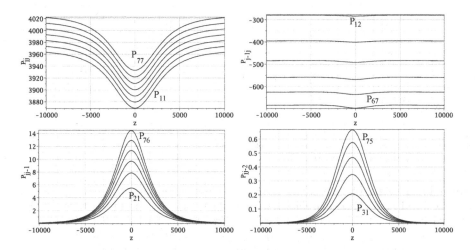

Fig. 4. Transverse dipole matrix elements $P_{nn'}^{|m||m|-1}$ (subscripts n, n' run $0,1,2,3,4,5,6$) for $m = -200$, $q = 1$, $\gamma = 2.553191 \cdot 10^{-5}$

As an example, Fig. 3 shows the eigenvalues and effective potentials (37), which agree with those calculated numerically using ODPEVP [10] with the accuracy of the order of 10^{-10}. We used finite element grid on the interval $\rho \in [\rho_{\min} = 2000, \rho_{\max} = 6000]$ with the Lagrange elements of fourth order. Expanding (37) into the Taylor series at $|z|/\rho_s \gg 1$, we arrive at perturbation expansion in powers of $1/z$ [8].

4   Calculations of the Transversal Dipole Matrix Elements

Using the scaled variable x defined by Eq. (18) one can express the transverse dipole matrix elements $P_{ij}^{|m|,|m|\mp1}(z) = \left\langle |m|, n \middle| \rho e^{\pm i\varphi} \middle| |m| \mp 1, n' \right\rangle$ and $P_{ij}^{-|m|,-|m|\pm1}(z) = \left\langle -|m|, n \middle| \rho e^{\mp i\varphi} \middle| -|m| \pm 1, n' \right\rangle$ possessing the property

$$\left\langle |m|, n \middle| \rho \exp(\pm i\varphi) \middle| |m| \mp 1, n' \right\rangle^* = \left\langle |m| \mp 1, n' \middle| \rho \exp(\mp i\varphi) \middle| |m|, n \right\rangle,$$

where $i = n + 1$ and $j = n' + 1$, in the following form

$$P_{ij}^{-|m|,-|m|\pm1}(z) = P_{ij}^{|m|,|m|\mp1}(z) = \sqrt{\frac{2}{\gamma^3}} \int\limits_0^\infty dx B_i^{|m|}(x; z) \sqrt{x} B_j^{|m|\mp1}(x; z). \quad (38)$$

According to Eqs. (22.7.12), (33.7.30), and (22.7.31) of [14], the dipole moment matrix elements calculated with normalized basis functions $||m|, n\rangle = B_{n|m|}^{(0)}(x) e^{i|m|\varphi}/\sqrt{2\pi}$ by means of Eq. (23) are expressed as

$$P_{ij}^{(0);|m||m|\mp1} = \sqrt{\frac{2}{\gamma^3}} \left\langle |m|, n \middle| \sqrt{x} e^{\pm i\varphi} \middle| |m| \mp 1, n' \right\rangle$$

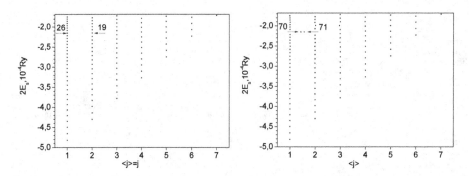

Fig. 5. Energy eigenvalues $2E_s$ for even ($\sigma = +1$) lower eigenstates vs the state number $\langle j \rangle$ calculated in the diagonal adiabatic approximation (left) and in the Kantorovich approximation at $j_{max} = 6$ with given accuracy (right). Here $m = -200$, $\gamma = 2.553191 \cdot 10^{-5}$, $q = 1$, $\sigma = +1$. The quantity $\langle j \rangle = \sum_j \int j \chi_{j,s}(z)^2 dz$ is the averaged quantum number, s is the eigenvalue number in the ascending energy sequence $E_1 < E_2 < ... < E_s < ... < \gamma/2$, corresponding to the number v of the eigenvalue $E_{j1} < E_{j2} < ... < E_{jv} < ... < \gamma/2$ counted at each $\langle j \rangle = j$ in diagonal approximation (17) of Eqs. (12)

$$= \sqrt{\frac{2}{\gamma^3}} \frac{1}{2\pi} \int_0^{2\pi} d\varphi \int_0^\infty e^{-i|m|\varphi} B_{i,|m|}^{(0)}(x) e^{\pm i\varphi} \sqrt{x} e^{i(|m|\mp 1)\varphi} B_{i,|m|\mp 1}^{(0)}(x) dx$$

$$= \sqrt{\frac{2}{\gamma}} \left[\delta_{nn'} \sqrt{n + |m| + 1/2 \mp 1/2} - \delta_{n\mp 1,n'} \sqrt{n + 1/2 \mp 1/2} \right]. \tag{39}$$

As a result of substituting Eqs. (34) and (39) into Eq. (38), the matrix elements take the following *analytic* form ($j = n + 1$)

$$P_{jj+t}^{|m|,|m|-1}(z) = \sum_{k=0}^{k_{max}} P_{jj+t}^{(k);|m||m|-1}(z),$$

$$P_{jj+t}^{(k);|m||m|-1}(z) = \sqrt{\frac{2}{\gamma}} \sum_{k'=0}^{k} \sum_{s=\max(-k,k'-k-t)}^{\min(k,k-k'-t)} \left[b_{n;n+s;|m|}^{(k')}(z) b_{n+t;n+s;|m|-1}^{(k-k')}(z) \right.$$

$$\left. \times \sqrt{n+s+|m|+1} - b_{n;n+s;|m|}^{(k')}(z) b_{n+t;n+s+1;|m|-1}^{(k-k')}(z) \sqrt{n+s+1} \right]. \tag{40}$$

Successful run of the Maple-implemented algorithm was performed up to $k_{max} = 6$ (run time 90 s with Intel Core i5, 3.36 GHz, 4 GB). A few first nonzero coefficients derived *in the analytic form* are presented below ($j = n + 1$):

$$P_{jj}^{(0);|m||m|-1}(z) = +\frac{\sqrt{2}\sqrt{n+|m|+1}}{\sqrt{\gamma}}, P_{jj}^{(1);|m||m|-1}(z) = -\frac{\sqrt{2}\sqrt{n+|m|}q}{\gamma^{5/2}(\rho_s^2 + z^2)^{3/2}},$$

$$P_{j-1j}^{(0);|m||m|-1}(z) = -\frac{\sqrt{n}\sqrt{2}}{\sqrt{\gamma}}, \tag{41}$$

$$P_{j-1j}^{(1);|m||m|-1}(z) = -\frac{\sqrt{n}\sqrt{2}\sqrt{n+|m|}(\sqrt{n+|m|-1} - \sqrt{n+|m|+1})q}{(\rho_s^2 + z^2)^{3/2}\gamma^{5/2}},$$

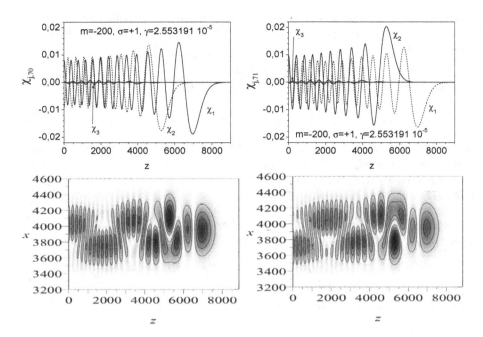

Fig. 6. Upper panels: the first three components of the eigenfunctions $\chi_{j,70}$ and $\chi_{j,71}$ ($j = 1, 2, 3$). The dominant components are $j = 1$ ($\langle j \rangle = 1.43$) with $v - 1 = 25$ nodes and $j = 2$ ($\langle j \rangle = 1.56$) with $v - 1 = 18$ nodes, respectively. Lower panels: the profile of the wave function $\Psi^{m=-200,\sigma=+1}_{s=70}(\rho, z)$ and $\Psi^{m=-200,\sigma=+1}_{s=71}(\rho, z)$ of the *resonance states* in the zx plane with the energies $2E^{m=-200,\sigma=+1}_{s=70} = -2.151832 \cdot 10^{-4}$Ry and $2E^{m=-200,\sigma=+1}_{s=71} = -2.150977 \cdot 10^{-4}$Ry pointed by arrows in the right panel of Fig. 5

$$P^{(1);|m||m|-1}_{jj-1}(z) = \frac{\sqrt{n}\sqrt{2}(\sqrt{n+|m|-1}\sqrt{n+|m|+1}-n-|m|)q}{(\rho_s^2 + z^2)^{3/2}\gamma^{5/2}}.$$

The comparison of our analytical numerical results with those obtained numerically using the program ODPEVP [10] shows the convergence of the perturbation series expansion up to $k_{\max} = 6$ with four significant digits. Expanding (40) into a Taylor series at $|z|/\rho_s \gg 1$, we arrive at the inverse power series for the dipole matrix elements. To obtain the leading terms at $|z| \to \infty$ it is sufficiently to put $\rho_s = 0$ in (41).

5   Calculations of Rydberg States and Decay Rates

In Fig. 5 we present an example of the lower part of discrete spectrum calculated in the diagonal adiabatic and Kantorovich approximations with the effective potentials (37) by means of the program KANTBP2 [1]. In numerical calculations at $q = -1$, $\gamma = 2.553191 \cdot 10^{-5}$ for $|m| \sim 200$, we use finite element grid on the interval $z \in [0, z_{\max} = 11000]$ with the Lagrange elements of fourth order. In

Fig. 6, we show an example of resonance states formed by coupling of the quasi-degenerate states with the energies $2E_{j=1,v=26}^{m=-200,\sigma=+1} = -2.151260 \cdot 10^{-4}$Ry and $2E_{j=2,v=19}^{m=-200,\sigma=+1} = -2.151202 \cdot 10^{-4}$Ry in the diagonal adiabatic approximation (17) pointed by arrows in the left panel of Fig. 5.

The partial transition decay rates $\Gamma_{\tilde{s} \to \tilde{s}'}$ are calculated as

$$\Gamma_{\tilde{s} \to \tilde{s}'} = \frac{4}{3} \frac{e^2 \omega_{\tilde{s}\tilde{s}'}^3}{4\pi\varepsilon_0 \hbar c^3} |\langle \tilde{s}'|\bar{\mathbf{r}}|\tilde{s}\rangle|^2, \quad \omega_{\tilde{s}\tilde{s}'} = (\bar{E}_{\tilde{s}'} - \bar{E}_{\tilde{s}})/\hbar. \tag{42}$$

In the above expressions, $\varepsilon_0 = 8.854187817 \cdot 10^{-12}$ F/m is the dielectric constant, the energy $\bar{E}_{\tilde{s}'} = E_{\tilde{s}'} E_B$ and the dipole moment $\langle \tilde{s}'|\bar{\mathbf{r}}|\tilde{s}\rangle = a_B \langle \tilde{s}'|\mathbf{r}|\tilde{s}\rangle$ are expressed in the atomic units $E_B = 2Ry = 4.35974434 \cdot 10^{-18}$ J, $a_B = 0.52917721092 \cdot 10^{-10}$ m, i.e.

$$\Gamma_{\tilde{s} \to \tilde{s}'} = 2.142 \cdot 10^{10} (E_{\tilde{s}'} - E_{\tilde{s}})^3 |\langle \tilde{s}'|\mathbf{r}|\tilde{s}\rangle|^2 \times \text{s}^{-1}. \tag{43}$$

Here $|\langle \tilde{s}'|\mathbf{r}|\tilde{s}\rangle|^2$ defined by the expression

$$|\langle \tilde{s}'|\mathbf{r}|\tilde{s}\rangle|^2 = (1/2)|\langle \tilde{s}'|\rho e^{-i\varphi}|\tilde{s}\rangle|^2 + |\langle \tilde{s}'|z|\tilde{s}\rangle|^2 + (1/2)|\langle \tilde{s}'|\rho e^{+i\varphi}|\tilde{s}\rangle|^2, \tag{44}$$

where $\langle \tilde{s}'|z|\tilde{s}\rangle$ and $\langle \tilde{s}'|\rho e^{\pm i\varphi}|\tilde{s}\rangle$ are the longitudinal and transverse dipole moment, respectively. As follows from Eq. (40),

$$\langle \tilde{s}'|z|\tilde{s}\rangle = \delta_{m'm}\delta_{-\sigma'\sigma} \sum_{i,j=1}^{j_{\max}} \int_{z_{\min}}^{z_{\max}} dz \chi_{i\tilde{s}'}^{m\sigma'}(z) z \chi_{j\tilde{s}}^{m\sigma}(z), \tag{45}$$

$$\langle \tilde{s}'|\rho e^{\pm i\varphi}|\tilde{s}\rangle = \delta_{m'm\mp 1}\delta_{\sigma'\sigma} \sum_{i,j=1}^{j_{\max}} \int_{z_{\min}}^{z_{\max}} dz \chi_{i\tilde{s}'}^{m'\sigma'}(z) P_{ij}^{m',m}(z) \chi_{j\tilde{s}}^{m\sigma}(z). \tag{46}$$

In Table 1 we show our present results for partial decay rates (43) and dipole moments (45) and (46). The results were obtained numerically by means of the program KANTBP 2.0 [1] using the analytically derived effective potentials (37) and matrix elements of transversal dipole moments (40), i.e., $M_{\tilde{s}'\tilde{s}} = \langle \tilde{s}'|\rho e^{-i\varphi}|\tilde{s}\rangle$ for cyclotron decay (C) ($q \to q' = q$, where $q = j - m$ is magnetron quantum number, $m \to m' = m - 1$, $\sigma \to \sigma' = \sigma$, $j \to j' = j - 1$, $v \to v' = v$); $M_{\tilde{s}'\tilde{s}} = \langle \tilde{s}'|z|\tilde{s}\rangle$ for the bounce decay (B) ($q \to q' = q$, $m \to m' = m$, $\sigma \to \sigma' = -\sigma$, $j \to j' = j$, $v \to v' = v - 1$), and $M_{\tilde{s}'\tilde{s}} = \langle \tilde{s}'|\rho e^{+i\varphi}|\tilde{s}\rangle$ for the magnetron decay (M) ($q \to q' = q - 1$, $m \to m' = m + 1$, $\sigma \to \sigma' = \sigma$, $j \to j' = j$, $v \to v' = v$). The results agree with the numerical ones from [12] within the required accuracy.

In Table 1 we also show the energy values $2E_{|\tilde{s}\rangle}$ calculated in the Kantorovich approximation (K) at $j_{\max} = 6$, and obtained by the aid of the diagonal approximation (17) in the *analytical form*

$$2E_{|\tilde{s}\rangle} \approx 2E_{i,v}^{m,\sigma} = U_{ii}^{(0)} + \mathcal{E}_{i;v}^{(0)} + \sum_{\kappa=2}^{\kappa_{\max}} \mathcal{E}_{i;v}^{(\kappa-1)}, \tag{47}$$

$$\mathcal{E}_{i;v}^{(0)} = \omega_{z,i}(2v+1), \quad \mathcal{E}_{i;v}^{(1)} = \frac{3U_i^{(4)}(2v^2 + 2v + 1)}{4\omega_{z,i}^2},$$

$$\mathcal{E}_{i;v}^{(2)} = -\frac{(2v+1)(17v^2 + 17v + 21)(U_i^{(4)})^2}{16\omega_{z,i}^5} + \frac{5(2v+1)(2v^2 + 2v + 3)U_i^{(6)}}{8\omega_{z,i}^3}.$$

Table 1. The partial transition decay rates $\Gamma_{\tilde{s}\to\tilde{s}'}$ evaluated using Eq. (43) from the state $|\tilde{s}\rangle = |j, v, \sigma, m\rangle$ to $|\tilde{s}'\rangle = |j', v', \sigma', m'\rangle$ with energies $2E_{|\tilde{s}\rangle}$ and $2E_{|\tilde{s}'\rangle}$ calculated using the Kantorovich approximation (K) at $j_{max} = 6$ and the corresponding dipole moments $M_{\tilde{s}'\tilde{s}}$. In square brackets, numerical results of [12] are given. The energies calculated in *analytical form* using the crude diagonal approximation with the Taylor series of $U_{ii}(z) = E_i(z)$ up to harmonic (H) and anharmonic (A) terms of order of z^2 and z^{10}, respectively. The corresponding energies in the diagonal approximation with Taylor series of $U_{ii}(z) = E_i(z) + H_{ii}(z)$ differing only in two last digits, are shown in parentheses.

	$\tilde{s}$	$\tilde{s}'$	$|j, v, \sigma, m\rangle$	$|j', v', \sigma', m'\rangle$	$\Gamma_{\tilde{s}\to\tilde{s}'}$, s^{-1}	$M_{\tilde{s}'\tilde{s}}$, a_B		$2E_{\tilde{s}}$, 10^{-4}Ry	$2E_{\tilde{s}'}$, 10^{-4}Ry
C	5	1	$|2,1,+1,-200\rangle$	$|1,1,+1,-201\rangle$	13.1	276.4	K	−4.29933	−4.80384
					[13.7]	[283]	H	−4.29978(76)	−4.80384(83)
							A	−4.30019(18)	−4.80424(23)
C	13	5	$|3,1,+1,-200\rangle$	$|2,1,+1,-201\rangle$	26.3	390.9	K	−3.78171	−4.28632
					[27.5]	[401]	H	−3.78299(95)	−4.28688(86)
							A	−3.78342(38)	−4.28729(27)
B	1	1	$|1,2,-1,-200\rangle$	$|1,1,+1,-200\rangle$	0.180	349.4	K	−4.73499	−4.81688
					[0.178]	[350]	H	−4.73329(27)	−4.81683(83)
							A	−4.73531(29)	−4.81724(23)
B	2	1	$|1,3,+1,-200\rangle$	$|1,2,-1,-200\rangle$	0.345	499.0	K	−4.65469	−4.73499
					[0.342]	[500]	H	−4.64974(71)	−4.73329(27)
							A	−4.65497(94)	−4.73531(29)
M	1	1	$|1,1,+1,-200\rangle$	$|1,1,+1,-199\rangle$	0.045	3870	K	−4.81688	−4.83003
					[0.044]	[3872]	H	−4.81683(83)	−4.82993(93)
							A	−4.81724(23)	−4.83034(33)

The latter was obtained using SNA like in Section 3, but for a perturbed 1D oscillator with *adiabatic frequency* $\omega_{z,i}$. It was accomplished with the help of a Taylor expansion up to $z^{2\kappa_{max}}$ of effective potentials $U_{ii}(z) = E_i(z) + H_{ii}(z)$ from Eq. (37) for the harmonic (H) and anharmonic (A) terms, i.e., $2\kappa_{max} = 2$ and $2\kappa_{max} = 10$, respectively,

$$U_{ii}(z) = U_{ii}(0) + \omega_{z,i}^2 z^2 + \sum_{\kappa=2}^{\kappa_{max}} U_i^{(2\kappa)} z^{2\kappa}. \tag{48}$$

Moreover, in Table 1 we present also the results for the energies (47) in the crude and adiabatic approximations obtained without and with the diagonal potential H_{ii}, respectively. One can see that the energies in crude adiabatic and adiabatic approximations differ only in two last significant figures, i.e., are the same within the accuracy of $\sim 10^{-8}$. One can see from Table 1 that the adiabatic harmonic (H) diagonal approximation and the crude anharmonic (A) one provide the *upper and lower estimations of the energy values of low-excited Rydberg states* with $j = 1$, respectively.

Remark 2. In the expansions (47) and (48), the coefficients are calculated using $U_{ii}^{(0)} = U_{ii}(0)$, $\omega_{z,i}^2 = (d^2 U_{ii}(z)/dz^2)_{z=0}/2$, $U_i^{(2\kappa)} = (d^{2\kappa} U_{ii}(z)/dz^{2\kappa})_{z=0}/((2\kappa)!)$.

In the harmonic approximation $\omega_{z,i}^2 = \sum_{k=1}^{k_{max}} \omega_{z,i,E}^{(k)} + \sum_{k=2}^{k_{max}} \omega_{z,i,H}^{(k)}$, where $\omega_{z,i,E}^{(k)} = (d^2 E_i^{(k)}(z)/dz^2)_{z=0}/2$ and $\omega_{z,i,H}^{(k)} = (d^2 H_{ii}^{(k)}(z)/dz^2)_{z=0}/2$, the leading terms are:

$$\omega_{z,i,E}^{(1)} = \frac{5q}{2\rho_s^3} - \frac{3q(2n+|m|+1)}{\gamma\rho_s^5}, \quad \omega_{z,i,H}^{(2)} = \frac{9q^2(2n^2+2n|m|+2n+|m|+1)}{\rho_s^{10}\gamma^4},$$

$$\omega_{z,i,E}^{(2)} = \frac{15q}{8\rho_s^3} - \frac{15q(2n+|m|+1)}{2\gamma\rho_s^5} + \frac{15q(6n^2+6n|m|+6n+m^2+3|m|+2)}{2\gamma^2\rho_s^7}$$

$$+ \frac{6q^2(2n+|m|+1)}{\gamma^3\rho_s^8}.$$

The substitution of $\rho_s = \sqrt{2|m|/\gamma}$ into the leading term $\omega_{z,i}^2 \approx \omega_{z,i,E}^{(1)}$ at $n=0$ yields $\omega_{z,i}^2 \approx (q\sqrt{\gamma}\gamma(2|m|-3))/(4m^2\sqrt{2|m|})$. At $q=1$ we obtain the *adiabatic parameter* $(\omega_\rho/\omega_{z,i=1})^{4/3} = |m|\gamma^{1/3}$, where $\omega_\rho = \gamma/2$, in agreement with [13].

6   Conclusions

A new efficient method to calculate wave functions and decay rates of high-$|m|$ Rydberg states of a hydrogen atom in a magnetic field is developed. It is based on the KM application to parametric eigenvalue problems in cylindrical coordinates. The results are in a good agreement with the calculations executed in spherical coordinates at fixed $|m| > 140$ for $\gamma \sim 2.553 \cdot 10^{-5}$. The elaborated SNA for calculation of the effective potentials, dipole moment matrix elements, and the perturbation solutions *in analytic form* allows us to generate effective approximations for a finite set of longitudinal equations. This provides benchmark calculations for the new version KANTBP3 of our earlier program KANTBP2 [1] announced in [9]. The developed approach is a useful tool for calculating the threshold phenomena in formation, decay, and ionization of (anti)hydrogen-like atoms and ions in magneto-optical traps [11,12,13], and channelling of ions in thin films [4].

The authors thank Prof. V.L. Derbov for valuable discussions.

References

1. Chuluunbaatar, O., Gusev, A.A., Vinitsky, S.I., Abrashkevich, A.G.: KANTBP 2.0: New version of a program for computing energy levels, reaction matrix and radial wave functions in the coupled-channel hyperspherical adiabatic approach. Phys. Commun. 179, 685–693 (2008)
2. Gusev, A., Gerdt, V., Kaschiev, M., Rostovtsev, V., Samoylov, V., Tupikova, T., Vinitsky, S.: A Symbolic-Numerical Algorithm for Solving the Eigenvalue Problem for a Hydrogen Atom in Magnetic Field. In: Ganzha, V.G., Mayr, E.W., Vorozhtsov, E.V. (eds.) CASC 2006. LNCS, vol. 4194, pp. 205–218. Springer, Heidelberg (2006)
3. Chuluunbaatar, O., Gusev, A.A., Derbov, V.L., Kaschiev, M.S., Melnikov, L.A., Serov, V.V., Vinitsky, S.I.: Calculation of a hydrogen atom photoionization in a strong magnetic field by using the angular oblate spheroidal functions. J. Phys. A 40, 11485–11524 (2007)

4. Gusev, A.A., Derbov, V.L., Krassovitskiy, P.M., Vinitsky, S.I.: Channeling problem for charged particles produced by confining environment. Phys. At. Nucl. 72, 768–778 (2009)

5. Chuluunbaatar, O., Gusev, A.A., Gerdt, V.P., Rostovtsev, V.A., Vinitsky, S.I., Abrashkevich, A.G., Kaschiev, M.S., Serov, V.V.: POTHMF: A program for computing potential curves and matrix elements of the coupled adiabatic radial equations for a hydrogen-like atom in a homogeneous magnetic field. Comput. Phys. Commun. 178, 301–330 (2008)

6. Gusev, A.A., Chuluunbaatar, O., Gerdt, V.P., Rostovtsev, V.A., Vinitsky, S.I., Derbov, V.L., Serov, V.V.: Symbolic-Numeric Algorithms for Computer Analysis of Spheroidal Quantum Dot Models. In: Gerdt, V.P., Koepf, W., Mayr, E.W., Vorozhtsov, E.V. (eds.) CASC 2010. LNCS, vol. 6244, pp. 106–122. Springer, Heidelberg (2010); arXiv:1104.2292

7. Vinitsky, S.I., Chuluunbaatar, O., Gerdt, V.P., Gusev, A.A., Rostovtsev, V.A.: Symbolic-Numerical Algorithms for Solving Parabolic Quantum Well Problem with Hydrogen-Like Impurity. In: Gerdt, V.P., Mayr, E.W., Vorozhtsov, E.V. (eds.) CASC 2009. LNCS, vol. 5743, pp. 334–349. Springer, Heidelberg (2009)

8. Chuluunbaatar, O., Gusev, A., Gerdt, V., Kaschiev, M., Rostovtsev, V., Samoylov, V., Tupikova, T., Vinitsky, S.: A Symbolic-Numerical Algorithm for Solving the Eigenvalue Problem for a Hydrogen Atom in the Magnetic Field: Cylindrical Coordinates. In: Ganzha, V.G., Mayr, E.W., Vorozhtsov, E.V. (eds.) CASC 2007. LNCS, vol. 4770, pp. 118–133. Springer, Heidelberg (2007)

9. Gusev, A.A., Vinitsky, S.I., Chuluunbaatar, O., Gerdt, V.P., Rostovtsev, V.A.: Symbolic-Numerical Algorithms to Solve the Quantum Tunneling Problem for a Coupled Pair of Ions. In: Gerdt, V.P., Koepf, W., Mayr, E.W., Vorozhtsov, E.V. (eds.) CASC 2011. LNCS, vol. 6885, pp. 175–191. Springer, Heidelberg (2011)

10. Chuluunbaatar, O., Gusev, A.A., Vinitsky, S.I., Abrashkevich, A.G.: ODPEVP: A program for computing eigenvalues and eigenfunctions and their first derivatives with respect to the parameter of the parametric self-adjoined Sturm-Liouville problem. Comput. Phys. Commun. 180, 1358–1375 (2009)

11. Chuluunbaatar, O., Gusev, A.A., Vinitsky, S.I., Derbov, V.L., Melnikov, L.A., Serov, V.V.: Photoionization and recombination of a hydrogen atom in a magnetic field. Phys. Rev. A 77, 034702-1–034702-4 (2008)

12. Guest, J.R., Choi, J.-H., Raithel, G.: Decay rates of high-$|m|$ Rydberg states in strong magnetic fields. Phys. Rev. A 68, 022509-1–022509-9 (2003)

13. Guest, J.R., Raithel, G.: High-$|m|$ Rydberg states in strong magnetic fields. Phys. Rev. A 68, 052502-1–052502-9 (2003)

14. Abramovits, M., Stegun, I.A.: Handbook of Mathematical Functions. Dover, New York (1972)

Quasi-stability versus Genericity

Amir Hashemi[1], Michael Schweinfurter[2], and Werner M. Seiler[2]

[1] Department of Mathematical Sciences, Isfahan University of Technology,
Isfahan, 84156-83111, Iran
Amir.Hashemi@cc.iut.ac.ir
[2] Institut für Mathematik, Universität Kassel
Heinrich-Plett-Straße 40, 34132 Kassel, Germany
{michael.schweinfurter,seiler}@mathematik.uni-kassel.de

Abstract. Quasi-stable ideals appear as leading ideals in the theory of
Pommaret bases. We show that quasi-stable leading ideals share many of
the properties of the generic initial ideal. In contrast to genericity, quasi-
stability is a characteristic independent property that can be effectively
verified. We also relate Pommaret bases to some invariants associated
with local cohomology, exhibit the existence of linear quotients in Pommaret bases and prove some results on componentwise linear ideals.

1 Introduction

The generic initial ideal of a polynomial ideal $0 \neq \mathcal{I} \unlhd \mathcal{P} = \Bbbk[\mathcal{X}] = \Bbbk[x_1, \ldots, x_n]$
was defined by Galligo [10] for the reverse lexicographic order and char $\Bbbk = 0$;
the extension to arbitrary term orders and characteristics is due to Bayer and
Stillman [5]. Extensive discussions can be found in [9, Sect. 15.9], [17, Chapt. 4]
and [13]. A characteristic feature of the generic initial ideal is that it is Borel-
fixed, a property depending on the characteristics of $\Bbbk$.

Quasi-stable ideals are known under many different names like ideals of nested
type [6], ideals of Borel type [19] or weakly stable ideals [7]. They appear naturally as leading ideals in the theory of Pommaret bases [25], a special class of
Gröbner bases with additional combinatorial properties. The notion of quasi-
stability is characteristic independent.

The generic initial ideal has found quite some interest, as many invariants
take the same value for $\mathcal{I}$ and $\mathrm{gin}\,\mathcal{I}$, whereas arbitrary leading ideals generally
lead to larger values. However, there are several problems with $\mathrm{gin}\,\mathcal{I}$: it depends
on char $\Bbbk$; there is no effective test known to decide whether a given leading
ideal is $\mathrm{gin}\,\mathcal{I}$ and thus one must rely on expensive random transformations for
its construction. The main point of the present work is to show that quasi-stable
leading ideals enjoy many of the properties of $\mathrm{gin}\,\mathcal{I}$ and can nevertheless be
effectively detected and deterministically constructed.

Throughout this article, $\mathcal{P} = \Bbbk[\mathcal{X}]$ denotes a polynomial ring in the variables
$\mathcal{X} = \{x_1, \ldots, x_n\}$ over an infinite field $\Bbbk$ of arbitrary characteristic and $0 \neq$
$\mathcal{I} \lhd \mathcal{P}$ a proper homogeneous ideal. When considering bases of $\mathcal{I}$, we will always
assume that these are homogeneous, too. $\mathfrak{m} = \langle \mathcal{X} \rangle \lhd \mathcal{P}$ is the homogeneous

V.P. Gerdt et al. (Eds.): CASC 2012, LNCS 7442, pp. 172–184, 2012.

maximal ideal. In order to be consistent with [24,25], we will use a non-standard convention for the reverse lexicographic order: given two arbitrary terms x^μ, x^ν of the same degree, $x^\mu \prec_{\mathrm{revlex}} x^\nu$ if the first non-vanishing entry of $\mu - \nu$ is positive. Compared with the usual convention, this corresponds to a reversion of the numbering of the variables $\mathcal{X}$.

2 Pommaret Bases

Pommaret bases are a special case of *involutive bases*; see [24] for a general survey. The algebraic theory of Pommaret bases was developed in [25] (see also [26, Chpts. 3-5]). Given an exponent vector $\mu = [\mu_1, \ldots, \mu_n] \neq 0$ (or the term x^μ or a polynomial $f \in \mathcal{P}$ with $\mathrm{lt}\, f = x^\mu$ for some fixed term order), we call $\min\{i \mid \mu_i \neq 0\}$ the *class* of μ (or x^μ or f), denoted by $\mathrm{cls}\,\mu$ (or $\mathrm{cls}\,x^\mu$ or $\mathrm{cls}\,f$). Then the *multiplicative variables* of x^μ or f are $\mathcal{X}_P(x^\mu) = \mathcal{X}_P(f) = \{x_1, \ldots, x_{\mathrm{cls}\,\mu}\}$. We say that x^μ is an *involutive divisor* of another term x^ν, if $x^\mu \mid x^\nu$ and $x^{\nu-\mu} \in \Bbbk[x_1, \ldots, x_{\mathrm{cls}\,\mu}]$. Given a finite set $\mathcal{F} \subset \mathcal{P}$, we write $\deg \mathcal{F}$ for the maximal degree and $\mathrm{cls}\,\mathcal{F}$ for the minimal class of an element of $\mathcal{F}$.

Definition 1. *Assume first that the finite set $\mathcal{H} \subset \mathcal{P}$ consists only of terms. $\mathcal{H}$ is a* Pommaret basis *of the monomial ideal $\mathcal{I} = \langle \mathcal{H} \rangle$, if as a $\Bbbk$-linear space*

$$\bigoplus_{h \in \mathcal{H}} \Bbbk[\mathcal{X}_P(h)] \cdot h = \mathcal{I} \tag{1}$$

(in this case each term $x^\nu \in \mathcal{I}$ has a unique involutive divisor $x^\mu \in \mathcal{H}$). A finite polynomial set $\mathcal{H}$ is a Pommaret basis *of the polynomial ideal $\mathcal{I}$ for the term order $\prec$, if all elements of $\mathcal{H}$ possess distinct leading terms and these terms form a Pommaret basis of the leading ideal $\mathrm{lt}\,\mathcal{I}$.*

Pommaret bases can be characterised similarly to Gröbner bases. However, involutive standard representations are unique. Furthermore, the existence of a Pommaret basis implies a number of properties that usually hold only generically.

Proposition 2 ([24, Thm. 5.4]). *The finite set $\mathcal{H} \subset \mathcal{I}$ is a Pommaret basis of the ideal $\mathcal{I} \lhd \mathcal{P}$ for the term order $\prec$, if and only if every polynomial $0 \neq f \in \mathcal{I}$ possesses a unique involutive standard representation $f = \sum_{h \in \mathcal{H}} P_h h$ where each non-zero coefficient $P_h \in \Bbbk[\mathcal{X}_P(h)]$ satisfies $\mathrm{lt}\,(P_h h) \preceq \mathrm{lt}\,(f)$.*

Proposition 3 ([24, Cor. 7.3]). *Let $\mathcal{H}$ be a finite set of polynomials and $\prec$ a term order such that no leading term in $\mathrm{lt}\,\mathcal{H}$ is an involutive divisor of another one. The set $\mathcal{H}$ is a Pommaret basis of the ideal $\langle \mathcal{H} \rangle$ with respect to $\prec$, if and only if for every $h \in \mathcal{H}$ and every non-multiplicative index $\mathrm{cls}\,h < j \leq n$ the product $x_j h$ possesses an involutive standard representation with respect to $\mathcal{H}$.*

Theorem 4 ([25, Cor. 3.18, Prop. 3.19, Prop. 4.1]). *Let $\mathcal{H}$ be a Pommaret basis of the ideal $\mathcal{I} \lhd \mathcal{P}$ for an order $\prec$.*

(i) *If $D = \dim(\mathcal{P}/\mathcal{I})$, then $\{x_1, \ldots, x_D\}$ is the unique maximal strongly independent set modulo $\mathcal{I}$ (and thus $\mathrm{lt}\,\mathcal{I} \cap \Bbbk[x_1, \ldots, x_D] = \{0\}$).*

(ii) *The restriction of the canonical map $\mathcal{P} \to \mathcal{P}/\mathcal{I}$ to the subring $\Bbbk[x_1, \ldots, x_D]$ defines a Noether normalisation.*

(iii) *If $d = \min_{h \in \mathcal{H}} \operatorname{cls} h$ is the minimal class of a generator in $\mathcal{H}$ and $\prec$ is the reverse lexicographic order, then $x_1, \ldots, x_{d-1}$ is a maximal $\mathcal{P}/\mathcal{I}$-regular sequence and thus depth $\mathcal{P}/\mathcal{I} = d - 1$.*

The involutive standard representations of the non-multiplicative products $x_j h$ appearing in Proposition 3 induce a basis of the first syzygy module. This observation leads to a stronger version of Hilbert's syzygy theorem.

Theorem 5 ([25, Thm. 6.1]). *Let $\mathcal{H}$ be a Pommaret basis of the ideal $\mathcal{I} \subseteq \mathcal{P}$. If we denote by $\beta_0^{(k)}$ the number of generators $h \in \mathcal{H}$ with $\operatorname{cls} \operatorname{lt} h = k$ and set $d = \operatorname{cls} \mathcal{H}$, then $\mathcal{I}$ possesses a finite free resolution*

$$0 \longrightarrow \mathcal{P}^{r_{n-d}} \longrightarrow \cdots \longrightarrow \mathcal{P}^{r_1} \longrightarrow \mathcal{P}^{r_0} \longrightarrow \mathcal{I} \longrightarrow 0 \qquad (2)$$

of length $n - d$ where the ranks of the free modules are given by

$$r_i = \sum_{k=d}^{n-i} \binom{n-k}{i} \beta_0^{(k)} . \qquad (3)$$

We denote by $\operatorname{reg} \mathcal{I}$ the *Castelnuovo-Mumford regularity* of $\mathcal{I}$ (considered as a graded module) and by $\operatorname{pd} \mathcal{I}$ its *projective dimension*. The *satiety* $\operatorname{sat} \mathcal{I}$ is the lowest degree from which on the ideal $\mathcal{I}$ and its *saturation* $\mathcal{I}^{\operatorname{sat}} = \mathcal{I} : \mathfrak{m}^\infty$ coincide. These objects can be easily read off from a Pommaret basis for $\prec_{\operatorname{revlex}}$.

Theorem 6 ([25, Thm. 8.11, Thm. 9.2, Prop. 10.1, Cor. 10.2]). *Let $\mathcal{H}$ be a Pommaret basis of the ideal $\mathcal{I} \lhd \mathcal{P}$ for the order $\prec_{\operatorname{revlex}}$. We denote by $\mathcal{H}_1 = \{h \in \mathcal{H} \mid \operatorname{cls} h = 1\}$ the subset of generators of class 1.*

(i) $\operatorname{reg} \mathcal{I} = \deg \mathcal{H}$.

(ii) $\operatorname{pd} \mathcal{I} = n - \operatorname{cls} \mathcal{H}$.

(iii) *Let $\tilde{\mathcal{H}}_1 = \{h/x_1^{\deg_{x_1} \operatorname{lt} h} \mid h \in \mathcal{H}_1\}$. Then the set $\bar{\mathcal{H}} = (\mathcal{H} \setminus \mathcal{H}_1) \cup \tilde{\mathcal{H}}_1$ is a weak Pommaret basis[1] of the saturation $\mathcal{I}^{\operatorname{sat}}$. Thus $\mathcal{I}^{\operatorname{sat}} = \mathcal{I} : x_1^\infty$ and the ideal $\mathcal{I}$ is saturated, if and only if $\mathcal{H}_1 = \emptyset$.*

(iv) $\operatorname{sat} \mathcal{I} = \deg \mathcal{H}_1$.

Remark 7. Bayer et al. [3] call a non-vanishing Betti number β_{ij} *extremal*, if $\beta_{k\ell} = 0$ for all $k \geq i$ and $\ell > j$. In [25, Rem. 9.7] it is shown how the positions and the values of all extremal Betti numbers can be obtained from the Pommaret basis $\mathcal{H}$ for $\prec_{\operatorname{revlex}}$. Let $h_{\gamma_1} \in \mathcal{H}$ be of minimal class among all generators of maximal degree in $\mathcal{H}$ and set $i_1 = n - \operatorname{cls} h_{\gamma_1}$ and $q_1 = \deg h_{\gamma_1}$. Then $\beta_{i_1, q_1 + i_1}$ is an extremal Betti number and its value is given by the number of generators of degree q_1 and class $n - i_1$. If $\operatorname{cls} h_{\gamma_1} = \operatorname{depth} \mathcal{I}$, it is the only one. Otherwise let h_{γ_2} be of minimal class among all generators of maximal degree in $\{h \in \mathcal{H} \mid \operatorname{cls} h < \operatorname{cls} h_{\gamma_1}\}$. Defining i_2, q_2 analogous to above, $\beta_{i_2, q_2 + i_2}$ is a further extremal Betti number and its value is given by the number of generators of degree q_2 and class $n - i_2$ and so on.

---

[1] Thus elimination of redundant generators yields a Pommaret basis [24, Prop. 5.7].

3 δ-Regularity and Quasi-stable Ideals

Not every ideal $\mathcal{I} \lhd \mathcal{P}$ possesses a finite Pommaret basis. One can show that this is solely a problem of the chosen variables $\mathcal{X}$; after a suitable linear change of variables $\tilde{\mathcal{X}} = A\mathcal{X}$ with a non-singular matrix $A \in \mathbb{k}^{n \times n}$ the transformed ideal $\tilde{\mathcal{I}} \lhd \tilde{\mathcal{P}} = \mathbb{k}[\tilde{\mathcal{X}}]$ has a finite Pommaret basis (for the same term order which we consider as being defined on exponent vectors) [25, Sect. 2].

Definition 8. *The variables $\mathcal{X}$ are δ-regular for $\mathcal{I} \lhd \mathcal{P}$ and the order $\prec$, if $\mathcal{I}$ has a finite Pommaret basis for $\prec$.*

In [25, Sect. 2] a method is presented to detect effectively whether given variables are δ-singular and, if this is the case, to produce deterministically δ-regular variables. Furthermore, it is proven there that generic variables are δ-regular so that one can also employ probabilistic approaches although these are usually computationally disadvantageous.

It seems to be rather unknown that Serre implicitly presented already in 1964 a version of δ-regularity. In a letter appended to [14], he introduced the notion of a *quasi-regular* sequence and related it to Koszul homology.[2] Let $\mathcal{V}$ be a finite-dimensional vector space, $S\mathcal{V}$ the symmetric algebra over $\mathcal{V}$ and $\mathcal{M}$ a finitely generated graded $S\mathcal{V}$-module. A vector $v \in \mathcal{V}$ is called quasi-regular at degree q for $\mathcal{M}$, if $vm = 0$ for an $m \in \mathcal{M}$ implies $m \in \mathcal{M}_{<q}$. A sequence $(v_1, \ldots, v_k)$ of vectors $v_i \in \mathcal{V}$ is quasi-regular at degree q for $\mathcal{M}$, if each v_i is quasi-regular at degree q for $\mathcal{M}/\langle v_1, \ldots, v_{i-1} \rangle \mathcal{M}$.

Given a basis $\mathcal{X}$ of $\mathcal{V}$, we can identify $S\mathcal{V}$ with the polynomial ring $\mathcal{P} = \mathbb{k}[\mathcal{X}]$. Then it is shown in [15, Thm. 5.4] that the variables $\mathcal{X}$ are δ-regular for a homogeneous ideal $\mathcal{I} \lhd \mathcal{P}$ and the reverse lexicographic order, if and only if they form a quasi-regular sequence for the module $\mathcal{P}/\mathcal{I}$ at degree reg $\mathcal{I}$.

Our first result describes the degrees appearing in the Pommaret basis for the reverse lexicographic order in an intrinsic manner and generalises [29, Lemma 2.3] where only Borel-fixed monomial ideals for char $\mathbb{k} = 0$ are considered.

Proposition 9. *Let the variables $\mathcal{X}$ be δ-regular for the ideal $\mathcal{I}$ and the reverse lexicographic order. If $\mathcal{H}$ denotes the corresponding Pommaret basis and $\mathcal{H}_i \subseteq \mathcal{H}$ the subset of generators of class i, then the integer*

$$q_i = \max \left\{ q \in \mathbb{N}_0 \mid (\langle \mathcal{I}, x_1, \ldots, x_{i-1} \rangle : x_i)_q \neq \langle \mathcal{I}, x_1, \ldots, x_{i-1} \rangle_q \right\} \qquad (4)$$

satisfies $q_i = \deg \mathcal{H}_i - 1$ (with the convention that $\deg \emptyset = \max \emptyset = -\infty$).

Proof. Set $\tilde{\mathcal{P}} = \mathbb{k}[x_i, \ldots, x_n]$ and $\tilde{\mathcal{I}} = \mathcal{I}|_{x_1 = \cdots = x_{i-1} = 0} \lhd \tilde{\mathcal{P}}$. Then it is easy to see that $q_i = \max \{ q \mid (\tilde{\mathcal{I}} : x_i)_q \neq \tilde{\mathcal{I}}_q \}$. Furthermore, the variables $x_i, \ldots, x_n$ are δ-regular for $\tilde{\mathcal{I}}$ and the reverse lexicographic order—the Pommaret basis of $\tilde{\mathcal{I}}$ is given by $\tilde{\mathcal{H}} = \bigcup_{k \geq i} \tilde{\mathcal{H}}_k$ with $\tilde{\mathcal{H}}_k = \mathcal{H}_k|_{x_1 = \cdots = x_{i-1} = 0}$ (cf. [27, Lemma 3.1]).

Assume first that $\tilde{\mathcal{H}}_i = \emptyset$. In this case $x_i f \in \tilde{\mathcal{I}}$ implies $f \in \tilde{\mathcal{I}}$, as one can immediately see from the involutive standard representation of $x_i f$ with respect

---

[2] Quasi-regular sequences were rediscovered by Schenzel et al. [23] under the name *filter-regular* sequences and by Aramova and Herzog [1] as *almost regular* sequences.

to $\tilde{\mathcal{H}}$ (all coefficients must lie in $\langle x_i \rangle$). If $\tilde{\mathcal{H}}_i \neq \emptyset$, then we choose a generator $\tilde{h}_{\max} \in \tilde{\mathcal{H}}_i$ of maximal degree. By the properties of $\prec_{\mathrm{revlex}}$, we find $\tilde{h}_{\max} \in \langle x_i \rangle$ and hence may write $\tilde{h}_{\max} = x_i \tilde{g}$. By definition of a Pommaret basis, $\tilde{g} \notin \tilde{\mathcal{I}}$ and thus $q_i \geq \deg \tilde{g} = \deg \mathcal{H}_i - 1$.

Assume now that $q_i > \deg \mathcal{H}_i - 1$. Then there exists a polynomial $\tilde{f} \in \tilde{\mathcal{P}} \setminus \tilde{\mathcal{I}}$ with $\deg \tilde{f} = q_i$ and $x_i \tilde{f} \in \tilde{\mathcal{I}}$. Consider the involutive standard representation $x_i \tilde{f} = \sum_{\tilde{h} \in \tilde{\mathcal{H}}} P_{\tilde{h}} \tilde{h}$ with respect to $\tilde{\mathcal{H}}$. If $\mathrm{cls}\,\tilde{h} > i$, then we must have $P_{\tilde{h}} \in \langle x_i \rangle$. If $\mathrm{cls}\,\tilde{h} = i$, then by definition $P_{\tilde{h}} \in \Bbbk[x_i]$. Since $\deg(x_i \tilde{f}) > \deg \tilde{\mathcal{H}}_i$, any non-vanishing coefficient $P_{\tilde{h}}$ must be of positive degree in this case. Thus we can conclude that all non-vanishing coefficients $P_{\tilde{h}}$ lie in $\langle x_i \rangle$. But then we may divide the involutive standard representation of $x_i \tilde{f}$ by x_i and obtain an involutive standard representation of $\tilde{f}$ itself so that $\tilde{f} \in \tilde{\mathcal{I}}$ in contradiction to the assumptions we made. □

Consider the following invariants related to the local cohomology of $\mathcal{P}/\mathcal{I}$ (with respect to the maximal graded ideal $\mathfrak{m} = \langle x_1, \ldots, x_n \rangle$):

$$
\begin{aligned}
a_i(\mathcal{P}/\mathcal{I}) &= \max \left\{ q \mid H^i_{\mathfrak{m}}(\mathcal{P}/\mathcal{I})_q \neq 0 \right\}, & 0 &\leq i \leq \dim(\mathcal{P}/\mathcal{I}), \\
\mathrm{reg}_t(\mathcal{P}/\mathcal{I}) &= \max \left\{ a_i(\mathcal{P}/\mathcal{I}) + i \mid 0 \leq i \leq t \right\}, & 0 &\leq t \leq \dim(\mathcal{P}/\mathcal{I}), \\
a_t^*(\mathcal{P}/\mathcal{I}) &= \max \left\{ a_i(\mathcal{P}/\mathcal{I}) \mid 0 \leq i \leq t \right\}, & 0 &\leq t \leq \dim(\mathcal{P}/\mathcal{I}).
\end{aligned}
$$

Trung [29, Thm. 2.4] related them for monomial Borel-fixed ideals and char $\Bbbk = 0$ to the degrees of the minimal generators. We can now generalise this result to arbitrary homogeneous polynomial ideals.

Corollary 10. *Let the variables $\mathcal{X}$ be δ-regular for the ideal $\mathcal{I} \lhd \mathcal{P}$ and the reverse lexicographic order. Denote again by $\mathcal{H}_i$ the subset of the Pommaret basis $\mathcal{H}$ of $\mathcal{I}$ consisting of the generators of class i and set $q_i = \deg \mathcal{H}_i - 1$. Then*

$$
\begin{aligned}
\mathrm{reg}_t(\mathcal{P}/\mathcal{I}) &= \max \left\{ q_1, q_2, \ldots, q_{t+1} \right\}, & 0 &\leq t \leq \dim(\mathcal{P}/\mathcal{I}), \\
a_t^*(\mathcal{P}/\mathcal{I}) &= \max \left\{ q_1, q_2 - 1, \ldots, q_{t+1} - t \right\}, & 0 &\leq t \leq \dim(\mathcal{P}/\mathcal{I}).
\end{aligned}
$$

Proof. This follows immediately from [29, Thm. 1.1] and Proposition 9. □

For monomial ideals it is in general useless to transform to δ-regular variables, as the transformed ideal is no longer monomial. Hence it is a special property of a monomial ideal to possess a finite Pommaret basis: such an ideal is called *quasi-stable*. The following theorem provides several purely algebraic characterisations of quasi-stability independent of Pommaret bases. It combines ideas and results from [4, Def. 1.5], [6, Prop. 3.2/3.6], [19, Prop. 2.2] and [25, Prop. 4.4].

Theorem 11. *Let $\mathcal{I} \lhd \mathcal{P}$ be a monomial ideal and $D = \dim(\mathcal{P}/\mathcal{I})$. Then the following statements are equivalent.*

(i) *$\mathcal{I}$ is quasi-stable.*
(ii) *The variable x_1 is not a zero divisor for $\mathcal{P}/\mathcal{I}^{\mathrm{sat}}$ and for all $1 \leq k < D$ the variable x_{k+1} is not a zero divisor for $\mathcal{P}/\langle \mathcal{I}, x_1, \ldots, x_k \rangle^{\mathrm{sat}}$.*

(iii) *We have* $\mathcal{I} : x_1^\infty \subseteq \mathcal{I} : x_2^\infty \subseteq \cdots \subseteq \mathcal{I} : x_D^\infty$ *and for all* $D < k \leq n$ *an exponent* $e_k \geq 1$ *exists such that* $x_k^{e_k} \in \mathcal{I}$.

(iv) *For all* $1 \leq k \leq n$ *the equality* $\mathcal{I} : x_k^\infty = \mathcal{I} : \langle x_k, \ldots, x_n \rangle^\infty$ *holds.*

(v) *For every associated prime ideal* $\mathfrak{p} \in \mathrm{Ass}(\mathcal{P}/\mathcal{I})$ *an integer* $1 \leq j \leq n$ *exists such that* $\mathfrak{p} = \langle x_j, \ldots, x_n \rangle$.

(vi) *If* $x^\mu \in \mathcal{I}$ *and* $\mu_i > 0$ *for some* $1 \leq i < n$, *then for each* $0 < r \leq \mu_i$ *and* $i < j \leq n$ *an integer* $s \geq 0$ *exists such that* $x_j^s x^\mu / x_i^r \in \mathcal{I}$.

The terminology "quasi-stable" stems from a result of Mall. The minimality assumption is essential here, as the simple example $\langle x^2, y^2 \rangle \lhd \Bbbk[x,y]$ shows.

Lemma 12 ([21, Lemma 2.13], [26, Prop. 5.5.6]). *A monomial ideal is stable,*[3] *if and only if its minimal basis is a Pommaret basis.*

Thus already in the monomial case Pommaret bases are generally not minimal. The following result of Mall characterises those polynomial ideals for which the reduced Gröbner basis is simultaneously a Pommaret basis. We provide here a much simpler proof due to a more suitable definition of Pommaret bases.

Theorem 13 ([21, Thm. 2.15]). *The reduced Gröbner basis of the ideal* $\mathcal{I} \lhd \mathcal{P}$ *is simultaneously a Pommaret basis, if and only if* $\mathrm{lt}\,\mathcal{I}$ *is stable.*

Proof. By definition, the leading terms $\mathrm{lt}\,\mathcal{G}$ of a reduced Gröbner basis $\mathcal{G}$ form the minimal basis of $\mathrm{lt}\,\mathcal{I}$. The assertion is now a trivial corollary to Lemma 12 and the definition of a Pommaret basis. □

4   The Generic Initial Ideal

If we fix an order $\prec$ and perform a linear change of variables $\tilde{\mathcal{X}} = A\mathcal{X}$ with a non-singular matrix $A \in \Bbbk^{n \times n}$, then, according to Galligo's Theorem [10,5], for almost all matrices A the transformed ideal $\tilde{\mathcal{I}} \lhd \tilde{\mathcal{P}} = \Bbbk[\tilde{\mathcal{X}}]$ has the same leading ideal, the *generic initial ideal* $\mathrm{gin}\,\mathcal{I}$ for the used order. By a further result of Galligo [11,5], $\mathrm{gin}\,\mathcal{I}$ is Borel fixed, i. e. invariant under the natural action of the Borel group. For char $\Bbbk = 0$, the Borel fixed ideals are precisely the stable ones; in positive characteristics the property of being Borel fixed has no longer such a simple combinatorial interpretation.

We will show in this section that many properties of the generic initial ideal $\mathrm{gin}\,\mathcal{I}$ also hold for the ordinary leading ideal $\mathrm{lt}\,\mathcal{I}$—provided the used variables are δ-regular. This observation has a number of consequences. While there does not exist an effective criterion for deciding whether a given leading ideal is actually $\mathrm{gin}\,\mathcal{I}$, δ-regularity is simply proven by the existence of a finite Pommaret basis. Furthermore, $\mathrm{gin}\,\mathcal{I}$ can essentially be computed only by applying a random change of variables which has many disadvantages from a computational point of view. By contrast, [25, Sect. 2] presents a deterministic approach for the

---

[3] In our "reverse" conventions, a monomial ideal $\mathcal{I}$ is called *stable*, if for every term $t \in \mathcal{I}$ and every index $k = \mathrm{cls}\,t < i \leq n$ also $x_i t / x_k \in \mathcal{I}$.

construction of δ-regular variables which in many case will lead to fairly sparse transformations.

From a theoretical point of view, the following trivial lemma which already appeared in [5,10] implies that proving a statement about quasi-stable leading ideals immediately entails the analogous statement about $\mathrm{gin}\,\mathcal{I}$.

Lemma 14. *The generic initial ideal* $\mathrm{gin}\,\mathcal{I}$ *is quasi-stable.*

Proof. For char $\Bbbk = 0$, the assertion is trivial, since then $\mathrm{gin}\,\mathcal{I}$ is even stable, as mentioned above. For arbitrary char $\Bbbk$, it follows simply from the fact that generic variables[4] are δ-regular and thus yield a quasi-stable leading ideal. $\square$

The next corollary is a classical result [13, Cor. 1.33] for which we provide here a simple alternative proof. The subsequent theorem extends many well-known statements about $\mathrm{gin}\,\mathcal{I}$ to the leading ideal in δ-regular variables (for $\prec_{\mathrm{revlex}}$); they are all trivial consequences of the properties of a Pommaret basis.

Corollary 15. *Let* $\mathcal{I} \lhd \mathcal{P}$ *be an ideal and* char $\Bbbk = 0$. *Then all bigraded Betti numbers satisfy the inequality* $\beta_{i,j}(\mathcal{P}/\mathcal{I}) \leq \beta_{i,j}(\mathcal{P}/\mathrm{gin}\,\mathcal{I})$.

Proof. We choose variables $\mathcal{X}$ such that $\mathrm{lt}\,\mathcal{I} = \mathrm{gin}\,\mathcal{I}$. By Lemma 14, these variables are δ-regular for the given ideal $\mathcal{I}$. As char $\Bbbk = 0$, the generic initial ideal is stable and hence the bigraded version of (3) applied to $\mathrm{lt}\,\mathcal{I}$ yields the bigraded Betti number $\beta_{i,j}(\mathcal{P}/\mathrm{gin}\,\mathcal{I})$. Now the claim follows immediately from analysing the resolution (2) degree by degree. $\square$

Theorem 16. *Let the variables* $\mathcal{X}$ *be* δ-*regular for the ideal* $\mathcal{I} \lhd \mathcal{P}$ *and the reverse lexicographic order* $\prec_{\mathrm{revlex}}$.

(i) $\mathrm{pd}\,\mathcal{I} = \mathrm{pd}\,\mathrm{lt}\,\mathcal{I}$.
(ii) $\mathrm{sat}\,\mathcal{I} = \mathrm{sat}\,\mathrm{lt}\,\mathcal{I}$.
(iii) $\mathrm{reg}\,\mathcal{I} = \mathrm{reg}\,\mathrm{lt}\,\mathcal{I}$.
(iv) $\mathrm{reg}_t\,\mathcal{I} = \mathrm{reg}_t\,\mathrm{lt}\,\mathcal{I}$ *for all* $0 \leq t \leq \dim(\mathcal{P}/\mathcal{I})$.
(v) $a_t^*(\mathcal{I}) = a_t^*(\mathrm{lt}\,\mathcal{I})$ *for all* $0 \leq t \leq \dim(\mathcal{P}/\mathcal{I})$.
(vi) *The extremal Betti numbers of* $\mathcal{I}$ *and* $\mathrm{lt}\,\mathcal{I}$ *occur at the same positions and have the same values.*
(vii) $\mathrm{depth}\,\mathcal{I} = \mathrm{depth}\,\mathrm{lt}\,\mathcal{I}$.
(viii) $\mathcal{P}/\mathcal{I}$ *is Cohen-Macaulay, if and only if* $\mathcal{P}/\mathrm{lt}\,\mathcal{I}$ *is Cohen-Macaulay.*

Proof. The assertions (i-v) are trivial corollaries of Theorem 6 and Corollary 10, respectively, where it is shown for all considered quantities that they depend only on the leading terms of the Pommaret basis of $\mathcal{I}$. Assertion (vi) is a consequence of Remark 7 and the assertions (vii) and (viii) follow from Theorem 4. $\square$

Remark 17. In view of Part (viii), one may wonder whether a similar statement holds for Gorenstein rings. In [27, Ex. 5.5] the ideal $\mathcal{I} = \langle z^2 - xy, yz, y^2, xz, x^2 \rangle \lhd$

---

[4] Recall that we assume throughout that $\Bbbk$ is an infinite field, although a sufficiently large finite field would also suffice [26, Rem. 4.3.19].

$\Bbbk[x, y, z]$ is studied. The used coordinates are δ-regular for $\prec_{\text{revlex}}$, as a Pommaret basis is obtained by adding the generator $x^2 y$. It follows from [27, Thm. 5.4] that $\mathcal{P}/\mathcal{I}$ is Gorenstein, but $\mathcal{P}/\operatorname{lt}\mathcal{I}$ not. A computation with CoCoA [8] gives here $\operatorname{gin}\mathcal{I} = \langle z^2, yz, y^2, xz, xy, x^3 \rangle$ (assuming $\operatorname{char}\Bbbk = 0$) and again one may conclude with [27, Thm. 5.4] that $\mathcal{P}/\operatorname{gin}\mathcal{I}$ is not Gorenstein.

5   Componentwise Linear Ideals

Given an ideal $\mathcal{I} \lhd \mathcal{P}$, we denote by $\mathcal{I}_{\langle d \rangle} = \langle \mathcal{I}_d \rangle$ the ideal generated by the homogeneous component $\mathcal{I}_d$ of degree d. Herzog and Hibi [16] called $\mathcal{I}$ componentwise linear, if for every degree $d \geq 0$ the ideal $\mathcal{I}_{\langle d \rangle} = \langle \mathcal{I}_d \rangle$ has a linear resolution. For a connection with Pommaret bases, we need a refinement of δ-regularity.

Definition 18. *The variables $\mathcal{X}$ are* componentwise δ-regular *for the ideal $\mathcal{I}$ and the order $\prec$, if all ideals $\mathcal{I}_{\langle d \rangle}$ for $d \geq 0$ have finite Pommaret bases for $\prec$.*

It follows from the proof of [25, Thm. 9.12] that for the definition of componentwise δ-regularity it suffices to consider the finitely many degrees $d \leq \operatorname{reg}\mathcal{I}$. Thus trivial modificiations of any method for the construction of δ-regular variables allow to determine effectively componentwise δ-regular variables.

Theorem 19 ([25, Thm. 8.2, Thm. 9.12]). *Let the variables $\mathcal{X}$ be componentwise δ-regular for the ideal $\mathcal{I} \lhd \mathcal{P}$ and the reverse lexicographic order. If $\mathcal{I}$ is componentwise linear, then the free resolution (2) of $\mathcal{I}$ induced by the Pommaret basis $\mathcal{H}$ is minimal and the Betti numbers of $\mathcal{I}$ are given by (3). Conversely, if the resolution (2) is minimal, then the ideal $\mathcal{I}$ is componentwise linear.*

The following corollary generalises the analogous result for stable ideals to componentwise linear ideals (Aramova et al. [2, Thm. 1.2(a)] noted a version for $\operatorname{gin}\mathcal{I}$). It is an immediate consequence of the linear construction of the resolution (2) in [25, Thm. 6.1] and its minimality for componentwise linear ideals.

Corollary 20. *Let $\mathcal{I} \lhd \mathcal{P}$ be componentwise linear. If the Betti number $\beta_{i,j}$ does not vanish, then also all Betti numbers $\beta_{i',j}$ with $i' < i$ do not vanish.*

As a further corollary, we obtain a simple proof of an estimate given by Aramova et al. [2, Cor. 1.5] (based on [18, Thm. 2]).

Corollary 21. *Let $\mathcal{I} \lhd \mathcal{P}$ be a componentwise linear ideal with $\operatorname{pd}\mathcal{I} = p$. Then the Betti numbers satisfy $\beta_i \geq \binom{p+1}{i+1}$.*

Proof. Let $\mathcal{H}$ be the Pommaret basis of $\mathcal{I}$ for $\prec_{\text{revlex}}$ in componentwise δ-regular variables and $d = \operatorname{cls}\mathcal{H}$. By Theorem 19, (2) is the minimal resolution of $\mathcal{I}$ and hence (3) gives us β_i. By Theorem 4, $p = n - d$. We also note that δ-regularity implies that $\beta_0^{(k)} > 0$ for all $d \leq k \leq n$. Now we compute

$$\beta_i = \sum_{k=d}^{n-i} \binom{n-k}{i} \beta_0^{(k)} = \sum_{\ell=i}^{p} \binom{\ell}{i} \beta_0^{(n-\ell)} \geq \sum_{\ell=i}^{p} \binom{\ell}{i} = \binom{p+1}{i+1}$$

by a well-known identity for binomial coefficients. $\square$

Example 22. The estimate in Corollary 21 is sharp. It is realised by any componentwise linear ideal whose Pommaret basis satisfies $\beta_0^{(i)} = 0$ for $i < d$ and $\beta_0^{(i)} = 1$ for $i \geq d$. As a simple monomial example consider the ideal $\mathcal{I}$ generated by the d terms $h_1 = x_n^{\alpha_n+1}$, $h_2 = x_n^{\alpha_n} x_{n-1}^{\alpha_{n-1}+1}, \ldots, h_d = x_n^{\alpha_n} \cdots x_{d+1}^{\alpha_{d+1}} x_d^{\alpha_d+1}$ for arbitrary exponents $\alpha_i \geq 0$. One easily verifies that $\mathcal{H} = \{h_1, \ldots, h_d\}$ is indeed simultaneously the Pommaret and the minimal basis of $\mathcal{I}$.

Recently, Nagel and Römer [22, Thm. 2.5] provided some criteria for componentwise linearity based on $\mathrm{gin}\,\mathcal{I}$ (see also [2, Thm 1.1] where the case char $\Bbbk = 0$ is treated). We will now show that again $\mathrm{gin}\,\mathcal{I}$ may be replaced by $\mathrm{lt}\,\mathcal{I}$, if one uses componentwise δ-regular variables. Furthermore, our proof is considerably simpler than the one by Nagel and Römer.

Theorem 23. *Let the variables $\mathcal{X}$ be componentwise δ-regular for the ideal $\mathcal{I} \lhd \mathcal{P}$ and the reverse lexicographic order. Then the following statements are equivalent:*

(i) *$\mathcal{I}$ is componentwise linear.*
(ii) *$\mathrm{lt}\,\mathcal{I}$ is stable and all bigraded Betti numbers β_{ij} of $\mathcal{I}$ and $\mathrm{lt}\,\mathcal{I}$ coincide.*
(iii) *$\mathrm{lt}\,\mathcal{I}$ is stable and all total Betti numbers β_i of $\mathcal{I}$ and $\mathrm{lt}\,\mathcal{I}$ coincide.*
(iv) *$\mathrm{lt}\,\mathcal{I}$ is stable and $\beta_0(\mathcal{I}) = \beta_0(\mathrm{lt}\,\mathcal{I})$.*

Proof. The implication "(i) $\Rightarrow$ (ii)" is a simple consequence of Theorem 19. Since our variables are componentwise δ-regular, the resolution (2) is minimal. This implies immediately that $\mathrm{lt}\,\mathcal{I}$ is stable. Applying Theorem 5 to the Pommaret basis $\mathrm{lt}\,\mathcal{H}$ of $\mathrm{lt}\,\mathcal{I}$ yields the minimal resolution of $\mathrm{lt}\,\mathcal{I}$. In both cases, the leading terms of all syzygies are determined by $\mathrm{lt}\,\mathcal{H}$ and hence the bigraded Betti numbers of $\mathcal{I}$ and $\mathrm{lt}\,\mathcal{I}$ coincide.

The implications "(ii) $\Rightarrow$ (iii)" and "(iii) $\Rightarrow$ (iv)" are trivial. Thus there only remains to prove "(iv) $\Rightarrow$ (i)". Let $\mathcal{H}$ be the Pommaret basis of $\mathcal{I}$. Since $\mathrm{lt}\,\mathcal{I}$ is stable by assumption, $\mathrm{lt}\,\mathcal{H}$ is its minimal basis by Lemma 12 and $\beta_0(\mathrm{lt}\,\mathcal{I})$ equals the number of elements of $\mathcal{H}$. The assumption $\beta_0(\mathcal{I}) = \beta_0(\mathrm{lt}\,\mathcal{I})$ implies that $\mathcal{H}$ is a minimal generating system of $\mathcal{I}$. Hence, none of the syzygies obtained from the involutive standard representations of the non-multiplicative products yh with $h \in \mathcal{H}$ and $y \in \overline{\mathcal{X}}_P(h)$ may contain a non-vanishing constant coefficients. By [25, Lemma 8.1], this observation implies that the resolution (2) induced by $\mathcal{H}$ is minimal and hence the ideal $\mathcal{I}$ is componentwise linear by Theorem 19.   □

6   Linear Quotients

Linear quotients were introduced by Herzog and Takayama [20] in the context of constructing iteratively a free resolution via mapping cones. As a special case, they considered monomial ideals where certain colon ideals defined by an ordered minimal basis are generated by variables. Their definition was generalised by Sharifan and Varabaro [28] to arbitrary ideals.

Definition 24. *Let $\mathcal{I} \lhd \mathcal{P}$ be an ideal and $\mathcal{F} = \{f_1, \ldots, f_r\}$ an ordered basis of it. Then $\mathcal{I}$ has linear quotients with respect to $\mathcal{F}$, if for each $1 < k \leq r$ the ideal $\langle f_1, \ldots, f_{k-1} \rangle : f_k$ is generated by a subset $\mathcal{X}_k \subseteq \mathcal{X}$ of variables.*

We show first that in the monomial case this concept captures the essence of a Pommaret basis. For this purpose, we "invert" some notions introduced in [25]. We associate with a monomial Pommaret basis $\mathcal{H}$ a directed graph, its *P-graph*. Its vertices are the elements of $\mathcal{H}$. Given a non-multiplicative variable $x_j \in \overline{\mathcal{X}}_P(h)$ for a generator $h \in \mathcal{H}$, there exists a unique involutive divisor $\bar{h} \in \mathcal{H}$ of $x_j h$ and we include a directed edge from h to $\bar{h}$.

An ordering of the elements of $\mathcal{H}$ is called an *inverse P-ordering*, if $\alpha > \beta$ whenever the P-graph contains a path from h_α to h_β. It is straightforward to describe explicitly an inverse P-ordering: we set $\alpha > \beta$, if $\operatorname{cls} h_\alpha < \operatorname{cls} h_\beta$ or if $\operatorname{cls} h_\alpha = \operatorname{cls} h_\beta$ and $h_\alpha \prec_{\text{lex}} h_\beta$, i.e. we sort the generators h_α first by their class and then within each class lexicographically (according to our reverse conventions!). One easily verifies that this defines an inverse P-ordering.

Example 25. Consider the monomial ideal $\mathcal{I} \subset \Bbbk[x, y, z]$ generated by the six terms $h_1 = z^2$, $h_2 = yz$, $h_3 = y^2$, $h_4 = xz$, $h_5 = xy$ and $h_6 = x^2$. One easily verifies that these terms form a Pommaret basis of $\mathcal{I}$. The P-graph in (5) shows that the generators are already inversely P-ordered, namely according to the description above.

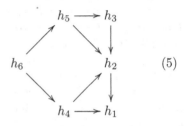

(5)

Proposition 26. *Let $\mathcal{H} = \{h_1, \ldots, h_r\}$ be an inversely P-ordered monomial Pommaret basis of the quasi-stable monomial ideal $\mathcal{I} \lhd \mathcal{P}$. Then the ideal $\mathcal{I}$ possesses linear quotients with respect to the basis $\mathcal{H}$ and*

$$\langle h_1, \ldots, h_{k-1} \rangle : h_k = \langle \overline{\mathcal{X}}_P(h_k) \rangle \qquad k = 1, \ldots r \,. \tag{6}$$

Conversely, assume that $\mathcal{H} = \{h_1, \ldots, h_r\}$ is a monomial generating set of the monomial ideal $\mathcal{I} \lhd \mathcal{P}$ such that (6) is satisfied. Then $\mathcal{I}$ is quasi-stable and $\mathcal{H}$ its Pommaret basis.

Proof. Let $y \in \overline{\mathcal{X}}_P(h_k)$ be a non-multiplicative variable for $h_k \in \mathcal{H}$. Since $\mathcal{H}$ is a Pommaret basis, the product $y h_k$ possesses an involutive divisor $h_i \in \mathcal{H}$ and, by definition, the P-graph of $\mathcal{H}$ contains an edge from k to i. Thus $i < k$ for an inverse P-ordering, which proves the inclusion "$\supseteq$".

The following argument shows that the inclusion cannot be strict. Consider a term $t \in \Bbbk[\mathcal{X}_P(h_k)]$ consisting entirely of multiplicative variables and assume that $t h_k \in \langle h_1, \ldots, h_{k-1} \rangle$, i.e. $t h_k = s_1 h_{i_1}$ for some term $s_1 \in \Bbbk[\mathcal{X}]$ and some index $i_1 < k$. By definition of a Pommaret basis, s_1 must contain at least one non-multiplicative variable y_1 of h_{i_1}. But now we may rewrite $y_1 h_{i_1} = s_2 h_{i_2}$ with $i_2 < i_1$ and $s_2 \in \Bbbk[\mathcal{X}_P(h_{i_2})]$. Since this implies $\operatorname{cls} h_2 \geq \operatorname{cls} h_1$, we find $\mathcal{X}_P(h_{i_1}) \subseteq \mathcal{X}_P(h_{i_2})$. Hence after a finite number of iterations we arrive at a representation $t h_k = s h_i$ where $s \in \Bbbk[\mathcal{X}_P(h_i)]$ which is, however, not possible for a Pommaret basis.

For the converse, we show by a finite induction over k that every non-multiplicative product $y h_k$ with $y \in \overline{\mathcal{X}}_P(h_k)$ possesses an involutive divisor h_i with $i < k$ which implies our assertion by Proposition 3. For $k = 1$ nothing is to be shown, since (6) implies in this case that all variables are multiplicative for h_1 (and thus this generator is of the form $h_1 = x_n^\ell$ for some $\ell > 0$), and $k = 2$ is trivial. Assume that our claim was true for $h_1, h_2, \ldots, h_{k-1}$. Because of (6), we may write $y h_k = t_1 h_{i_1}$ for some $i_1 < k$. If $t_1 \in \mathbb{k}[\mathcal{X}_P(h_{i_1})]$, we set $i = i_1$ and are done. Otherwise, t_1 contains a non-multiplicative variable $y_1 \in \overline{\mathcal{X}}_P(h_{i_1})$. By our induction assumption, $y_1 h_{i_1}$ has an involutive divisor h_{i_2} with $i_2 < i_1$ leading to an alternative representation $y h_k = t_2 h_{i_2}$. Now we iterate and find after finitely many steps an involutive divisor h_i of $y h_k$, since the sequence $i_1 > i_2 > \cdots$ is strictly decreasing and h_1 has no non-multiplicative variables.   □

Remark 27. As we are here exclusively concerned with Pommaret bases, we formulated and proved the above result only for this special case. However, Proposition 26 remains valid for any involutive basis with respect to a *continuous* involutive division L (and thus for all divisions of practical interest). The continuity of L is needed here for two reasons. Firstly, it guarantees the existence of an L-ordering, as for such divisions the L-graph is always acyclic [25, Lemma 5.5]. Secondly, the above argument that finitely many iterations lead to a representation $t h_k = s h_i$ where s contains only multiplicative variables for h_i is specific for the Pommaret division and cannot be generalised. However, the very definition of continuity [12, Def. 4.9] ensures that for continuous divisions such a rewriting cannot be done infinitely often.

In general, we cannot expect that the second part of Proposition 26 remains true, when we consider arbitrary polynomial ideals. However, for the first part we find the following variation of [28, Thm. 2.3].

Proposition 28. *Let $\mathcal{H}$ be a Pommaret basis of the polynomial ideal $\mathcal{I} \lhd \mathcal{P}$ for the term order $\prec$ and $h' \in \mathcal{P}$ a polynomial with $\operatorname{lt} h' \notin \operatorname{lt} \mathcal{H}$. If $\mathcal{I} : h' = \langle \overline{\mathcal{X}}_P(h') \rangle$, then $\mathcal{H}' = \mathcal{H} \cup \{h'\}$ is a Pommaret basis of $\mathcal{J} = \mathcal{I} + \langle h' \rangle$. If furthermore $\mathcal{I}$ is componentwise linear, the variables $\mathcal{X}$ are componentwise δ-regular and $\mathcal{H}'$ is a minimal basis of $\mathcal{J}$, then $\mathcal{J}$ is componentwise linear, too.*

Proof. If $\mathcal{I} : h' = \langle \overline{\mathcal{X}}_P(h') \rangle$, then all products of h' with one of its non-multiplicative variables lie in $\mathcal{I}$ and hence possess an involutive standard representation with respect to $\mathcal{H}$. This immediately implies the first assertion.

In componentwise δ-regular variables all syzygies obtained from the involutive standard representations of products $y h$ with $h \in \mathcal{H}$ and $y \in \overline{\mathcal{X}}_P(h)$ are free of constant coefficients, if $\mathcal{I}$ is componentwise linear. If $\mathcal{H}'$ is a minimal basis of $\mathcal{J}$, the same is true for all syzygies obtained from products $y h'$ with $y \in \overline{\mathcal{X}}_P(h')$. Hence we can again conclude with [25, Lemma 8.1] that the resolution of $\mathcal{J}$ induced by $\mathcal{H}'$ is minimal and $\mathcal{J}$ componentwise linear by Theorem 19.   □

References

1. Aramova, A., Herzog, J.: Almost regular sequences and Betti numbers. Amer. J. Math. 122, 689–719 (2000)
2. Aramova, A., Herzog, J., Hibi, T.: Ideals with stable Betti numbers. Adv. Math. 152, 72–77 (2000)

3. Bayer, D., Charalambous, H., Popescu, S.: Extremal Betti numbers and applications to monomial ideals. J. Alg. 221, 497–512 (1999)
4. Bayer, D., Stillman, M.: A criterion for detecting m-regularity. Invent. Math. 87, 1–11 (1987)
5. Bayer, D., Stillman, M.: A theorem on refining division orders by the reverse lexicographic orders. Duke J. Math. 55, 321–328 (1987)
6. Bermejo, I., Gimenez, P.: Saturation and Castelnuovo-Mumford regularity. J. Alg. 303, 592–617 (2006)
7. Caviglia, G., Sbarra, E.: Characteristic-free bounds for the Castelnuovo-Mumford regularity. Compos. Math. 141, 1365–1373 (2005)
8. CoCoATeam: CoCoA: a system for doing Computations in Commutative Algebra, http://cocoa.dima.unige.it
9. Eisenbud, D.: Commutative Algebra with a View Toward Algebraic Geometry. Graduate Texts in Mathematics, vol. 150. Springer, New York (1995)
10. Galligo, A.: A propos du théorème de préparation de Weierstrass. In: Norguet, F. (ed.) Fonctions de Plusieurs Variables Complexes. Lecture Notes in Mathematics, vol. 409, pp. 543–579. Springer, Berlin (1974)
11. Galligo, A.: Théorème de division et stabilité en géometrie analytique locale. Ann. Inst. Fourier 29(2), 107–184 (1979)
12. Gerdt, V., Blinkov, Y.: Involutive bases of polynomial ideals. Math. Comp. Simul. 45, 519–542 (1998)
13. Green, M.: Generic initial ideals. In: Elias, J., Giral, J., Miró-Roig, R., Zarzuela, S. (eds.) Six Lectures on Commutative Algebra. Progress in Mathematics, vol. 166, pp. 119–186. Birkhäuser, Basel (1998)
14. Guillemin, V., Sternberg, S.: An algebraic model of transitive differential geometry. Bull. Amer. Math. Soc. 70, 16–47 (1964), (With a letter of Serre as appendix)
15. Hausdorf, M., Sahbi, M., Seiler, W.: δ- and quasi-regularity for polynomial ideals. In: Calmet, J., Seiler, W., Tucker, R. (eds.) Global Integrability of Field Theories, pp. 179–200. Universitätsverlag Karlsruhe, Karlsruhe (2006)
16. Herzog, J., Hibi, T.: Componentwise linear ideals. Nagoya Math. J. 153, 141–153 (1999)
17. Herzog, J., Hibi, T.: Monomial Ideals. Graduate Texts in Mathematics, vol. 260. Springer, London (2011)
18. Herzog, J., Kühl, M.: On the Bettinumbers of finite pure and linear resolutions. Comm. Alg. 12, 1627–1646 (1984)
19. Herzog, J., Popescu, D., Vladoiu, M.: On the Ext-modules of ideals of Borel type. In: Commutative Algebra. Contemp. Math, vol. 331, pp. 171–186. Amer. Math. Soc., Providence (2003)
20. Herzog, J., Takayama, Y.: Resolutions by mapping cones. Homol. Homot. Appl. 4, 277–294 (2002)
21. Mall, D.: On the relation between Gröbner and Pommaret bases. Appl. Alg. Eng. Comm. Comp. 9, 117–123 (1998)
22. Nagel, U., Römer, T.: Criteria for componentwise linearity. Preprint arXiv:1108.3921 (2011)
23. Schenzel, P., Trung, N., Cuong, N.: Verallgemeinerte Cohen-Macaulay-Moduln. Math. Nachr. 85, 57–73 (1978)
24. Seiler, W.: A combinatorial approach to involution and δ-regularity I: Involutive bases in polynomial algebras of solvable type. Appl. Alg. Eng. Comm. Comp. 20, 207–259 (2009)

25. Seiler, W.: A combinatorial approach to involution and δ-regularity II: Structure analysis of polynomial modules with Pommaret bases. Appl. Alg. Eng. Comm. Comp. 20, 261–338 (2009)

26. Seiler, W.: Involution — The Formal Theory of Differential Equations and its Applications in Computer Algebra. Algorithms and Computation in Mathematics, vol. 24. Springer, Berlin (2009)

27. Seiler, W.: Effective genericity, δ-regularity and strong Noether position. Comm. Alg. (to appear)

28. Sharifan, L., Varbaro, M.: Graded Betti numbers of ideals with linear quotients. Matematiche 63, 257–265 (2008)

29. Trung, N.: Gröbner bases, local cohomology and reduction number. Proc. Amer. Math. Soc. 129, 9–18 (2001)

Invariant Theory:
Applications and Computations
(Invited Talk)

Gregor Kemper

Zentrum Mathematik M11,
Technische Universität München
Boltzmannstr. 3, 85748 Garching, Germany
Kemper@ma.tum.de

Abstract. Being at the crossroads of several mathematical disciplines, invariant theory has a wide range of applications. Many of these depend on computing generating or at least separating subsets of rings of invariants. This talk gives some examples in which invariant theory is applied to graph theory, computer vision, and coding theory. We also give an overview of the state of the art of algorithmic invariant theory.

V.P. Gerdt et al. (Eds.): CASC 2012, LNCS 7442, p. 185, 2012.
© Springer-Verlag Berlin Heidelberg 2012

Local Generic Position for Root Isolation of Zero-Dimensional Triangular Polynomial Systems

Jia Li[1], Jin-San Cheng[2], and Elias P. Tsigaridas[3]

[1] Beijing Electronic Science and Technology Institute
[2] KLMM, AMSS, Chinese Academy of Sciences
[3] POLSYS project, INRIA, LIP6/CNRS
jcheng@amss.ac.cn, lijia@besti.edu.cn, elias@polsys.lip6.fr

Abstract. We present an algorithm to isolate the real roots, and compute their multiplicities, of a zero-dimensional triangular polynomial system, based on the local generic position method. We also present experiments that demonstrate the efficiency of the method.

1 Introduction

Solving polynomial systems is a basic problem in the fields of computational sciences, engineering, etc. A usual technique is to transform the input polynomial system to a triangular one using well known algebraic elimination methods, such as Gröbner bases, characteristic sets, CAD, and resultants. In most of the cases we have to deal with zero-dimensional systems. For example, for computing the topology of a real algebraic curve or surface with CAD based methods [2,6,13] we need to isolate the real roots of a zero-dimensional triangular system and also know their multiplities.

A (zero-dimensional) triangular system has the form $\Sigma_n = \{f_1, \ldots, f_n\}$, where $f_i \in \mathbb{Q}[x_1, \ldots, x_i]$ $(i = 1, \ldots, n)$, and $\mathbb{Q}$ is the field of rational numbers. Our aim is to isolate the zeros $\boldsymbol{\xi}^n = (\xi_1, \ldots, \xi_n) \in \mathbb{C}^n$ (or $\mathbb{R}^n$) of Σ_n, where $\mathbb{C}, \mathbb{R}$ are the fields of complex and real numbers, respectively.

The local generic position method (shortly LGP) was introduced in [4]. It was used to solve bivariate polynomial systems and the experiments show that it is competitive in practice. The method has been extended to solve general zero-dimensional systems using Gröbner basis computations and linear univariate representation [5]. In this paper, we extend LGP to solve general zero-dimensional triangular systems using only resultant computations.

We will explain how to isolate the roots of a zero-dimensional polynomial system as $\Sigma = \{f(x), g(x, y), h(x, y, z)\}$. The case with more variables are similar. At first, we isolate the roots of $f(x) = 0$ and compute the root separation bound as r_1. Then we compute the root bound of $g(x, y) = f(x) = 0$ on y and denote as R_2. Choose a rational number s_1 such that $0 < s_1 < r_1/R_2$. The roots of $f(x - s_1 y) = g(x - s_1 y, y) = 0$ are in a generic position. Let $h_1 = \mathrm{Res}_y(f(x - s_1 y), g(x - s_1 y, y))$. And the roots of $f = g = 0$ corresponding

V.P. Gerdt et al. (Eds.): CASC 2012, LNCS 7442, pp. 186–197, 2012.

to α, where $f(\alpha) = 0$, are uniquely projected to the neighborhood of α with radius r_1 by $h_1 = 0$, say β_i. Thus we can recover the y-coordinate of these roots from $y = (\beta_i - \alpha)/s_1$. Thus we get the roots of $f = g = 0$. And we get a root separation bound for these y roots for a fixed α, choose the smallest one from all roots of $f = 0$ as r_2. Similarly, compute a root bound R_3 for z coordinates of the roots of $f = g = h = 0$. Choose a rational s_2 as $0 < s_2 < r_2/R_3$. Let $g_1 = \mathrm{Res}_z(g(x, y - s_2 z), h(x, y - s_2 z, z))$. Isolate the roots of $f = g_1 = 0$ as for isolating the roots of $f = g = 0$ (We can use the same s_1). For each root $P = (\alpha, \beta)$ of $f = g = 0$, we can recover z-coordinates of the roots of $f = g = h = 0$ from the roots $P_i = (\alpha, \gamma_i)$ of $f = g_1 = 0$ in P's neighborhood with radius r_2 by $z = (\gamma_i - \beta)/s_2$. Thus we get all the roots of $f = g = h = 0$. And we also get an algebraic representation of the zeros of the system Σ: each coordinate of each zero is a linear combination of roots of several univariate polynomials. Using this representation we can compute the zeros of the system up to any desired precision. Our method is **complete** in the sense that Σ_n can be any zero-dimensional triangular system.

There is an extensive bibliography for isolating the roots of zero-dimensional triangular polynomial systems. However, most of the methods can not apply to triangular systems with multiple zeros directly [8,9,15,3,19]. Usually, they decompose the system into triangular systems without multiple zeros and then isolate the real zeros. Cheng et al [7] provided a direct method which does not compute an algebraic representation of the real zeros and can not compute their multiplicities. In [22] a method for computing the multiplicities is presented in the case where the zeros have already been computed. Concerning the algebraic representation of the roots of a polynomial system, let us mention Rouillier [17] that used the rational univariate representation, and Gao and Chou [12] that presented a representation for the zeros of a radical characteristic set. Using Gröbner basis computations, [1] presented a representation of the zeros of a system that depends on the multiplicities.

The rest of the paper is structured as follows: In Section 2 we present the theory of isolating the roots of a zero-dimensional triangular polynomial system. In Section 3, we give the algorithm, present an example, and compare our method with other methods. We conclude in Section 4.

2   Zero-Dimensional Triangular System Solving

Let $\Sigma_i = \{f_1(x_1), f_2(x_1, x_2), \ldots, f_i(x_1, x_2, \ldots, x_i)\} \in \mathbb{Q}[x_1, x_2, \ldots, x_i](i = 1, \ldots, n)$ be a general zero-dimensional triangular system. $\boldsymbol{\xi}^i = (\xi_1, \ldots, \xi_i) \in \mathrm{Zero}(\Sigma_i)$, where $\mathrm{Zero}(t)$ represents the zero set of $t = 0$. And t can be a polynomial or a polynomial system.

Let $f \in \mathbb{C}[x]$. Then the separation bound $\mathrm{sep}(f)$ and root bound $\mathrm{rb}(f)$ of f are defined as follows: $\mathrm{sep}(f) := \min\{\Delta(\alpha, \beta)|\forall \alpha, \beta \in \mathbb{C} \, s.t. f(\alpha) = f(\beta) = 0, \alpha \neq \beta\}$, where $\Delta(\alpha, \beta) := \max\{|\mathrm{Re}(\alpha - \beta)|, |\mathrm{Im}(\alpha - \beta)|\}$, $\mathrm{Re}(\alpha - \beta), \mathrm{Im}(\alpha - \beta)$ are the real part and imaginary part of $\alpha - \beta$ respectively. We also need the definition of the root bound: $\mathrm{rb}(f) := \max\{|\alpha||\forall \alpha \in \mathbb{C} \, s.t. f(\alpha) = 0\}$.

Assume that we have solved the system $\Sigma_i (1 \leq i \leq n-1)$. The assumption is reasonable since we can solve Σ_1 directly with many existing tools, such as [18,21]. And we can get a separation bound r_1 of $f_1(x_1)$. Based on the roots of $f_1 = 0$, we can estimate the root bound R_2.

Let $r_j (1 \leq j \leq i)$ be a positive rational number, such that

$$r_j \leq \frac{1}{2} \min_{\boldsymbol{\xi}^{j-1} \in \mathrm{Zero}(\Sigma_{j-1})} \mathrm{sep}(f_j(\boldsymbol{\xi}^{j-1}, x_j)). \tag{1}$$

We can compute r_j after we get the roots of $f_j(\boldsymbol{\xi}^{j-1}, x_j) = 0$.

Based on the zeros of Σ_j, we can estimate the root bound on x_{j+1} (we will show how to estimate the bound later) to get a positive rational number R_{j+1}, such that

$$R_{j+1} \geq \max_{\boldsymbol{\xi}^j \in \mathrm{Zero}(\Sigma_j)} \mathrm{rb}(f_{j+1}(\boldsymbol{\xi}^j, x_{j+1})). \tag{2}$$

We usually add a previously estimated value, say r'_{j+1}, for r_{j+1} to the above root bound to ensure that after shearing and projection, the fixed neighborhoods of the zeros of $T_{i,i}(X_{i,i})$ (see definition below) are disjoint. Then when we compute r_{j+1}, we choose the one no larger than r'_{j+1}.

We say two plane curves defined by $f, g \in \mathbb{C}[x, y]$ s.t. $\gcd(f, g) = 1$ are in a **generic position** w.r.t. y if (1) The leading coefficients of f and g w.r.t. y have no common factors, and (2) If h is the resultant of f and g w.r.t. y, then any $\alpha \in \mathbb{C}$ such that $h(\alpha) = 0$, $f(\alpha, y), g(\alpha, y)$ have only one common zero in $\mathbb{C}$.

Now we introduce **local generic position** [4,5]. Given $f, g \in \mathbb{Q}[x, y]$, not necessarily in generic position, we consider the the mapping $\phi : (x, y) \to (x + sy, y), s \in \mathbb{Q}$, with the following properties: (i) $\phi(f), \phi(g)$ are in a generic position w.r.t. y, and (ii) Let $\bar{h}, h$ be the resultants of $\phi(f), \phi(g)$ and f, g w.r.t. y, respectively. Each root α of $h(x) = 0$ has a neighbor interval H_α such that $H_\alpha \cap H_\beta = \emptyset$ for roots $\beta \neq \alpha$ of $h = 0$. And any root (γ, η) of $f = g = 0$ which has a same x-coordinate γ, is mapped to $\gamma' = \gamma + s\eta \in H_\gamma$, where $h(\gamma) = 0, \bar{h}(\gamma') = 0$. Thus we can recover $\eta = \frac{\gamma' - \gamma}{s}$.

2.1   Basic Theory and Method

For each $\boldsymbol{\xi}^i = (\xi_1, \ldots, \xi_i) \in \mathrm{Zero}(\Sigma_i)$, the roots of $f_{i+1}(\boldsymbol{\xi}^i, x_{i+1}) = 0$ are bounded by R_{i+1}. We can take a shear mapping on $f_{i+1}(x_1, \ldots, x_{i+1})$ such that when projected to i-D space, all the roots of $f_{i+1}(\boldsymbol{\xi}^i, x_{i+1}) = 0$ are projected into the fixed neighborhood of ξ_i (centered at ξ_i and radius bounded by r_i). This can be achieved by take the following shear mapping on (x_i, x_{i+1}).

$$X_{2,i+1} = x_i + \frac{r_i}{R_{i+1}} x_{i+1}, \; X_{1,i+1} = x_{i+1}. \tag{3}$$

Applying (3) to the system Σ_{i+1}, we derive a new system $\Sigma'_{i+1} = \{f_1(x_1), \ldots, f_{i-1}(x_1, \ldots, x_{i-1}), f_i(x_1, \ldots, x_{i-1}, X_{2,i+1} \quad - \quad \frac{r_i}{R_{i+1}} X_{1,i+1}), f_{i+1}(x_1, \ldots, x_{i-1},$

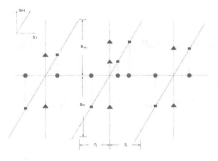

Fig. 1. Local generic position

$X_{2,i+1} - \frac{r_i}{R_{i+1}} X_{1,i+1}, X_{1,i+1})\}$. Let

$$T_{2,i+1}(x_1, \ldots, x_{i-1}, X_{2,i+1}) = \mathrm{Res}_{X_{1,i+1}} (f_i(x_1, \ldots, x_{i-1}, X_{2,i+1} -$$
$$\frac{r_i}{R_{i+1}} X_{1,i+1}), f_{i+1}(x_1, \ldots, x_{i-1}, X_{2,i+1} \quad - \quad \frac{r_i}{R_{i+1}} X_{1,i+1}, X_{1,i+1})),$$

where $\mathrm{Res}_t(f, g)$ is the resultant of f and g w.r.t. t. There is only one root of $f_{i+1}(\xi_1, \ldots, \xi_{i-1}, \theta_2 - \frac{r_i}{R_{i+1}} X_{1,i+1}, X_{1,i+1}) = 0$ corresponding to each i-D root $(\xi_1, \ldots, \xi_{i-1}, \theta_2) \in \mathrm{Zero}(\Sigma_i^*)$, where $\Sigma^* = \Sigma_{i-1} \cup \{T_{2,i+1}\}$. As is shown in Figure 1, θ_2 is some dot point on x_i-axis. Corresponding to each dot point, there is only one box point. Each box point corresponds to one triangle point. We will further study the relationship between the zeros of Σ_{i+1} and $\Sigma_{i-1} \cup \{T_{2,i+1}\}$ below. Considering the multiplicities of the zeros, we give the following lemma.

Lemma 1. *For each zero $\boldsymbol{\xi}^i$ of Σ_{i-1}, there exists a one to one correspondence between the roots of $\{f_i(\xi_1, \ldots, \xi_{i-1}, x_i), f_{i+1}(\xi_1, \ldots, \xi_{i-1}, x_i, x_{i+1})\} = 0$ and the roots of $T_{2,i+1}(\xi_1, \ldots, \xi_{i-1}, X_{2,i+1}) = 0$, and the multiplicities of corresponding zeros in their equation(s) are the same.*

Proof. Note that we derive the system $\Theta_2 := \{f_i(\xi_1, \ldots, \xi_{i-1}, X_{2,i+1} - \frac{r_i}{R_{i+1}} X_{1,i+1}), f_{i+1}(\xi_1, \ldots, \xi_{i-1}, X_{2,i+1} - \frac{r_i}{R_{i+1}} X_{1,i+1}, X_{1,i+1})\}$ from the system $\Theta_1 := \{f_i(\xi_1, \ldots, \xi_{i-1}, x_i), f_{i+1}(\xi_1, \ldots, \xi_{i-1}, x_i, x_{i+1})\}$ by coordinate system transformation. So there exists a one to one correspondence between their zeros, including the multiplicities of the zeros by the properties of local generic position. And the coordinate system transformation ensures that for any zero (ξ_i, ξ_{i+1}), when projected to x_i-axis by LGP method, the zero is in the fixed neighborhood of ξ_i (centered at ξ_i and radius bounded by r_i). This ensures that all the zeros of Θ_2, when projected to x_i-axis, do not overlap, which means any root of $T_{2,i+1}(\xi_1, \ldots, \xi_{i-1}, X_{2,i+1}) = 0$ corresponds to one zero of Θ_2. So there exists a one to one correspondence between roots of $T_{2,i+1}(\xi_1, \ldots, \xi_{i-1}, X_{2,i+1}) = 0$ and the zeros of Θ_1. It is not difficult to find that the degree of the polynomial $f_i(\xi_1, \ldots, \xi_{i-1}, X_{2,i+1} - \frac{r_i}{R_{i+1}} X_{1,i+1})$ w.r.t. $X_{1,i+1}$ is equal to its total degree. And $T_{2,i+1}(\xi_1, \ldots, \xi_{i-1}, X_{2,i+1})$ is the resultant of the two polynomials in Θ_2 w.r.t. $X_{1,i+1}$. Based on the theory in Section 1.6 in [11], we can conclude that the multiplicities of the roots in $T_{2,i+1}(\xi_1, \ldots, \xi_{i-1}, X_{2,i+1}) = 0$ equals the multiplicities of the corresponding zeros of Θ_2, and then Θ_1. So the lemma is true.

Lemma 2. *There exists a one to one correspondence between the zeros of triangular systems Σ_{i+1} and $\Sigma_{i-1} \cup \{T_{2,i+1}(x_1, \ldots, x_{i-1}, X_{2,i+1})\}$. And the corresponding zeros have the same multiplicities in their system.*

Proof. Since both the systems have a same sub-system Σ_{i-1}, we can derive that the lemma is correct by Lemma 1.

Lemma 3. *For $(\xi_1, \ldots, \xi_i) \in \text{Zero}(\Sigma_i)$, the roots of $f_{i+1}(\xi_1, \ldots, \xi_i, x_{i+1})$ are:*

$$x_{i+1} = \frac{R_{i+1}}{r_i}(\zeta_2 - \xi_i), \text{ where } |\zeta_2 - \xi_i| < r_i, \ T_{2,i+1}(\xi_1, \ldots, \xi_{i-1}, \zeta_2) = 0. \tag{4}$$

Proof. The first formula is directly derived from (3). Since the first formula just holds for ζ_2, corresponding zeros having ξ_i as coordinate, the inequality holds.

The above lemma tells us how to derive the roots of $f_{i+1}(\xi_1, \ldots, \xi_i, x_{i+1}) = 0$ from the roots of $T_{2,i+1}(\xi_1, \ldots, \xi_{i-1}, X_{2,i+1}) = 0$. From (1) and (2), the corollary below is obvious.

Corollary 1. *All the roots of $T_{2,i+1}(\xi_1, \ldots, \xi_{i-1}, X_{2,i+1}) = 0$ are inside the fixed neighborhood of 0 (centered at 0 bounded by R_i) for all $(\xi_1, \ldots, \xi_{i-1}) \in \text{Zero}(\Sigma_{i-1})$.*

We apply the previous procedure on the triangular system $\Sigma_{i-1} \cup \{T_{2,i+1}\}$ with the mapping

$$X_{3,i+1} = x_{i-1} + \frac{r_{i-1}}{R_i}X_{2,i+1}, \ X_{2,i+1} = X_{2,i+1}, \tag{5}$$

$T_{3,i+1}(x_1, \ldots, x_{i-2}, X_{3,i+1}) = \text{Res}_{X_{2,i+1}}(f_{i-2}(x_1, \ldots, x_{i-3}, X_{3,i+1} - \frac{r_{i-1}}{R_i}X_{2,i+1}), T_{2,i+1}(x_1, \ldots, x_{i-3}, X_{3,i+1} - \frac{r_{i-1}}{R_i}X_{2,i+1}, X_{2,i+1}))$.

So, we have a triangular system $\Sigma_{i-2} \cup \{T_{3,i+1}\}$. Since Corollary 1 holds, the results in Lemma 2 still hold on $\Sigma_{i-1} \cup \{T_{2,i+1}\}$ and $\Sigma_{i-2} \cup \{T_{3,i+1}\}$. By (5), and similarly as (4), we derive

$$\zeta_2 = \frac{R_i}{r_{i-1}}(\zeta_3 - \xi_{i-1}), \ |\zeta_3 - \xi_{i-1}| < r_{i-1}, \quad T_{3,i+1}(\xi_1, \ldots, \xi_{i-2}, \zeta_3) = 0. \tag{6}$$

Then we have $x_{i+1} = \frac{R_{i+1}}{r_i}(\frac{R_i}{r_{i-1}}(\zeta_3 - \xi_{i-1}) - \xi_i)$, where $|\zeta_3 - \xi_{i-1}| < r_{i-1}$, $|\frac{R_i}{r_{i-1}}(\zeta_3 - \xi_{i-1}) - \xi_i| < r_i$, $T_{3,i+1}(\xi_1, \ldots, \xi_{i-2}, \zeta_3) = 0$.

The above formula means that we can get the roots of $f_{i+1}(\xi_1, \ldots, \xi_i, x_{i+1}) = 0$ by solving $T_{3,i+1}(\xi_1, \ldots, \xi_{i-2}, X_{3,i+1}) = 0$ directly.

Step by step, we can derive a univariate polynomial $T_{i+1,i+1}(X_{i+1,i+1})$. It holds $\zeta_i = \frac{R_2}{r_1}(\zeta_{i+1} - \xi_1)$ and $|\zeta_{i+1} - \xi_1| < r_1$. Now we can represent $\text{Zero}(f_{i+1}(\xi_1, \ldots, \xi_i, x_{i+1}))$ by $\xi_1, \ldots, \xi_i$ and the roots of $T_{i+1,i+1}(X_{i+1,i+1})$, where $(\xi_1, \ldots, \xi_i) \in \text{Zero}(\Sigma_i)$.

Lemma 4. *For any zero $(\xi_1, \ldots, \xi_i) \in \text{Zero}(\Sigma_i)$, each root ξ_{i+1} of $f_{i+1}(\xi_1, \ldots, \xi_i, x_{i+1}) = 0$ is mapped to a root of $T_{i+1,i+1}(X_{i+1,i+1}) = 0$. And we can derive ξ_{i+1} by $T_{i+1,i+1}(X_{i+1,i+1}) = 0$ as follows.*

$$\xi_{i+1} = \frac{R_{i+1}}{r_i}(\zeta_2 - \xi_i), \zeta_2 = \frac{R_i}{r_{i-1}}(\zeta_3 - \xi_{i-1}), \ldots, \zeta_i = \frac{R_2}{r_1}(\zeta_{i+1} - \xi_1), T_{i+1,i+1}(X_{i+1,i+1}) = 0,$$

where $|\zeta_2 - \xi_i| < r_i, |\zeta_3 - \xi_{i-1}| < r_{i-1}, \ldots, |\zeta_{i+1} - \xi_1| < r_1$.

Proof. Using Lemma 3 recursively, we can derive the above formula.

Lemma 5. *For any* $(\xi_1, \ldots, \xi_i) \in \text{Zero}(\Sigma_i)$, *each distinct root* ξ_{i+1} *of* $f_{i+1}(\xi_1, \ldots, \xi_i, x_{i+1}) = 0$ *is mapped to a root of* $T_{i+1,i+1}(X_{i+1,i+1}) = 0$. *And we can derive:*

$$\xi_{i+1} = (\prod_{j=1}^{i} \frac{R_{j+1}}{r_j})(\eta_{i+1} - \eta_i), \tag{7}$$

where $\eta_{i+1} \in \text{Zero}(T_{i+1,i+1})$, $\eta_i \in \text{Zero}(T_{i,i})$, *and* $|\eta_{i+1} - \eta_i| < (\prod_{j=1}^{i-1} \frac{r_j}{R_{j+1}})r_i$.

Proof. According to Lemma 4, we know

$$\xi_i = \frac{R_i}{r_{i-1}}(\zeta_2 - \xi_{i-1}) = \cdots = (\prod_{j=1}^{i-1} \frac{R_{j+1}}{r_j})\eta_i - \sum_{k=1}^{i-1}((\prod_{j=k}^{i-1} \frac{R_{j+1}}{r_j})\xi_k).$$

Note that here $\zeta_i = \eta_i$. Similarly, we have

$$\begin{aligned}
\xi_{i+1} &= (\prod_{j=1}^{i} \frac{R_{j+1}}{r_j})\eta_{i+1} - \sum_{k=1}^{i}[(\prod_{j=k}^{i} \frac{R_{j+1}}{r_j})\xi_k] \\
&= (\prod_{j=1}^{i} \frac{R_{j+1}}{r_j})\eta_{i+1} - \sum_{k=1}^{i-1}[(\prod_{j=k}^{i} \frac{R_{j+1}}{r_j})\xi_k] - \frac{R_{i+1}}{r_i}\xi_i \\
&= (\prod_{j=1}^{i} \frac{R_{j+1}}{r_j})\eta_{i+1} - \frac{R_{i+1}}{r_i}\sum_{k=1}^{i-1}[(\prod_{j=k}^{i-1} \frac{R_{j+1}}{r_j})\xi_k] - \frac{R_{i+1}}{r_i}\xi_i \\
&= (\prod_{j=1}^{i} \frac{R_{j+1}}{r_j})\eta_{i+1} - \frac{R_{i+1}}{r_i}\sum_{k=1}^{i-1}[(\prod_{j=k}^{i-1} \frac{R_{j+1}}{r_j})\xi_k] \\
&\quad - \frac{R_{i+1}}{r_i}((\prod_{j=1}^{i-1} \frac{R_{j+1}}{r_j})\eta_i - \sum_{k=1}^{i-1}[(\prod_{j=k}^{i-1} \frac{R_{j+1}}{r_j})\xi_k]) \\
&= (\prod_{j=1}^{i} \frac{R_{j+1}}{r_j})(\eta_{i+1} - \eta_i).
\end{aligned}$$

$$|\eta_{i+1} - \eta_i| = \prod_{j=1}^{i} \frac{r_j}{R_{j+1}}|\xi_{i+1}| < \prod_{j=1}^{i} \frac{r_j}{R_{j+1}}R_{i+1} = (\prod_{j=1}^{i-1} \frac{r_j}{R_{j+1}})r_i.$$

The lemma has been proved.

Lemma 6. *The multiplicity of the zero* $(\xi_1, \ldots, \xi_i, \xi_{i+1})$ *of* Σ_{i+1} *is equal to the multiplicity of the corresponding root in* $T_{i+1,i+1}(X_{i+1,i+1}) = 0$.

Proof. Using Lemma 2 recursively, we can derive the lemma.

Theorem 1. *With the notations above, we have the following representation for a general zero-dimensional triangular system* $\Sigma_n : \{\{T_{1,1}, \ldots, T_{n,n}\}, \{r_1, \ldots, r_{n-1}\}, \{R_2, \ldots, R_n\}\}$, *such that the zeros of* Σ_n *can be derived as follows.*

$$\begin{aligned}
\xi_1 &= \eta_1, \eta_1 \in \text{Zero}(T_{1,1}), \\
\xi_2 &= \frac{R_2}{r_1}(\eta_2 - \eta_1), \eta_2 \in \text{Zero}(T_{2,2}), |\eta_2 - \eta_1| < r_1, \\
&\cdots \\
\xi_i &= (\prod_{j=1}^{i-1} \frac{R_{j+1}}{r_j})(\eta_i - \eta_{i-1}), \eta_i \in \text{Zero}(T_{i,i}), |\eta_{i+1} - \eta_i| < (\prod_{j=1}^{i-1} \frac{r_j}{R_{j+1}})r_i, \\
&\cdots \\
\xi_n &= (\prod_{j=1}^{n-1} \frac{R_{j+1}}{r_j})(\eta_n - \eta_{n-1}), \eta_n \in \text{Zero}(T_{n,n}), |\eta_n - \eta_{n-1}| < (\prod_{j=1}^{n-2} \frac{r_j}{R_{j+1}})r_{n-1},
\end{aligned}$$

where $T_{j,j}(j = 1, \ldots, n)$ *are univariate polynomials,* $T_{1,1} = f_1$. *For each zero* $(\xi_1, \ldots, \xi_i)$ $(1 \leq i \leq n)$ *of the system* Σ_i, *the multiplicity of the zero in the system is the multiplicity of the corresponding zero* η_i *in the univariate polynomial* $T_{i,i}$.

Remark: From the second part of the theorem, we can compute the multiplicities of the roots of $f_{i+1}(\xi_1, \ldots, \xi_i, x_{i+1}) = 0$, it is the multiplicity of the zero $(\xi_1, \ldots, \xi_{i+1})$ in Σ_{i+1} dividing the multiplicity of the zero $(\xi_1, \ldots, \xi_i)$ in Σ_i.

2.2   Estimation of Bounds r_i, R_{i+1}

To estimate the bounds r_i, R_{i+1}, we can directly derive the bound by the method in [10]. But the derived bounds r_i is tiny and R_{i+1} is huge. We prefer to use direct methods to get the bounds.

The methods to estimate the bound for r_i, R_{i+1} can be used both for complex and real roots isolation. We focus on real roots isolation in this paper.

For R_{i+1}, we have two methods to estmate it. One of them is computing

$$S(x_{i+1}) = \text{Res}_{x_1}(\text{Res}_{x_2}(\cdots \text{Res}_{x_i}(f_{i+1}, f_i), \cdots, f_2), f_1) \tag{8}$$

first, estimating R_{i+1} by estimating the root bound of $S(x_{i+1})$.

Now, we introduce the second method. we at first estimate the root bound on $f_{i+1}(\xi_1, \ldots, \xi_i, x_{i+1}) = 0$ for a fixed zero $(\xi_1, \ldots, \xi_i)$. Doing so, we need to use the definition of sleeve (see [7,14,15] for details). Given $g \in \mathbb{Q}[x_1, \ldots, x_n]$, we decompose it uniquely as $g = g^+ - g^-$, where $g^+, g^- \in \mathbb{Q}[x_1, \ldots, x_n]$ each has only positive coefficients and with minimal number of monomials. Given an isolating box $\square \boldsymbol{\xi}^i = [a_1, b_1] \times \cdots \times [a_i, b_i]$ for $\boldsymbol{\xi}^i = (\xi_1, \ldots, \xi_i)$, we assume that $a_j, b_j, \xi_j \geq 0, 1 \leq j \leq i$ since we can take a coordinate system transformation to satisfy the condition when $\xi_j < 0$. Then we define

$$f^u(x) = f^u_{i+1}(\square\boldsymbol{\xi}^i; x) = f^+_{i+1}(\boldsymbol{b}_i, x) - f^-_{i+1}(\boldsymbol{a}_i, x),$$
$$f^d(x) = f^d_{i+1}(\square\boldsymbol{\xi}^i; x) = f^+_{i+1}(\boldsymbol{a}_i, x) - f^-_{i+1}(\boldsymbol{b}_i, x), \tag{9}$$

where $\boldsymbol{a}_i = (a_1, \ldots, a_i)$, $\boldsymbol{b}_i = (b_1, \ldots, b_i)$. Then (f^u, f^d) is a **sleeve** of $f_{i+1}(\boldsymbol{\xi}^i, x_{i+1})$. When considering $x \geq 0$, we have (see [7]): $f^d(x) \leq f_{i+1}(\boldsymbol{\xi}^i, x) \leq f^u(x)$.

If the leading coefficients of f_u and f_d have the same signs, then we can find that the root bound of $f_{i+1}(\boldsymbol{\xi}^i, x)$ is bounded by the root bounds of f_u and f_d.

Lemma 7. *[20] Let a polynomial of degree d be $f(x) = a_d x^d + a_{d-1} x^{d-1} + \ldots + a_0 \in \mathbb{R}[x], a_d \neq 0$. Let $R = 1 + \max_{0 \leq k \leq d-1} |\frac{a_k}{a_d}|$, then all zeros of $f(x)$ lie inside the circle of radius R about the origin.*

If the considered triangular system is not regular, the leading coefficients of f_u and f_d always have different signs. But the absolute value of the leading coefficients are very close to zero. So usually, the root bound of $f_{i+1}(\boldsymbol{\xi}^i, x)$ is also bounded by the larger of the root bound of f_u and f_d. Then we can get R_{i+1} by the lemma above.

For r_i, we can directly compute the bound on the zeros of Σ_i using (1). It is for complex roots. Since we focus on real roots, we compute r_i after we get the real roots of $\Sigma_i = 0$ with the following formula.

$$\text{sep}(f) := \min\{|\alpha - \beta| | \forall \alpha, \beta \in \mathbb{R} \, s.t. f(\alpha) = f(\beta) = 0, \alpha \neq \beta\}. \tag{10}$$

If we use (1) to compute r_i, the roots of the system are in a local generic position after a shear mapping. When we use (10) to compute r_i, though all the real roots of the system have some local property, the complex roots may not in a generic position. Since a random shear mapping usually puts the system into a generic position, we can get the real roots of the given system.

2.3   Precision Control

When we compute the approximating zeros of a given zero-dimensional triangular system with the method we provided, the errors of the zeros will cumulate. So we need to control the error under a wanted precision. This is what we want to discuss in this subsection.

Consider the coordinate ξ_i of the zero $\pmb{\xi}^n = (\xi_1, \ldots, \xi_n)$ of the triangular system Σ_n in Theorem 1. Assume that we derive the coordinate ξ_j under the precision $\rho_j (> 0)$, and we isolate the roots of $T_{j,j}(X_{j,j}) = 0$ under the precision $\epsilon_j (> 0)$, Note that $\rho_1 = \epsilon_1$.

From (7), the following lemma is clear.

Lemma 8. *With the symbol above, we can derive that the root precision ρ_i for ξ_i is defined as follows: $\rho_i = (\prod_{j=1}^{i-1} \frac{R_{j+1}}{r_j})(\epsilon_i + \epsilon_{i-1})$.*

From Lemma 8, we can compute the zeros of Σ_n under any given precision by controlling the precisions $\epsilon_i (1 \le i \le n)$. For example, we can set them as follows if we require the precision of the output zeros to be ϵ.

$$\epsilon_i = \Pi_{j=1}^i \frac{r_j}{R_{j+1}} \frac{\epsilon}{2} (1 \le i \le n-1), \epsilon_n = \Pi_{j=1}^{n-1} \frac{r_j}{R_{j+1}} \frac{\epsilon}{2}. \tag{11}$$

In order to avoid refining the roots when we want to control the precision under a given ϵ, we can assume $\frac{R_{i+1}}{r_i}$ to be less than a number previously, such as $10, 2^3$, before we solve the system. This help us to previously estimate the precisions that should be used to get the roots of $T_{i,i}(X_{i,i}) = 0 (1 \le i \le n)$.

For root isolation, we require not only the roots satisfying the given precision, but the isolating boxes being disjoint for distinct roots. We will show how to ensure that the isolating boxes are disjoint. For real numbers α and β, $\alpha < \beta$ in $\mathbb{R}$, if we use intervals $[a, b]$ and $[c, d]$ to represent them respectively. Denote

$$|\alpha| = |b - a|, \mathrm{Dis}(\alpha, \beta) = \begin{cases} c - b, & b < c, \\ 0, & b \ge c. \end{cases}$$

Let $\xi^i = (\xi_1^i, \ldots, \xi_n^i) \in [a_1^i, b_1^i] \times \ldots \times [a_n^i, b_n^i] \subset \mathbb{R}^n, i = 1, 2$. Denote

$$|\xi^i| = \max_{j=1,\ldots,n} \{b_j^i - a_j^i\}, \mathrm{Dis}(\xi^1, \xi^2) = \min_{j=1,\ldots,n} \{\mathrm{Dis}(\xi_j^1, \xi_j^2)\}.$$

If $\mathrm{Dis}(\xi^1, \xi^2) > 0$, we say ξ^1 and ξ^2 are disjoint.

Theorem 2. *Use the notations above. We use intervals to represent real numbers and use boxes to represent real points, if for any $\eta_i^j \in \mathrm{Zero}(T_{i,i})$, $\eta_{i-1} \in \mathrm{Zero}(T_{i-1,i-1})$, $|\eta_i^j - \eta_{i-1}| < (\prod_{j=1}^{i-2} \frac{r_j}{R_{j+1}}) r_i$, $i = 2, \ldots, n; j = 1, 2$,*

$$\mathrm{Dis}(\eta_i^1, \eta_i^2) > |\eta_{i-1}|, \tag{12}$$

then any two real zeros $\pmb{\xi}^1 = (\xi_1^1, \ldots, \xi_n^1)$ and $\pmb{\xi}^2 = (\xi_1^2, \ldots, \xi_n^2)$ of Σ_n are disjoint.

Proof. We need only to consider the case η_i^1, η_i^2 are in the neighborhood of η_{i-1}. Otherwise, they are obviously disjoint. According to (7), for any $i = 2, \ldots, n$,

$$\xi_i^1 = (\textstyle\prod_{j=1}^{i-1} \frac{R_{j+1}}{r_j})(\eta_i^1 - \eta_{i-1}), \quad \xi_i^2 = (\textstyle\prod_{j=1}^{i-1} \frac{R_{j+1}}{r_j})(\eta_i^2 - \eta_{i-1}).$$

$$\mathrm{Dis}(\xi_i^1, \xi_i^2) = (\prod_{j=1}^{i-1} \frac{R_{j+1}}{r_j})\mathrm{Dis}(\eta_i^1 - \eta_{i-1}, \eta_i^2 - \eta_{i-1}) \geq (\prod_{j=1}^{i-1} \frac{R_{j+1}}{r_j})(\mathrm{Dis}(\eta_i^1, \eta_i^2) - |\eta_{i-1}|) > 0 \quad (13)$$

if (12) is satisfied. Thus $\mathrm{Dis}(\boldsymbol{\xi}^1, \boldsymbol{\xi}^2) > 0$.

3   The Algorithm and Experiments

Algorithm 3. *Isolate the real roots of a 0-dim. triangular system.*
Input: *A zero-dimensional triangular system Σ_n, a precision ϵ.*
Output: *The solutions of the system in isolating interval representation.*

1. Isolate the real roots of $f_1(x_1) = 0$ under the precision $\rho = \frac{\epsilon}{20}$. Let $T_{1,1} = f_1$.
2. For i from 2 to n,
   (a) Estimate r_{i-1} with method in Section 2.2.
   (b) Estimate R_i with method in Section 2.2.
   (c) Compute $T_{i,i}(X_{i,i})$ with method in Section 2.1.
   (d) Isolate the real roots of $T_{i,i}(X_{i,i}) = 0$ with precision $\Pi_{j=1}^{i-1} \frac{r_j}{R_{j+1}} \frac{\epsilon}{20}$ (if $i = n$, take $\Pi_{j=1}^{n-1} \frac{r_j}{R_{j+1}} \frac{\epsilon}{2}$). Compute the multiplicities of the roots if needed when $i = n$.
   (e) If (12) is not satisfied, then refine the real roots of $T_{i-1}^{i-1}(X_{i-1}^{i-1}) = 0$ until (12) is satisfied.
   (f) Recover the real zeros of Σ_i from $T_{i,i}(X_{i,i})$ and Σ_{i-1} by Theorem 1.
3. Get the solutions:$\{\{T_{1,1}(X_{1,1}), \ldots, T_{n,n}(X_{n,n})\}, \{r_1, \ldots, r_{n-1}\}, \{R_2, \ldots, R_n\}\}$, or numeric solutions and their corresponding multiplicities.

Example 4. *Consider the system $\{x^2 - 6, 5\,x^2 + 10\,xy + 6\,y^2 - 5, x^2 + 2\,xy + 2\,y^2 + 4\,yz + 5\,z^2 - 1\}$. We derive a symbolic representation of the roots, as well as a floating point approximation up to precision $\frac{1}{10^3}$. We isolate the roots of $f_1 = 0$ using precision $\frac{1}{2 \cdot 10^4}$ and we derive the zero set: $H = \{\xi_1^1 = -2.449490070, \xi_1^2 = 2.449490070\}$. Let $r_1 = 2$. Consider $\xi_1 \approx -2.449490070 \in [-2.45, -2.44]$. We can use $-2.45, -2.44$ to construct $f^u(y), f^d(y)$ for $f_2(\xi_1, y)$. We compute a root bound for $f^u(y), f^d(y)$. For both it is ≤ 6. Similarly, we compute a root bound for the other root in H. we notice that all the root bounds are less than 6. We have computed $r_2 = 2$, so we set $R_2 = 6 + 2 = 2^3$. By considering a coordinate system transformation, we derive a system Σ_2' as follows $\{X_{2,2}^2 - \frac{1}{2} X_{2,2} X_{2,1} + \frac{1}{16} X_{2,1}^2 - 6, 5 X_{2,2}^2 + \frac{15}{2} X_{2,2} X_{2,1} + \frac{61}{16} X_{2,1}^2 - 5\}$. Hence we can compute $T_{2,2} = 36 X_{2,2}^4 - \frac{1083}{4} X_{2,2}^2 + \frac{130321}{256}$. Solve $T_{2,2}(X_{2,2}) = 0$ under the precision $\frac{1}{8 \cdot 10^4}$, we have its real roots and multiplicities (the number in each bracket is the multiplicity of the root in the system): $G = \{\eta_2^1 = -1.939178944 \,[2], \eta_2^2 = 1.939178944 \,[2]\}$.*

For each root η_2 in G, if it satisfies $|\eta_2 - \xi_1| < r_1 = 2$, then it corresponds to ξ_1, where ξ_1 is a root in H. And the multiplicity of (ξ_1, η_2) in the given system

is the corresponding multiplicity of η_2 in $T_2^2 = 0$. In this way, we can get the approximating roots of the subsystem Σ_2:
$$\{[-2.449490070\,[1], 2.041244504\,[2]], [2.449490070\,[1], -2.041244504\,[2]]\}.$$
With the method of Section 2.2, we estimate $r_3 = 2$, and we derive that 3 is a bound for the z coordinate. Let $R_3 = 2+2 = 4$ and $r_2 = 2$ and consider a coordinate system transformation as mentioned above. By computing the resultant, we can get $T_{3,3} = 810000\,x^8 - 13500000\,x^6 + 84375000\,x^4 - 234375000\,x^2 + 244140625.$

Then, we get the solution of the given triangular system as follows.
$$\{\{X_{1,1}{}^2 - 6, 36\,X_{2,2}{}^4 - \tfrac{1083}{4}\,X_{2,2}{}^2 + \tfrac{130321}{256}, 810000\,x^8 - 13500000\,x^6$$
$$+ 84375000\,x^4 - 234375000\,x^2 + 244140625\}, \{2,2\}, \{8,4\}\}.$$

We solve $T_{3,3}$ using precision $\frac{1}{16\cdot10^4}$, and derive its roots and multiplicities:
$$J = \{\eta_3^1 = -2.041241452\,[4], \eta_3^2 = 2.041241452\,[4]\}.$$

For each root η_3 in J, if it satisfies $|\eta_3 - \eta_2| < \frac{r_1}{R_2}r_2 = \frac{1}{2}$, then it corresponds to the same (ξ_1, ξ_2) with η_2, where (ξ_1, ξ_2) is a root in Σ_2. And the multiplicity of (ξ_1, ξ_2, η_3) in the given system is the corresponding multiplicity of η_3. In this way, we can get the approximating roots of the system:

$$\{[-2.449490070\,[1], 2.041244504\,[2], -0.816497800\,[4]], [2.449490070\,[1],$$
$$-2.041244504\,[2], 0.816497800\,[4]]\}.$$

Using Lemma 8, the precision of roots is $4(\frac{1}{8\cdot10^4} + \frac{1}{16\cdot10^4}) < \frac{1}{10^3}$.

In the below, we illustrate the function of our algorithm by some examples. The timings are collected on a computer running Maple 15 with 2.29GHz CPU, 2G memory and Windows XP by using the time command in Maple.

We compare our method with Discoverer, Isolate, EVB and Vincent-Collins-Akritas algorithm. Discoverer is a tool for solving problems about polynomial equations and inequalities [19]. Isolate is a tool to solve general equation systems based on Realsolving C library by Rouillier. EVB is developed by Cheng et al in [7]. Vincent-Collins-Akritas(VCA) algorithm which isolates real roots for univariate polynomials uses techniques which are very close to the ones used by Rioboo in [16]. Sqf is the method in [7] for zero-dimensional triangular system without multiple roots. All the required precision are 0.001.

Table 1. Timing of Real Root Isolation of System without Multiple Roots (Seconds)

Degree(k)	Vars	LGP	Dis	Iso	VCA	EVB	Sqf
1-11	2	0.104	0.228	0.187	0.861	0.577	0.014
12-30	2	23.790	17.269	175.684	28.362	210.295	0.140
1-4	3	25.614	19.290	202.681	29.727	246.102	0.156
5-9	3	55.579	41.769	453.217	62.407	512.930	0.223
1-3	4	11.947	8.748	87.936	14.612	105.436	0.077
4-5	4	40.597	30.529	327.949	46.067	379.516	0.190

In Table 1, we compare different methods by computing some zero-dimensional triangular polynomial systems without multiple roots. All the tested systems are generated randomly and dense. They have the form $(f_1, f_2, \ldots, f_n)$ in which $\deg(f_1) = \deg(f_2) = \ldots = \deg(f_n) = k$ are the total degrees of the

Table 2. Timing of Real Root Isolate of surfaces(Seconds)

Example	Degree	LGP	Iso	EVB
f_1	6	84.000	394.344	118.922
f_2	4	0.031	0.078	0.078
f_3	4	0.250	0.375	0.453
f_4	4	0.218	0.578	0.86

polynomials and their coefficients are random integers from -9 to 9. For each $i = 1, 2, \ldots, 30; n = 2, 3, 4$, we compute five systems. After all, we divide those systems into serval groups by degree and the number of variables, and take average timings for different group.

In Table 2, we find four famous surfaces. We generate the systems as follow. Let f be the defining polynomial of a surface in $\mathbb{R}^3$. We compute the resultant of f and $\frac{\partial f}{\partial z}$ with respect to z. Denote its squarefree part as g. Then we compute the resultant of g and $\frac{\partial g}{\partial y}$ with respect to y and denote the squarefree part as h. Thus we get a triangular polynomial system $\{h, g, f\}$. We decompose them into sub-system. This kind of system is usually zero-dimensional and have multiple roots. The total timing for each surface are collected in table 2 for the methods which can deal with multiple roots directly. They are Isolate, EVB and LGP.

f_1 is Barth's sextic surface with 65 singularities. f_2 is A squared off sphere. f_3 is a deformation of quartics. f_4 is a Kummer surface with 16 singular points.

$$f_1 = 4(2.618x^2 - y^2)(2.618y^2 - z^2)(2.618z^2 - x^2) - 4.236(x^2 + y^2 + z^2 - 1)^2;$$
$$f_2 = x^4 + y^4 + z^4 - 1; f_3 = (x^2 - 1)^2 + (y^2 - 1)^2 + (z^2 - 1)^2 - s, s = 1;$$
$$f_4 = x^4 + y^4 + z^4 - (0.5x + 1)^2(x^2 + y^2 + z^2) - (x^2y^2 + x^2z^2 + y^2z^2) + (0.5x + 1)^4.$$

From the data, we can find that LGP works well for system with multiple roots comparing to the existing direct method. For the systems without multiple roots, Sqf is the most efficient method. LGP works well for system with fewer roots. For the systems with higher degrees or more variables, that is, systems with more roots, LGP will slow down comparing to other methods. The reason is that $\prod_{j=1}^{i-1} \frac{r_j}{R_{j+1}}$ becomes small, thus the resultant computations take much more time.

4   Conclusion and Future Work

We present an algorithm to isolate the real roots and to count the multiplicities of a zero-dimensional triangular system directly. It is effective and efficient, especially comparing to other direct methods. We will analyze the complexity of the algorithm in our full version.

Acknowledgement. The authors would like to thank the anonymous referees for their valuable suggestions. The work is partially supported by NKBRPC (2011CB302400), NSFC Grants (11001258, 60821002, 91118001), and China-France project EXACTA (60911130369) and the French National Research Agency (ANR-09-BLAN-0371-01).

References

1. Alonso, M.-E., Becker, E., Roy, M.-F., Wörmann, T.: Multiplicities and idempotents for zero dimensional systems. In: Algorithms in algebraic Geometry and Applications. Progress in Mathematics, vol. 143, pp. 1–20. Birkhäuser (1996)
2. Berberich, E., Kerber, M., Sagraloff, M.: Exact Geometric-Topological Analysis of Algebraic Surfaces. In: Teillaud, M. (ed.) Proc. of the 24th ACM Symp. on Computational Geometry (SoCG), pp. 164–173. ACM Press (2008)
3. Boulier, F., Chen, C., Lemaire, F., Moreno Maza, M.: Real Root Isolation of Regular Chains. In: ASCM 2009, pp. 1–15 (2009)
4. Cheng, J.S., Gao, X.S., Li, J.: Root isolation for bivariate polynomial systems with local generic position method. In: ISSAC 2009, pp. 103–110 (2009)
5. Cheng, J.S., Gao, X.S., Guo, L.: Root isolation of zero-dimensional polynomial systems with linear univariate representation. J. of Symbolic Computation (2011)
6. Cheng, J.S., Gao, X.S., Li, M.: Determining the Topology of Real Algebraic Surfaces. In: Martin, R., Bez, H.E., Sabin, M.A. (eds.) IMA 2005. LNCS, vol. 3604, pp. 121–146. Springer, Heidelberg (2005)
7. Cheng, J.S., Gao, X.S., Yap, C.K.: Complete Numerical Isolation of Real Roots in 0-dimensional Triangular Systems. JSC 44(7), 768–785 (2009)
8. Collins, G.E., Johnson, J.R., Krandick, W.: Interval arithmetic in cylindrical algebraic decomposition. Journal of Symbolic Computation 34, 145–157 (2002)
9. Eigenwillig, A., Kettner, L., Krandick, W., Mehlhorn, K., Schmitt, S., Wolpert, N.: A Descartes Algorithm for Polynomials with Bit-Stream Coefficients. In: Ganzha, V.G., Mayr, E.W., Vorozhtsov, E.V. (eds.) CASC 2005. LNCS, vol. 3718, pp. 138–149. Springer, Heidelberg (2005)
10. Emiris, I.Z., Mourrain, B., Tsigaridas, E.P.: The DMM bound: Multivariate (aggregate) separation bounds. In: ISSAC 2010, pp. 243–250. ACM, Germany (2010)
11. Fulton, W.: Introduction to intersection theory in algebraic geometry. CBMS Regional Conference Series in Mathematics, vol. 54. Conference Board of the Mathematical Sciences, Washington, DC (1984)
12. Gao, X.S., Chou, S.C.: On the theory of resolvents and its applications. Mathematics and Systems Science (1997)
13. Hong, H.: An Efficient Method for Analyzing the Topology of Plane Real Algebraic Curves. Mathematics and Computers in Simulation 42, 571–582 (1996)
14. Hong, H., Stahl, V.: Safe start region by fixed points and tightening. Computing 53(3-4), 323–335 (1994)
15. Lu, Z., He, B., Luo, Y., Pan, L.: An Algorithm of Real Root Isolation for Polynomial Systems. In: SNC 2005 (2005)
16. Rioboo, R.: Computation of the real closure of an ordered field. In: ISSAC 1992. Academic Press, San Francisco (1992)
17. Rouillier, F.: Solving zero-dimensional systems through the rational univariate representation. AAECC 9, 433–461 (1999)
18. Sagraloff, M.: When Newton meets Descartes: A Simple and Fast Algorithm to Isolate the Real Roots of a Polynomial. CoRR abs/1109.6279 (2011)
19. Xia, B., Zhang, T.: Real Solution Isolation Using Interval Arithmetic. Computers and Mathematics with Applications 52, 853–860 (2006)
20. Yap, C.: Fundamental Problems of Algorithmic Algebra. Oxford University Press, New York (2000)
21. Yap, C., Sagraloff, M.: A simple but exact and efficient algorithm for complex root isolation. In: ISSAC 2011, pp. 353–360 (2011)
22. Zhang, Z.H., Fang, T., Xia, B.C.: Real solution isolation with multiplicity of 0-dimensional triangular systems. Science China: Information Sciences 54(1), 60–69 (2011)

On Fulton's Algorithm for Computing Intersection Multiplicities

Steffen Marcus[1], Marc Moreno Maza[2], and Paul Vrbik[2]

[1] Department of Mathematics, University of Utah
[2] Department of Computer Science, University of Western Ontario

Abstract. As pointed out by Fulton in his *Intersection Theory*, the intersection multiplicities of two plane curves $V(f)$ and $V(g)$ satisfy a series of 7 properties which *uniquely* define $I(p; f, g)$ at each point $p \in V(f, g)$. Moreover, the proof of this remarkable fact is constructive, which leads to an algorithm, that we call Fulton's Algorithm. This construction, however, does not generalize to n polynomials $f_1, \ldots, f_n$. Another practical limitation, when targeting a computer implementation, is the fact that the coordinates of the point p must be in the field of the coefficients of $f_1, \ldots, f_n$. In this paper, we adapt Fulton's Algorithm such that it can work at any point of $V(f, g)$, rational or not. In addition, we propose algorithmic criteria for reducing the case of n variables to the bivariate one. Experimental results are also reported.

1 Introduction

Intuitively, the intersection multiplicity of two plane curves counts the number of times these curves intersect. There are more formal ways to define this number. The following one is commonly used, see for instance [9,11,12,6,18]. Given an arbitrary field k and two bivariate polynomials $f, g \in k[x, y]$, consider the affine algebraic curves $C := V(f)$ and $D := V(g)$ in $\mathbb{A}^2 = \overline{k}^2$, where $\overline{k}$ is the algebraic closure of k. Let p be a point in the intersection. The *intersection multiplicity* of p in $V(f, g)$ is defined to be

$$I(p; f, g) := \dim_{\overline{k}}(\mathcal{O}_{\mathbb{A}^2, p} / \langle f, g \rangle)$$

where $\mathcal{O}_{\mathbb{A}^2, p}$ and $\dim_{\overline{k}}(\mathcal{O}_{\mathbb{A}^2, p} / \langle f, g \rangle)$ are the local ring at p and the dimension of the vector space $\mathcal{O}_{\mathbb{A}^2, p} / \langle f, g \rangle$. The intersection multiplicity of two plane curves at a point admits many properties. Among them are the seven below, which are proved in [9, Section 3-3] as well as in [11,12].

(2-1) $I(p; f, g)$ is a non-negative integer for any C, D, and p such that C and D have no common component at p. We set $I(p; f, g) = \infty$ if C and D have a common component at p.

(2-2) $I(p; f, g) = 0$ if and only if $p \notin C \cap D$.

(2-3) $I(p; f, g)$ is invariant under affine change of coordinates on $\mathbb{A}^2$.

(2-4) $I(p; f, g) = I(p; g, f)$.

V.P. Gerdt et al. (Eds.): CASC 2012, LNCS 7442, pp. 198–211, 2012.

(2-5) $I(p; f, g)$ is greater or equal to the product of the multiplicity (see [9, §3.1]) of p in f and g, with equality occurring if and only if C and D have no tangent lines in common at p.

(2-6) $I(p; f, gh) = I(p; f, g) + I(p; f, h)$ for all $h \in k[x, y]$.

(2-7) $I(p; f, g) = I(p; f, g + hf)$ for all $h \in k[x, y]$.

Remarkably, Properties (2-1) through (2-7) uniquely determine $I(p; f, g)$. This observation is made by Fulton in [9, Section 3-3] where he exhibits an algorithm for computing $I(p; f, g)$ using (2-1) through (2-7) as rewrite rules.

In order to obtain a practical implementation of this algorithm, a main obstacle must be overcome. To understand it, let us first recall that computer algebra systems efficiently manipulate multivariate polynomials whenever their coefficients are in the field of rational numbers or in a prime field. In particular, popular algorithms for decomposing the algebraic variety $V(f_1, \ldots, f_n)$ with $f_1, \ldots, f_n \in k[x_1, \ldots, x_n]$ rely only on operations in the field k, thus avoiding to manipulate non-rational numbers, that is, elements of $\overline{k} \setminus k$. For instance, algorithms such as those of [4] represent the variety $V(f_1, \ldots, f_n)$ (which is a subset of $\overline{k}^n$) with finitely many regular chains $T_1, \ldots, T_e$ of $k[x_1, \ldots, x_n]$ such that we have

$$V(f_1, \ldots, f_n) = V(T_1) \cup \cdots \cup V(T_e). \tag{1}$$

Now, observe that the intersection multiplicity $I(p; f_1, \ldots, f_n)$ of $f_1, \ldots, f_n$ at a point p is truly a local notion, while each of the $V(T_i)$ may consist of more than one point, even if T_i generates a maximal ideal of $k[x_1, \ldots, x_n]$. Therefore, in order to use regular chains for computing intersection multiplicities, one needs to be able to compute "simultaneously" all the $I(p; f_1, \ldots, f_n)$ for $p \in V(T_i)$

In Section 5 we propose an algorithm achieving the following task in the bivariate case: given $\mathcal{M} \subset k[x, y]$ a maximal ideal, compute the common value of all $I(p; f, g)$ for $p \in V(\mathcal{M})$. In Section 6, we relax the assumption of $\mathcal{M}$ being maximal and require only that a zero-dimensional regular chain $T \subset k[x, y]$ generates $\mathcal{M}$. However, in this case, the values of $I(p; f, g)$ for $p \in V(T)$ may not be all the same. This situation is handled via splitting techniques as in [4].

Thus, for $n = 2$, we obtain a procedure TriangularizeWithMultiplicity$(f_1, \ldots, f_n)$ which returns finitely many pairs $(T_1, m_1), \ldots, (T_e, m_e)$ where $T_1, \ldots, T_e \subset k[x_1, \ldots, x_n]$ are regular chains and $m_1, \ldots, m_e$ are non-negative integers satisfying Equation (1) and for each $i = 1, \ldots, e$, we have

$$(\forall p \in V(T_i)) \ I(p; f_1, \ldots, f_n) = m_i. \tag{2}$$

We are also interested in generalizing Fulton's Algorithm to n multivariate polynomials in n variables—our ultimate goal being an algorithm that realizes the above specification for $n \geq 2$.

We denote by $\mathbb{A}^n$ the n-dimensional affine space over $\overline{k}$. Let $f_1, \ldots, f_n \in k[x_1, \ldots, x_n]$ be n polynomials generating a zero-dimensional ideal with (necessarily finite) zero set $V(f_1, \ldots, f_n) \subset \mathbb{A}^n$. Let p be a point in the intersection $V(f_1) \cap \cdots \cap V(f_n)$, that is, $V(f_1, \ldots, f_n)$. The *intersection multiplicity* of p in $V(f_1, \ldots, f_n)$ is the generalization of the 2-variable case (as in [6,18])

$$I(p; f_1, \ldots, f_n) := \dim_{\overline{k}} (\mathcal{O}_{\mathbb{A}^n, p} / \langle f_1, \ldots, f_n \rangle),$$

where $\mathcal{O}_{\mathbb{A}^n,p}$ and $\dim_{\overline{k}}(\mathcal{O}_{\mathbb{A}^n,p}/\langle f_1, \ldots, f_n \rangle)$ are (respectively) the local ring at the point p and the dimension of the vector space $\mathcal{O}_{\mathbb{A}^n,p}/\langle f_1, \ldots, f_n \rangle$.

Among the key points in the proof of Fulton's algorithmic construction is that $k[x_1]$ is a principal ideal domain. Fulton uses Property (2-7) in an elimination process similar to that of the Euclidean Algorithm. Since $k[x_1, \ldots, x_{n-1}]$ is no longer a PID for $n \geq 3$, there is no natural generalization of (2-1) through (2-7) to the n-variate setting (up to our knowledge) that would lead to an algorithm for computing $I(p; f_1, \ldots, f_n)$.

To overcome this obstacle, at least for some practical examples, we propose an algorithmic criterion to reduce the n-variate case to that of $n-1$ variables. This reduction requires two hypotheses: $V(f_n)$ is non-singular at p, and the tangent cone of $V(f_1, \ldots, f_{n-1})$ at p and the tangent hyperplane of $V(f_n)$ at p meet only at the point p. The second hypothesis ensures that each component of the curve $V(f_1, \ldots, f_{n-1})$ meets the hypersurface $V(f_n)$ without tangency at p. This *transversality* assumption yields a reduction from n to $n-1$ variables proved with Theorem 1.

In Section 7, we discuss this reduction in detail. In particular, we propose a technique which, in some cases, replaces $f_1, \ldots, f_n$ by polynomials $g_1, \ldots, g_n$ generating the same ideal and for which the hypotheses of the reduction hold. Finally, in Section 8 we give details on implementing the algorithms herein and in Section 9 we report on our experimentation for both the bivariate case and the techniques of Section 7.

We conclude this introduction with a brief review of related works. In [5], the Authors report on an algorithm with the same specification as the above TriangularizeWithMultiplicity$(f_1, \ldots, f_n)$. Their algorithm requires, however, that the number of input polynomials is 2. In [17], the Authors outline an algorithm with similar specifications as ours. However, this algorithm is not complete, even in the bivariate case, in the sense that it may not compute the intersection multiplicities of all regular chains in a triangular decomposition of $V(f_1, \ldots, f_n)$.

In addition, our approach is novel thanks to an important feature which makes it more attractive in terms of performance. We first compute a triangular decomposition of $V(f_1, \ldots, f_n)$ (by *any* available method) thus without trying to "preserve" any multiplicity information. Then, once $V(f_1, \ldots, f_n)$ is decomposed we work "locally" at each regular chain. This enables us to quickly discover points p of intersection multiplicity one by checking whether the Jacobian matrix of f_1, $\ldots, f_n$ is invertible at p. We have observed experimentally that this strategy leads to massive speedup.

2   Regular Chains

In this section, we recall the notions of a regular chain. From now on we assume that the variables of the polynomial ring $k[x_1, \ldots, x_n]$ are ordered as $x_n > \cdots > x_1$. For a non-constant $f \in k[x_1, \ldots, x_n]$, the *main variable* of f is the largest variable appearing in f, while the *initial* of f is the leading coefficient of f w.r.t. the main variable of f. Let $T \subset k[x_1, \ldots, x_n]$ be a set of n non constant polynomials. We say that T is *triangular* if the main variables of the elements of T are pairwise different. Let t_i be the polynomial of T with main variable x_i. We say that T is a (zero-dimensional) *regular chain* if, for $i = 2, \ldots, n$ the initial of t_i is invertible modulo the ideal $\langle t_1, \ldots, t_{i-1} \rangle$. Regular chains are also defined in positive dimension, see [1,15].

For any maximal ideal $\mathcal{M}$ of $k[x_1, \ldots, x_n]$ there exists a regular chain T generating $\mathcal{M}$, see [14]. Therefore, for any zero-dimensional ideal $\mathcal{I}$ of $k[x_1, \ldots, x_n]$ there exist finitely many regular chains $T_1, \ldots, T_e \subset k[x_1, \ldots, x_n]$ such that we have $V(\mathcal{I}) = V(T_1) \cup \cdots \cup V(T_e)$. Various algorithms, among them those published in [20,10,14,19,4], compute such decompositions. The Triangularize command of the RegularChains library [16] in MAPLE implements the decomposition algorithm of [4]. This library also implements another algorithm of [4] that we will use in this paper and which is specified hereafter. For a regular chain $T \subset k[x_1, \ldots, x_n]$ and a polynomial $p \in k[x_1, \ldots, x_n]$, the operation Regularize(p, T) returns regular chains $T_1, \ldots, T_e \subset k[x_1, \ldots, x_n]$ such that we have $V(T) = V(T_1) \cup \cdots \cup V(T_e)$ and for all $i = 1, \ldots, e$ we have either $V(p) \cap V(T_i) = \emptyset$ or $V(T) \subset V(p)$. We will make use of the following result which can easily be derived from [4]: if Regularize(p, T) returns $T_1, \ldots, T_e$, then we have

$$(\forall p \in V(T_i)) \ \text{Regularize}(p, T_i) = T_i. \tag{3}$$

3   Intersection Multiplicity

As above, let $f_1, \ldots, f_n \in k[x_1, \ldots, x_n]$ be n polynomials in n variables such that the ideal $\langle f_1, \ldots, f_n \rangle$ they generate is zero-dimensional. Let $p \in V(f_1, \ldots, f_n)$ and denote the maximal ideal at p by $\mathcal{M}_p$. When needed, denote the coordinates of p by $(\alpha_1, \ldots, \alpha_n)$, so that we have $\mathcal{M}_p = \langle x_1 - \alpha_1, \ldots, x_n - \alpha_n \rangle$.

Definition 1. *The* intersection multiplicity *of p in $V(f_1, \ldots, f_n)$ is given by the length of $\mathcal{O}_{\mathbb{A}^n, p} / \langle f_1, \ldots, f_n \rangle$ as an $\mathcal{O}_{\mathbb{A}^n, p}$-module.*

Since we consider $\mathbb{A}^n$ as defined over the algebraically closed field $\overline{k}$, we know (see, for instance, [8]) that the length of this module is equal to its dimension as a $\overline{k}$ vector space, which is precisely the definition of Section 1. Our algorithm depends on the fact that the intersection multiplicity satisfies a generalized collection of properties similar to (2-1) through (2-7) for the bi-variate case. They are the following:

(n-1) $I(p; f_1, \ldots, f_n)$ is a non-negative integer.
(n-2) $I(p; f_1, \ldots, f_n) = 0$ if and only if $p \notin V(f_1, \ldots, f_n)$.
(n-3) $I(p; f_1, \ldots, f_n)$ is invariant under affine change of coordinates on $\mathbb{A}^n$.
(n-4) $I(p; f_1, \ldots, f_n) = I(p; f_{\sigma(1)}, \ldots, f_{\sigma(n)})$ for any $\sigma \in \mathfrak{S}_n$.
(n-5) $I(p; (x_1 - \alpha_1)^{m_1}, \ldots, (x_n - \alpha_n)^{m_n}) = m_1 \cdots m_n$, for all non-negative integers $m_1, \ldots, m_n$.
(n-6) If $g, h \in k[x_1, \ldots, x_n]$ make $f_1, \ldots, f_{n-1}, gh$ a zero-dimensional, then $I(p; f_1, \ldots, f_{n-1}, gh) = I(p; f_1, \ldots, f_{n-1}, g) + I(p; f_1, \ldots, f_{n-1}, h)$ holds.
(n-7) $I(p; f_1, \ldots, f_{n-1}, g) = I(p; f_1, \ldots, f_{n-1}, g + h)$ for all $h \in \langle f_1, \ldots, f_{n-1} \rangle$.

In order to reduce the case of n variables (and n polynomials) to that of $n-1$ variables (see Section 7) we require an additional property when $n > 2$. Of course, the assumptions necessary for this property may not hold for every polynomial system. However, we discuss in Section 7 a technique that can overcome this limitation for some practical examples.

(n-8) Assume the hypersurface $h_n = V(f_n)$ is non-singular at p. Let v_n be its tangent hyperplane at p. Assume furthermore that h_n meets each component of the curve $C = V(f_1, \ldots, f_{n-1})$ transversely, that is, the tangent cone $TC_p(C)$ intersects v_n only at the point p. Let $h \in k[x_1, \ldots, x_n]$ be the degree 1 polynomial defining v_n. Then, we have

$$I(p; f_1, \ldots, f_n) = I(p; f_1, \ldots, f_{n-1}, h_n).$$

Recall that the tangent cone $TC_p(C)$ can be thought of as the set of tangents given by limiting the secants to C passing through p. If $g_1, \ldots, g_s \in k[x_1, \ldots, x_n]$ are polynomials generating the radical of the ideal $\langle f_1, \ldots, f_{n-1} \rangle$, then $TC_p(C)$ is also given by $TC_p(C) = \langle \text{in}(g_1), \ldots, \text{in}(g_s) \rangle$ where $\text{in}(g_i)$, for $i = 1, \ldots, s$, is the initial form of g_i, that is, the homogeneous component of g_i of the lowest degree.

Theorem 1. $I(p; f_1, \ldots, f_n)$ *satisfies the properties (n-1) through (n-8).*

Proof. For the first seven properties, adapting the proofs of [9,12] is routine, except for (n-6), and we omit them for space consideration. For (n-6) and (n-8), as well as the others, the reader is refered to our technical report with the same title and available in the Computing Research Repository (CoRR).

4   Expansion of a Polynomial Family about at an Algebraic Set

The tools introduced herein help build an algorithm for computing the intersection multiplicity of $f_1, \ldots, f_n$ at any point of $V(f_1, \ldots, f_n)$, whenever the ideal $\langle f_1, \ldots, f_n \rangle$ is zero-dimensional and when, for $n > 2$, certain hypothesis are met.

Let $y_1, \ldots, y_n$ be n new variables with ordering $y_n > \cdots > y_1$. Let $F^1, \ldots, F^n \in k[x_1, \ldots, x_n, y_1, \ldots, y_n]$ be polynomials in $x_1, \ldots, x_n, y_1, \ldots, y_n$ with coefficients in k. We order the monomials in $y_1, \ldots, y_n$ (resp. $x_1, \ldots, x_n$) with the lexicographical term order induced by $y_n > \cdots > y_1$ (resp. $x_n > \cdots > x_1$). We denote by $S_{F^1}, \ldots, S_{F^n}$ the respective monomial supports (i.e. the set of monomials with non-zero coefficients) of $F^1, \ldots, F^n$, regarded as polynomials in the variables $y_1, \ldots, y_n$ and with coefficients in $k[x_1, \ldots, x_n]$. Let i be any integer index in $1, \ldots, n$. Write

$$F^i = \sum_{\mu \in S_{F^i}} F^i_\mu \mu, \qquad (4)$$

where all F^i_μ are polynomials of $k[x_1, \ldots, x_n]$. In particular, the F^i_1 represent F^i_μ when $\mu = y^0_1 \cdots y^0_n = 1$. Denote by $F^i_{<y_n}$ the polynomial of $k[x_1, \ldots, x_n][y_1, \ldots, y_{n-1}]$ defined by

$$F^i_{<y_n} = \sum_{\substack{\mu \in S^i_F \\ \deg(\mu, y_n) = 0}} F^i_\mu \mu.$$

Let $\mathcal{I}$ be a (proper) ideal of $k[x_1, \ldots, x_n]$. We denote by $\mathsf{NF}(f, \mathcal{I})$ the normal form of f w.r.t. the reduced lexicographical Gröbner basis of $\mathcal{I}$ for $x_n > \cdots > x_1$.

Let $p \in \mathbb{A}^n$ with coordinates $\alpha = (\alpha_1, \ldots, \alpha_n)$. For a monomial $\mu = y^{e_1}_1 \cdots y^{e_n}_n$, we denote by $\mathrm{shift}(\mu, \alpha)$ the polynomial of $\overline{k}[x_1, \ldots, x_n]$ defined by

$$\mathrm{shift}(\mu, \alpha) = (x_1 - \alpha_1)^{e_1} \cdots (x_n - \alpha_n)^{e_n}.$$

We denote by $\mathcal{M}_\alpha$ the maximal ideal of $\overline{k}[x_1, \ldots, x_n]$ generated by $x_1 - \alpha_1, \ldots, x_n - \alpha_n$. When no confusion is possible, we simply write F and f instead of F^i and f_i. We denote by $\mathrm{eval}(F, \alpha)$ the polynomial

$$\mathrm{eval}(F, \alpha) = \sum_{\mu \in S_F} \mathsf{NF}(F_\mu, \mathcal{M}_\alpha) \,\mathrm{shift}(\mu, \alpha) \tag{5}$$

in $\overline{k}[x_1, \ldots, x_n]$. We call this the *specialization of F at α*. Let $W \subset \mathbb{A}^n$ be an algebraic set over k, that is, the zero set $V(P)$ in $\mathbb{A}^n$ of some $P \subset k[x_1, \ldots, x_n]$. Finally, consider a family $(f_\alpha, \alpha \in W)$ of polynomials of $\overline{k}[x_1, \ldots, x_n]$.

We say that F is an *expansion of f about W* if for every point α of W we have $f = \mathrm{eval}(F, \alpha)$. More generally, we say that F is an *expansion of the polynomial family* $(f_\alpha, \alpha \in W)$ *about W* if for every point α of W we have $f_\alpha = \mathrm{eval}(F, \alpha)$. We conclude this section with a fundamental example of the concepts introduced below. For $\mu = y^{e_1} \cdots y^{e_n}$, we denote by $c(f, \mu)$ the polynomial of $k[x_1, \ldots, x_n]$ defined by $c(f, \mu) = \frac{1}{e_1! \cdots e_n!} \frac{\partial^{e_1 + \cdots + e_n} f}{\partial x^{e_1}_1 \cdots \partial x^{e_n}_n}$. (One should recognize these as the coefficients in a Taylor expansion.) Let $S_C(f)$ be the set of the $y^{e_1} \cdots y^{e_n}$ monomials such that $e_i \leq \deg(f, x_i)$ holds for all $i = 1, \ldots, e$. Then, the polynomial $C(f) = \sum_{\mu \in S_C(f)} c(f, \mu) \mu$ is an expansion of f about W.

5 Computing Intersection Multiplicities of Bivariate Systems: Irreducible Case

We follow the notations introduced in Section 4. Let $F^1, \ldots, F^n$ be the expansions of $f_1, \ldots, f_n$ about an algebraic set $W \subset \mathbb{A}^n$. In this section, we assume $W = V(\mathcal{M})$ holds for a maximal ideal $\mathcal{M}$ of $k[x_1, \ldots, x_n]$ and that $n = 2$ holds.

Theorem 2. *The intersection multiplicity of f_1, f_2 is the same at any point of $V(\mathcal{M})$; we denote it by $I(\mathcal{M}; f_1, f_2)$. Moreover, Algorithm 1 computes this multiplicity from F^1, F^2 by performing arithmetic operations in $k[x_1, x_2]$ only.*

This first claim in Theorem 2 should not surprise the expert reader. The length of the module $\mathcal{O}_{\mathbb{A}^n, p} / \langle f_1, \ldots, f_n \rangle$ over a non-algebraically closed field is not necessarily equal to the dimension as a k vector space, though length equals dimension

---

Algorithm 1. $\mathsf{IM}_2(\mathcal{M}; F^1, F^2)$

Input: $F^1, F^2 \in k[x_1, x_2, y_1, y_2]$ and $\mathcal{M} \subset k[x_1, x_2]$ maximal such that F^1, F^2 are expansions of $f_1, f_2 \in k[x_1, x_2]$ about $V(\mathcal{M})$ and $\langle f_1, f_2 \rangle$ is a zero-dimensional ideal.

Output: $I(\mathcal{M}; f_1, f_2)$.

1 **if** $\mathsf{NF}(F_1^1, \mathcal{M}) \neq 0$ **then**
2 $\quad$ **return** 0;

3 **if** $\mathsf{NF}(F_1^2, \mathcal{M}) \neq 0$ **then**
4 $\quad$ **return** 0;

5 $r := \deg(F^1_{<y_2} \bmod \mathcal{M}, y_1)$;
6 $s := \deg(F^2_{<y_2} \bmod \mathcal{M}, y_1)$;
7 **if** $r = 0$ **then**
8 $\quad$ **return** $\mathrm{tdeg}(F^2_{<y_2} \bmod \mathcal{M}, y_1) + \mathsf{IM}_2(\mathcal{M}; \frac{F^1 - F^1_{<y_2}}{y_2}, F^2)$;

9 **if** $s = 0$ **then**
10 $\quad$ **return** $\mathrm{tdeg}(F^1_{<y_2} \bmod \mathcal{M}, y_1) + \mathsf{IM}_2(\mathcal{M}; F^1, \frac{F^2 - F^2_{<y_2}}{y_2})$;

11 $a_1 := \mathrm{lc}(F^1_{<y_2} \bmod \mathcal{M}, y_1)$;
12 $a_2 := \mathrm{lc}(F^2_{<y_2} \bmod \mathcal{M}, y_1)$;
13 **if** $r \leq s$ **then**
14 $\quad$ let $b_1 \in k[x_1, x_2]$ such that $a_1 b_1 \equiv 1 \bmod \mathcal{M}$;
15 $\quad$ $H := F^2 - a_2 b_1 y_1^{s-r} F^1$;
16 $\quad$ **return** $\mathsf{IM}_2(\mathcal{M}; F^1, H)$;

17 let $b_2 \in k[x_1, x_2]$ such that $a_2 b_2 \equiv 1 \bmod \mathcal{M}$;
18 $H := F^1 - a_1 b_2 y_1^{r-s} F^2$;
19 **return** $\mathsf{IM}_2(\mathcal{M}; H, F^2)$;

---

when the field is algebraically closed. The dimension, however, remains the same over both k and $\overline{k}$.

Proof. We show that $\mathsf{IM}_2(\mathcal{M}; F^1, F^2)$, as returned by Algorithm 1, computes $I(p; f_1, f_2)$ uniformly for all $p \in V(\mathcal{M})$ and performs operations in $k[x_1, x_n]$ only. Algorithm correctness and termination follows from three claims.

Claim 1: If $I(p; f_1, f_2) = 0$ holds for some $p \in V(\mathcal{M})$, then $\mathsf{IM}_2(\mathcal{M}; F^1, F^2)$ correctly returns 0.

Claim 2: If $I(p; f_1, f_2) > 0$ holds for all $p \in V(\mathcal{M})$, and if either $\deg(F^1_{<y_2} \bmod \mathcal{M}, y_1) = 0$ or $\deg(F^2_{<y_2} \bmod \mathcal{M}, y_1) = 0$ holds, then $\mathsf{IM}_2(\mathcal{M}; F^1, F^2)$ correctly invokes $\mathsf{IM}_2(\mathcal{M}; G^1, G^2)$ where each $G^i \in k[x_1, x_2, y_1, y_2]$ is an expansion of a polynomial family about $V(\mathcal{M})$ such that $\min(\deg(G^1, y_2), \deg(G^2, y_2)) < \min(\deg(F^1, y_2), \deg(F^2, y_2))$.

Claim 3: If $I(p; f_1, f_2) > 0$ holds for all point $p \in V(\mathcal{M})$, and if $\deg(F^1_{<y_2} \bmod \mathcal{M}, y_1) > 0$ and $\deg(F^2_{<y_2} \bmod \mathcal{M}, y_1) > 0$ both hold, then the call $\mathsf{IM}_2(\mathcal{M}; F^1, F^2)$ correctly invokes $\mathsf{IM}_2(\mathcal{M}; G^1, G^2)$ where each $G^i \in$

$k[x_1, x_2, y_1, y_2]$ is an expansions of a polynomial family about $V(\mathcal{M})$ such that $\min(\deg(G^1_{<y_2}, y_1), \deg(G^2_{<y_2}, y_1)$ is strictly less than $\min(\deg(F^1_{<y_2}, y_1),$ $\deg(F^2_{<y_2}, y_1)$.

Proof (of Claim 1). Assume that there is $p \in V(\mathcal{M})$ such that $I(p; f_1, f_2) = 0$ holds. From (2-2), this implies that we have $p \notin V(f_1, f_2)$. Since $\mathcal{M}$ is maximal, we deduce that $W \cap V(f) = \emptyset$ holds. Thus, the intersection multiplicity of f_1, f_2 is null at any point of $V(\mathcal{M})$. Moreover, deciding whether this latter fact holds amounts to testing whether one of $\mathsf{NF}(F^1_1, \mathcal{M})$, $\mathsf{NF}(F^2_1, \mathcal{M})$ is zero or not, which can be computed in $k[x_1, x_2]$ with a regular chain generating $\mathcal{M}$.

Remark 1. From now on, we assume that $I(p; f_1, f_2) > 0$ holds for all $p \in V(\mathcal{M})$. Since $\mathcal{M}$ is maximal, this implies that $W \subseteq V(F^1_1)$ and $W \subseteq V(F^2_1)$ both hold. Besides, the ideal $\mathcal{M}$ is one of the associated primes of $\langle f_1, f_2 \rangle \subset k[x_1, x_2]$.

Proof (of Claim 2). Assume that either

$$\deg\left(F^1_{<y_2} \bmod \mathcal{M}, y_1\right) = 0 \quad \text{or} \quad \deg\left(F^2_{<y_2} \bmod \mathcal{M}, y_1\right) = 0$$

holds. Since the role of f_1 and f_2 can be exchanged, using (2-4), we assume that $\deg(F^1_{<y_2} \bmod \mathcal{M}, y_1) = 0$ holds. Consider any point $\alpha = (\alpha_1, \alpha_2)$ of $V(\mathcal{M})$. Since F^1_1 is null modulo $\mathcal{M}$, the relation $\deg(F^1_{<y_2} \bmod \mathcal{M}, y_1) = 0$ implies that the whole polynomial $F^1_{<y_2}$ is actually null modulo $\mathcal{M}$. Thus, the specialization $\mathrm{eval}(F^1, \alpha)$ can be divided by $x_2 - \alpha_2$. Applying (2-6), we have

$$I(p; f_1, f_2) = I(p; x_2 - \alpha_2, f_2) + I(p; \tfrac{f_1}{x_2 - \alpha_2}, f_2), \tag{6}$$

where $I(p; x_2 - \alpha_2, f_2)$ is the trailing degree of f_2 evaluated at $x_2 = \alpha_2$ (via (2-5)). Since F^1, F^2 are expansions of f_1, f_2 about $V(\mathcal{M})$, Equation (6) yields

$$\mathsf{IM}_2(\mathcal{M}; F^1, F^2) = \mathrm{tdeg}(F^2_{<y_2} \bmod \mathcal{M}, y_1) + \mathsf{IM}_2(\mathcal{M}; \tfrac{F^1 - F^1_{<y_2}}{y_2}, F^2) \tag{7}$$

where $\mathrm{tdeg}(F^1_{<y_2} \bmod \mathcal{M}, y_1)$ is the trailing degree of $F^1_{<y_2}$ regarded as a polynomial in y_1 with coefficients in the field $k[x_1, x_2]/\mathcal{M}$.

Proof (of Claim 3). We assume that

$$\deg(F^1_{<y_2} \bmod \mathcal{M}, y_1) > 0 \text{ and } \deg(F^2_{<y_2} \bmod \mathcal{M}, y_1) > 0$$

both hold. Since the role of f_1 and f_2 can be exchanged, using (2-4),

$$\deg(F^1_{<y_2} \bmod \mathcal{M}, y_1) \le \deg(F^2_{<y_2} \bmod \mathcal{M}, y_1)$$

is assumed to hold. Let $a_1, a_2 \in k[x_1, x_2]$ be polynomials and $r \le s$ be positive integers such that $a_1 y_1^r$ and $a_2 y_1^s$ are the leading terms of $F^1_{<y_2}$ and $F^2_{<y_2}$ regarded as polynomials in y_1 with coefficients in $k[x_1, x_2]/\mathcal{M}$. Since $W \cap V(a_1) = \emptyset$ holds there exists a polynomial $b_1 \in k[x_1, x_2]$ such that we have $a_1 b_1 \equiv 1 \bmod \mathcal{M}$. Define $H := F^2 - a_2 b_1 y_1^{s-r} F^1$. Clearly, this an expansion of a polynomial family $(h_\alpha, \alpha \in V(\mathcal{M}))$ about $V(\mathcal{M})$ such that we have $\mathrm{eval}(H, \alpha) = h_\alpha$ where

$$h_\alpha := f_2 - a_2(\alpha) b_1(\alpha)(x_1 - \alpha_1)^{s-r} f_1. \tag{8}$$

Using (2-7), we have $I(p; f_1, f_2) = I(p; f_1, h_\alpha)$, for all $p \in V(\mathcal{M})$, yielding

$$\mathsf{IM}_2(\mathcal{M}; F^1, F^2) = \mathsf{IM}_2(\mathcal{M}; F^1, H). \tag{9}$$

6   Computing Intersection Multiplicities of Bivariate Systems: Zero-Dimensional Case

The generalization from *irreducible* zero-dimensional algebraic sets $V(\mathcal{M})$ to *arbitrary* ones relies on standard techniques for computing triangular decomposition of polynomial systems (see for instance [20,10,14,19,4]).

Algorithm 2 is the adaptation of Algorithm 1 for $n = 2$ variables. In this algorithm we use two yet unmentioned methods: LT and Tdeg, and one yet unmentioned language construct: **output**. Similar to Regularize, the call $\mathsf{LT}(F^i, T)$, or leading term of F^i modulo $\langle T \rangle$, returns a list of pairs, (C, a_{F^i}), where $C \subset k[x_1, x_2]$ is a regular chain and a_{F^i} is the lexicographical leading term of F^i when viewed as a polynomial in $y_1 < y_2$ with coefficients in $k[x_1, x_2]/\langle C \rangle$; moreover the union of $V(C)$'s form a partition of $V(T)$. The specification for TDeg "trailing degree" is analogue. Finally, as we are returning a *sequence* we use the language construct **output**(x, y) to indicate that (x, y) has been added to the sequence that will ultimately be returned.

Theorem 3. *Algorithm 2 terminates and works correctly.*

Proof. We distinguish two cases: Algorithm 2 *does not split* the computations and *does split* the computations. In this proof, $C_1, \ldots, C_e \subset$ designate regular chains of $k[x_1, \ldots, x_n]$ such that $V(T)$ is the disjoint union of $V(C_1), \ldots, V(C_e)$. *Non-splitting case:* Assume that $\mathsf{IM}_2(T; F^1, F^2)$ computed by Algorithm 2 does not split the computation, thus returning a single pair (T, m). Using Relation (3), one can check that $\mathsf{IM}_2(C_i; F^1, F^2)$ returns (C_i, m), for each $i = 1, \ldots, e$. Assume that $C_1, \ldots, C_e$ generate maximal ideals. One can check that, when it does not split, Algorithm 2 performs the same computation as Algorithm 1. By virtue of Theorem 2, Algorithm 1 works correctly with input maximal ideals, thus each call $\mathsf{IM}_2(C_i; F^1, F^2)$ correctly returns (C_i, m). Consequently, $\mathsf{IM}_2(T; F^1, F^2)$ correctly returns (T, m) also, since is the disjoint union of $V(C_1), \ldots, V(C_e)$.

Splitting case: From now on, assume now that the call $\mathsf{IM}_2(T; F^1, F^2)$ computed by Algorithm 2 splits and returns pairs $(C_1, m_1), \ldots, (C_e, m_e)$, where we no longer assume that $C_1, \ldots, C_e$ generate maximal ideals. From the non-splitting case and Relation (3), we know that each call $\mathsf{IM}_2(C_i; F^1, F^2)$ correctly returns (C_i, m). We conclude again with the fact that $V(T)$

7   Reduction to the Bivariate Case

We return to the n-variate case, using the same notations as in Sections 3. We discuss how this n-variate case can be reduced to the bivariate one, for which Algorithm 2 computes the intersection multiplicity of two plane curves (without common components) at any point of their intersection.

We start by considering Property (n-8) of Section 3. Let $p \in V(f_1, \ldots, f_n)$. Assume the hypersurface $h_n = V(f_n)$ is non-singular at p. Let v_n be its tangent hyperplane at p. Assume furthermore that the tangent cone $TC_p(\mathcal{C})$ intersects

---

Algorithm 2. $\mathsf{IM}_2(T; F^1, F^2)$

Input: F^1 and F^2 as given in Algorithm 1
Output: Finitely many pairs (T_i, m_i) where $T_i \subset k[x_1, \ldots, x_n]$ are regular
chains and $m_i \in \mathbb{Z}^+$ such that Equation (1) holds and for all
$p \in V(T^i)$ we have $I(p; f_1, \ldots, f_n) = m_i$.

1 **for** $T \in \mathsf{Regularize}\left(F_1^1, T\right)$ **do**
2   **if** $F_1^1 \notin \langle T \rangle$ **then**
3   ⌊ **output**$(T, 0)$;
4   **else**
5   **for** $T \in \mathsf{Regularize}\left(F_1^2, T\right)$ **do**
6   **if** $F_1^2 \notin \langle T \rangle$ **then**
7   ⌊ **output**$(T, 0)$;
8   **else**
9   **for** $(T, a_{F^1}) \in \mathsf{LT}\left(F_{<y_2}^1, T\right)$ **do**
10   **for** $(T, a_{F^2}) \in \mathsf{LT}\left(F_{<y_2}^2, T\right)$ **do**
               /* Wlog $deg(F_{<y_2}^1) \le deg(F_{<y_2}^2)$ */
11   **if** $a_{F^1} \in \langle T \rangle$ **then**
12   **for** $(T, d) \in \mathsf{TDeg}\left(F_{<y_2}^2, T\right)$ **do**
13   **for** $(T, i) \in \mathsf{IM}_2(T, \frac{F^1 - F_{<y_2}^1}{y_2}, F^2)$ **do**
14   ⌊ **output**$(T, (d + i))$;
15   **else**
16               $H \leftarrow F^2 - a_{F^2} \cdot \mathsf{Inverse}\left(a_F^1, T\right) \cdot F^1$;
17   **output**$\left(\mathsf{IM}_2(T, F^1, H)\right)$;

---

v_n only at the point p. Let $h \in k[x_1, \ldots, x_n]$ be the degree 1 polynomial defining v_n. Finally, recall (Theorem 1) that $I(p; f_1, \ldots, f_n) = I(p; f_1, \ldots, f_{n-1}, h)$ holds.

Up to re-numbering the variables, we can assume that the coefficient of x_n in h is non-zero, thus $h = x_n - h'$, where $h' \in k[x_1, \ldots, x_{n-1}]$. Hence, we can rewrite the ideal $\langle f_1, \ldots, f_{n-1}, h \rangle$ as $\langle g_1, \ldots, g_{n-1}, h \rangle$ where g_i is obtained from f_i by substituting x_n with h'. If instead of a point p, we have a zero-dimensional regular chain $T \subset k[x_1, \ldots, x_n]$, we use the techniques developed in Sections 5 and 6 to reduce to the case of a point. Assuming $x_1 < \cdots < x_n$, this leads to $I(p; f_1, \ldots, f_n) = I(T \cap k[x_1, \ldots, x_{n-1}]; g_1, \ldots, g_{n-1})$.

In practice, this reduction from n to $n - 1$ variables does not always apply. For instance, this is the case for *Ojika 2* $\subseteq k[x, y, z]$:

$$x^2 + y + z - 1 = x + y^2 + z - 1 = x + y + z^2 - 1 = 0. \tag{10}$$

However, using the equation $x^2 + y + z - 1 = 0$ to eliminate z from the other two, we obtain two bivariate polynomials $f, g \in k[x, y]$. At any point of $p \in V(h, f, g)$ the tangent cone of the curve $V(f, g)$ is independent of z; in some sense it is "vertical". Moreover, at any point of $p \in V(h, f, g)$ the tangent space of $V(h)$ is *not* vertical. Thus, the reduction applies without computing *any* tangent cones.

We conclude this section by explaining how the tangent cone $TC_p(\mathcal{C})$ is computed when the above trick does not apply. For simplicity, assume $\overline{k} = \mathbb{C}$ and assume that none of the $V(f_i)$ are singular at p. For each component $\mathcal{G}$ through p of $\mathcal{C} = V(f_1, \ldots, f_{n-1})$, we proceed as follows: There exists a neighborhood B of p such that $V(f_i)$ is not singular at all $q \in (B \cap \mathcal{G}) \setminus \{p\}$, for $i = 1, \ldots, n-1$. Let $v_i(q)$ be the tangent hyperplane of $V(f_i)$ at q. Regard $v_1(q) \cap \cdots \cap v_{n-1}(q)$ as a parametric variety with the coordinates of q as parameters. Then, we have $TC_p(\mathcal{G}) = v_1(q) \cap \cdots \cap v_{n-1}(q)$ when q approaches p, which we compute by a variable elimination process. Finally, $TC_p(\mathcal{C})$ is the union of all the $TC_p(\mathcal{G})$. This approach avoids standard basis computation and extends easily for working with the zero set $V(T)$ of a zero-dimensional regular chain T instead of a point p.

8   Implementation

We have done an implementation in MAPLE that depends heavily on the Regular-Chains library. As this implementation is sufficiently different from the theoretical algorithm it is meaningful to discuss how we realized it.

These differences can be traced back to a common origin: the data structure simulating the expansions F^i defined in Section 4 for the purpose of the algorithms of Sections 5 and 6. Recall that the expansions $F^1, \ldots, F^n$ belong to $k[x_1, \ldots, x_n, y_1, \ldots, y_n]$ where $x_1, \ldots, x_n$ are the variables of the input polynomials $f_1, \ldots, f_n$ and where $y_1, \ldots, y_n$ are essentially "placeholders". But our algorithms fundamentally treat $F^1, \ldots, F^n$ as vectors, performing only additions and subtractions on them.

While these expansions $F^1, \ldots, F^n$ are a nice trick to manipulate "simultaneously" Taylor expansions at several points of a variety, a naïve implementation could suffer from performance bottleneck (hardly surprisingly when doubling the number of variables). In particular, we observe that during the execution of the algorithms, all the partial derivatives of $f_1, \ldots, f_n$ may not be needed. Therefore, one may wish to take advantage of *lazy* or *delayed* evaluation.

A structure utilizing delayed computation is well suited for this. To demonstrate why, suppose that $\mathcal{F}^i$ is a data structure implementing F^i such that $\mathcal{F}^i(a_1, \ldots, a_n) = F^i_\mu$ for $\mu = y_1^{a_1} \cdots y_n^{a_n}$. To determine $\mathcal{F}^i(a_1, \ldots, a_n + 1)$ one must only compute $\frac{1}{a_n+1} \frac{\partial \mathcal{F}^i(a_i, \ldots, a_n)}{\partial x_n}$. Combining this rule with $\mathcal{F}^i(a_1, \ldots, a_{n-1}, 0) = \mathcal{F}^i(a_1, \ldots, a_{n-1})$ and $\mathcal{F}^i(0) = f_i$ gives a recursive function whose output matches our specification. We call these "lazy Taylor expansions" (LTEs).

Moreover these LTEs have a very useful property: $\mathcal{F}^i(a_1, \ldots, a_{n-1}) \equiv F^i_{<y_n}$. They are also surprisingly straightforward to implement in MAPLE.

Notice that the "data structure" for the LTEs are in fact procedures. Therefore any method processing LTEs, like Subtract for instance, will take as input procedures and return a procedure. This notion may be unusual but requires very little overhead (practically undetectable in our experiments). We outline the remaining important methods for our algorithms:

Division by y_n:

$$\frac{\mathcal{F}^i(a_1,\ldots,a_n)}{y_n} = \mathcal{F}^i(a_1,\ldots,a_n+1)$$

Multiplication by μ: Let $\mathcal{F}^i(a_1,\ldots,a_n) = 0$ if there is i for which $a_i < 0$, then

$$\mathcal{F}^i(a_1,\ldots,a_n) \cdot \left(y_1^{b_1} \cdots y_n^{b_n}\right) = \mathcal{F}^i(a_1 - b_1,\ldots,a_n - b_n)$$

Substitute $y_n = h_1 y_1 + \cdots + h_{n-1} y_{n-1}$. For every $b_1,\ldots,b_n$ with $b_n > 0$, $\mathcal{F}(b_1,\ldots,b_n) \leftarrow 0$ and

$$\mathcal{F}(a_1 + k_1,\ldots,a_{n-1} + k_{n-1}) \leftarrow \mathcal{F}(a_1,\ldots,a_{n-1})+$$

$$\sum_{k_1+\cdots+k_{n-1}=b_n} \binom{b_n}{k_1,\ldots,k_{n-1}} h_1^{k_1} \cdots h_{n-1}^{k_{n-1}}.$$

Using these LTEs along with careful, and repeated, invocations of the Regular-Chains[Regularize] command, our algorithms can be realized.

9   Experiments

We have fully implemented the bivariate case, that is, Algorithm 2, on top of the RegularChains library in MAPLE. As this is the base case for the n-variate algorithm it is of paramount importance that it runs fast and correctly. The n-variate implementation is a work in progress and there is large room for improvements.

We choose to study systems taken from [2] and [13]—a suite of examples used for benchmarking and testing bivariate system solvers. All timings are given in seconds and the base field has characteristic 962592769 in all cases. It should be noted that, despite 962592769 being a so-called FFT-prime, we are *not* using the FastArithmeticTools package of the RegularChains library. This is because our current implementation is only generic and works in any characteristic. However, some of the systems in [13] are too challenging for being directly solved in characteristic zero without using an approach based on modular, or other advanced, techniques. Results are in Table 1.

We are happy with the results of these experiments for two reasons. First, we could not find an instance where Triangularize produced regular chains for which our algorithm IM_2 could not correctly and expeditiously determine the intersection multiplicities. Secondly, applying Property (2-5) from Section 1 to our bivariate code admits a speedup factor in the hundreds. Indeed this property enables us to determine if the intersection multiplicity is one simply by checking the invertibility of the Jacobian of f_1, f_2 modulo the current regular chain.

Our n-variate implementation is based on the techniques discussed in Section 7. As with the bivariate case, our experiments are done in characteristic 962592769. We have taken examples from [7] (a paper on intersection multiplicity) and from [3] (a test suite for benchmarking homotopy solvers). Observe that the reduction techniques of Section 7 apply successfully for 3 examples and partially for 2 examples. We also note that tangent cone computations are currently a bottleneck. A new algorithm for this task is work in progress.

Table 1. (LEFT) Input Polynomials (after specialization to bivariate). **(RIGHT)** Experimental results for the bivariate case. Dimension is calculated by MAPLE's PolynomialIdeals:-NumberOfSolutions command which gives the number of solutions counted *with* multiplicity. Time($\triangle$ize) is time required by RegularChains:-Triangularize to decompose the system into N=#rc's many regular chains and Time(rc_im) = Time(rc_im(rc$_1$)) +$\cdots$+ Time(rc_im(rc$_N$)): the total time for rc_im, our implementation of Algorithm 3, to determine intersection multiplicities of an entire system.

Label	Name	terms	degree
1	hard_one	30	37
2	L6_circles	4	24
3	spiral29_24	63	52
4	tryme	38	59
5	challenge_12	49	30
6	challenge_12_1	64	40
7	compact_surf	52	18
8	degree_6_surf	467	42
9	mignotte_xy	81	64
10	SA_4_4_eps	63	33
11	spider	292	36

System	Dim	Time($\triangle$ize)	#rc's	Time(rc_im)
$\langle 1,3\rangle$	888	9.7	20	19.2
$\langle 1,4\rangle$	1456	226.0	8	9.023
$\langle 1,5\rangle$	1595	169.4	8	25.4
$\langle 3,5\rangle$	1413	22.5	27	28.6
$\langle 4,5\rangle$	1781	218.4	9	13.9
$\langle 5,1\rangle$	1759	113.0	10	15.8
$\langle 6,8\rangle$	1680	99.7	12	37.6
$\langle 6,9\rangle$	2560	299.3	10	22.9
$\langle 6,10\rangle$	1320	131.9	7	8.4
$\langle 6,11\rangle$	1440	59.8	17	27.5
$\langle 7,8\rangle$	1152	32.8	12	16.2
$\langle 7,9\rangle$	756	18.5	16	11.2
$\langle 8,9\rangle$	1984	374.5	10	11.3
$\langle 8,10\rangle$	1362	232.5	7	9.3
$\langle 8,11\rangle$	1256	49.6	17	45.7
$\langle 9,11\rangle$	1792	115.1	16	17.2
$\langle 10,11\rangle$	1180	40.9	17	21.3

Table 2. Experimental results for the n-variate case. Dimension is again the dimension of the vector space $k[x_1,\ldots,x_n]/\langle f_1,\ldots,f_n\rangle$ and Points is the degree of the variety $V(f_1,\ldots,f_n)$. $\triangle$ize and rc_im are the same as in Table 1. Cones and COV give (respectively) the time to calculate the tangent cones or to do a change of variables of the system. Finally, Total is the sum of the previous three columns and Success is the number of points (counted *with* multiplicity) for which the bivariate reduction was success full over the dimension of of the vector space $k[x_1,\ldots,x_n]/\langle f_1,\ldots,f_n\rangle$.

Name	Dim	Points	$\triangle$ize	Cones	COV	rc_im	Total	Success
Nbody5	99	49	1.60	0.00	0.06	1.90	2.00	51/99
mth191	27	18	0.56	5400.00	0.04	0.01	5400.00	23/27
ojika2	8	5	0.20	8.20	0.13	0.47	8.80	8/8
E-Arnold1	45	30	0.89	1100.00	0.01	1800.00	2900.00	45/45
ShiftedCubes	27	25	0.66	0.00	0.00	0.52	0.52	27/27

Acknowledgements. This project has benefited from useful conversations with Dr. Roi Docampo and Dr. Noah Giansiracusa, and was funded, in part, by grants from Maplesoft, MITACS and NSERC of Canada.

References

1. Aubry, P., Lazard, D., Moreno Maza, M.: On the theories of triangular sets. J. Symb. Comp. 28(1-2), 105–124 (1999)
2. Berberich, E., Emeliyanenko, P., Sagraloff, M.: An elimination method for solving bivariate polynomial systems: Eliminating the usual drawbacks. CoRR, abs/1010.1386 (2010)
3. Bini, D., Mourrain, B.: Polynomial test suite, http://www-sop.inria.fr/saga/POL/ (accessed: April 1, 2012)
4. Chen, C., Moreno Maza, M.: Algorithms for computing triangular decompositions of polynomial systems. In: Proc. ISSAC 2011, pp. 83–90. ACM (2011)
5. Cheng, J.-S., Gao, X.-S.: Multiplicity preserving triangular set decomposition of two polynomials. CoRR, abs/1101.3603 (2011)
6. Cox, D., Little, J., O'Shea, D.: Using Algebraic Geometry. Graduate Text in Mathematics, vol. 185. Springer, New York (1998)
7. Dayton, B.H., Zeng, Z.: Computing the multiplicity structure in solving polynomial systems. In: Proceedings of ISSAC 2005, pp. 116–123. ACM (2005)
8. Fulton, W.: Introduction to intersection theory in algebraic geometry. CBMS Regional Conference Series in Mathematics, vol. 54. Conference Board of the Mathematical Sciences, Washington, DC (1984)
9. Fulton, W.: Algebraic curves. Advanced Book Classics. Addison-Wesley (1989)
10. Kalkbrener, M.: A generalized euclidean algorithm for computing triangular representations of algebraic varieties. J. Symb. Comp. 15, 143–167 (1993)
11. Kirwan, F.: Complex algebraic curves. London Mathematical Society Student Texts, vol. 23. Cambridge University Press, Cambridge (1992)
12. Knapp, A.W.: Cornerstones. In: Advanced algebra. Birkhäuser Boston Inc., Boston (2007), Along with a companion volume ıt Basic algebra
13. Labs, O.: A list of challenges for real algebraic plane curve visualization software. In: Emiris, I.Z., Sottile, F., Theobald, T. (eds.) Nonlinear Computational Geometry, pp. 137–164. Springer, New York (2010)
14. Lazard, D.: Solving zero-dimensional algebraic systems. J. Symb. Comp. 15, 117–132 (1992)
15. Lemaire, F., Moreno Maza, M., Pan, W., Xie, Y.: When does (T) equal Sat(T)? In: Proc. ISSAC 2008, pp. 207–214. ACM Press (2008)
16. Lemaire, F., Moreno Maza, M., Xie, Y.: The RegularChains library. In: Ilias, S. (ed.) Maple Conference 2005, pp. 355–368 (2005)
17. Li, Y.L., Xia, B., Zhang, Z.: Zero decomposition with multiplicity of zero-dimensional polynomial systems. CoRR, abs/1011.1634 (2010)
18. Shafarevich, I.R.: Basic algebraic geometry 1, 2nd edn. Springer, Berlin (1994)
19. Wang, D.M.: Elimination Methods. Springer (2000)
20. Wu, W.T.: A zero structure theorem for polynomial equations solving. MM Research Preprints 1, 2–12 (1987)

A Note on the Space Complexity
of Fast D-Finite Function Evaluation

Marc Mezzarobba

Inria, AriC, LIP (UMR 5668 CNRS-ENS Lyon-Inria-UCBL)
marc@mezzarobba.net

Abstract. We state and analyze a generalization of the "truncation trick" suggested by Gourdon and Sebah to improve the performance of power series evaluation by binary splitting. It follows from our analysis that the values of D-finite functions (i.e., functions described as solutions of linear differential equations with polynomial coefficients) may be computed with error bounded by 2^{-p} in time $O(p(\lg p)^{3+o(1)})$ and space $O(p)$. The standard fast algorithm for this task, due to Chudnovsky and Chudnovsky, achieves the same time complexity bound but requires $\Theta(p \lg p)$ bits of memory.

1 Introduction

Binary splitting is a well-known and widely applicable technique for the fast multiple precision numerical evaluation of rational series. For any series $\sum_n s_n$ with $\limsup_n |s_n|^{1/n} < 1$ whose terms s_n obey a linear recurrence relation with polynomial coefficients, e.g.,

$$\ln 2 = \sum_{n=0}^{\infty} s_n, \qquad s_n = \frac{1}{(n+1)2^{n+1}}, \qquad 2(n+2)s_{n+1} - (n+1)s_n = 0,$$

the binary splitting algorithm allows one to compute the partial sum $\sum_{n=0}^{N-1} s_n$ in $O(M(N(\lg N)^2))$ bit operations [5,3]. Here $M(n)$ stands for the complexity of multiple precision integer multiplication, and $\lg$ denotes the binary logarithm. As $N = O(p)$ terms of the series are enough to make the approximation error less than 2^{-p}, the complexity of the algorithm is softly linear in the precision p, assuming $M(n) = O(n(\lg n)^{O(1)})$.

Methods based on binary splitting tend to be favored in practice even in cases when asymptotically faster algorithms (typically AGM iterations [2]) would apply. One high-profile example is the computation of billions of digits of classical constants such as π, $\zeta(3)$ or γ. Basically all record computation in recent years were achieved by evaluating suitable series using variants of binary splitting [9,28].

A drawback of the classical binary splitting algorithm, both from the complexity point of view and in practice, is its comparatively large memory usage. Indeed, the algorithm amounts to the computation of a product tree of matrices derived from the recurrence—see Sect. 3 below for details. The intermediate

---

V.P. Gerdt et al. (Eds.): CASC 2012, LNCS 7442, pp. 212–223, 2012.

results are matrices of rational numbers whose bit sizes roughly double from one level to the next. Near the root, their sizes can (and in general do) reach $\Theta(p \lg p)$, even though the output has size $\Theta(p)$.

However, the space complexity can be lowered to $O(p)$ using a slight variation of the classical algorithm. The basic idea is to truncate the intermediate results to a precision $O(p)$ when they start taking up more space than the final result. Of course, these truncations introduce errors. To make the trick into a genuine algorithm, we need to analyze the errors, add a suitable number of "guard digits" at each step and check that the space and time complexity of the resulting process stay within the expected bounds.

The opportunity to improve the practical behavior of binary splitting using truncations has been noticed by authors of implementations on several occasions over the last decade or so. Gourdon and Sebah [10] describe truncation as a "crucial" optimization. Besides the expected drop of memory usage, they report running time improvements by an "appreciable" constant factor. Cheng et al. [4] compare truncation with alternative (less widely applicable but sometimes more efficient) approaches. Most recently, Kreckel [14] explicitly asks how to make sure that the new roundoff errors do not affect the correctness of the result.

Indeed, the above-mentioned error analysis did not appear in the literature until very recently. An article by Yakhontov [26,27] now provides the required bounds in the case of the generalized hypergeometric series $_pF_q$, which covers all examples where the truncation trick had been used before. But the applicability of the method is actually much wider.

The purpose of this note is to present a more general and arguably simpler analysis. Our version is more general in two main respects. First, besides hypergeometric series, it applies to the solutions of linear ordinary differential equations with rational coefficients, also known as *D-finite* (or holonomic) series [21]. D-finite series are exactly those whose coefficients obey a linear recurrence relation with rational coefficients, while hypergeometric series correspond to recurrences of the first order. Second, we take into account the coefficient size of the recurrence that generates the series to be computed. Allowing the size of the coefficients to vary with the target precision p makes it possible to use the modified binary splitting procedure as part of the "bit burst" algorithm [5] to handle evaluations at general real or complex points approximated by rationals of size $\Theta(p)$.

Additionally, our analysis readily adapts to other applications of binary splitting. The simplicity and generality of the proof are direct consequences of viewing the algorithm primarily as the computation of a product tree. See Gosper [8] and Bernstein [1, §12–16] for further comments on this point of view.

The remainder of this note is organized as follows. Section 2 contains some notations and assumptions. In Sect. 3, we recall the standard binary splitting algorithm, which will serve as a subroutine in the linear-space version. Then, in Sect. 4, we state and analyze the "truncated" variant that achieves the linear space complexity for general D-finite functions. Finally, Sect. 5 offers a few comments on other variants of the binary splitting method and possible extensions of the analysis.

2 Setting

The performance of the binary splitting algorithm crucially depends on that of integer multiplication. Following common usage, we denote by $M(n)$ a bound on the time needed to multiply two integers of at most n bits. Currently the best theoretical bound [7] is $M(n) = O(n(\lg n)\exp O(\lg^* n))$, where $\lg^* n = \min\{k \lg^{\circ k} n \leqslant 1\}$. In practice, implementations such as GMP [11] use variants of the Schönhage-Strassen algorithm of complexity $O(n(\lg n)(\lg\lg n))$. We make the usual assumption [25] that the function $n \mapsto M(n)/n$ is nondecreasing. It follows that $M(n) + M(m) \leqslant M(n+m)$. We also assume that the *space* complexity of integer multiplication is linear, which is true for the standard algorithms.

Write $\mathbb{K} = \mathbb{Q}(i)$, and define the *bit size* of a number $(x+iy)/w \in \mathbb{K}$ (where $w, x, y \in \mathbb{Z}$) as $\lceil \lg w \rceil + \lceil \lg x \rceil + \lceil \lg y \rceil + 1$. Consider a linear differential equation with coefficients in $\mathbb{K}(z)$. It will prove convenient to clear all denominators (both polynomial and integer) and multiply the equation by a power of z to write it as

$$\left(a_r(z)\left(z\frac{\mathrm{d}}{\mathrm{d}z}\right)^r + \cdots + a_1(z)z\frac{\mathrm{d}}{\mathrm{d}z} + a_0(z) \right) \cdot y(z) = 0, \qquad a_k \in \mathbb{Z}[i][z]. \quad (1)$$

Let $s = \max_k \deg a_k$, and let h_1 denote the maximum bit size of the coefficients of the a_k. Although our complexity estimates depend on r and h_1, we do not consider more general dependencies on the equation. Thus, the a_k are assumed to vary only in ways that can be described in terms of these two parameters. Specifically, we assume that $s = O(1)$ and that the coefficients of $a_k(z)/a_r(0)$ are all restricted to some bounded domain.

We also assume that 0 is an ordinary (i.e. nonsingular) point of (1). This implies that $a_r(0) \neq 0$ and $s \geqslant r$. The case of *regular singular* points (those for which we still have $a_r(0) \neq 0$ but possibly $s < r$ [13, Chap. 9]) is actually similar [23,17]; we focus on ordinary points to avoid cumbersome notations.

Let $\rho = \min\{|z| : a_r(z) = 0\} \in (0, \infty]$. Then any formal series solution $y(z) = \sum_{n \geqslant 0} y_n z^n$ of (1) converges on the disk $|z| < \rho$. We select a particular solution (say, by specifying initial values $y(0), \ldots, y^{(r-1)}(0)$ in some fixed, bounded domain), and an evaluation point $\zeta \in \mathbb{K}$ with $|\zeta| < \rho$. Let h_2 denote the bit size of ζ, and let $h = h_1 + h_2$. Again, h_2 is allowed to grow to infinity, but we assume that $|\zeta|$ is bounded away from ρ.

Given $p \geqslant 0$, our goal is to compute a complex number $\omega \in \mathbb{K}$ such that $|\omega - y(\zeta)| \leqslant 2^{-p}$. By a classical argument, which can be reconstructed by substituting a series with indeterminate coefficients into (1), the sequence (y_n) obeys a recurrence relation of the form

$$b_0(n)y_{n+r} + b_1(n)y_{n+r-1} + \cdots + b_s(n)y_{n+r-s} = 0, \qquad b_j \in \mathbb{K}[n]. \quad (2)$$

Writing $a_k(z) = a_{k,0} + a_{k,1}z + \cdots + a_{k,s}z^s$, the b_j are given explicitly by

$$b_j(n) = \sum_{k=0}^{r} a_{k,j}(n+r-j)^k. \quad (3)$$

Based on the matrix form of the recurrence (2), set

$$B(n) = \begin{pmatrix} \zeta C(n) & 0 \\ R & 1 \end{pmatrix} \in \mathbb{K}(n)^{(s+1)\times(s+1)} \tag{4}$$

where

$$C(n) = \begin{pmatrix} 1 & & & \\ & \ddots & & \\ & & 1 & \\ -\frac{b_s(n)}{b_0(n)} & \cdots & \cdots & -\frac{b_1(n)}{b_0(n)} \end{pmatrix}, \qquad R = \begin{pmatrix} \underbrace{0 \quad \cdots \quad 0}_{s-r \text{ zeroes}} & 1 & \underbrace{0 \quad \cdots \quad 0}_{r-1 \text{ zeroes}} \end{pmatrix}.$$

Let $P(a,b) = B(b-1)\cdots B(a+1)B(a)$ for all $a \leqslant b$. (In particular, $P(a,a)$ is the identity matrix.)

One may check that $b_0(n) \neq 0$ for $n \geqslant 0$, due to the fact that 0 is an ordinary point of (1). Thus the computation of a partial sum $S_N = \sum_{n=0}^{N-1} y_n \zeta^n$ reduces to that of the matrix product $P(0,N)$. Indeed, we have

$$(y_{n+r-s}\zeta^n, \ldots, y_{n+r-1}\zeta^n, S_n)^{\mathrm{T}} = P(0,n)\,(y_{r-s}, \ldots, y_{r-1}, 0)^{\mathrm{T}}$$

where $y_{r-s} = 0, \ldots, y_{-1} = 0, y_0, \ldots, y_{r-1}$ are easily determined from the initial values of the differential equation.

3 Review of the Classical Binary Splitting Algorithm

Since the entries of the matrix $B(n)$ are rational functions of n, the bit size of $P(a,b)$ grows as $O((b-a)\lg b)$ when $b, (b-a) \to \infty$. This bound is sharp in the sense that it is reached for some (in fact, most) differential equations. Computing $P(a,b)$ as $B(b-1) \cdot [B(b-2) \cdot [\cdots B(a)]]$ then takes time at least quadratic in $b - a$, as can be seen from the combined size of the intermediate results. The term "binary splitting" refers to the technique of reorganizing the product into a *balanced tree* of subproducts, using the relation $P(a,b) = P(m,b) \cdot P(a,m)$ with $m = \lfloor \frac{1}{2}(a+b) \rfloor$, and so on recursively.

A slight complication stems from the fact that removing common divisors between the numerators and denominators of the fractions appearing in the intermediate $P(a,b) \in \mathbb{K}^{r\times r}$ would in general be too expensive. Multiplying the numerators and denominators separately and doing a single final division yields better complexity bounds. Let

$$\hat{B}(n) = b_0(n)\check{\zeta}B(n) \in \mathbb{Z}[i][n]^{(s+1)\times(s+1)}, \quad \zeta = \hat{\zeta}/\check{\zeta} \ (\hat{\zeta} \in \mathbb{Z}[i], \check{\zeta} \in \mathbb{Z}). \tag{5}$$

The entries of $\hat{B}(n)$ are polynomials of degree at most r and bit size $O(h)$. To compute $P(a,b)$ by binary splitting, we multiply the $\hat{B}(n)$ for $a \leqslant n < b$ using Algorithm 1, and then divide the resulting matrix by its bottom right entry. The general algorithm considered here was first published by Chudnovsky and Chudnovsky [5], with (up to minor details) the analysis summarized in Prop. 1. The idea of binary splitting was known long before [8,1].

Algorithm 1. BinSplit(a, b)
1 If $b - a \leqslant$ (some threshold)
    2 Return $\hat{B}(b-1) \cdots \hat{B}(a)$ where $\hat{B}$ is defined by (5)
3 else
    4 Return BinSplit($\lfloor \frac{a+b}{2} \rfloor, b$) · BinSplit($a, \lfloor \frac{a+b}{2} \rfloor$)

Proposition 1. [5] As $b, N = b - a, h, r \to \infty$ with $r = \mathrm{O}(N)$, Algorithm 1 computes an unreduced fraction equal to $P(a, b)$ in $\mathrm{O}(\mathrm{M}(N(h + r \lg b)) \lg N)$ operations, using $\mathrm{O}(N(h + r \lg b))$ bits of memory. Assuming $\mathrm{M}(n) = n(\lg n)(\lg \lg n)^{\mathrm{O}(1)}$, both bounds are sharp.

Proof (sketch). The bit sizes of the matrices that get multiplied together at any given depth $0 \leqslant \delta < \lceil \lg N \rceil$ in the recursive calls are at most $C2^{-\delta}N(h + d \lg b)$ for some C. Since there are at most 2^{δ} such products and the multiplication function $\mathrm{M}(\cdot)$ was assumed to be subadditive, the contribution of each level is bounded by $\mathrm{M}(C(b-a)(h + d \lg b))$, whence the total time complexity. See [5,17] for details. The intermediate results stored or multiplied together at any stage of the computation are disjoint subproducts of $B(b-1) \cdots B(a)$, and we assumed the space complexity of n-bit integer multiplication to be $\mathrm{O}(n)$, so the space required by the algorithm is linear in the combined size of the $B(n)$. Finally, it is not hard to construct examples of differential equations that reach these bounds.

Remark 1. The link between our setting and the more common description of the algorithm for hypergeometric series is as follows. In the notation of Haible and Pananikolaou [12] also used in Yakhontov's article, the partial sums of the hypergeometric series are related to its defining parameters a, b, p, q by

$$\begin{pmatrix} \tilde{s}(i+1) \\ S(i) \end{pmatrix} = \begin{pmatrix} \frac{p(i)}{q(i)} & 0 \\ \frac{a(i)}{b(i)} \frac{p(i)}{q(i)} & b(i)q(i) \end{pmatrix} \begin{pmatrix} \tilde{s}(i) \\ S(i-1) \end{pmatrix}, \qquad \tilde{s}(i) = \frac{b(i)}{a(i)} s(i).$$

This equation becomes $\begin{pmatrix} B(i) \\ T(0,i) \end{pmatrix} = \begin{pmatrix} b(i)p(i) & 0 \\ a(i)p(i) & b(i)q(i) \end{pmatrix} \begin{pmatrix} B(i-1)P(i-1) \\ T(0,i-1) \end{pmatrix}$ upon clearing denominators. The standard recursive algorithm for hypergeometric series may be seen an "inlined" computation of the associated product tree. Each recursive step is equivalent to the computation of the matrix product $\begin{pmatrix} B_r P_r & 0 \\ T_r & B_r Q_r \end{pmatrix} \begin{pmatrix} B_l P_l & 0 \\ T_l & B_l Q_l \end{pmatrix}$.

We return to the evaluation of a D-finite power series within its disk of convergence. From the differential equation (1), suitable initial conditions, the evaluation point ζ and a target precision p, one can *compute* [18] a truncation order N such that $|S_N - y(\zeta)| \leqslant 2^{-p}$ and

$$\begin{cases} N \sim Kp = \big(\lg(|\zeta|/\rho)\big)^{-1}p, & \text{if } \rho < \infty, \\ N = \Theta(p/\lg p), & \text{if } \rho = \infty. \end{cases} \tag{6}$$

Combined with these estimates, Proposition 1 implies the following.

Corollary 1. *Write $\ell = h + r \lg p$. Under the assumptions of Proposition 1, one can compute $y(\zeta)$ in $O(M(\ell p \lg p))$ bit operations, using $O(\ell p)$ bits of memory. The complexity goes down to $O(M(\ell p))$ operations and $O(\ell p / \lg p)$ bits of memory when $a_r(z)$ is a constant.*

This result is the basis of more general evaluation algorithms for D-finite functions [5]. Indeed, binary splitting can be used to compute the required series sums at each step when solving a differential equation of the form (1) by the so-called method of Taylor series [15]. Corollary 1 thus extends to the evaluation of y outside the disk $|z| < \rho$. Chudnovsky and Chudnovsky further showed how to reduce the cost of evaluation from $\Omega(hp) = \Omega(p^2)$ to softly linear in p when $h = \Theta(p)$. This last situation is very natural since it covers the case where the point ζ is itself a $O(p)$-digits approximation resulting from a previous computation. The method, known as the *bit burst* algorithm, consists in solving the differential equation along a path made of approximations of ζ of exponentially increasing precision. Its time complexity is $O(M(p(\lg p)^2))$ [16]. The improvements from the next section apply to all these settings. See also [24] for an overview of more sophisticated applications.

4 "Truncated" Binary Splitting

The superiority of binary splitting over alternatives like summing the series in floating-point arithmetic results from the controlled growth of intermediate results. Indeed, in the product tree computed by Algorithm 1, the exact representations of most subproducts $P(a, b)$ are much more compact than $\Theta(p)$-digits approximations would be. However, as already mentioned, the bit sizes of the $P(a, b)$ also grow larger than p near the root of the tree. The size of a subproduct appearing at depth δ is roughly $2^{-\delta} N(h + r \lg N)$. Assuming $N = \Theta(p)$, this means that the intermediate results get significantly larger than the output in the top $\Theta(\lg \lg p)$ levels of the tree.

A natural remedy is to use a hybrid of binary splitting and naive summation. More precisely, we split the full product $P(0, N)$ into $\Delta = \Theta(\ln N)$ subproducts of $O(p)$ bits each, which are computed by binary splitting. The results are accumulated by successive multiplications at precision $O(p)$.

We make use of the following notations to state and analyze the algorithm. In Equations (7) to (11) below, the coefficients of a general matrix $A \in \mathbb{C}^{k \times k}$ are denoted $a_{p,q} = x_{p,q} + i y_{p,q}$ $(1 \leqslant p, q \leqslant k)$ with $x_{p,q}, y_{p,q} \in \mathbb{R}$. Let $\|\cdot\|$ be a submultiplicative norm on $\mathbb{C}^{k \times k}$, and let $\beta_k > 0$ be such that

$$\|A\| \leqslant \beta_k \mathcal{N}(A), \qquad \mathcal{N}(A) = \max\{|x_{i,j}|, |y_{i,j}|\}_{1 \leqslant i,j \leqslant k}. \tag{7}$$

For definiteness, assume for now that $\|\cdot\| = \|\cdot\|_1$ is the matrix norm induced by the vector 1-norm. (We will discuss this choice later.) Then it holds that

$$\mathcal{N}(A) \leqslant \|A\|_1 = \max_{j=1}^{k} \sum_{i=1}^{k} |a_{i,j}| \leqslant \sqrt{2} k \mathcal{N}(A) \tag{8}$$

Algorithm 2. TruncBinSplit(p)

The notation $X^{(q)}$, $q = 0, 1, \ldots$ refers to a single memory location X at different points q of the computation.

1. Set $\varepsilon = 2^{-p}$
2. Compute N such that $|S_N - y(\zeta)| \leqslant \varepsilon$ [22,18]
3. Set $\Delta = \lceil \frac{N}{p}(h + r \lg N) \rceil$, where h and r are given following Eq. (1)
4. Compute M such that $\max_{q=0}^{\Delta-1} \|P(\lfloor \frac{q}{\Delta}N \rfloor, \lfloor \frac{q+1}{\Delta}N \rfloor)\| + \varepsilon \leqslant M \leqslant C^{N/\Delta}$, where C does not depend on p, h, r [say, by approximating the right-hand side of (9) from above with $O(\lg p)$ bits of precision]
5. Initialize $\tilde{P}^{(0)} := \mathrm{id} \in \mathbb{K}^{(s+1)\times(s+1)}$
6. For $q = 0, 1, \ldots, \Delta - 1$
7. $\quad \hat{Q} = (\hat{Q}_{i,j}) := \mathrm{BinSplit}(\lfloor \frac{q}{\Delta}N \rfloor, \lfloor \frac{q+1}{\Delta}N \rfloor)$ (Algorithm 1)
8. $\quad \tilde{Q}^{(q)} := \mathrm{Trunc}(\hat{Q}_{s+1,s+1}^{-1} \cdot \hat{Q}, \frac{1}{2\Delta}M^{-\Delta+1}\varepsilon)$
9. $\quad \tilde{P}^{(q+1)} := \mathrm{Trunc}(\tilde{Q}^{(q)} \cdot \tilde{P}^{(q)}, \frac{1}{2\Delta}M^{-\Delta+q+1}\varepsilon)$
10. Return $\tilde{P}^{(\Delta)}$

---

and

$$\|P(a,b)\| \leqslant \prod_{n=a}^{b-1} \|B(n)\| \leqslant \prod_{n=a}^{b-1} \left(1 + |\zeta| + |\zeta| \max_{k=1}^{s} \left| \frac{b_k(n)}{b_0(n)} \right| \right). \tag{9}$$

Observe that, since 1 is an eigenvalue of $B(n)$ and the norm $\|\cdot\|$ is assumed to be submultiplicative, we have $\|B(n)\| \geqslant 1$ for all n. Besides, it is clear from (3) that $\|B(n)\|$ is bounded.

Given $a \in \mathbb{Q}$ and $\varepsilon < 1$, let

$$\mathrm{Trunc}(a, \varepsilon) = \mathrm{sgn}(a) \lfloor 2^e |a| \rfloor 2^{-e}, \quad e = \lceil \lg \varepsilon^{-1} \rceil. \tag{10}$$

We have $|\mathrm{Trunc}(a, \varepsilon) - a| \leqslant \varepsilon$; the size of $\mathrm{Trunc}(a, \varepsilon)$ is $O(\lg \varepsilon^{-1})$ for bounded a; and $\mathrm{Trunc}(a, \varepsilon)$ may be computed in $O(\mathsf{M}(h + e))$ bit operations where h is the bit size of a. We extend the definition to matrices $A \in \mathbb{K}^{k\times k}$ by

$$\mathrm{Trunc}(A, \varepsilon) = \left(\mathrm{Trunc}(x_{p,q}, \beta_k^{-1}\varepsilon) + i\,\mathrm{Trunc}(y_{p,q}, \beta_k^{-1}\varepsilon) \right)_{1\leqslant p,q\leqslant k}, \tag{11}$$

so that again $\|\mathrm{Trunc}(A, \varepsilon) - A\| \leqslant \varepsilon$. Note that we often write expressions of the form $\mathrm{Trunc}(a \star b, \varepsilon)$ for some operator $\star$. Though this does not affect our complexity bounds, it is usually better to compute the approximate value of $a \star b$ directly instead of starting with an exact computation and truncating the result. See Brent and Zimmermann [3] for some relevant algorithms.

The complete binary splitting algorithm with truncations is stated as Algorithm 2. Its key properties are summarized in the following propositions.

Proposition 2. *The output $\tilde{P} = \mathrm{TruncBinSplit}(p)$ of Algorithm 2 is such that $\|\tilde{P} - P(0, N)\| \leqslant 2^{-p}$.*

Proof. Set $P^{(q)} = P(0, \lfloor \frac{q}{\Delta}N \rfloor)$ and $Q^{(q)} = P(\lfloor \frac{q}{\Delta}N \rfloor, \lfloor \frac{q+1}{\Delta}N \rfloor)$. Then, for $0 \leqslant q \leqslant \Delta$, it holds that

$$\|\tilde{P}^{(q)} - P^{(q)}\| \leqslant \frac{q}{\Delta} \frac{\varepsilon}{M^{\Delta-q}}. \tag{12}$$

Indeed, this is true for $q = 0$. After Step 8 of each loop iteration, we have the bound $\|\tilde{Q}^{(q)} - Q^{(q)}\| \leqslant \frac{1}{2\Delta}M^{-\Delta+1}\varepsilon \leqslant \varepsilon$ since $\|B(n)\| \geqslant 1$ for all n. Using (12) and the inequality $\|\tilde{Q}^{(q)}\| \leqslant M$ from Step 3, it follows that

$$\|\tilde{Q}^{(q)}\tilde{P}^{(q)} - Q^{(q)}P^{(q)}\| \leqslant \|\tilde{Q}^{(q)} - Q^{(q)}\|\,\|P^{(q)}\| + \|\tilde{Q}^{(q)}\|\,\|\tilde{P}^{(q)} - P^{(q)}\|$$
$$\leqslant \frac{2q+1}{2\Delta}\frac{\varepsilon}{M^{\Delta-q-1}}.$$

After taking into account the truncation error from Step 9, we obtain

$$\|\tilde{P}^{(q+1)} - P^{(q+1)}\| = \|\tilde{P}^{(q+1)} - Q^{(q)}P^{(q)}\| \leqslant \frac{q+1}{\Delta}\frac{\varepsilon}{M^{\Delta-q-1}}.$$

which concludes the induction.

Proposition 3. *Not counting the cost of Step 2, Algorithm 2 runs in time*

$$\begin{cases} O\big(M(p)(h+r\lg p)\lg p\big), & \text{if } \rho < \infty, \\ O\big(M(p)(h+r\lg p)\big), & \text{if } \rho = \infty, \end{cases} \tag{13}$$

as $p, h, r \to \infty$ with $r = O(\lg p)$ and $h = O(p)$. In both cases, it uses $O(p)$ bits of memory (where the hidden constant is independent of h and r, under the same growth assumptions).

We neglect the cost of finding N to avoid a lengthy discussion of the complexity of the corresponding bound computation algorithms. It could actually be checked to be polynomial in r and $\lg p$.

Proof. Computing the bound M using Equation (9) as suggested is more than enough to ensure that $\lg M = O(N/\Delta)$. It requires $O(N)$ arithmetic operations on $O(\lg p)$-bit numbers, that is, $o(N(\lg N)^2)$ bit operations.

By Proposition 1, each of the Δ calls to BinSplit requires

$$O\big(M(\tfrac{N}{\Delta}(h+r\lg N))\lg N\big) = O\big(M(p)\lg p\big)$$

bit operations. The resulting matrices $Q^{(p)}$ all have size $O(p)$, hence the divisions from Step 8 can be done in $O(M(p))$ operations using Newton's method [25, Chap. 9]. The truncations in Steps 8 and 9 ensure that the bit sizes of $\tilde{P}$ and $\tilde{Q}$ are always at most

$$\lg \varepsilon^{-1} + \Delta\lg M + \lg \Delta + O(1) = O(p). \tag{14}$$

It follows that the matrix multiplications from Step 9 take $O(M(p))$ operations each. Summing up, each iteration of the loop from Step 6 can be performed in $O(M(p)\lg p)$ operations, for a total of $O(\Delta M(p)\lg p)$. Equation (13) follows upon setting $N = O(p)$ or $N = O(p/\lg p)$ according to (6).

The required memory comprises space for the current values of $\tilde{P}^{(q)}$ and $Q^{(q)}$, any temporary storage used by the operations from Steps 7 to 9, and an additional $O(\lg p)$ bits to manipulate auxiliary variables such as M and q. We

Table 1. Complexity of some D-finite function evaluation algorithms based on binary splitting. The rows labeled "BinSplit" summarize the cost of computing a single sum by binary splitting, with or without truncations. Those labeled "BitBurst" refer to the computation of $y(\zeta)$ by the "bit burst" method, using either of Algorithm 1 and Algorithm 2 at each step. All entries are asymptotic bounds as $p, h \to \infty$ with $h = O(p)$. In the "BinSplit" case, we also let r tend to infinity under the assumption that $r = O(\lg p)$. The whole point of the "bit burst" method is to get rid the dependency on h.

		Time	Space (classical)	Space (trunc.)
$\rho < \infty$	BinSplit	$O(M(p(h + r \lg p) \lg p))$	$O(p(h + r \lg p))$	$O(p)$
	BitBurst	$O(M(p(\lg p)^2))$	$O(p \lg p)$	$O(p)$
$\rho = \infty$	BinSplit	$O(M(p(h + r \lg p)))$	$O(p(r + h/\lg p))$	$O(p)$
	BitBurst	$O(M(p(\lg p)^2))$	$O(p)$	$O(p)$

have seen that $\tilde{P}^{(q)}$ and $Q^{(q)}$ have bit size $O(p)$. Besides, our assumption that fast integer multiplication could be performed in linear space implies the same property for division by Newton's method. Thus, Steps 8 and 9 use $O(p)$ bits of auxiliary storage. Finally, again by Proposition 1, the calls to Algorithm 1 use $O((N/\Delta)(h + r \lg p)) = O(p)$ bits of memory.

Plugging Algorithm 2 into the numerical evaluation algorithms mentioned at the end of Sect. 3 yields corresponding improvements for the evaluation of D-finite functions at more general points. Table 1 summarizes the complexity bounds we obtain. The omitted proofs are direct adaptations of those that apply without truncations [5,22,17]. There would be much to say on the hidden constant factors. The main result may be stated more precisely as follows.

Theorem 1. *Let $U \subset \mathbb{C}$ be a simply connected domain such that $0 \in U$ and $a_r(z) \neq 0$ for all $z \in U$. Fix $\ell_0, \ldots, \ell_{r-1} \in \mathbb{C}$ and $\zeta \in U$. Assume that 0 is an ordinary point of (1), and let y be the unique solution of (1) defined on U and such that $y^{(k)}(0) = \ell_k$, $0 \leqslant k < r$. Then, the value $y(\zeta)$ may be computed with error bounded by 2^{-p} in time $O(M(p)(\lg p)^2)$ and space $O(p)$, not counting the resources needed to approximate the ℓ_k or ζ to precision $O(p)$ or to find suitable truncation orders for the Taylor series involved.*

Finally, some comments are in order regarding the "working precision", that is, the size p' of the entries of $\tilde{P}$ and $\tilde{Q}$ in Algorithm 2. Equation (14) suggests a number of "guard digits" $p' - p = \Theta(p)$. Moreover, if the bound M is computed using (9), the hidden constant depends on the choice of $\|\cdot\|$.

Let $B_\infty = \lim_{n \to \infty} B(n)$. For the norm $\|\cdot\|_{\mathrm{opt}}$ given by Lemma 1 below, we have

$$\lg \|P(a, b)\|_{\mathrm{opt}} \leqslant \sum_{n=a}^{b-1} \lg \big(\|B_\infty\|_{\mathrm{opt}} + O(n^{-1})\big) = O\big(\lg(b - a)\big),$$

and hence $\lg \|P(a, b)\| = O(\lg(b - a))$ for any norm $\|\cdot\|$.

Lemma 1. *There exists a matrix norm* $\|\cdot\|_{\mathrm{opt}}$ *such that* $\|B_\infty\|_{\mathrm{opt}} = 1$.

Proof. We mimic the classical proof of Householder's theorem [20, Sect. 4.2]. By (3), the limit $C_\infty = \lim_{n\to\infty} C(n)$ is the companion matrix of the polynomial $z^s a_r(1/z)$. The eigenvalues of ζC_∞ are strictly smaller than 1 in absolute value since $|\zeta| < \rho$. Let Γ be such that $\Gamma^{-1} C_\infty \Gamma$ is in (lower) Jordan normal form. Let $\lambda > 0$, and set $\Pi = \mathrm{diag}(1, \lambda, \dots, \lambda^s) \cdot \mathrm{diag}(\Gamma, 1)$. Then $\Pi^{-1} B_\infty \Pi$ is lower triangular, with off-diagonal entries tending to zero as $\lambda \to 0$. Hence we have $\|\Pi^{-1} B_\infty \Pi\|_1 = 1$ for λ small enough. We choose such a λ (e.g., $\lambda = \frac{1 - |\zeta|/\rho}{2 \max(1, |\zeta|)}$) and set $\|A\|_{\mathrm{opt}} = \|\Pi^{-1} A \Pi\|_1$.

One way to eliminate the overestimation in the algorithm is to compute approximations of the matrices $P(\lfloor \frac{q}{\Delta} N \rfloor, \lfloor \frac{q+1}{\Delta} N \rfloor)$ with $O(\lg p)$ digits of precision before doing the computation at full precision. One then uses the norms of these approximate products instead of those of the individual $B(n)$ to determine M. We can also explicitly construct an approximation $\tilde{\Pi}$ of the matrix Π from the proof of Lemma 1 precise enough that $\|\tilde{\Pi}^{-1} B_\infty \tilde{\Pi}\|_1 = 1$, and use the corresponding norm instead of $\|\cdot\|_1$ in (9). (Compare [22, Algorithm B].) Other options include computing symbolic bounds on the coefficients of $P(a, b)$ as a function of a and b [18] or finding an explicit integer n_0 such that $n \geqslant n_0 \Rightarrow \|B(n)\|_{\mathrm{opt}} = 1$ based on the symbolic expression of n. Which variant to use in practice depends on the features of the implementation platform.

In any case, replacing the $O(\cdot)$ in the space complexity bound by an explicit constant would also require more specific assumptions on the memory representation of the objects we work with, as well as finer control on the space complexity of integer multiplication and division (see, e.g., Roche [19]).

5   Final Remarks

What we lose and what we retain. The price we pay for the reduced memory usage is the ability to easily extend the computation to higher precision. Indeed, the classical algorithm computes the exact value of the matrix $P(0, N)$, from which we can deduce $P(0, N')$ for any $N' > N$ in time roughly proportional to $N' - N$. This is no longer true with the linear-space variant. In some "lucky" cases where $P(0, N)$ can be represented exactly in linear space, it is possible to get the memory usage down to $O(N)$ while preserving restartability: see Cheng et al. [4] and the references therein. Additionally, the resulting running time is reportedly lower than using truncations, probably owing to the fact that the size of the subproducts in the $\lg(N/\Delta)$ lower levels of the tree is reduced as well. Unfortunately, the applicability of the technique is limited to very special cases.

Two other traditional selling points of the binary splitting method are its easy parallelization and good memory locality. Nothing is lost in this respect, except that the memory bound grows to $\Theta(t \cdot p)$ when using $t = o(\lg N)$ parallel tasks in the approximate part of the computation.

Generalizations. The idea of binary splitting "with truncations" and the outline of its analysis adapt to various settings not covered here. For instance, we may consider systems of linear differential equations instead of scalar equations [5]. Product trees of matrices over number fields $\mathbb{K}' = \mathbb{Q}(\alpha)$ other than $\mathbb{Q}(i)$ or over rings of truncated power series $\mathbb{K}' [[\varepsilon]] / \langle \varepsilon^k \rangle$ are also useful, respectively, to evaluate limits of D-finite functions at regular singular points of their defining equations, and to make the analytic continuation process more efficient for equations of large order [22,17]. It is not essential either that the coefficients of the recurrence relation satisfied by the y_n are rational functions of n: all we really ask is that they have suitable growth properties and can be computed fast.

Implementation. We are working on an implementation of the algorithm from Sect. 4 in an experimental branch of the software package NumGfun [16]. The current state of the code is available from

http://marc.mezzarobba.net/supporting-material/trunc-CASC2012/.

A comparison (updated periodically) with the implementation of binary splitting without truncations used in previous releases of NumGfun is also included.

Acknowledgments. I would like to thank Nicolas Brisebarre and Bruno Salvy for encouraging me to write this note and offering useful comments, and Anne Vaugon for proofreading parts of it.

References

1. Bernstein, D.J.: Fast multiplication and its applications. In: Buhler, J., Stevenhagen, P. (eds.) Algorithmic Number Theory, pp. 325–384. Cambridge University Press (2008), http://www.msri.org/communications/books/Book44/
2. Borwein, J.M., Borwein, P.B.: Pi and the AGM. Wiley (1987)
3. Brent, R.P., Zimmermann, P.: Modern Computer Arithmetic. Cambridge University Press (2010), http://www.loria.fr/~zimmerma/mca/mca-cup-0.5.7.pdf
4. Cheng, H., Hanrot, G., Thomé, E., Zima, E., Zimmermann, P.: Time- and space-efficient evaluation of some hypergeometric constants. In: Wang, D. (ed.) ISSAC 2007, pp. 85–91. ACM (2007),
   http://www.cs.uleth.ca/~cheng/papers/issac2007.pdf
5. Chudnovsky, D.V., Chudnovsky, G.V.: Computer algebra in the service of mathematical physics and number theory. In: Chudnovsky and Jenks [6], pp. 109–232
6. Chudnovsky, D.V., Jenks, R.D. (eds.): Computers in Mathematics, Stanford University. Lecture Notes in Pure and Applied Mathematics, vol. 125 (1986), Dekker (1990)
7. Fürer, M.: Faster integer multiplication. SIAM Journal on Computing 39(3), 979–1005 (2009), http://www.cse.psu.edu/~furer/Papers/mult.pdf
8. Gosper, W.: Strip mining in the abandoned orefields of nineteenth century mathematics. In Chudnovsky and Jenks [6], pp 261–284
9. Gourdon, X., Sebah, P.: Constants and records of computation. Updated August 12 (2010), http://numbers.computation.free.fr/Constants/constants.html
10. Gourdon, X., Sebah, P.: Binary splitting method (2001),
    http://numbers.computation.free.fr/Constants/Algorithms/splitting.ps

11. Granlund, T., et al.: GNU Multiple Precision Arithmetic Library,
    http://gmplib.org/
12. Haible, B., Papanikolaou, T.: Fast multiprecision evaluation of series of rational
    numbers (1997), http://www.informatik.tu-darmstadt.de/TI/Mitarbeiter/
    papanik/ps/TI-97-7.ps.gz
13. Hille, E.: Ordinary differential equations in the complex domain. Wiley (1976),
    Dover reprint (1997)
14. Kreckel, R.B.: decimal(γ) =~ "0.57721566[0-9]{1001262760}39288477" (2008),
    http://www.ginac.de/~kreckel/news.html#EulerConstantOneBillionDigits
15. Mathews, J.H.: Bibliography for Taylor series method for D.E.'s (2003),
    http://math.fullerton.edu/mathews/n2003/taylorde/TaylorDEBib/Links/
    TaylorDEBib_lnk_3.html
16. Mezzarobba, M.: NumGfun: a package for numerical and analytic computation
    with D-finite functions. In: Koepf, W. (ed.) ISSAC 2010, pp. 139–146. ACM (2010),
    http://arxiv.org/abs/1002.3077, doi:10.1145/1837934.1837965
17. Mezzarobba, M.: Autour de l'évaluation numérique des fonctions D-finies. Thèse
    de doctorat, École polytechnique (November 2011),
    http://tel.archives-ouvertes.fr/pastel-00663017/
18. Mezzarobba, M., Salvy, B.: Effective bounds for P-recursive sequences. Journal of
    Symbolic Computation 45(10), 1075–1096 (2010),
    http://arxiv.org/abs/0904.2452, doi:10.1016/j.jsc.2010.06.024
19. Roche, D.S.: Efficient Computation with Sparse and Dense Polynomials. PhD the-
    sis, University of Waterloo (2011),
    http://uspace.uwaterloo.ca/handle/10012/5869
20. Serre, D.: Matrices. Graduate Texts in Mathematics, vol. 216. Springer (2002)
21. Stanley, R.P.: Differentiably finite power series. European Journal of Combina-
    torics 1(2), 175–188 (1980)
22. van der Hoeven, J.: Fast evaluation of holonomic functions. Theoretical Computer
    Science 210(1), 199–216 (1999),
    http://www.texmacs.org/joris/hol/hol-abs.html
23. van der Hoeven, J.: Fast evaluation of holonomic functions near and in regular
    singularities. Journal of Symbolic Computation 31(6), 717–743 (2001),
    http://www.texmacs.org/joris/singhol/singhol-abs.html
24. van der Hoeven, J.: Transséries et analyse complexe effective. Habilitation à diriger
    des recherches, Université Paris-Sud, Orsay, France (2007),
    http://www.texmacs.org/joris/hab/hab-abs.html
25. von Zur Gathen, J., Gerhard, J.: Modern Computer Algebra, 2nd edn. Cambridge
    University Press (2003)
26. Yakhontov, S.V.: Calculation of hypergeometric series with quasi-linear time and
    linear space complexity. Vestnik Samarskogo Gosudarstvennogo Tekhnicheskogo
    Universiteta. Seriya: Fiziko-Matematicheskie Nauki 24, 149–156 (2011)
27. Yakhontov, S.V.: A simple algorithm for the evaluation of the hypergeometric series
    using quasi-linear time and linear space. Preprint 1106.2301v1, arXiv (June 2011),
    English version of [26], http://arxiv.org/abs/1106.2301
28. Yee, A.J.: Mathematical constants – billions of digits. Updated March 7 (2011),
    http://www.numberworld.org/digits/

Inversion Modulo Zero-Dimensional Regular Chains

Marc Moreno Maza, Éric Schost, and Paul Vrbik

Department of Computer Science, Western University
{moreno,eschost,pvrbik}@csd.uwo.ca

Abstract. We consider the questions of inversion modulo a regular chain in dimension zero and of matrix inversion modulo such a regular chain. We show that a well-known idea, Leverrier's algorithm, yields new results for these questions.

1 Introduction

Triangular sets, and more generally regular chains, constitute a useful data structure for encoding the solutions of algebraic systems. Among the fundamental operations used by these objects, one finds a few low-level operations, such as multiplication and division in dimension zero. Higher-level algorithms can then be built upon these subroutines: for instance, the authors of [8] outline a probabilistic and modular algorithm for solving zero-dimensional polynomial systems with rational coefficients. Their algorithm requires matrix inversion modulo regular chains.

Despite a growing body of work, the complexity of several basic questions remains imperfectly understood. In this article, we consider the question of inversion modulo a triangular set in dimension zero, and by extension, matrix inversion modulo such a triangular set. We show that a well-known idea, Leverrier's algorithm, surprisingly admits new results for these questions.

Triangular sets. We adopt the following convention: a *triangular set* is a family of polynomials $\mathbf{T} = (T_1, \ldots, T_n)$ in $k[X_1, \ldots, X_n]$, where k is a field. We require that for all i, $T_i \in k[X_1, \ldots, X_i]$ is monic in X_i and reduced with respect to $\langle T_1, \ldots, T_{i-1} \rangle$. Note that the slightly more general notion of a *regular chain* allows for non necessarily monic T_i; in that case, the requirement is that the leading coefficient of T_i be invertible modulo $\langle T_1, \ldots, T_{i-1} \rangle$. These regular chains may be called "zero-dimensional", since they encode finitely many points. Note that we *do not* require that the ideal $\langle \mathbf{T} \rangle$ be radical.

Multiplication modulo triangular sets. In the context of triangular sets, the first non-trivial algorithmic question is modular multiplication. For this, and for the question of inversion in the following paragraph, the input and output are polynomials reduced modulo $\langle \mathbf{T} \rangle$. We thus denote by $R_{\mathbf{T}}$ the residue class ring $k[X_1, \ldots, X_n]/\langle T_1, \ldots, T_n \rangle$. For all $i \leq n$, let us write $d_i = \deg(T_i, X_i)$; the n-tuple $(d_1, \ldots, d_n)$ is the *multi-degree* of $\mathbf{T}$. Then, the set of monomials

V.P. Gerdt et al. (Eds.): CASC 2012, LNCS 7442, pp. 224–235, 2012.
© Springer-Verlag Berlin Heidelberg 2012

$$M_{\mathbf{T}} = \left\{ X_1^{e_1} \cdots X_n^{e_n} \mid 0 \le e_i < d_i \text{ for all } i \right\}$$

is the canonical basis of the k-vector space $R_{\mathbf{T}}$; its cardinality is the integer $\delta_{\mathbf{T}} = d_1 \cdots d_n$, which we call the *degree* of $\mathbf{T}$. In all our algorithms, elements of $R_{\mathbf{T}}$ are represented on this basis.

As of now, the best known algorithm for modular multiplication features the following running time [15]. For $x \ge 1$, write $\lg(x) = \log(\max(x, 2))$. Then, there exists a universal constant K such that given A, B in $R_{\mathbf{T}}$, one can compute $AB \in R_{\mathbf{T}}$ using at most $K4^n \delta_{\mathbf{T}} \lg(\delta_{\mathbf{T}}) \lg \lg(\delta_{\mathbf{T}})$ operations in k.

Inversion modulo triangular sets. For inversion, several questions can be posed. In this paper we consider the problem: *given $A \in R_{\mathbf{T}}$, decide whether A is invertible, and if so, compute its inverse.* We are also interested in its the generalization to matrices over $R_{\mathbf{T}}$: *given a $(d \times d)$ matrix $A \in \mathscr{M}_d(R_{\mathbf{T}})$, decide whether it is invertible, and if so, compute its inverse.* We simply call this the problem of *invertibility test / inversion* in $R_{\mathbf{T}}$ (or in $\mathscr{M}_d(R_{\mathbf{T}})$).

This question should be contrasted with the following one: given $A \in R_{\mathbf{T}}$, decompose the ideal $\langle \mathbf{T} \rangle$ into a product of pairwise coprime ideals of the form $\langle \mathbf{T}_1 \rangle \cap \cdots \cap \langle \mathbf{T}_r \rangle$, all $\mathbf{T}_i$ being triangular sets, such that for all $i \le r$, A is either a unit modulo $\langle \mathbf{T}_i \rangle$, or zero modulo $\langle \mathbf{T}_i \rangle$; we also compute the inverse of A modulo all $\langle \mathbf{T}_i \rangle$ that are among the first category. A similar, albeit more complex, question could be raised for matrices over $R_{\mathbf{T}}$. To distinguish it from the previous problem, we call this question the *quasi-inverse* computation.

When the ideal $\langle \mathbf{T} \rangle$ is maximal, so $R_{\mathbf{T}}$ is a field, the two questions are the same. Without this assumption the question of computing quasi-inverses is more complex than the inversion problem: when A is a zero-divisor modulo $\langle \mathbf{T} \rangle$, the first approach would just return "not invertible"; the second approach would actually require us to do some extra work.

As of now, most known algorithms naturally handle the second, more general problem. Indeed, the natural approach is the following: to compute an inverse in the residue class ring $R_{\mathbf{T}} = k[X_1, \ldots, X_n]/\langle T_1, \ldots, T_n \rangle$, we see it as $R_{\mathbf{T}'}[X_n]/\langle T_n \rangle$, where $\mathbf{T}'$ is the triangular set $(T_1, \ldots, T_{n-1})$ in $k[X_1, \ldots, X_{n-1}]$. Then, testing if $A \in R_{\mathbf{T}}$ is invertible, and inverting it when possible, is usually done by computing its extended GCD with T_n in $R_{\mathbf{T}'}[X_n]$, see [12,18,7,15]. This approach requires several quasi-inverse computations in $R_{\mathbf{T}'}$ (namely those of all leading terms that arise during the extended GCD algorithm). Even if A is invertible in $R_{\mathbf{T}'}$, some of these leading terms may be zero-divisors, thus we may have to decompose $\mathbf{T}$.

Main results. Our two main results concern the inversion problem, first for elements of $R_{\mathbf{T}}$, then for matrices over $R_{\mathbf{T}}$.

In what follows, in addition to $\delta_{\mathbf{T}}$, let $s_{\mathbf{T}} = \max(d_1, \ldots, d_n)$. Our theorems also involve the quantity ω, which denotes the exponent of matrix multiplication [4, Ch. 15]: explicitly, this means that ω is such that over any ring $\mathbb{A}$, matrices of size d can be multiplied in d^ω operations $(+, \times)$ in $\mathbb{A}$. We take $2 < \omega \le 3$, the best known value being $\omega \le 2.3727$ [24].

Theorem 1. *There exists a constant C such that: If $1, \ldots, s_\mathbf{T}$ are units in k, then one can perform an invertibility test / inversion in $R_\mathbf{T}$ using*

$$C 4^n n \, \delta_\mathbf{T} \, s_\mathbf{T}^{(\omega-1)/2} \lg(\delta_\mathbf{T}) \lg \lg(\delta_\mathbf{T})$$

operations in k.

Dropping logarithmic factors, we see that the cost of inversion modulo $\langle \mathbf{T} \rangle$ grows like $4^n \delta_\mathbf{T} \, s_\mathbf{T}^{(\omega-1)/2}$, whereas the cost of multiplication modulo $\langle \mathbf{T} \rangle$ grows like $4^n \delta_\mathbf{T}$. In other words, the overhead for inversion grows like $s_\mathbf{T}^{(\omega-1)/2}$, which is between $s_\mathbf{T}^{1/2}$ and $s_\mathbf{T}$, depending on ω.

The second theorem describes the cost of matrix invertibility test and inversion.

Theorem 2. *There exists a constant C such that: If $1, \ldots, s_\mathbf{T}$ are units in k, then one can perform an invertibility test / inversion in $\mathscr{M}_d(R_\mathbf{T})$ using*

$$C 4^n \delta_\mathbf{T} \left(d^{\omega+1/2} + n s_\mathbf{T}^{(\omega-1)/2} \right) \lg(\delta_\mathbf{T}) \lg \lg(\delta_\mathbf{T})$$

operations in k.

Previous work. As stated above, most previous works on the invertibility question in $R_\mathbf{T}$ actually give algorithms for quasi-inverses, using *dynamic evaluation* techniques [9]. Unfortunately, managing the decompositions induced in quasi-inverse computations in an efficient manner leads to very complex algorithms: as of now, the fastest algorithm for quasi-inverse follows from [7,6], and features a running time of the form $\lambda^n \prod_{i \leq n} d_i \lg(d_i)^4 \lg \lg(d_i)$, for some non-explicit constant λ (conservative estimates give $\lambda \geq 60$).

Dynamic evaluation techniques carry over to matrix inversion, and make it possible to implement Gaussian elimination with coefficients in $R_\mathbf{T}$, handling decompositions of $\mathbf{T}$ when zero-divisors are met. The complexity of such a process seems quite complex to analyze; to our knowledge, this has not been done yet.

The algorithms from [7,6] apply half-GCD techniques in a recursive manner, together with fast Chinese remaindering techniques to handle splitting. We mention here another approach from [16]: using evaluation / interpolation techniques, the Authors extend it in [17] to an algorithm with cost growing like $2^n \sum_{i=1}^{n} \left(i^2 d_1 \cdots d_i d_i^{i+1} \right)$.

The main ingredient in our theorems is Leverrier's algorithm [13], a method for computing the characteristic polynomial of a matrix by means of the computation of the traces of its powers. Once the characteristic polynomial is known, it can be used to express the inverse of a matrix A as a polynomial in A — we still refer to this extension to inverse computation as Leverrier's algorithm, somewhat inappropriately.

This algorithm has been rediscovered, extended and improved in work by (among others) Souriau [23], Faddeev [10], Csanky [5], and Preparata and Sarwate [19]. The latter reference introduces the "baby steps / giant steps" techniques that are used herein; note on the other hand that the focus in these references is on the parallel complexity of characteristic polynomial or the inverse, which is not our main interest here.

Similar "baby steps / giant steps" techniques have been discovered in other contexts (algorithms on polynomials and power series) by Brent and Kung [3] and Shoup [21,22]. In these references, though, no mention was made of applications to modular inversion.

2   Leverrier's Algorithm

In this paper, we are interested in inversion algorithms which:

1. invert dense $(d \times d)$ matrices with entries in a ring $\mathbb{A}$;
2. invert elements in the $\mathbb{A}$-algebra $\mathbb{A}[X]/\langle T \rangle$, for some degree d monic polynomial T in $\mathbb{A}[X]$.

When we use these results, we take $\mathbb{A}$ of the form $R_{\mathbf{T}}$, for some triangular set $\mathbf{T}$. Our goal is to perform as little invertibility tests / inversions in $\mathbb{A}$ as possible: we thus rely on Leverrier's algorithm, which only does one. With $\mathbb{A}$ of the form $R_{\mathbf{T}}$, this allows us to avoid unnecessary splittings of $\mathbf{T}$.

Since both scenarios share many similarities, we strive to give a unified presentation, at the cost of a slight increase in notational burden.

2.1   Setup and Main Result

The following setup enables us to handle both cases above at once. Let $\mathbb{A}$ be our base ring and let $\mathscr{M}_d(\mathbb{A})$ be the free $\mathbb{A}$-algebra of $(d \times d)$ matrices over $\mathbb{A}$. We consider an $\mathbb{A}$-algebra $\mathbb{B}$ that is free of rank e as an $\mathbb{A}$-module, and which admits an $\mathbb{A}$-algebra embedding $\phi : \mathbb{B} \to \mathscr{M}_d(\mathbb{A})$; we assume $d \leq e$. The two above scenarios fit into this description:

1. In the first case, $\mathbb{B}$ is the whole $\mathbb{A}$-algebra $\mathscr{M}_d(\mathbb{A})$ and ϕ is the identity; here, $e = d^2$;
2. In the second case, $\mathbb{B}$ is the $\mathbb{A}$-algebra $\mathbb{A}[X]/\langle T \rangle$. It can be identified to a subalgebra of $\mathscr{M}_d(\mathbb{A})$ by means of the mapping ϕ that maps $A \in \mathbb{B} = \mathbb{A}[X]/\langle T \rangle$ to the $(d \times d)$ matrix of multiplication by A. In this case, the rank of $\mathbb{B}$ is $e = d$.

To any element $A \in \mathbb{B}$, we associate its *trace* $\mathrm{tr}(A) \in \mathbb{A}$, defined as the trace of the matrix $\phi(A) \in \mathscr{M}_d(\mathbb{A})$, and its *characteristic polynomial* $\chi_A \in \mathbb{A}[X]$, defined as the characteristic polynomial of the matrix $\phi(A)$; the latter is a monic polynomial of degree d in $\mathbb{A}[X]$. Finally, the *determinant* $\det(A)$ of A is defined similarly, as the determinant of $\phi(A)$.

For our computations, we suppose that a basis B of the $\mathbb{A}$-module $\mathbb{B}$ is known. In both cases above, we have a canonical choice: matrices with a single non-zero entry, equal to one, in the first case, and the monomial basis $1, X, \ldots, X^{d-1}$ in the second case.

An addition in $\mathbb{B}$ then takes e operations $(+, \times)$ in $\mathbb{A}$. For multiplication, things are less straightforward: we let $M(\mathbb{B})$ be such that one multiplication in $\mathbb{B}$ can be done using $M(\mathbb{B})$ operations $(+, \times)$ in $\mathbb{A}$. The other black-box we need is for determining the trace: we let $T(\mathbb{B})$ be such that the traces of all basis elements of $\mathbb{B}$ can be computed in $T(\mathbb{B})$ operations $(+, \times)$ in $\mathbb{A}$. We give details below on $M(\mathbb{B})$ and $T(\mathbb{B})$ for our two main cases of interest.

Then, Leverrier's algorithm, combined with baby steps / giant steps techniques, yields the following result.

Proposition 1. *Suppose that* $1, \ldots, d$ *are units in* $\mathbb{A}$*. Given* $A \in \mathbb{B}$*, one can decide whether* A *is invertible, and if so compute its inverse, using*

$$T(\mathbb{B}) + O\left(\sqrt{d}\, M(\mathbb{B}) + d^{(\omega-1)/2} e\right)$$

operations $(+, \times)$ *in* $\mathbb{A}$*, and one invertibility test / inversion in* $\mathbb{A}$*.*

We will prove this result explicitly. Still, although this result may not have appeared before in this exact form, its specializations to our two cases of interest are not exactly new. As we said in the introduction, when $\mathbb{B} = \mathcal{M}_d(\mathbb{A})$, this approach is essentially Preparata and Sarwate's algorithm [19]. When $\mathbb{B} = \mathbb{A}[X]/\langle T \rangle$, this is in essence a combination of results of Brent and Kung [3] and Shoup [21,22], although these references do not explicitly discuss inverse computation, but respectively modular composition and minimal polynomial computation.

Our first case of interest is $\mathbb{B} = \mathcal{M}_d(\mathbb{A})$, with rank $e = d^2$. In this case, computing the traces of all basis elements is straightforward, so $T(\mathbb{B})$ takes linear time $O(e) = O(d^2)$. Matrix multiplication takes time $M(\mathbb{B}) = d^\omega$, so that we end up with a total of

$$O\left(d^{\omega+1/2}\right)$$

operations $(+, \times)$ in $\mathbb{A}$, as is well-known.

Our second case of interest is $\mathbb{B} = \mathbb{A}[X]/\langle T \rangle$, with rank $e = d$. In this case, computing the traces of all basis elements requires some work (namely, computing the Taylor series expansion of a rational function), and can be done in $O(M(d))$ operations $(+, \times)$ in $\mathbb{A}$, see [20] — here, and in what follows, $M(d)$ is a multiplication time function, such that we can multiply degree d polynomials in $M(d)$ base ring operations [11, Ch. 9]. Multiplication in $\mathbb{B}$ takes time $O(M(d))$ as well, so we end up with a total of

$$O\left(\sqrt{d}\, M(d) + d^{(\omega+1)/2}\right) = O\left(d^{(\omega+1)/2}\right)$$

operations $(+, \times)$ in $\mathbb{A}$.

Other cases could be considered along these lines, such as taking $\mathbb{B}$ of the form $\mathbb{A}[X_1, X_2]/\langle T_1, T_2 \rangle$, with $\langle T_1, T_2 \rangle$ a triangular set of degree d, but we do not need this here.

2.2   Outline of the Algorithm

In essence, Leverrier's algorithm relies on two facts: for A in $\mathbb{B}$, *(i)* the traces of the powers of A are the Newton sums of χ_A (A's characteristic polynomial) and *(ii)* Cayley-Hamilton's theorem, which says that A cancels χ_A.

Fact *(i)* above is made explicit in the following folklore lemma; see e.g. [1] for essentially the same arguments, in the case where $\mathbb{B} = \mathscr{M}_d(\mathbb{A})$.

Lemma 1. *Let* $\mathrm{rev}(\chi_A) = X^d \chi_A(1/X)$ *be the reverse polynomial of* χ_A. *Then the following holds in* $\mathbb{A}[[X]]$:

$$\frac{\mathrm{rev}(\chi_A)'}{\mathrm{rev}(\chi_A)} = -\sum_{i \geq 0} \mathrm{tr}(A^{i+1}) X^i. \tag{1}$$

Proof. This equality is well-known when $\mathbb{A}$ is a field and when $\mathbb{B} = \mathscr{M}_d(\mathbb{A})$. We use this fact to prove the lemma in our slightly more general setting.

Let $\mu_{1,1}, \dots, \mu_{d,d}$ be d^2 indeterminates over $\mathbb{Z}$, and let μ be the $(d \times d)$ matrix with entries $(\mu_{i,j})$. It is sufficient to prove we have

$$\frac{\mathrm{rev}(\chi_\mu)'}{\mathrm{rev}(\chi_\mu)} = -\sum_{i \geq 0} \mathrm{tr}(\mu^{i+1}) X^i, \tag{2}$$

where $\mathrm{tr}(\mu)$, χ_μ and $\mathrm{rev}(\chi_\mu)$ are defined as previously. Indeed, starting from the equality for μ, we can deduce it for $A \in \mathbb{B}$ by applying the evaluation morphism $\mu_{i,j} \mapsto \phi(A)_{i,j}$, where $\phi(A)_{i,j}$ is the (i,j)-th entry of the matrix $\phi(A) \in \mathscr{M}_d(\mathbb{A})$.

To prove our equality for μ, we can see the variables $\mu_{i,j}$ over $\mathbb{Q}$, so that we are left to prove (2) over the field $\mathbb{L} = \mathbb{Q}(\mu_{1,1}, \dots, \mu_{d,d})$. Since $\mathbb{L}$ is a field, it is sensible to introduce the roots $\gamma_1, \dots, \gamma_d$ of χ_μ in $\overline{\mathbb{L}}$, which are thus the eigenvalues of μ. Then, (2) is a well-known restatement of the Newton-Girard identities (see for instance Lemma 2 in [2]). $\square$

Let us write

$$\chi_A = X^d - a_1 X^{d-1} - \cdots - a_d.$$

Then, extracting coefficients in (1) shows that knowing the values $s_k = \mathrm{tr}(A^k)$, for $k = 1, \dots, d$, enables us to obtain the coefficients a_k in a successive manner using the formula

$$a_k = \frac{1}{k} \left(s_k - \sum_{i=1}^{k-1} s_{k-i} a_i \right). \tag{3}$$

(Note our assumption that $1, \dots, d$ are units in $\mathbb{A}$ makes this identity well-defined.) Computing all a_k in this manner takes a quadratic number of operations in $\mathbb{A}$. Using Newton iteration to solve the differential equation (1), which essentially boils down to computing a power series exponential, one can compute $a_1, \dots, a_d$ from $s_1, \dots, s_d$ in $O(\mathsf{M}(d))$ operations $(+, \times)$ in $\mathbb{A}$ [3,20].

Thus, we now assume we know the characteristic polynomial χ_A of A. Fact *(ii)* above then amounts to the following. Cayley-Hamilton's theorem implies that $\chi_A(\phi(A)) = 0$ in $\mathscr{M}_d(\mathbb{A})$, and thus that $\chi_A(A) = 0$ in $\mathbb{B}$; in other words,

$$A^d - a_1 A^{d-1} - \cdots - a_{d-1} A - a_d = 0.$$

This can be rewritten as

$$A(A^{d-1} - a_1 A^{d-2} - \cdots - a_{d-1}) = a_d.$$

Thus, if $a_d = \det(A)$ is invertible in $\mathbb{A}$, A is invertible in $\mathbb{B}$, with inverse

$$A^{-1} = a_d^{-1}(A^{d-1} - a_1 A^{d-2} - \cdots - a_{d-1}); \qquad (4)$$

conversely, if A is invertible in $\mathbb{B}$, $\phi(A)$ is invertible in $\mathscr{M}_d(\mathbb{A})$, and thus a_d is invertible in $\mathbb{A}$.

To summarize this outline, Leverrier's algorithm can decide if A is invertible (and if so compute its inverse) by means of the following steps:

1. compute the traces $s_1, \ldots, s_d$ of the powers of A
2. deduce χ_A using (1), using $O(\mathsf{M}(d))$ operations $(+, \times)$ in $\mathbb{A}$
3. A is invertible in $\mathbb{B}$ if and only if a_d is invertible in $\mathbb{A}$; if so, we deduce A^{-1} by means of (4).

2.3   Baby-Steps / Giant Steps Techniques

The direct implementation of Step 1 of Leverrier's algorithm consists of computing the powers $A^1, \ldots, A^d$, then taking their traces; this requires $O(d)$ multiplications in $\mathbb{B}$. Similarly, the direct approach to Step 3 by means of Horner's scheme requires $O(d)$ multiplications in $\mathbb{B}$. As is well-known, the baby steps / giant steps techniques allows for the reduction of the number of multiplications for both steps, from $O(d)$ to $O(\sqrt{d})$. We review this idea here, and analyze it in our general setup.

The dual of $\mathbb{B}$. As a preliminary, we say a few words about linear forms over $\mathbb{B}$. Let $\mathbb{B}^* = \mathrm{Hom}_{\mathbb{A}}(\mathbb{B}, \mathbb{A})$ be the dual of $\mathbb{B}$, that is, the set of $\mathbb{A}$-linear forms $\mathbb{B} \to \mathbb{A}$. For instance, the trace $\mathrm{tr} : \mathbb{B} \to \mathbb{A}$ is in $\mathbb{B}^*$.

Since we assume we have an $\mathbb{A}$-basis B of $\mathbb{B}$, it is natural to represent elements of $\mathbb{B}^*$ by means of their values on the basis B. Since we assume that $\mathbb{B}$ has rank e, its elements can be seen as column-vectors of size e, and the elements of $\mathbb{B}^*$ as row-vectors of size e. Then, applying a linear form to an element takes $O(e)$ operations $(+, \times)$ in $\mathbb{A}$.

There exists a useful operation on $\mathbb{B}^*$, the *transposed product*. The $\mathbb{A}$-module $\mathbb{B}^*$ can be turned into a $\mathbb{B}$-module: to any $A \in \mathbb{B}$, and to any $\lambda \in \mathbb{B}^*$, we can associate the linear form $A \circ \lambda : \mathbb{B} \to \mathbb{A}$ defined by $(A \circ \lambda)(B) = \lambda(AB)$. A general algorithmic theorem, the transposition principle [4, Th. 13.20], states that given A and λ, one can compute the linear form $A \circ \lambda$ using $\mathsf{M}(\mathbb{B})$ operations in $\mathbb{A}$ (that is, for the same cost as multiplication in $\mathbb{B}$).

Step 1. Using transposed products, we now explain how to implement the first step of Leverrier's algorithm. As a preliminary, we "compute the trace", that is, its values on the basis B. As per our convention, this takes $\mathsf{T}(\mathbb{B})$ operations $(+, \times)$ in $\mathbb{A}$.

Let $m = \lfloor \sqrt{d} \rfloor$ and $m' = \lceil (d+1)/m \rceil$, so that both m and m' are $O(\sqrt{d})$. The baby steps / giant steps version of Step 1 first computes the sequence of "baby steps"

$$M_0, M_1, M_2, \ldots, M_m = A^0, A^1, A^2, \ldots, A^m,$$

by means of repeated multiplications by A. Then, by repeated transposed multiplications by M_m, we compute the "giant steps" (which are here linear forms)

$$\lambda_0, \lambda_1, \lambda_2, \ldots, \lambda_{m'} = \mathrm{tr}, \ M_m \circ \mathrm{tr}, \ M_m^2 \circ \mathrm{tr}, \ \ldots, \ M_m^{m'} \circ \mathrm{tr}.$$

Computing all M_i and λ_j takes $O(\sqrt{d})$ multiplications and transposed multiplications in $\mathbb{B}$, for a total of $O(\sqrt{d}\,\mathsf{M}(\mathbb{B}))$ operations $(+, \times)$ in $\mathbb{A}$.

Knowing the M_i and λ_j, we can compute the required traces as $\lambda_j(M_i)$, for $0 \le i < m$ and $0 \le j < m'$, since they are given by

$$\lambda_j(M_i) = \mathrm{tr}(M_i M_m^j) = \mathrm{tr}(A^i A^{mj}) = \mathrm{tr}(A^{i+mj}),$$

and the exponent $i + mj$ cover all of $0, \ldots, d$. As we saw above, computing each $\lambda_j(M_i)$ amounts to doing a dot-product in size e, so a direct approach would give a cost of $O(de)$ operations in $\mathbb{A}$.

Better can be done, though. Consider the $(e \times m)$ matrix Γ whose columns give the coefficients of $M_0, \ldots, M_{m-1}$ on the basis B, and the $(m' \times e)$ matrix Λ whose rows give the coefficients of $\lambda_0, \ldots, \lambda_{m'-1}$ on the dual basis of B. Then, the (j, i)-th entry of $\Lambda\Gamma$ is precisely the value $\lambda_j(M_i)$. Since m and m' are both equivalent to $\sqrt{d}$, a naive matrix multiplication algorithm computes the product $\Lambda\Gamma$ in $O(de)$ operations in $\mathbb{A}$, as above. However, by doing a block product, with $O(e/\sqrt{d})$ blocks of size $O(\sqrt{d})$, we obtain $\Lambda\Gamma$ using $O(d^{(\omega-1)/2}e)$ operations $(+, \times)$ in $\mathbb{A}$.

Step 3. In order to perform Step 3, we have to evaluate $A^{d-1} - a_1 A^{d-2} - \cdots - a_{d-1}\mathbf{I}$, then divide by a_d if possible. Let us write $a_0 = -1$, and define $\alpha_i = -a_{d-1-i}$ for $i = 0, \ldots, d-1$; then, the quantity to compute is

$$p(A) = \sum_{i=0}^{d-1} \alpha_i A^i.$$

We extend the sum, by adding dummy coefficients α_i set to zero, to write

$$p(A) = \sum_{i=0}^{mm'-1} \alpha_i A^i;$$

this is valid, since by construction $mm' - 1 \ge d$. For $k \ge 0$, let us then define

$$\sigma_k = \sum_{i=km}^{(k+1)m-1} \alpha_i M_{i-km} = \sum_{i=0}^{m-1} \alpha_{i+km} M_i;$$

then, we see that we have

$$p(A) = (\cdots(\sigma_{m'-1}M_m + \sigma_{m'-2})\,M_m + \cdots)\,M_m + \sigma_0. \tag{5}$$

Using this formula, the algorithm to compute $p(A)$ first requires the computation of all M_i, for $i = 0, \ldots, m$, using $O(\sqrt{d}\,\mathsf{M}(\mathbb{B}))$ operations $(+, \times)$ in $\mathbb{A}$.

Next, we have to compute $\sigma_0, \ldots, \sigma_{m'-1}$. As for Step 1, let $\varGamma$ denote the $(e \times m)$ matrix whose columns give the coefficients of $A^0 = M_0, \ldots, A^{m-1} = M_m$. Then, σ_k is obtained by right-multiplying the matrix $\varGamma$ by the size m column vector $[\alpha_{km} \cdots \alpha_{(k+1)m-1}]^t$. Joining all these column vectors in a $(m \times m')$ matrix $\varDelta$, we obtain all σ_k by computing the product $\varGamma\varDelta$. As for Step 1, the cost is $O(d^{(\omega-1)/2}e)$ operations $(+, \times)$ in $\mathbb{A}$.

Finally, once all σ_k are known, we obtain $p(A)$ by means of m' products and additions in $\mathbb{B}$; the cost is $O(\sqrt{d}\,\mathsf{M}(\mathbb{B}))$ operations $(+, \times)$ in $\mathbb{A}$. Putting all costs seen before together, we obtain the cost announced in Proposition 1.

3   Proof of the Main Theorems

Using Proposition 1, it becomes straightforward to prove Theorems 1 and 2. Let $\mathbf{T} = (T_1, \ldots, T_n)$ be a triangular set of multidegree $(d_1, \ldots, d_n)$ in $k[X_1, \ldots, X_n]$. First, we deal with invertibility test and inversion in $R_{\mathbf{T}}$, assuming that all integers from 1 to $s_{\mathbf{T}} = \max(d_1, \ldots, d_n)$ are units in k.

Let A be in $R_{\mathbf{T}}$. As in the introduction, we view $R_{\mathbf{T}}$ as $R_{\mathbf{T}'}[X_n]/\langle T_n\rangle$, where $\mathbf{T}'$ is the triangular set $(T_1, \ldots, T_{n-1})$ in $k[X_1, \ldots, X_{n-1}]$. Applying Proposition 1, and referring to the discussion just after it, we see that we can decide whether A is invertible in $R_{\mathbf{T}}$, and if so compute its inverse, using

1. $O\left(d_n^{(\omega+1)/2}\right)$ operations $(+, \times)$ in $R_{\mathbf{T}'}$; and
2. one invertibility test / inversion in $R_{\mathbf{T}'}$.

As recalled in the introduction, multiplications in $R_{\mathbf{T}'}$ can be done for the cost of $K4^{n-1}\delta_{\mathbf{T}'}\lg(\delta_{\mathbf{T}'})\lg\lg(\delta_{\mathbf{T}'})$ operations in k, for some constant K. The same holds for additions in $R_{\mathbf{T}'}$, since additions can be done in optimal time $\delta_{\mathbf{T}'}$. Let K' be a constant such that the big-Oh estimate in the first item above is bounded by $K'd_n^{(\omega+1)/2}$.

Notice $\delta_{\mathbf{T}'} = d_1 \cdots d_{n-1}$, and that it admits the obvious upper bound: $\delta_{\mathbf{T}'} \leq \delta_{\mathbf{T}}$. Then, the total running time $\mathsf{I}(d_1, \ldots, d_n)$ of the invertibility test / inversion algorithm follows the recurrence

$$\mathsf{I}(d_1, \ldots, d_n) \leq KK'4^{n-1}d_1\cdots d_{n-1}d_n^{(\omega+1)/2}\lg(\delta_{\mathbf{T}})\lg\lg(\delta_{\mathbf{T}}) + \mathsf{I}(d_1, \ldots, d_{n-1}),$$

which can be simplified as

$$\mathsf{I}(d_1, \ldots, d_n) \leq C4^n\delta_{\mathbf{T}}d_n^{(\omega-1)/2}\lg(\delta_{\mathbf{T}})\lg\lg(\delta_{\mathbf{T}}) + \mathsf{I}(d_1, \ldots, d_{n-1}),$$

with $C = KK'/4$. Unrolling the recurrence, we obtain

$$\mathsf{I}(d_1, \ldots, d_n) \leq C4^n\delta_{\mathbf{T}}\left(d_1^{(\omega-1)/2} + \cdots + d_n^{(\omega-1)/2}\right)\lg(\delta_{\mathbf{T}})\lg\lg(\delta_{\mathbf{T}}).$$

With $s_{\mathbf{T}} = \max(d_1, \cdots, d_n)$, this admits the upper bound

$$\mathsf{I}(d_1, \ldots, d_n) \leq C 4^n n\, \delta_{\mathbf{T}}\, s_{\mathbf{T}}^{(\omega-1)/2} \lg(\delta_{\mathbf{T}}) \lg \lg(\delta_{\mathbf{T}}),$$

which proves Theorem 1.

Table 1. Experimental results (in seconds)

n	d	$\delta_{\mathbf{T}}$	m	Traces	CharPoly	Inverse	Horner	Total	MatrixInverse Time
1	2	2	3	0.03	0.00	0.00	0.01	0.04	0.2
1	2	2	6	0.03	0.00	0.00	0.02	0.05	0.07
1	2	2	9	0.07	0.00	0.00	0.06	0.13	0.15
1	2	2	12	0.19	0.00	0.00	0.12	0.31	0.34
1	2	2	15	0.26	0.00	0.00	0.23	0.49	0.54
1	2	2	18	0.47	0.00	0.00	0.45	0.92	0.72
1	10	10	3	0.02	0.00	0.00	0.01	0.03	0.12
1	10	10	6	0.10	0.01	0.00	0.09	0.20	0.39
1	10	10	9	0.43	0.01	0.00	0.21	0.65	1.09
1	10	10	12	0.96	0.01	0.00	0.63	1.60	2.26
1	10	10	15	1.67	0.02	0.00	1.29	2.98	4.09
1	10	10	18	3.17	0.02	0.00	2.09	5.28	6.67
1	18	18	3	0.02	0.01	0.00	0.03	0.06	0.22
1	18	18	6	0.33	0.01	0.00	0.20	0.54	0.87
1	18	18	9	0.93	0.02	0.00	0.50	1.45	2.28
1	18	18	12	2.30	0.02	0.00	1.51	3.83	4.60
1	18	18	15	4.22	0.03	0.00	3.36	7.61	8.02
1	18	18	18	8.07	0.05	0.00	5.43	13.56	13.14
3	3	27	3	0.14	0.02	0.08	0.22	0.46	7.7
3	3	27	6	1.75	0.07	0.07	1.46	3.35	10.4
3	3	27	9	5.68	0.11	0.08	3.58	9.45	15.5
3	3	27	12	13.47	0.16	0.07	9.18	22.8	24
3	3	27	15	22.9	0.27	0.08	19.4	42.8	35.7
3	3	27	18	42.67	0.27	0.07	30	73	52.2
3	4	64	3	0.88	0.22	0.58	1.6	3.28	54.5
3	4	64	6	10.6	0.43	0.63	9.80	21.4	100
3	4	64	9	32.8	0.77	0.62	22.5	56.7	184
3	4	64	12	74.9	1.07	0.63	55.1	132	324
3	4	64	15	121	1.38	0.65	111	233	524
3	4	64	18	213	1.67	0.58	163	379	840
3	5	125	3	0.75	0.08	0.63	0.74	2.20	159
3	5	125	6	7.07	0.22	0.63	5.07	14	299
3	5	125	9	22.5	0.38	0.55	12.6	36.0	548
3	5	125	12	53.7	0.65	0.54	33.2	88.1	960
3	5	125	15	94.1	0.84	0.54	72.1	167	1582
3	5	125	18	175.08	1.08	0.57	112	288	2462

Theorem 2 then follows from the combination of Proposition 1 and Theorem 1. To invert a $(d \times d)$ matrix A with entries in $R_{\mathbf{T}}$, we apply Leverrier's algorithm in Proposition 1, over the ring $\mathbb{A} = R_{\mathbf{T}}$. As explained after Proposition 1, the cost is $O(d^{\omega+1/2})$ operations $(+, \times)$ in $R_{\mathbf{T}}$, followed by the invertibility test / inversion of the determinant of A in $R_{\mathbf{T}}$. The cost reported in Theorem 2 then follows easily from the bounds on the cost of multiplication and invertibility test in $R_{\mathbf{T}}$.

4   Experimental Results

In this section, we compare Maple implementations of two approaches: our own recursive Leverrier algorithm and the existing (Gauss-Bareiss based) method from the `RegularChains` Maple library [14]. Our implementation uses the `RegularChains` library for normal forms, multiplication, etc, so we believe that this is a fair comparison.

Letting $p = 962592769$, we choose a random dense regular chain $\mathbf{T}$ in $\mathbb{F}_p[X_1, \ldots, X_n]$, with varying n, with and multidegree $(d, \ldots, d)$ for some varying d. We invert a random (and thus invertible) $m \times m$ matrix A with random entries in $R_{\mathbf{T}}$. We compare our results to the `MatrixInverse` function from `RegularChains`.

Table 1 gives the results of our experiments on a AMD Athlon running Linux, using Maple 15. For our algorithm, we detail the timings for trace computation (Step 1 of the algorithm), reconstituting the characteristic polynomial χ_A (Step 2), the inverse of the determinant of A, and the computation of the inverse of A itself (Step 3). As was to be expected, Step 1 and Step 3 take comparable times. For $n = 1$, our algorithm behaves very similarly to the built-in `MatrixInverse`. Already for $n = 3$, our implementation usually gives better results.

Acknowledgments. We acknowledge the support of the Canada Research Chairs Program and of NSERC.

References

1. Abdeljaoued, J., Lombardi, H.: Méthodes matricielles: introduction à la complexité algébrique. Mathématiques & Applications, vol. 42. Springer (2004)
2. Bostan, A., Flajolet, P., Salvy, B., Schost, É.: Fast computation of special resultants. J. Symb. Comp. 41(1), 1–29 (2006)
3. Brent, R.P., Kung, H.T.: Fast algorithms for manipulating formal power series. Journal of the ACM 25(4), 581–595 (1978)
4. Bürgisser, P., Clausen, M., Shokrollahi, A.: Algebraic Complexity Theory. Springer (1997)
5. Csanky, L.: Fast parallel matrix inversion algorithms. SIAM J. Comput. 5(4), 618–623 (1976)
6. Dahan, X., Jin, X., Moreno Maza, M., Schost, É.: Change of ordering for regular chains in positive dimension. Theoretical Computer Science 392(1-3), 37–65 (2008)
7. Dahan, X., Moreno Maza, M., Schost, É., Xie, Y.: On the complexity of the D5 principle. Transgressive Computing, 149–168 (2006)

8. Dahan, X., Moreno Maza, M., Schost, É., Wu, W., Xie, Y.: Lifting techniques for triangular decompositions. In: ISSAC 2005, pp. 108–115. ACM Press (2005)
9. Della Dora, J., Discrescenzo, C., Duval, D.: About a New Method for Computing in Algebraic Number Fields. In: Caviness, B.F. (ed.) EUROCAL 1985. LNCS, vol. 204, pp. 289–290. Springer, Heidelberg (1985)
10. Faddeev, D., Sominskii, I.: Collected problems in higher algebra. Freeman (1949)
11. von Zur Gathen, J., Gerhard, J.: Modern Computer Algebra. Cambridge University Press (1999)
12. Langemyr, L.: Algorithms for a multiple algebraic extension. In: Effective Methods in Algebraic Geometry. Progr. Math, vol. 94, pp. 235–248. Birkhäuser (1991)
13. Le Verrier, U.J.J.: Sur les variations séculaires des éléments elliptiques des sept planètes principales : Mercure, Venus, La Terre, Mars, Jupiter, Saturne et Uranus. J. Math. Pures Appli. 4, 220–254 (1840)
14. Lemaire, F., Moreno Maza, M., Xie, Y.: The RegularChains library. In: Kotsireas, I.S. (ed.) Maple Conference 2005, pp. 355–368 (2005)
15. Li, X., Moreno Maza, M., Schost, É.: Fast arithmetic for triangular sets: from theory to practice. Journal of Symbolic Computation 44(7), 891–907 (2009)
16. Li, X., Maza, M.M., Pan, W.: Computations modulo regular chains. In: ISSAC 2009, pp. 239–246. ACM Press (2009)
17. Li, X., Moreno Maza, M., Pan, W.: Gcd computations modulo regular chains. Technical report, Univ. Western Ontario, 30 pages (2009) (submitted)
18. Moreno Maza, M., Rioboo, R.: Polynomial GCD Computations over Towers of Algebraic Extensions. In: Giusti, M., Cohen, G., Mora, T. (eds.) AAECC 1995. LNCS, vol. 948, pp. 365–382. Springer, Heidelberg (1995)
19. Preparata, F.P., Sarwate, D.V.: An improved parallel processor bound in fast matrix inversion. Information Processing Letters 7(2), 148–150 (1978)
20. Schönhage, A.: The fundamental theorem of algebra in terms of computational complexity. Technical report, Univ. Tübingen, 73 pages (1982)
21. Shoup, V.: Fast construction of irreducible polynomials over finite fields. Journal of Symbolic Computation 17(5), 371–391 (1994)
22. Shoup, V.: Efficient computation of minimal polynomials in algebraic extensions of finite fields. In: ISSAC 1999, pp. 53–58. ACM Press (1999)
23. Souriau, J.-M.: Une méthode pour la décomposition spectrale et l'inversion des matrices. Comptes rendus des Séances de l'Académie des Sciences 227, 1010–1011 (1948)
24. Vassilevska Williams, V.: Breaking the Coppersmith-Winograd barrier (2011)

Sparse Polynomial Powering Using Heaps

Michael Monagan and Roman Pearce

Department of Mathematics, Simon Fraser University, Burnaby, B.C., Canada
{mmonagan,rpearcea}@cecm.sfu.ca

Abstract. We modify an old algorithm for expanding powers of dense polynomials to make it work for sparse polynomials, by using a heap to sort monomials. It has better complexity and lower space requirements than other sparse powering algorithms for dense polynomials. We show how to parallelize the method, and compare its performance on a series of benchmark problems to other methods and the Magma, Maple and Singular computer algebra systems.

Keywords: Sparse Polynomials, Powers, Heaps, Parallel Algorithms.

1 Introduction

Expanding powers of sparse polynomials is an elementary function of computer algebra systems. Despite receiving a lot of attention in the 1970's, a fragmented situation exists today where the fastest sparse methods make time and memory tradeoffs that improve one case at the expense of others. Thus, programmers of computer algebra systems must implement multiple routines and carefully select among them to obtain good performance.

For an introduction to this problem and current methods it is hard improve on the papers by Richard Fateman [1,2]. He characterizes the relative performance of the algorithms by counting coefficient operations. We briefly discuss these results. Let f be a polynomial with t terms to be raised to a power $k > 1$. We use f_i to refer to the ith term of f and $\#f$ to refer to the number of terms of f. We consider two cases: *sparse* and *dense*.

In the sparse case, the terms of f interact as if they were algebraically independent, e.g. as in $f = x_1 + x_2 + \cdots + x_t$. Expanding f^k creates $\binom{k+t-1}{k}$ terms, the most possible. In the dense case the terms of f combine as much as possible, e.g. as in $f = 1 + x + x^2 + \cdots + x^{t-1}$. If there are no cancellations, f^k will have $k(t-1) + 1$ terms.

We want a sparse algorithm to have good performance in the dense case, to allow for a smooth transition to dense methods inside a general purpose routine. The literature suggested that current sparse methods do an order of magnitude too much work in the dense case, so we developed new methods to address this. This in turn forced us to reassess sparse and dense algorithms for powering, as the consensus heavily favors dense algorithms.

Our contribution is two methods for powering sparse polynomials. The first, Sparse SUMS, has the best performance in the dense case. The second method, which we call FPS, is a modification to improve performance in the sparse case.

V.P. Gerdt et al. (Eds.): CASC 2012, LNCS 7442, pp. 236–247, 2012.

The methods in the literature are as follows.

RMUL computes $f^i = f \cdot f^{i-1}$ for $i = 2 \ldots k$. The memory taken by f^{i-2} may be reused to hold f^i so that total storage is at most twice the result.

RSQR computes $f^i = (f^{i/2})^2$ for $i = 2 \ldots \lfloor \log_2 k \rfloor$, with extra multiplication by f at each 1 in the binary expansion of k. E.g. $f^{13} = f^{1101_2} = (((f)^2 \cdot f)^2)^2 \cdot f$.

Gentleman and Heindel note in [4,5] that *RSQR* is vastly inferior to *RMUL* in the sparse case. *RSQR* also requires asymptotically fast dense multiplication to improve on *RMUL* in the dense case. Therefore, *RSQR* is a dense algorithm. The best feature of *RMUL* is that it aggressively combines like terms. This can be of great importance on large problems which "fill-in". Its weakness is sparse problems and high powers.

BINA selects $f_1 \in f$ and expands $g = (f_1 + 1)^k$ using the binomial theorem. It expands $(f - f_1)^i$ for $i = 2 \ldots k$ using *RMUL* and merges $f^k = \sum_{i=0}^{k} g_i \cdot (f - f_1)^i$.

BINB is similar to *BINA* except that f is split into equal-sized parts $f = g + h$. It expands and merges $f^k = \sum_{i=0}^{k} \binom{k}{i} \cdot g^i \cdot h^{k-i}$.

Binomial methods originate with Fateman in [1], who shows that *BINB* is nearly optimal in the sparse case. Alagar and Probst [11] improve on this using recursion, and Rowan [16] expands the set of powers $\{g^i\}$ more efficiently, both for the sparse case only. For the dense case, Fateman in [2] shows that *BINA* is comparable to *RMUL* and much faster than *BINB*. The tradeoff made in *BINB* assumes that few like terms combine. This makes it unsuitable for our purpose. In *BINA*, we avoid unbalanced merging by storing all $(f - f_1)^i$ and performing a simultaneous n-ary merge that multiplies by each g_i inline. This makes *BINA* extremely fast in most cases, at the cost of extra memory.

MNE generates all combinations of terms with multinomial coefficients, see [6]. This quickly becomes infeasible in the dense case.

FFT performs fast multipoint evaluation at roots of unity modulo primes, uses modular exponentiation on the values, then performs fast interpolation. Over $\mathbb{Z}$ it uses multiple primes and Chinese remaindering.

As noted by Ponder in [10], the FFT can be competitive in practice because high powers of sparse polynomials tend to fill in. For multivariate polynomials, one can use the Kronecker substitution as suggested by Moenck [9], however this separates the variables with very high degrees and thus limits gains from fill-in. A weakness of the FFT is that small polynomials raised to high powers over $\mathbb{Z}$ require many large FFTs. For that case the following classical method is faster, a crucial fact which was brought to our attention by Greg Fee.

SUMS is a dense method. Let $f = \sum_{i=0}^{d} f_i x^i$. To compute $g = f^k = \sum_{i=0}^{kd} g_i x^i$ we compute $g_0 = f_0^k$ and use the formula $g_i = \frac{1}{i f_0} \sum_{j=1}^{\min(d,i)} ((k+1)j - i) f_j g_{i-j}$ for $i = 1 \ldots kd$.

The *SUMS* algorithm is originally due to Euler and is used to exponentiate power series, see [2,3,8]. The algorithm is extremely fast for small polynomials raised to large powers, as it is linear in k and quadratic in d.

Two features of the *SUMS* formula recall the sparse multiplication algorithm of Johnson [7]. First, it computes each new term of the result in order. Second, it merges pairwise products $f_j g_{i-j}$ of equal degree, but scaled by $((k+1)j - i)$. Our starting point was to make a sparse method by skipping over products that a sparse representation omits, that is, where f_j or g_{i-j} equals zero.

What methods do computer algebra systems presently use for this problem? Singular 3.1 uses *RMUL*. Magma 2.17 uses *RSQR*. Maple 16 selects among our implementations of *RMUL*, *BINA*, and *RSQR*. For univariate powering, Maple estimates when *RSQR* will beat *BINA*. For multivariate powers, Maple bounds the extra memory needed for *BINA* and uses *RMUL* when this is too large.

For the underlying multiplications, Magma and Maple use dense algorithms for univariate polynomials over $\mathbb{Z}$. Magma uses the Schönhage-Strassen method with a single modulus of the form $2^{2^k} + 1$. Maple evaluates at a large integer of the form 2^{64i} to leverage the FFT from integer multiplication. For multivariate multiplications, Maple, Magma, and Singular all use classical sparse algorithms and distributed polynomial representations. Maple uses our codes from [12,14].

Our paper is organized as follows. Section 2 develops the Sparse SUMS and FPS algorithms and describes our implementation. The complexity of powering is discussed in Section 2.1. Section 2.2 describes our approach to parallelization which we also used successfully for sparse polynomial division in [15]. Section 3 compares the performance of the algorithms on benchmark problems.

2   Sparse Sums

For completeness we briefly derive *SUMS*. Let $f = \sum_{i=0}^d f_i x^i \in \mathbb{Q}[x]$ and $g = f^k$. Then $g' = k\,f^{k-1} \cdot f'$ and $f \cdot g' = k\,g \cdot f'$. Equating terms of degree $i-1$ in

$$(f_0 + f_1 x + \cdots)(g_1 + 2g_2 x + \cdots) = k(g_0 + g_1 x + \cdots)(f_1 + 2f_2 x + \cdots)$$

we obtain

$$\sum_{j=0}^{\min(d,i)} f_j x^j \cdot (g_{i-j} x^{i-j})' = \sum_{j=1}^{\min(d,i)} k g_{i-j} x^{i-j} \cdot (f_j x^j)'$$

from which we isolate g_i to obtain the formula for $i > 0$. $\square$

Algorithm Dense SUMS (descending order).

Input: dense polynomial $f = f_0 + f_1 x + \cdots + f_d x^d$, $f_d \neq 0$ stored as an
         array $[f_0, f_1, \ldots, f_d]$ indexed from zero, and a positive integer k.

Output: dense polynomial $g = f^k$.

1    $g :=$ an array with $kd + 1$ elements indexed from zero

2    $g_{kd} := f_d^k$

3   for i from $kd - 1$ to 0 by -1 do

4       $e := kd - i$

5       $c := \sum_{j=1}^{\min(d,e)} ((k+1)j - e) \cdot f_{d-j} \cdot g_{i+j}$

6       $g_i := c/(e \cdot f_d)$

7   return g

Our first task is to modify *SUMS* to produce the terms in descending order, dividing by the leading coefficient of f rather than the constant term f_0. This leads into the sparse version and solves the problem of what to do when $f_0 = 0$.

In algorithm Dense SUMS we identify i as the degree of the next term being computed for g. To compute g_i, we merge products of degree $i + d$, scaling by $((k + 1)j - e)$. To make our sparse algorithm, we express this scale factor using the terms' degrees. To merge $f_\alpha x^\alpha \times g_\beta x^\beta$ where $\alpha + \beta = i + d$, we scale by $((k + 1)j - e) = \beta - k\alpha$.

The sparse version of *SUMS* is presented below. It uses a heap of pointers into f and g to combine only non-zero products. The heap is used to merge the set of all pairwise term products $f_i \times g_j$ in descending order. We exploit the fact that the term $f_i \times g_j$ is strictly greater than $f_i \times g_{j+1}$ and $f_{i+1} \times g_j$ to reduce the size of the heap. In lines 12 and 13, we avoid having multiple f_i in the heap with the same g_j. Also note, because the coefficients of g are much larger than those of f, there is an advantage to multiplying $(\beta - k\alpha) \cdot cof(f_i)$ first.

Algorithm Sparse SUMS.
Input: sparse univariate polynomial $f = f_1 + f_2 + \cdots + f_t \in \mathbb{Z}[x]$
         with terms descending in degree, and a positive integer k.
Output: sparse polynomial $g = f^k$.
1     $H :=$ an empty heap ordered by degree with maximum element H_1
2     $g := f_1^k$
3   insert $f_2 \times g_1 = (2, 1, \deg(f_2) + \deg(g_1))$ into H
4   while $|H| > 0$ and $\deg(H_1) \geq \deg(f)$ do
5        $M := \deg(H_1); C := 0; Q := \{\};$
6   while $|H| > 0$ and $\deg(H_1) = M$ do
7          $(i, j, M) := extract\_max(H)$
8          $(\alpha, \beta) := (degree(f_i), degree(g_j))$
9          $C := C + (\beta - k\alpha) \cdot cof(f_i) \cdot cof(g_j)$
10        $Q := Q \cup \{(i, j)\}$
11   for all $(i, j) \in Q$ do
12   if $j < \#g$ and ($i = 1$ or $f_{i-1} \times g_{j+1}$ was merged) insert $f_i \times g_{j+1}$ into H
13   if $i < \#f$ and $f_{i+1} \times g$ not in H then insert $f_{i+1} \times g_j$ into H
14   if $C \neq 0$ then
15         $C := C/((\deg(g_1) - M) \cdot cof(f_1))$
16         $g := g + C\, x^{M - \deg(f_1)}$
17   if $f_2 \times g$ has no term in H then insert $f_2 \times g_{\#g}$ into H
18   return g

In computer memory, the heap is an array of size $O(\#f)$ with pointers into a second array for the products $f_i \times g_j$. For most inputs (1000 terms or fewer) these structures fit inside the L1 cache. For each $f_i \in f$, we maintain a pointer to the next term $g_j \in g$ for which we have yet to merge $f_i \times g_j$. This makes the test for whether $f_{i-1} \times g_j$ has been merged easy. We simply check if the pointer for f_{i-1} has advanced beyond g_j. We set a bit to indicate whether each product $f_i \times g_j$ is in the heap or not. For dense polynomials, we also use an optimization called *chaining* to combine products with equal monomials, see [13,14].

2.1 Complexity and Optimizations

Theorem 1. *Sparse sums expands $g = f^k \in \mathbb{Z}[x]$ using $(2\,\#f - 1)\,\#g + 2\log k$ coefficient multiplications, $\#g$ divisions, and $O(\#f\#g \log \#f)$ comparisons. It stores g and uses $O(\#f)$ additional memory.*

Proof. Binary powering $g_1 = f_1^k$ does at most $2\log k$ multiplications. We merge the set of all products $\{f_i \times g_j\}$ for $2 \le i \le \#f$ and $1 \le j \le \#g$ with the heap. Each product requires two multiplications in line 9 and $O(\log \#f)$ comparisons for the heap in lines 7, 10 and 13. We do not count the exponent multiplication in $\beta - k\alpha$. To construct each term of g, we perform one multiplication and one division in line 15. The objects stored other than g are the heap H and set Q which have at most $\#f$ entries. $\square$

For multivariate polynomials we use the Kronecker substitution to treat the problem as univariate. In general, one can use any invertible map of monomials to integers so long as monomial multiplications correspond to integer additions. The mapping has two caveats that do not occur in the other sparse algorithms. Because we multiply by the exponents, any padding in the map that increases the univariate degrees can also increase the cost of coefficient arithmetic in *Sparse SUMS*. And, because we divide by the exponents, we cannot run the algorithm mod p if the degree of g under the mapping is greater than or equal to p.

Our benchmarks revealed one case where *Sparse SUMS* is inefficient. When sparse polynomials, e.g. those arising from a Kronecker substitution, are raised to a low power, typically $\#f^k \gg \#f^{(k-1)}$. The cost of *RMUL* will be mostly in the final step which does $\#f \cdot \#f^{(k-1)}$ multiplications. But *Sparse SUMS* does $O(\#f \times \#f^k)$ coefficient operations, which could be far more in total.

We note that *Sparse SUMS* could construct f^{k+1} almost for free, because it already multiplies every term of f^k by every term of f except for f_1. To exploit this we created a variant that we call *FPS*. It uses the *Sparse SUMS* algorithm to compute f^{k-1} and outputs f^k as a side effect.

We present *FPS* at the end of this section by adding lines to our description of *Sparse SUMS*. To reduce the number of coefficient operations, lines 9 and 11 should reuse $cof(f_i) \cdot cof(g_j)$, and lines 17 and 18 should update C and S with $C := C/(\deg(g_1) - M)$; $S := S + C$; $C := C/cof(f_1)$.

Table 1 counts coefficient multiplications to compare the cost of the sparse algorithms. The sparse result has $(k + t - 1)!/(k!(t - 1)!)$ terms, so *BINB* is nearly optimal. *RMUL* is more expensive by a factor of k, slowing it down on high powers, and *BINA* by a factor of $kt/(k+t-1)$, which balances contributions from k and t. *Sparse SUMS* adds a factor of $(2t - 1)$ and *FPS* adds a factor of $(2t - 1)k/(k + t - 1)$. Those methods also do many divisions in the sparse case, however their cost does not dominate.

The FFT is inefficient for sparse problems. One may assume these problems have distinct variables, e.g. $(1 + x + y + z)^{50}$, and Kronecker substitution must separate variables in the result. For t terms to the power k, we must replace the ith term by at least $x^{(k+1)^{i-2}}$ for $i > 2$, so the degree of f^k is $d = k\,(k + 1)^{t-2}$.

An FFT does about $\frac{1}{2}n \log_2 n$ multiplications, where n is the first power of 2 greater than d. For example, $(1 + x + y + z)^{50}$ will have $d = 50 \cdot 51^2 = 130050$

Table 1. Coefficient multiplications to power $(t \text{ terms})^k$

	sparse case	dense case
RMUL	$\dfrac{(k+t-1)!}{(t-1)!(k-1)!} - t$	$t(k-1)(kt-k+2)/2 \in O(k^2 t^2)$
BINA	$\dfrac{t \cdot (k+t-2)!}{(t-1)!(k-1)!} + 2k$	$t(k-1)(kt-2k+4)/2 + 2 \in O(k^2 t^2)$
BINB	$\dfrac{(k+t-1)!}{k!(t-1)!} + \cdots$	$k^2(k-1)(t-2)^2/24 + \cdots \in O(k^3 t^2)$
SUMS	$\dfrac{(2t-1)(k+t-1)!}{k!(t-1)!}$	$(2t-1)((t-1)k+1) \in O(kt^2)$
FPS	$\dfrac{(2t-1)(k+t-2)!}{(k-1)!(t-1)!}$	$(2t-1)((t-1)(k-1)+1) \in O(kt^2)$

and $n = 2^{14}$. The two FFT calls do about $n\log_2 n = 2.29 \times 10^6$ multiplications, but *SUMS* and *FPS* compute the result in 1.64×10^5 and 1.55×10^5 operations. In the dense case, *SUMS* and *FPS* are $O(kt^2)$ and the other sparse algorithms are $O(k^2 t^2)$. The FFT is $O(d\log d)$ where $d = ((t-1)k+1)$ is now the size of the result, however, *SUMS* can still win if $\log d > 2t$, that is, *SUMS* is the best method for raising small dense polynomials to high powers.

Algorithm FPS.
Input: sparse univariate polynomial $f = f_1 + f_2 + \cdots + f_t \in \mathbb{Z}[x]$
        with terms descending in degree, and a positive integer k.
Output: sparse polynomial $h = f^k$.
1   $H :=$ an empty heap ordered by degree with maximum element H_1
2   $g := f_1^{k-1};\ \ h := f_1^k$
3   insert $f_2 \times g_1 = (2, 1, \deg(f_2) + \deg(g_1))$ into H
4   while $|H| > 0$ do
5      $M := \deg(H_1);\ C := 0;\ S := 0;\ Q := \{\};$
6   while $|H| > 0$ and $\deg(H_1) = M$ do
7         $(i, j, M) := extract\_max(H)$
8         $(\alpha, \beta) := (degree(f_i), degree(g_j))$
9         $S := S + cof(f_i) \cdot cof(g_j)$
10   if $M \geq \deg(f_1)$ and $\beta \neq (k-1)\alpha$
11   then $C := C + (\beta - (k-1)\alpha) \cdot cof(f_i) \cdot cof(g_j)$
12       $Q := Q \cup \{(i, j)\}$
13   for all $(i, j) \in Q$ do
14   if $j < \#g$ and ($i = 1$ or $f_{i-1} \times g_{j+1}$ was merged) insert $f_i \times g_{j+1}$ into H
15   if $i < \#f$ and $f_{i+1} \times g$ not in H then insert $f_{i+1} \times g_j$ into H
16   if $C \neq 0$ then
17         $C := C/((\deg(g_1) - M) \cdot cof(f_1))$
18         $S := S + C \cdot cof(f_1)$
19         $g := g + C\, x^{M - \deg(f_1)}$
20   if $f_2 \times g$ has no term in H then insert $f_2 \times g_{\#g}$ into H
21   if $S \neq 0$ then $h := h + S\, x^M$
22   return h

2.2   Parallelization

Our design for the parallel algorithm follows the approach used for polynomial division in [15]. Both problems have a tight data-dependency among the terms in the result. That is, each new term of g can depend on any subset of previous terms with no predictable pattern. To create parallelism we split the work into dynamically interacting pieces and exploit structure to hide latencies.

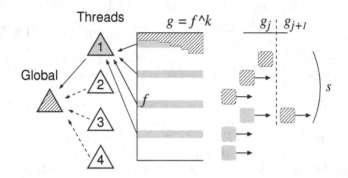

Fig. 1. Threads multiply strips of f by all of g. A global function merges the results from the threads and the first strip, while computing new terms of g.

Figure 1 shows features common to all our parallel algorithms. The work of merging products $f_i \times g_j$ is divided into strips along the terms of f, so threads are given subsets of f to multiply by g. A global function combines their results and computes new terms of g. This function is protected by a lock and may be called by any thread, which allows them to cooperatively balance the load [12].

Another feature from our earlier work on division [15] is used to resolve the data-dependency. The first strip of f is assigned to the global function, so that as new terms g_j are computed there is no delay in merging $f_2 \times g_j$. Recall that this term must be compared to all others immediately as it could be used next.

The global strip is also used to resolve the nasty problem of blocked threads. Threads block when they merge $f_i \times g_j$ and go to insert $f_i \times g_{j+1}$ in their heap only to find that g_{j+1} does not exist. The reason could be a delay, but perhaps $f_i \times g_j$ was merged by the global function and no new term of g was computed. In that case, the global function now needs $f_{i+1} \times g_j$ to progress. Our solution is for the global function to steal rows from the threads when this happens.

To implement stealing, we have two shared variables that are read by all of the threads. The first variable t is the number of terms computed in the result. The variable s is the number of rows stolen by the global function. To ensure a valid state, threads must read s before t, and the global function must update t before incrementing s. We enforce this with memory barriers.

Incrementing t means that a new term of g was computed, and alongside its monomial and coefficient the global function stores the current value of s. This tells the threads what products involving g_t are stolen and must not be merged. When threads block waiting for t to be incremented, they attempt to enter the

Table 2. Timings for completely sparse $(t \text{ terms})^k$

input		result			C code				Magma		Singular
t	k	terms	degree	bits	SUMS	FPS	RMUL	BINA	FFT	RSQR	RMUL
3	100	5151	10100	152	0.001	0.001	0.026	0.001	0.01	0.25	0.05
3	250	31626	62750	388	0.007	0.013	0.484	0.011	0.45	12.84	1.04
3	500	125751	250500	784	0.035	0.069	4.560	0.055	3.48	278.13	12.75
3	1000	501501	1001000	1575	0.208	0.414	45.664	0.333	31.38	–	125.29
3	2500	3128751	6252500	3951	2.328	4.770	–	5.667	(*)	–	–
4	50	23426	130050	92	0.005	0.007	0.033	0.005	0.12	1.34	0.18
4	100	176851	1020100	191	0.040	0.073	0.763	0.055	3.10	98.71	2.49
4	200	1373701	8080200	389	0.373	0.714	13.151	0.521	74.36	–	44.61
4	400	10827401	64320400	788	3.636	7.405	247.743	5.144	–	–	889.79
6	20	53130	$3.89 \cdot 10^6$	42	0.008	0.008	0.021	0.006	3.25	0.77	0.07
6	30	324632	$2.77 \cdot 10^7$	67	0.056	0.057	0.173	0.039	63.27	26.00	1.17
6	40	1221759	$1.13 \cdot 10^8$	91	0.332	0.531	1.471	0.222	–	460.42	6.67
6	50	3478761	$3.38 \cdot 10^8$	117	1.000	1.682	6.547	0.838	–	–	26.89
6	70	17259390	$1.78 \cdot 10^9$	167	5.123	9.256	49.476	5.029	–	–	176.80
8	15	170544	$2.51 \cdot 10^8$	34	0.031	0.027	0.052	0.023	(*)	0.95	0.10
8	20	888030	$1.71 \cdot 10^9$	47	0.179	0.162	0.337	0.117	–	36.20	1.84
8	25	3365856	$7.72 \cdot 10^9$	62	0.677	0.649	1.504	0.452	–	284.64	10.70
8	30	10295472	$2.66 \cdot 10^{10}$	76	2.838	3.135	6.143	1.546	–	–	42.92
8	35	26978328	$7.62 \cdot 10^{10}$	90	9.042	13.828	28.342	5.927	–	–	148.97
12	10	352716	$2.59 \cdot 10^{11}$	22	0.088	0.055	0.074	0.050	–	1.61	0.18
12	12	1352078	$1.65 \cdot 10^{12}$	29	0.364	0.231	0.330	0.199	–	11.84	0.89
12	14	4457400	$8.07 \cdot 10^{12}$	35	1.222	0.864	1.220	0.672	–	78.81	4.06
12	16	13037895	$3.22 \cdot 10^{13}$	41	3.538	2.631	3.970	1.982	–	500.20	21.99
12	18	34597290	$1.10 \cdot 10^{14}$	47	9.339	7.166	11.468	5.402	–	–	(*)
12	20	84672315	$2.15 \cdot 10^{14}$	54	22.537	18.071	29.922	13.360	–	–	–

– Not attempted. (*) Ran out of memory.

global function and then they update their local copies of s and t. The global function can steal rows with impunity. We do this whenever it is blocked.

3   Benchmarks

Our benchmarks were performed on a 2.66 GHz Intel Core i7 920 with 6 GB of RAM running Linux. This is a 64 bit 4 core processor. Timings are the median time in seconds of 3 runs. Magma timings are for version 2.17. Singular timings are for version 3.10. Timings for *SUMS*, *FPS*, *RMUL*, and *BINA* are real times from our C library. For Magma and Singular we report CPU timings, which we found to be less precise.

3.1   Sparse Problems

To create polynomials with t terms whose powers up to k are completely sparse, we used Kronecker's substitution on $F = 1 + x_1 + x_2 + \cdots + x_{t-1}$ to construct

$$f = 1 + x + x^{(k+1)} + x^{(k+1)^2} + \cdots + x^{(k+1)^{t-2}}.$$

This polynomial to the power k generates the largest possible number of terms. That is what is meant by sparse. Notice how we can not have too many terms t before the integer exponents become massive. This suggests that most practical problems (whose result can be stored) have $t \ll k$, so the extra factor of $2t - 1$ in the cost of *SUMS* is not as disadvantageous as it may first appear.

Table 2 compares *SUMS*, *FPS*, *RMUL* and *BINA*. The polynomials are too short to run our parallel algorithms. For Magma we give two times; FFT is the *RSQR* algorithm with Schönhage-Strassen multiplication. We also tried writing the problem as multivariate, which uses *RSQR* and sparse arithmetic. Singular uses sparse arithmetic and *RMUL* which is a sensible choice.

The timings show that *SUMS* is consistently faster than *RMUL*, and is the fastest method for higher powers in fewer variables. The *FPS* method becomes slower relative to *SUMS* as k increases but faster as t increases. *BINA* is highly competitive in all cases, and is the fastest method tested for 6 or more terms.

3.2 Dense Problems

Table 3 shows timings for expanding powers of the polynomial

$$f = 1 + x + x^2 + \cdots + x^{t-1}.$$

Dense problems are a strong case for *SUMS*. *RMUL* and *BINA* are competitive only for low powers of large polynomials, where the FFT is the fastest method. This implies *SUMS* is the best sparse method to complement the FFT. Higher powers benefit *SUMS* versus the FFT. For 500 terms, *SUMS* goes from 21 times slower at $k = 10$ down to 1.5 times slower at $k = 320$, breaking even at $k = 640$. Our parallel speedup appears to be limited to 3.8. The timings for *FPS* do not fit in the table, but they are slower than *SUMS* by a factor of 3. We think this ratio will improve with optimization.

Table 4 reports the time to power two dense multivariate polynomials. The data shows that conventional sparse methods (*RMUL* and *BINA*) beat the FFT as the number of variables increases. Because it has better complexity on dense

Table 3. Timings for completely dense $(t$ terms$)^k$

t	k	*SUMS*	*RMUL*	*BINA*	*FFT*	t	k	*SUMS* (4 cores)	*RMUL*	*FFT*
10	200	0.001	0.085	0.098	0.006	500	10	0.084	0.026	0.151	0.004
10	500	0.005	0.752	1.078	0.095	500	20	0.198	0.058	1.343	0.014
10	1000	0.015	4.474	8.178	0.501	500	40	0.476	0.131	6.944	0.057
10	1500	0.032	13.386	29.630	0.510	500	80	1.200	0.322	34.933	0.247
10	2000	0.055	29.808	–	2.640	500	160	3.351	0.921	192.162	1.352
10	2500	0.082	55.433	–	2.670	500	320	10.616	2.808	–	6.890
100	50	0.023	0.415	0.428	0.017	1000	3	0.045	0.015	0.034	0.001
100	100	0.057	2.087	2.165	0.056	1000	5	0.078	0.026	0.115	0.003
100	200	0.159	11.091	11.728	0.262	1000	10	0.361	0.102	0.797	0.013
100	400	0.497	66.643	71.487	1.360	1000	20	0.824	0.228	5.714	0.030
100	800	1.730	446.477	–	6.990	1000	40	1.951	0.525	29.393	0.130
100	1600	6.087	–	–	36.310	1000	80	5.035	1.325	149.326	0.570

Table 4. Timings for dense multivariate f^k

$f = (1+x+y)^{15}$			$t = 136$				Magma	Singular
k	$\#g$	SUMS	4 cores	FPS	RMUL	4 cores	BINA	FFT	RMUL
20	45451	0.536	0.149	0.685	1.514	0.429	1.553	0.49	12.33
40	180901	3.157	0.846	4.181	15.833	4.406	16.375	5.49	134.59
60	406351	9.263	2.478	12.552	65.276	17.927	66.790	27.27	522.59
80	721801	20.439	5.402	28.110	182.717	49.830	187.178	56.42	–
120	1622701	64.117	16.618	88.688	–	–	–	325.60	–

$f = (1+w+x+y+z)^4$			$t = 70$				Magma	Singular
k	$\#g$	SUMS	2 cores	FPS	RMUL	2 cores	BINA	FFT	RMUL
4	4845	0.005	0.005	0.003	0.003	0.003	0.003	0.30	0.01
8	58905	0.068	0.062	0.048	0.071	0.047	0.072	1.24	1.01
12	270725	0.711	0.440	1.021	0.955	0.589	0.995	10.84	10.40
16	814385	2.311	1.297	3.784	5.238	3.120	5.443	65.50	46.49
20	1929501	5.852	4.755	10.337	17.164	10.065	17.790	218.14	166.02
24	3921225	12.313	11.350	22.643	44.008	25.513	45.489	391.42	394.08
28	7160245	23.430	22.754	45.458	97.179	56.745	100.277	(*)	–

problems, SUMS has a much easier time beating the FFT. It gains more as the power k or the number of variables is increased.

The only case where SUMS loses to RMUL or BINA is $k = 4$ in the second problem. In that case, and also for $k = 8$, the FPS algorithm does much better. The parallel speedup for SUMS is good on the first problem but it deteriorates on the second problem as k increases. We suspect the routine is struggling with data dependencies because parallel division of f^k by f shows the same issue.

3.3 Real Examples

We were first motivated to investigate sparse powering by a post to the Sage development newsgroup by Tom Coates. He wanted to raise the polynomial

$$f = xy^3z^2 + x^2y^2z + xy^3z + xy^2z^2 + y^3z^2 + y^3z$$
$$+ 2y^2z^2 + 2xyz + y^2z + yz^2 + y^2 + 2yz + z$$

to high powers, but no computer algebra system could do it in reasonable time. This can now be done quickly. Table 5 shows that SUMS is the fastest method. Note, in order to get Magma to use the FFT, we explicitly converted $f(x, y, z)$ into a univariate polynomial using Kronecker's substitution. Otherwise Magma will use sparse RSQR, which takes 134.49 seconds for $k = 40$.

In [17], Zeilberger writes (in 1994):

"In my research on constant term conjectures, I often need to expand powers of polynomials P^m where m is very large and P is (usually) a polynomial of several variables. I was frustrated by the slowness of all the commercial computer algebra packages. For example, in Maple, it takes several days to expand $(1 + 3x + 2x^2)^{3000}$."

Table 5. Timings (in CPU seconds) to power f^k

	result	C code			Magma	Maple	Singular
k	#g	$SUMS$	$RMUL$	$BINA$	FFT	$RMUL$	$RMUL$
40	243581	0.159	0.968	0.941	1.47	1.36	5.50
70	1284816	0.941	10.833	10.624	28.26	13.97	62.85
100	3721951	3.026	48.932	51.670	93.64	59.37	316.11
150	12499176	10.880	276.320	–	(*)	324.00	–
250	57636126	68.626	–	–	–	–	–

– Not attempted. (*) Ran out of memory.

Zeilberger coded dense $SUMS$ in Maple and noted that it was theoretically faster than the FFT, although his analysis does not account for the coefficients which exceed 2300 decimal digits. At the time Maple was using $BINA$, which is a poor choice on this problem as it needs over 2 GB of memory to store all the expanded powers of $(3x + 1)^i$ for i up to 3000.

Table 6 shows that $SUMS$ is by far the fastest method on this example. The digits column shows the length in decimal digits of the largest coefficient in the result. By default, Maple 16 uses $RSQR$ and performs univariate multiplication by evaluating at a suitable power of 2 and leveraging the FFT from fast integer multiplication. This takes 1 second on our Intel Core i7 2.66 GHz machine. But $SUMS$ takes under 9 milliseconds! It does fewer than $2t^2k = 2 \cdot 9 \cdot 3000 = 54000$ coefficient multiplications; and because the coefficients of f are small, at most half of those are multiprecision.

Table 6. Timings (in CPU seconds) to power $(2x^2 + 3x + 1)^k$

		C code			Magma	Maple 16	Singular
k	digits	$SUMS$	$RMUL$	$BINA$	FFT	$RSQR$	$RMUL$
1000	777	0.00130	0.302	0.591	0.02	0.088	0.76
2000	1555	0.00418	1.858	6.562	0.08	0.419	4.62
3000	2333	0.00884	5.461	28.847	0.25	1.03	15.04
4000	3111	0.01540	12.202	83.870	0.41	2.13	35.57
5000	3889	0.02318	23.008	(*)	1.31	3.48	70.32

(*) $BINA$ ran out of space; it exceeded the 6 gigabytes available.

4   Conclusion

We adapted a classical method for powering dense series to make a new method for powering sparse polynomials. $SUMS$ has better complexity than other sparse algorithms in the dense case, which is important for general problems. It has reasonable performance in the completely sparse case.

In comparing $SUMS$ with $RMUL$, the larger the power and the smaller the polynomial, the better. We also compared it to the FFT and explained why the FFT struggles to power multivariate polynomials. It is due to the very high degrees that are needed in Kronecker substitution when powering. We conclude that $SUMS$ has a wide range of applicability. It performed extremely well on a benchmark problem coming from a real application.

Our effort to parallelize Sparse $SUMS$ was largely successful. For inputs with a large number of terms, 500 or more, we often obtained good parallel speedup. A problem with this approach is that it requires the input to have a lot of terms, at least 50, to conceal communication latencies.

One improvement that we can make is to generate the terms of the output $g = g_1 + g_2 + \cdots + g_m$ from both directions in parallel and meet in the middle. Our next task is to optimize and parallelize the FPS variant presented here. That algorithm should offer better performance in the cases where $SUMS$ loses to $RMUL$ or $BINA$, while retaining the best qualities of $SUMS$.

Acknowledgment. We thank the referees for their suggestions which have improved this paper.

References

1. Fateman, R.: On the computation of powers of sparse polynomials. Studies in Appl. Math. 53, 145–155 (1974)
2. Fateman, R.: Polynomial multiplication, powers, and asymptotic analysis: some comments. SIAM J. Comput. 3(3), 196–213 (1974)
3. Fettis, H.: Algorithm 158. Communications of the ACM 6, 104 (1963)
4. Gentleman, M.: Optimal multiplication chains for computing a power of a symbolic polynomial. Math Comp. 26(120), 935–939 (1972)
5. Heindel, L.: Computation of powers of multivariate polynomials over the integers. J. Comput. Syst. Sci. 6(1), 1–8 (1972)
6. Horowitz, E., Sahni, S.: The computation of powers of symbolic polynomials. SIAM J. Comput. 4(2), 201–208 (1975)
7. Johnson, S.C.: Sparse polynomial arithmetic. ACM SIGSAM Bulletin 8(3), 63–71 (1974)
8. Knuth, D.: The Art of Computer Programming, Seminumerical Algorithms, vol. 2. Addison-Wesley (1998)
9. Moenck, R.: Another Polynomial Homomorphism. Acta Informatica 6, 153–169 (1976)
10. Ponder, C.: Parallel multiplication and powering of polynomials. J. Symbolic. Comp. 11(4), 307–320 (1991)
11. Probst, D., Alagar, V.: A Family of Algorithms for Powering Sparse Polynomials. SIAM J. Comput. 8(4), 626–644 (1979)
12. Monagan, M., Pearce, R.: Parallel Sparse Polynomial Multiplication Using Heaps. In: Proc. of ISSAC 2009, pp. 295–315. ACM Press (2009)
13. Monagan, M., Pearce, R.: Polynomial Division Using Dynamic Arrays, Heaps, and Packed Exponent Vectors. In: Ganzha, V.G., Mayr, E.W., Vorozhtsov, E.V. (eds.) CASC 2007. LNCS, vol. 4770, pp. 295–315. Springer, Heidelberg (2007)
14. Monagan, M., Pearce, R.: Sparse Polynomial Division Using a Heap. J. Symbolic. Comp. 46(7), 807–922 (2011)
15. Monagan, M., Pearce, R.: Parallel Sparse Polynomial Division Using Heaps. In: Proc. of PASCO 2010, pp. 105–111. ACM Press (2010)
16. Rowan, W.: Efficient Polynomial Substitutions of a Sparse Argument. ACM Sigsam Bulletin 15(3), 17–23 (1981)
17. Zeilberger, D.: The J.C.P. Miller recurrence for exponentiating a polynomial, and its q-analog. J. Difference Eqns and Appls 1(1), 57–60 (1995), http://www.math.rutgers.edu/~zeilberg/mamarim/mamarimPDF/power.pdf

Stability Conditions of Monomial Bases and Comprehensive Gröbner Systems

Katsusuke Nabeshima

Institute of Socio-Arts and Sciences,
University of Tokushima,
1-1 Minamijosanjima, Tokushima, 770-8502, Japan
nabeshima@tokushima-u.ac.jp

Abstract. A new stability condition of monomial bases is introduced. This stability condition is stronger than Kapur-Sun-Wang's one. Moreover, a new algorithm for computing comprehensive Gröbner systems, is also introduced by using the new stability condition. A number of segments generated by the new algorithm is smaller than that of segments of in Kapur-Sun-Wang's algorithm.

1 Introduction

First, in this paper, we introduce a new stability condition of monomial bases which is enhanced and stronger than the previous results. Second, we construct an algorithm for computing comprehensive Gröbner systems by using the new stability condition.

Comprehensive Gröbner systems for parametric ideals were introduced, constructed, and studied by Weispfenning [16] in 1992. After Weispfenning's paper was published, Kapur introduced an algorithm [8] for parametric Gröbner bases and Dolzman-Sturm implemented and published the software [3]. There was, however, no big development about comprehensive Gröbner systems (or bases) for ten years. Last ten years, the big developments were made by Kapur-Sun-Wang, Montes, Nabeshima, Sato, Suzuki, and Weispfenning [9,10,11,12,14,15,17].

Some of algorithms for computing comprehensive Gröbner systems are based on stability of Gröbner bases of ideals under specializations (Kalkbrener's results [7]). Each algorithm of them has a different "stability condition" of monomial bases. In 2010, Kapur-Sun-Wang discovered a wonderful stability condition [9] and constructed an algorithm for computing comprehensive Gröbner systems by using the stability condition. As Kapur-Sun-Wang's stability condition is stronger than Suzuki-Sato's [15] and Nabeshima's [12] ones, Kapur-Sun-Wang's algorithm works more efficient than them.

In this paper, we improve Kapur-Sun-Wang's algorithm by using the new strong stability condition. The main advantage of the new algorithm is that, it generates fewer segments compared to Kapur-Sun-Wang's algorithm [9].

The paper is organized as follows. Section 2 gives notations and definitions that will be used in this paper. Section 3 reviews Kapur-Sun-Wang's stability

V.P. Gerdt et al. (Eds.): CASC 2012, LNCS 7442, pp. 248–259, 2012.

condition and gives the new stability condition which is the main result. Section 4 describes an algorithm for computing comprehensive Gröbner systems by using the new stability condition and compares the new algorithm with Kapur-Sun-Wang's one.

2   Preliminary

We use the notation X as the abbreviation of n variables $X_1, \ldots, X_n$ and the notation A as the abbreviation of m variables $A_1, \ldots, A_m$. Let K and $\bar{K}$ be fields such that $\bar{K}$ is an algebraic closure field of K. $\mathrm{pp}(X)$, $\mathrm{pp}(A)$ and $\mathrm{pp}(A, X)$ are the sets of power products of X, A and $A \cup X$, respectively. $\prec_{A,X}$ is an admissible block order on $\mathrm{pp}(A, X)$ such that $A \ll X$. $\prec_X$ and $\prec_A$ are the restriction of $\prec_{A,X}$ on $\mathrm{pp}(X)$ and $\mathrm{pp}(A)$, respectively.

For a polynomial $f \in K[A][X]$ (polynomial ring over $K[A]$ in the variables X), the leading power product, leading coefficient and leading monomial of f w.r.t. the order $\prec_X$ are denoted by $\mathrm{lpp}_X(f), \mathrm{lc}_X(f)$ and $\mathrm{lm}_X(f)$, respectively. Since f can be regarded as an element of $K[A, X]$, in this case, the leading power product, leading coefficient and leading monomial of f w.r.t. the order $\prec_{A,X}$ are denoted by $\mathrm{lpp}_{A,X}, \mathrm{lc}_{A,X}$ and $\mathrm{lm}_{A,X}$, respectively. Let F be a subset of $K[A][X]$. We define $\mathrm{lc}_X(F) := \{\mathrm{lc}_X(f) | f \in F\}$ and $\mathrm{lpp}_X(F) := \{\mathrm{lpp}_X(f) | f \in F\}$.

$\mathbb{Q}$ and $\mathbb{C}$ are the field of rational numbers and the field of complex numbers, respectively. Angle brackets $\langle \cdot \rangle$ are defined as follows: let $f_1, \ldots, f_l \in R$ where R is a commutative ring with identity. Then, $\langle f_1, \ldots, f_l \rangle := \{\sum_{i=1}^{s} h_i f_i | h_1, \ldots, h_s \in R\}$.

For every $\bar{a} \in \bar{K}^m$, we can define the canonical specialization homomorphism $\sigma_{\bar{a}} : K[A] \to \bar{K}$ induce by $\bar{a}$, and we can naturally extend it to $\sigma_{\bar{a}} : K[A][X] \to \bar{K}[X]$. The image under σ of an ideal $I \in K[A][X]$ generates the extension $\sigma(I) := \{\sigma(f) | f \in I\} \subseteq \bar{K}[X]$.

For example, let $f = abx^2y + xy + ax + by + 2 \in \mathbb{Q}[a, b][x, y]$ and $(a, b) = (-2, 3), \left(0, \frac{1}{3}\right) \in \mathbb{Q}^2$. Then, $\sigma_{(-2,3)}(f) = -6x^2y + xy - 2x + 3y + 2$ and $\sigma_{\left(0, \frac{1}{3}\right)}(f) = \frac{1}{3}y + 2$. That is, we can regard $\sigma_{\bar{a}}$ as substituting $\bar{a}$ into m variables A.

For $f_1, \ldots, f_k \in K[A]$, $\mathbb{V}(f_1, \ldots, f_k) \subseteq \bar{K}^m$ denotes the affine variety of $f_1, \ldots, f_k$, i.e., $\mathbb{V}(f_1, \ldots, f_k) = \{\bar{a} \in \bar{K}^m | f_1(\bar{a}) = \cdots = f_k(\bar{a}) = 0\}$ and $\mathbb{V}(0) := \bar{K}^m$. In this paper, we use an algebraically constructible set that has a form $\mathbb{V}(f_1, \ldots, f_k) \backslash \mathbb{V}(g_1, \ldots, g_l) \subseteq \bar{K}^m$ where $f_1, \ldots, f_k, g_1, \ldots, g_l \in K[A]$.

Definition 1. We call an ideal $I \subseteq K[A][X]$ **stable** under the ring homomorphism σ and a term order $\prec_X$ if it satisfies

$$\sigma(\mathrm{lm}_X(I)) = \mathrm{lm}_X(\sigma(I))$$

where $\sigma(\mathrm{lm}_X(I)) := \{\sigma(\mathrm{lm}_X(f)) | f \in I\}$ and $\mathrm{lm}_X(\sigma(I)) := \{\mathrm{lm}_X(f) | f \in \sigma(I)\}$.

In several papers [1,5,4,6,7], the stability of Gröbner bases under specialization was studied. Stability conditions of this paper are based on the following theorem.

Theorem 1 (Kalkbrener [7]). Let σ be a ring homomorphism from $K[A]$ to $\bar{K}$, I an ideal in $K[A][X]$ and $G = \{g_1, \ldots, g_s\}$ a Gröbner basis of I w.r.t. a term order $\prec_X$. We assume that the g_i's are ordered in such a way that there exists an $r \in \{1, \ldots, s\}$ with $\sigma(\mathrm{lc}_X(g_i)) \neq 0$ for $i \in \{1, \ldots, r\}$ and $\sigma(\mathrm{lc}_X(g_i)) = 0$ for $i \in \{r+1, \ldots, s\}$. Then, the following three conditions are equivalent.
(1) I is stable under σ and $\prec_X$.
(2) $\{\sigma(g_1), \ldots, \sigma(g_r)\}$ is a Gröbner basis of $\sigma(I)$ w.r.t. the term order $\prec_X$.
(3) For every $i \in \{r+1, \ldots, s\}$, $\sigma(g_i)$ is reducible to 0 modulo $\{\sigma(g_1), \ldots, \sigma(g_r)\}$ in $\bar{K}[X]$.

Definition 2. Let I be an ideal in $K[A][X]$. If a monomial set $M = \{m_1, \ldots, m_l\} \subset \mathrm{pp}(X)$ satisfies the next two properties:
(1) there exists $\bar{a} \in \bar{K}^m$ such that $\mathrm{lm}_X(\sigma_{\bar{a}}(I)) = \langle m_1, \ldots, m_l \rangle$, and
(2) for all $i, j \in \{1, \ldots, l\}$ $(i \neq j)$, $m_i \nmid m_j$,
then M is called **"a specialized minimal leading monomial basis of I"**.

Let $I = \langle ax^2 + x, by + 1 \rangle$ be an ideal in $\mathbb{C}[a, b][x, y]$. In case $a = b = 1$, then $\sigma_{\{a=1, b=1\}}(I) = \langle x^2 + x, y + 1 \rangle$. Thus, $\{x^2, y\}$ is a specialized minimal leading monomial basis of I. Moreover, for $a = 0$ and $b \neq 0$ but not for $a \neq 0$ or for $b = 0$, $\{x, y\}$ is a specialized minimal leading monomial basis of I.

Definition 3 (stability conditions). Let I be an ideal in $K[A][X]$, $M \subset \mathrm{pp}(X)$ be a specialized leading monomial basis of I. Moreover, let $\mathbb{A} \subseteq \bar{K}$ be an algebraically constructible set. If for all $\bar{a} \in \mathbb{A}$, the canonical specialization homomorphism $\sigma_{\bar{a}}$ satisfies $\sigma_{\bar{a}}(\mathrm{lm}_X(I)) = \mathrm{lm}_X(\sigma_{\bar{a}}(I)) = \langle M \rangle$, then we call $\mathbb{A}$ a **stability condition** of M.

Let $I = \langle ax^2 + x, by + 1 \rangle$ be an ideal in $\mathbb{C}[a, b][x, y]$. Then, $\mathbb{C}^2 \setminus \mathbb{V}(ab)$ is a stability condition of $\{x^2, y\}$. In general, a stability condition of a monomial basis is not unique.

3   Stability Conditions of Monomial Bases

Here, we describe stability conditions of monomial bases. Let us fix the term order $\prec_X$ in $K[A][X]$, and let I be an ideal in $K[A][X]$. Then, we have a question.

When Does a Specialization Homomorphism σ Make I Stable?
Some of answers were given by Kapur-Sun-Wang, Nabeshima and Suzuki-Sato [9,12,15]. As Kapur-Sun-Wang's one is stronger than Suzuki-Sato's and Nabeshima's, we review Kapur-Sun-Wang's stability condition, first. Second, we give a new stability condition which is stronger than Kapur-Sun-Wang's one.

3.1   Kapur-Sun-Wang's Stability Condition

Definition 4 (Kapur-Sun-Wang [9]). Given a set G of polynomials which are a subset of $K[A, X]$ and $\prec_{A,X}$, let Noncomparable(G) be a subset, called F, of G such that (i) every polynomial $g \in G$ is such that $\mathrm{lpp}_X(g)$ is a multiple of $\mathrm{lpp}_X(f)$ for some $f \in F$, and further (ii) for any two distinct $f_1, f_2 \in F$, neither $\mathrm{lpp}_X(f_1)$ is a multiple of $\mathrm{lpp}_X(f_2)$ nor $\mathrm{lpp}_X(f_2)$ is a multiple of $\mathrm{lpp}_X(f_1)$.

It is easy to see that $\langle \mathrm{lpp}_X(\mathrm{Noncomparable}(G)\rangle = \langle \mathrm{lpp}_X(G)\rangle$ and a set Noncomparable(G) may **NOT be unique**. Kapur-Sun-Wang have introduced the following theorem.

Theorem 2 (Kapur-Sun-Wang [9]). Let G be a Gröbner basis of an ideal $I \subset K[A, X]$ w.r.t. $\prec_{A,X}$. Let $G_r = G \cap K[A]$ and $G_m = \mathrm{Noncomparable}(G \backslash G_r)$. Assume $G_m = \{g_1, \ldots, g_s\}$. Then, for all $\bar{a} \in \mathbb{V}(G_r) \backslash (\mathbb{V}(\mathrm{lc}_X(g_1)) \cup \mathbb{V}(\mathrm{lc}_X(g_2)) \cup \cdots \cup \mathbb{V}(\mathrm{lc}_X(g_s)))$, $\sigma_{\bar{a}}(G_m)$ is a Gröbner basis of $\sigma_{\bar{a}}(I)$ w.r.t. $\prec_X$ in $\bar{K}[X]$. (Remark that $\mathbb{V}(\mathrm{lc}_X(g_1)) \cup \cdots \cup \mathbb{V}(\mathrm{lc}_X(g_s)) = \mathbb{V}(\mathrm{lc}_X(g_1) \cdots \mathrm{lc}_X(g_s))$.)

By Theorem 1 and Theorem 2, we can easily obtain the following corollary.

Corollary 1. With the same notations in Theorem 2, then, a stability condition of $\mathrm{lpp}(\mathrm{Noncomparable}(G \backslash G_r))$ is $\mathbb{V}(G_r) \backslash (\mathbb{V}(\mathrm{lc}_X(g_1)) \cup \mathbb{V}(\mathrm{lc}_X(g_2)) \cup \cdots \cup \mathbb{V}(\mathrm{lc}_X(g_s)))$. (Clearly, in $K[A][X]$, $\mathrm{lpp}(\mathrm{Noncomparable}(G \backslash G_r))$ is a specialized minimal leading monomial basis of I.)

3.2 A New Stability Condition

Here, we give a new stability condition of monomial bases. In order to introduce the new condition, we need the following definition.

Definition 5. Given a set $F \subset K[A, X]$ and $\prec_{A,X}$, **the minimal basis of** $\mathrm{lpp}_X(F)$ is denoted by $\mathrm{MBlpp}(F)$, i.e., $\mathrm{MBlpp}(F) := \{\mathrm{lpp}_X(f) | \mathrm{lpp}_X(g) \nmid \mathrm{lpp}_X(f)$, for any $g \in F$ such that $\mathrm{lpp}_X(g) \neq \mathrm{lpp}_X(f)\}$. ($\mathrm{MBlpp}(F)$ is also the reduced Gröbner basis of $\langle \mathrm{lpp}_X(F)\rangle$ w.r.t. $\prec_X$ and a specialized minimal leading monomial basis of $\langle F \rangle$ in $K[A][X]$.)

It is obvious that $\mathrm{MBlpp}(F)$ is unique and $\langle \mathrm{MBlpp}_X(F)\rangle = \langle \mathrm{lpp}_X(F)\rangle = \langle \mathrm{lpp}_X(\mathrm{Noncomparable}(F))\rangle$. We give a simple example.

Let $F = \{ax^2 - y, ay^2 - 1, ax - 1, (a+1)x - y, (a+1)y - 1\} \subset \mathbb{Q}[a, x, y]$ with the lexicographic order such that $x \succ y \gg a$. Then, $F_1 = \{ax - 1, (a+1)y - a\}$ and $F_2 = \{(a+1)x - y, (a+1)y - a\}$ are both Noncomparable(F). $\{x, y\}$ is $\mathrm{MBlpp}(F)$. It is easy to verify that $\langle \mathrm{lpp}_X(F)\rangle = \langle \mathrm{lpp}_X(F_1)\rangle = \langle \mathrm{lpp}_X(F_2)\rangle = \langle \mathrm{MBlpp}(F)\rangle = \langle x, y\rangle$.

Now, we are ready to introduce a new stability condition. The following theorem is the main result of this paper.

Theorem 3. Let G be a Gröbner basis of an ideal $I \subset K[A, X]$ w.r.t. $\prec_{A,X}$. Let $G_r = G \cap K[A]$ and $\mathrm{MBlpp}(G \backslash G_r) = \{p_1, p_2, \ldots, p_s\}$. Assume $G_{p_i} = \{f \in G | \mathrm{lpp}_X(f) = p_i\}$ for each $i = 1, \ldots, s$. Then, for all $\bar{a} \in \mathbb{V}(G_r) \backslash (\mathbb{V}(\mathrm{lc}_X(G_{p_1}) \cup \mathbb{V}(\mathrm{lc}_X(G_{p_2}) \cup \cdots \cup \mathbb{V}(\mathrm{lc}_X(G_{p_s})))$, $\sigma_{\bar{a}}(G_{p_1} \cup G_{p_2} \cup \cdots \cup G_{p_s})$ is a Gröbner basis of $\sigma_{\bar{a}}(I)$ w.r.t. $\prec_X$ in $\bar{K}[X]$. (Remark that, for $V := \mathbb{V}(f_1, \ldots, f_s)$ and $W := \mathbb{V}(g_1, \ldots, g_t)$, $V \cup W = \mathbb{V}(f_i g_j : 1 \leq i \leq s, 1 \leq j \leq t)$.)

Proof. Since $\bar{a}$ is from $\mathbb{V}(G_r) \backslash (\mathbb{V}(\mathrm{lc}_X(G_{p_1}) \cup \cdots \cup \mathbb{V}(\mathrm{lc}_X(G_{p_s})))$, there exists $h_j \in G_{p_j}$ such that $\sigma_{\bar{a}}(\mathrm{lm}_X(h_j)) \neq 0$ for each $j = 1, \ldots, s$. Set $G = \{g_1, \ldots, g_l\}$ and $H = \{h_1, \ldots, h_s\}$. Then, $H \subseteq G$ and $\mathrm{MBlpp}(H) = \mathrm{MBlpp}(G \backslash G_r) = \{p_1, \ldots, p_s\}$. First, we proof the following claim.

Claim 1

"For all $\bar{a} \in \mathbb{V}(G_r) \backslash (\mathbb{V}(\mathrm{lc}_X(G_{p_1}) \cup \cdots \cup \mathbb{V}(\mathrm{lc}_X(G_{p_s})))$, I is stable under $\sigma_{\bar{a}}$ and $\prec_X$. That is, $\sigma_{\bar{a}}(\mathrm{lm}_X(I)) = \mathrm{lm}_X(\sigma_{\bar{a}}(I))$."

Proof of the claim: It is obvious that $\sigma_{\bar{a}}(\mathrm{lm}_X(I)) \subset \mathrm{lm}_X(\sigma_{\bar{a}}(I))$. We show the reverse inclusion. As G is a Gröbner basis of I w.r.t. $\prec_X$ in $K[A][X]$, it is enough to show that $\langle \sigma_{\bar{a}}(\mathrm{lm}_X(g_1)), \ldots, \sigma_{\bar{a}}(\mathrm{lm}_X(g_l)) \rangle \supset \mathrm{lm}_X(\sigma_{\bar{a}}(I))$. Let $f \in I$ with $\sigma_{\bar{a}}(f) \neq 0$. It suffices to show

"there exists g_i such that $\mathrm{lm}_X(g_i)$ divides $\mathrm{lm}_X(\sigma_{\bar{a}}(f))$ and $\sigma_{\bar{a}}(\mathrm{lc}_X(g_i)) \neq 0$, where $1 \leq i \leq l$." $(*)$

We do the proof by induction on $\prec_X$.

(Induction basis:) Let $\mathrm{lpp}_X(f)$ be the smallest element in $\mathrm{lpp}_X(I)$ w.r.t. $\prec_X$. If $\sigma_{\bar{a}}(\mathrm{lc}_X(f)) \neq 0$, then $\mathrm{lc}_X(\sigma_{\bar{a}}(f)) = \sigma_{\bar{a}}(\mathrm{lc}_X(f))$ is obvious. Thus, $\mathrm{lpp}_X(\sigma_{\bar{a}}(f)) = \mathrm{lpp}_X(f)$ is divisible by some elements of $\{\mathrm{lpp}_X(g_1), \ldots, \mathrm{lpp}_X(g_l)\}$. We assume that $\sigma_{\bar{a}}(\mathrm{lc}_X(f)) = 0$. Since $f \in I$, there exists $h \in H$ such that $\mathrm{lpp}_X(h) | \mathrm{lpp}_X(f)$ and $\sigma_{\bar{a}}(\mathrm{lc}_X(h)) \neq 0$. Then, for the h, defining

$$f' = \mathrm{lc}_X(h)f - \mathrm{lc}_X(f) \cdot \frac{\mathrm{lpp}_X(f)}{\mathrm{lpp}_X(h)} h,$$

we obtain $\mathrm{lpp}_X(f') \prec_X \mathrm{lpp}_X(f)$, $f' \in I$ and $\mathrm{lpp}_X(\sigma_{\bar{a}}(f')) = \mathrm{lpp}_X(\sigma_{\bar{a}}(f))$. As $\mathrm{lpp}_X(f)$ is the smallest element in $\mathrm{lpp}_X(I)$, f' must be 0. Since $\sigma_{\bar{a}}(\mathrm{lc}_X(h)) \neq 0$, thus $\sigma_{\bar{a}}(f) = 0$. This is a contradiction. Therefore, if $\sigma_{\bar{a}}(f) \neq 0$, then we have always $\sigma_{\bar{a}}(\mathrm{lc}_X(f)) \neq 0$.

(Induction step:) We assume that $(*)$ holds for polynomials whose leading power products are smaller than $\mathrm{lpp}_X(f)$ w.r.t. $\prec_X$. If $\sigma_{\bar{a}}(\mathrm{lc}_X(f)) \neq 0$, then $\mathrm{lc}_X(\sigma_{\bar{a}}(f)) = \sigma_{\bar{a}}(\mathrm{lc}_X(f))$ is obvious. Thus, $\mathrm{lpp}_X(\sigma_{\bar{a}}(f)) = \mathrm{lpp}_X(f)$ is divisible by some elements of $\{\mathrm{lpp}_X(g_1), \ldots, \mathrm{lpp}_X(g_l)\}$. We assume that $\sigma_{\bar{a}}(\mathrm{lc}_X(f)) = 0$. Since $f \in I$, there exists $h \in H$ such that $\mathrm{lpp}_X(h) | \mathrm{lpp}_X(f)$ and $\sigma_{\bar{a}}(\mathrm{lc}_X(h)) \neq 0$. Then, for the h, defining

$$f' = \mathrm{lc}_X(h)f - \mathrm{lc}_X(f) \cdot \frac{\mathrm{lpp}_X(f)}{\mathrm{lpp}_X(h)} h,$$

we obtain $\mathrm{lpp}_X(\sigma_{\bar{a}}(f')) = \mathrm{lpp}_X(\sigma_{\bar{a}}(f))$ and $\mathrm{lpp}_X(f') \prec_X \mathrm{lpp}_X(f)$. Hence, by the induction hypothesis, $\mathrm{lpp}_X(\sigma_{\bar{a}}(f'))$ is divisible by some elements of $\{\mathrm{lpp}_X(g_1), \ldots, \mathrm{lpp}_X(g_l)\}$. That is, $\mathrm{lpp}_X(\sigma_{\bar{a}}(f))$ is also divisible by them. Therefore, $(*)$ holds. $\qquad\square$

By Claim 1 and Theorem 1, $\sigma_{\bar{a}}(H)$ is a Gröbner basis of $\sigma_{\bar{a}}(I)$ w.r.t. $\prec_X$ in $\bar{K}[X]$. Assume that $U = \{u \in G_{p_1} \cup \cdots \cup G_{p_s} | \sigma_{\bar{a}}(\mathrm{lm}_X(u)) \neq 0\}$. Then, clearly $H \subseteq U \subset I$ and $\langle \mathrm{lm}_X(\sigma_{\bar{a}}(H)) \rangle = \langle \mathrm{lm}_X(\sigma_{\bar{a}}(U)) \rangle$. By Theorem 1 (3), for all $f \in (G_{p_1} \cup \cdots \cup G_{p_s}) \backslash U$, $\sigma_{\bar{a}}(f)$ is reducible to 0 module $\sigma_{\bar{a}}(H)$. Thus, $\sigma_{\bar{a}}(G_{p_1} \cup \cdots \cup G_{p_s})$ is also a Gröbner basis of $\sigma_{\bar{a}}(I)$ w.r.t. $\prec_X$ in $\bar{K}[X]$. $\qquad\square$

By Theorem 1 and Theorem 3, it is easy to see the following corollary.

Corollary 2. With the same notations in Theorem 3, a stability condition of $\mathrm{MBlpp}(G \backslash G_r)$ is $\mathbb{V}(G_r) \backslash (\mathbb{V}(\mathrm{lc}_X(G_{p_1}) \cup \cdots \cup \mathbb{V}(\mathrm{lc}_X(G_{p_s})))$.

Let compare Corollary 1 with Corollary 2. It is clear that $\mathrm{MBlpp}(G \backslash G_r) = \mathrm{lpp}(\mathrm{Noncomparable}(G \backslash G_r))$. The stability condition of Corollary 1 is $V := \mathbb{V}(G_r) \setminus (\mathbb{V}(\mathrm{lc}_X(g_1)) \cup \cdots \cup \mathbb{V}(\mathrm{lc}_X(g_s)))$, and the stability condition of Corollary 2 is $W := \mathbb{V}(G_r) \setminus (\mathbb{V}(\mathrm{lc}_X(G_{p_1})) \cup \cdots \cup \mathbb{V}(\mathrm{lc}_X(G_{p_s})))$. As obviously $\mathrm{lc}_X(g_i) \subseteq \mathrm{lc}_X(G_{p_i})$ for each $1 \le i \le s$, we have $\mathbb{V}(\mathrm{lc}_X(G_{p_i})) \subseteq \mathbb{V}(\mathrm{lc}_X(g_i))$. Therefore, $V \subseteq W$. This means that the stability condition of Corollary 2, is stronger than the condition of Corollary 1. The algebraically constructible set W is the new stability condition.

4 Comprehensive Gröbner Systems

In this section, we introduce an algorithm for computing comprehensive Gröbner systems. It is well-known that some of algorithm are based on stability conditions of monomial bases. So that an application of stability conditions is to compute comprehensive Gröbner systems.

Definition 6. Let F be a subset of $K[A][X]$, $\mathbb{A}_1, .., \mathbb{A}_l$ algebraically constructible subsets of $\bar{K}^m$ and $G_1, \ldots, G_l$ subsets of $K[A][X]$. Let $\mathbb{S}$ be a subset of $\bar{K}^m$ such that $\mathbb{S} \subseteq \mathbb{A}_1 \cup \cdots \cup \mathbb{A}_l$. A finite set $\mathcal{G} = \{(\mathbb{A}_1, G_1), \ldots, (\mathbb{A}_l, G_l)\}$ of pairs is called a **comprehensive Gröbner system** on $\mathbb{S}$ for F if $\sigma_{\bar{a}}(G_i)$ is a Gröbner basis of the ideal $\langle \sigma_{\bar{a}}(F) \rangle$ in $\bar{K}[X]$ for each $i = 1, \ldots, l$ and $\bar{a} \in \mathbb{A}_i$. Each $(\mathbb{A}_i, G_i)$ is called a **segment** of $\mathcal{G}$. We simply say $\mathcal{G}$ is a comprehensive Gröbner system for F if $\mathbb{S} = \bar{K}^m$.

In [9], an algorithm for computing comprehensive Gröbner systems is constructed by using the stability condition of Corollary 1. Now, Corollary 2 allows us to design a new algorithm for computing comprehensive Gröbner systems.

Algorithm CGSsmall
Input: (E, N, F): E, N, finite subsets of $K[A]$; F, a finite subset of $K[A, X]$.
Output: a finite set of 3-tuples (E_i, N_i, G_i) such that $((\mathbb{V}(E_i) \setminus \mathbb{V}(N_i)), G_i)$ constitute a comprehensive Gröbner system of F on $\mathbb{V}(E) \setminus \mathbb{V}(N)$.
BEGIN
1: **if** $\mathbb{V}(E) \setminus \mathbb{V}(N) = \emptyset$ **then** return $\{\}$; **end-if**;
2: $G \leftarrow \mathsf{ReducedGröbnerBasis}(\langle F \cup E \rangle, \prec_{A,X})$;
3: **if** $1 \in G$ **then** return $\{(E, N, \{1\})\}$; **end-if**;
4: $G_r \leftarrow G \cap K[A]$;
5: **if** $(\mathbb{V}(E) \setminus \mathbb{V}(G_r)) \setminus \mathbb{V}(N) = \emptyset$ **then**
6: $\quad$ $\mathcal{PGB} \leftarrow \{\}$; **else** $\mathcal{PGB} \leftarrow \{(E, G_r \wedge N, \{1\})\}$;
7: **end-if**;
8: **if** $\mathbb{V}(G_r) \setminus \mathbb{V}(N) = \emptyset$ **then** return $\mathcal{PGB}$; **else**
9: $\quad$ $\{p_1, \ldots, p_s\} \leftarrow \mathsf{MBlpp}(G \backslash G_r)$;
10: $\quad$ **for** $j = 1$ **to** s **do**
11: $\quad\quad$ $G_{p_j} \leftarrow \{g \in G \mid \mathrm{lpp}_X(g) = p_j\}$; $j \leftarrow j + 1$;
12: $\quad$ **end-for**;
13: $\quad$ **if** $(\mathbb{V}(G_r) \setminus \mathbb{V}(N)) \setminus (\mathbb{V}(\mathrm{lc}_X(G_{p_1})) \cup \cdots \cup \mathbb{V}(\mathrm{lc}_X(G_{p_s}))) \ne \emptyset$ **then**

14:      $\mathcal{PGB} \leftarrow \mathcal{PGB} \cup \{(E, N \wedge \mathrm{lc}_X(G_{p_1}) \wedge \cdots \wedge \mathrm{lc}_X(G_{p_s}), G_{p_1} \cup \cdots \cup G_{p_s})\}$;
15:      **end-if;**
16:      $\mathcal{PGB} \leftarrow \mathcal{PGB} \cup \mathrm{CGSsmall}(E \cup \mathrm{lc}_X(G_{p_1}), N, G \backslash G_r) \cup$
17:      $\mathrm{CGSsmall}(E \cup \mathrm{lc}_X(G_{p_2}), N \wedge \mathrm{lc}_X(G_{p_1}), G \backslash G_r) \cup$
18:      $\mathrm{CGSsmall}(E \cup \mathrm{lc}_X(G_{p_3}), N \wedge \mathrm{lc}_X(G_{p_1}) \wedge \mathrm{lc}_X(G_{p_2}), G \backslash G_r) \cup$
19:      $\cdots \cdots$
20:          $\cup \mathrm{CGSsmall}(E \cup \mathrm{lc}_X(G_{p_s}), N \wedge \mathrm{lc}_X(G_{p_1}) \wedge \cdots \wedge \mathrm{lc}_X(G_{p_{s-1}}), G \backslash G_r)$;
21:      return $\mathcal{PGB}$;
22: **end-if;**
END
(**Note that** $A \wedge B = \{fg | f \in A, g \in B\}$.)

The proof of the correctness of this algorithm, follows Theorem 3 and Kapur-Sun-Wang's algorithm [9]. The proof of the termination is the same as the proof of Suzuki-Sato's algorithm [15]. In this algorithm, we have deliberately avoided tricks and optimizations. These techniques are applicable to obtain small and nice outputs of a comprehensive Gröbner system. (See [9,14,15].)

Example 1. Let $F = \{ax^2 - xy + y^2, bxy + y, ax^2 - y, (b+1)xy^2 + ax\} \subset \mathbb{C}[a, b][x, y]$, where a, b, x, y variables. Fix the block order $\prec_{\{a,b\},\{x,y\}}$ such that $\mathrm{pp}(a, b) \ll \mathrm{pp}(x, y)$, $y \prec_{lex} x$ and $a \prec_{tdr} b$ where $\prec_{lex}$ is the lexicographic order and $\prec_{tdr}$ is the total degree reverse lexicographic order. A new algorithm works as follows.

The reduced Gröbner basis of $\langle F \rangle$ w.r.t. $\prec_{\{a,b\},\{x,y\}}$ is $G = \{(a + b^2 + b)y, ((b - 2)a + 1)y, (a^2 + 6a - b - 3)y, -y^2 + (a + 2b + 1)y, ax + by, xy + (-a - 2b - 2)y\}$.

(1) Since $\mathrm{MBlpp}(G) = \{x, y\}$, we obtain $G_x = \{ax + by\}$ and $G_y = \{(a + b^2 + b)y, ((b - 2)a + 1)y, (a^2 + 6a - b - 3)y\}$. Thus, $\mathrm{lc}_{\{x,y\}}(G_x) = \{a\}$ and $\mathrm{lc}_{\{x,y\}}(G_y) = \{a + b^2 + b, (b - 2)a + 1, a^2 + 6a - b - 3\}$. By Theorem 3, $(\mathbb{C}^2 \backslash (\mathbb{V}(a) \cup \mathbb{V}(a + b^2 + b, (b - 2)a + 1, a^2 + 6a - b - 3)), G_x \cup G_y)$ is a segment. In order to compute the comprehensive Gröbner system, next we consider two cases $[a = 0]$ and $[a + b^2 + b = 0, (b - 2)a + 1 = 0, a^2 + 6a - b - 3 = 0, a \neq 0]$.

(2) Let consider the case $[a = 0]$ (i.e., vanish all elements of $\mathrm{lm}_{\{x,y\}}(G_x)$). The reduced Gröbner basis of $\langle \mathrm{lc}_{\{x,y\}}(G_x) \cup G \rangle$ w.r.t. $\prec_{\{a,b\},\{x,y\}}$, is $G_1 = \{a, y\}$. Then, $G_{11} := G_1 \backslash (G_1 \cap \mathbb{C}[a, b]) = \{y\}$ and $\mathbb{V}(\mathrm{lc}_{\{x,y\}}(G_{11})) = \mathbb{V}(1) = \emptyset$. By Theorem 3, for all $\alpha \in \mathbb{V}(a)$, $\sigma_\alpha(G_{11})$ is a Gröbner basis of $\langle \sigma_\alpha(F) \rangle$. Since $\mathbb{V}(\mathrm{lc}_{\{x,y\}}(G_{11})) = \emptyset$, this branch does not have any sub-branches.

(3) Next, we consider the case $[a + b^2 + b = 0, (b - 2)a + 1 = 0, a^2 + 6a - b - 3 = 0, a \neq 0]$ (i.e., vanish all elements of $\mathrm{lm}_{\{x,y\}}(G_y)$). The reduced Gröbner basis of $\langle G \cup \mathrm{lc}_{\{x,y\}}(G_y) \rangle$ w.r.t. $\prec_{\{a,b\},\{x,y\}}$, is $G_2 = \{a + b^2 + b, (b - 2)a + 1, a^2 + 6a - b - 3, y^2 - (a + 2b + 1)y, x + (a + 3b)y\}$. Then, $G_{22} := G_2 \backslash (G_2 \cap \mathbb{C}[a, b]) = \{y^2 - (a + 2b + 1)y, x + (a + 3b)y\}$ and $\mathbb{V}(\mathrm{lc}_{\{x,y\}}(G_{22})) = \mathbb{V}(1, 1) = \emptyset$. By Theorem 3, we obtain a segment $(\mathbb{V}(a + b^2 + b, (b - 2)a + 1, a^2 + 6a - b - 3), G_{22})$ of a comprehensive Gröbner system of F. Since $\mathbb{V}(\mathrm{lc}_{\{x,y\}}(G_{22})) = \emptyset$, this branch does not have any sub-branches.

Thus, a comprehensive Gröbner system of F is :
$$\{(\mathbb{C}^2 \setminus (\mathbb{V}(a) \cup \mathbb{V}(a + b^2 + b, (b-2)a + 1, a^2 + 6a - b - 3)), G_x \cup G_y\}),$$
$$(\mathbb{V}(a), G_{11}), \quad (\mathbb{V}(a + b^2 + b, (b-2)a + 1, a^2 + 6a - b - 3), G_{22})\}.$$
This output has three segments.

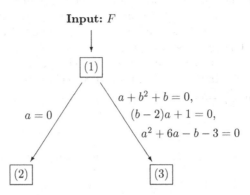

Fig. 1.

One of nice points of the new algorithm is to vanish all elements of $\mathrm{lm}_X(G_{p_i})$, if $\mathbb{V}(\mathrm{lc}_X(G_{p_i})) \neq \emptyset$. When the algorithm generates a new branch (see **Figure 1** (1) $\rightarrow$ (3)), by vanishing all elements of $\mathrm{lm}_X(G_{p_i})$, a polynomial whose leading power product is p_i, does not appear in the next reduced Gröbner basis, in many cases. Therefore, a number of reduced Gröbner basis computations (line 2) becomes small and a number of segments also becomes small.

In the algorithm, there are two main computation steps: one is Gröbner basis computation (line 2) and the other is checking consistency of parametric constraints $\mathbb{V}(E) \setminus \mathbb{V}(N)$ (line 1,5,8,13).

Let $N = \{f_1, \ldots, f_s\}$. Then, $\langle N \rangle \subseteq \sqrt{\langle E \rangle}$ if and only if $f_1, \ldots, f_s \in \sqrt{\langle E \rangle}$ where $\sqrt{\langle E \rangle}$ is a radical ideal of $\langle E \rangle$. If $\langle N \rangle \subseteq \sqrt{\langle E \rangle}$, then $\mathbb{V}(N) \supseteq \mathbb{V}(E)$. This means that $\mathbb{V}(E) \setminus \mathbb{V}(N)$ is an emptyset. By this fact, the next algorithm returns empty, if $\mathbb{V}(N) \supseteq \mathbb{V}(E)$, otherwise, returns nonempty.

(Checking consistency)

Input:(E, N): $E \subset K[A]$, $N = \{f_1, \ldots, f_s\} \subset K[A]$,
Output: empty or nonempty.
for each $i = 1, \ldots, s$ **do**
if $f_i \notin \sqrt{\langle E \rangle}$ **then** return nonempty; **end-if**;
end-for;
return empty;

In [9], Kapur-Sun-Wang gave great algorithms for checking "f is in the ideal generated by $\sqrt{\langle E \rangle}$". One can adapt these algorithms to check $f_i \notin \sqrt{\langle E \rangle}$.

In the new algorithm CGSsmall, checking consistency is costly in particular. This is because in many cases, the input N of CGSsmall has more than one polynomials. That is, in an algebraically constructible set $\mathbb{V}(E) \setminus \mathbb{V}(N)$, $\mathbb{V}(N)$ is defined by more than one polynomials. In Kapur-Sun-Wang's one, N has always one polynomial, essentially. (See [9].) Thus, in checking consistency of parametric constraints, Kapur-Sun-Wang's algorithm is more efficient than CGSsmall. Specially, line 13 of CGSsmall is quite costly. However, even if we remove line 13 and line 15 of CGSsmall (costly computation), then the algorithm still outputs a comprehensive Gröbner system. Note that this output probably has segments whose algebraically constructible sets are empty. This is one of techniques to obtain a comprehensive Gröbner system, quickly.

In order to avoid costly computation (for computation performance), we can also change line 13-15 of the algorithm CGSsmall as follows:

(Rough check)

Flag← 1;
for each $i = 1, \ldots, s$ **do**
if $\mathbb{V}(G_r) \backslash (\mathbb{V}(N) \cup \mathbb{V}(\mathrm{lc}_X(G_{p_i}))) = \emptyset$ **then** Flag← 0; break; **end-if**;
end-for;
if Flag=1 **then**
$\mathcal{PGB} \leftarrow \mathcal{PGB} \cup \{(G_r, N \wedge \mathrm{lc}_X(G_{p_1}) \wedge \cdots \wedge \mathrm{lc}_X(G_{p_s}), G_{p_1} \cup \cdots \cup G_{p_s})\}$;
end-if;

When we change line 13-15 of CGSsmall as the above, then we call the algorithm CGSsmall_R. The algorithm CGSsmall_R obviously outputs a comprehensive Gröbner system. Note that the outputs probably have segments whose algebraically constructible sets are empty. However, in fact, this "Rough check" works **quite well**. See Table 1 and Table 2.

Thus, in order to avoid costly computation, line 13-15 of CGSsmall can be changed into the following procedure. The following procedure outputs segments whose algebraically constructible sets are not empty.

(A new procedure in line 13-15)

Flag← 1;
for each $i = 1, \ldots, s$ **do**
if $\mathbb{V}(G_r) \backslash (\mathbb{V}(N) \cup \mathbb{V}(\mathrm{lc}_X(G_{p_i}))) = \emptyset$ **then** Flag← 0; break; **end-if**;
end-for;
if Flag=1 **then**
if $(\mathbb{V}(G_r) \backslash \mathbb{V}(N)) \backslash (\mathbb{V}(\mathrm{lc}_X(G_{p_1})) \cup \cdots \cup \mathbb{V}(\mathrm{lc}_X(G_{p_s}))) \neq \emptyset$ **then**
$\mathcal{PGB} \leftarrow \mathcal{PGB} \cup \{(G_r, N \wedge \mathrm{lc}_X(G_{p_1}) \wedge \cdots \wedge \mathrm{lc}_X(G_{p_s}), G_{p_1} \cup \cdots \cup G_{p_s})\}$;
end-if;
end-if;

When we change line 13-15 of CGSsmall as the above, then we call the algorithm CGSsmall, again. We adopt this algorithm CGSsmall for implementing.

The new algorithms CGSsmall and CGSsmall_R have been implemented in the computer algebra system Risa/Asir[13] by the author.

Table 1 shows a comparison of time of checking consistency on "line 13-15". This statistical data shows that how "Rough check" works well.

Table 2 shows a comparison of our implementation on Risa/Asir with Kapur-Sun-Wang's implementation on SINGULAR[2]. (One of [9]'s authors gave the SINGULAR implementation to the author.) Problems in Table 1 and Table 2, are picked from [9]. A list of problems is in the appendix.

A PC [CPU: Intel i7-2600 3.4 GHZ 3.4GHZ, Memory 4 GB RAM, OS: Windows 7 (64 bit)] was used. The time was given in second. The version of Risa/Asir and SINGULAR are 20110721 and 3-1-1, respectively.

Table 1. Time of checking consistency on line 13-15 (cpu sec.)

Problem	$F6$	$F8$	$E4$	$E5$	$S1$	$S2$	$S3$	$S4$	$S5$	$P3P$
with rough check	0	0	0	0.015	0	0.015	5.179	0	13.119	0.405
without rough check	0	5.226	0	0.015	0	0.015	17.674	0.062	23.774	32499.8

Table 2.

Problem	CA system	Algorithm	No. of segments	time (cpu sec.)
$F6$	SINGULAR	Kapur-Sun-Wang	7	0.187
	Risa/Asir	CGSsamll	4	0.015
	Risa/Asir	CGSsamll_R	4	0.015
$F8$	SINGULAR	Kapur-Sun-Wang	18	0.312
	Risa/Asir	CGSsmall	17	0.187
	Risa/Asir	CGSsmall_R	17	0.187
$E4$	SINGULAR	Kapur-Sun-Wang	8	0.078
	Risa/Asir	CGSsmall	7	0.015
	Risa/Asir	CGSsmall_R	7	0.015
$E5$	SINGULAR	Kapur-Sun-Wang	12	0.468
	Risa/Asir	CGSsmall	3	0.062
	Risa/Asir	CGSsmall_R	3	0.047
$S1$	SINGULAR	Kapur-Sun-Wang	7	0.078
	Risa/Asir	CGSsmall	6	0.016
	Risa/Asir	CGSsmall_R	6	0.015
$S2$	SINGULAR	Kapur-Sun-Wang	24	4.181
	Risa/Asir	CGSsmall	7	73.96
	Risa/Asir	CGSsmall_R	7	71.12
$S3$	SINGULAR	Kapur-Sun-Wang	26	2.901
	Risa/Asir	CGSsmall	12	9.173
	Risa/Asir	CGSsmall_R	12	3.495
$S4$	SINGULAR	Kapur-Sun-Wang	32	2.153
	Risa/Asir	CGSsmall	15	0.546
	Risa/Asir	CGSsmall_R	16	0.484
$S5$	SINGULAR	Kapur-Sun-Wang	36	50.248
	Risa/Asir	CGSsmall	21	28.69
	Risa/Asir	CGSsmall_R	21	17.58
$P3P$	SINGULAR	Kapur-Sun-Wang	36	2.418
	Risa/Asir	CGSsmall	12	4.337
	Risa/Asir	CGSsmall_R	12	4.119

As is evident from Table 2, the main advantage of the new algorithm is that, it generates fewer segments compared to Kapur-Sun-Wang's algorithm.

A set $\mathrm{MBlpp}(G \backslash G_i)$ in algorithm CGSsmall, has a lot of information of $\langle F \rangle$. For instance, when we compute dimensions of a parametric ideal, we directly need $\mathrm{MBlpp}(G_i)$ in each parametric constraint.

Remark that the new algorithm does not output minimal comprehensive Gröbner systems. (The definition of minimal comprehensive Gröbner systems is in [9].) After obtaining a comprehensive Gröbner system from the algorithm CGSsmall, it is possible to transform the comprehensive Gröbner system into a minimal comprehensive Gröbner system by applying the same idea of Kapur-Sun-Wang's algorithm. In this transformation, we do not need to compute a reduced Gröbner basis. We only need to compute reductions. Here, we do not describe the detail because of page restriction. The detail will appear elsewhere.

5   Concluding Remarks

A new stability condition of monomial bases was given. Moreover, a new algorithm for computing comprehensive Gröbner system was introduced.

If we adapt the "strong" stability condition (Corollary 2) to construct an algorithm for computing comprehensive Gröbner systems, then checking consistency of parametric constraints becomes costly. However, the number of reduced Gröbner bases computation become small. That is, the number of segments of a comprehensive Gröbner system is small. The new algorithm CGSsmall has these properties. In contrast, if we adapt Corollary 1 to construct the algorithm, then, checking consistency of parametric constraints is not costlier than the new algorithm CGSsmall. However, the number of reduced Gröbner bases computation, is bigger than the new algorithm. That is, the number of segments of a comprehensive Gröbner system is bigger than the new algorithm. This is Kapur-Sun-Wang's algorithm.

Acknowledgments. A part of this work has been supported by Grant-in-Aid for Young Scientists (B) (No. 22740065).

References

1. Becker, T.: On Gröbner bases under specialization. Applicable Algebra in Engineering, Communication and Computing 5, 1–8 (1994)
2. Decker, W., Greuel, G.-M., Pfister, G., Schönemann, H.: Singular 3-1-3, A computer algebra system for polynomial computations (2011)
3. Dolzmann, A., Sturm, T.: Redlog: Computer algebra meets computer logic. ACM SIGSAM Bulletin 31(2), 2–9 (1997)
4. Fortuna, E., Gianni, P., Trager, B.: Degree reduction under specialization. Journal of Pure and Applied Algebra 164(1), 153–163 (2001)
5. Gianni, P.: Properties of Gröbner bases under specializations. In: Davenport, J. (ed.) EUROCAL 1987, pp. 293–297. ACM Press (1987)
6. Kalkbrener, M.: Solving Systems of Algebraic Equations Using Gröbner Base. In: Davenport, J. (ed.) EUROCAL 1987. LNCS, vol. 378, pp. 293–297. Springer, Heidelberg (1987)

7. Kalkbrener, M.: On the stability of Gröbner bases under specializations. Journal of Symbolic Computation 24, 51–58 (1997)
8. Kapur, D.: An approach for solving systems of parametric polynomial equations. In: Saraswat, V., Hentenryck, P. (eds.) Principles and Practice of Constraint Programming, pp. 217–244. MIT Press (1995)
9. Kapur, D., Sun, Y., Wang, D.: A new algorithm for computing comprehensive Gröbner systems. In: Watt, S. (ed.) International Symposium on Symbolic and Algebraic Computation, pp. 29–36. ACM Press (2010)
10. Montes, A.: A new algorithm for discussing Gröbner basis with parameters. Journal of Symbolic Computation 33(1-2), 183–208 (2002)
11. Montes, A., Wibmer, M.: Gröbner bases for polynomial systems with parameters. Journal of Symbolic Computation 45(12), 1391–1425 (2010)
12. Nabeshima, K.: A speed-up of the algorithm for computing comprehensive Gröbner systems. In: Brown, C. (ed.) International Symposium on Symbolic and Algebraic Computation, pp. 299–306. ACM Press (2007)
13. Noro, M., Takeshima, T.: Risa/Asir- A computer algebra system. In: Wang, P. (ed.) International Symposium on Symbolic and Algebraic Computation, pp. 387–396. ACM Press (1992)
14. Suzuki, A., Sato, Y.: An alternative approach to comprehensive Gröbner bases. Journal of Symbolic Computation 36(3-4), 649–667 (2003)
15. Suzuki, A., Sato, Y.: A simple algorithm to compute comprehensive Gröbner bases using Gröbner bases. In: Dumas, J.-G. (ed.) International Symposium on Symbolic and Algebraic Computation, pp. 326–331. ACM Press (2006)
16. Weispfenning, V.: Comprehensive Gröbner bases. Journal of Symbolic Computation 14(1), 1–29 (1992)
17. Weispfenning, V.: Canonical comprehensive Gröbner bases. Journal of Symbolic Computation 36(3-4), 669–683 (2003)

Appendix

In the following sets of polynomials, x, y, z, w are main variables and $a, b, c, d, e, f,$ p, q, r are variables in coefficient rings. The term order is the total degree reverse lexicographic order such that $w \prec z \prec y \prec x$. (Theses are from [9].)

- $F6 = \{x^4 + ax^3 + bx^2 + cx + d, 4x^3 + 3ax^2 + 2bx + c\}$.
- $F8 = \{ax^2 + by, cw^2 + z, (x - z)^2 + (y - w)^2, 2dxw - 2by\}$.
- $E4 = \{(a - 1)z - b(x - 1), (a - 1)(x + 1) + bz, (a + 1)w - b(y + 1), (a + 1)(y - 1) + bw, (y - a)^2 + w^2 - (x - a)^2 - z^2\}$.
- $E5 = \{(x - a)^2 + (y - 1)^2 - a^2 - 1, (z + b)^2 + (w - 1)^2 - b^2 - 1, a(x - a) + (y - 1) + (a^2 + 1)d, -b(z + b) + (w - 1) + (1 + b^2)f, a(y - 1) - (x - a) + (a^2 + 1)c, -b(w - 1) - (z + b) + (b^2 + 1)e, xw - 2x - zy + 2z, d^2 + c^2 - 1, f^2 + e^2 - 1\}$.
- $S1 = \{ax^4 + cx^2 + y, bx^3 + x^2 + 2, cx^2 + dx + y\}$.
- $S2 = \{ax^3y + cxy^2 + bx + y, x^4y + 3dy, cx^2 + bxy, x^2y^2 + ax^2, x^5 + y^5\}$.
- $S3 = \{ax^2y + bx^2 + y^3, ax^2y + bxy + cy^2, ay^3 + bx^2y + cxy\}$.
- $S4 = \{x^4 + ax^3 + bx^2 + cxy + d, 4x^3 + 3ax^2y + 2bx + c + y\}$.
- $S5 = \{ax^2 + byz + czw, cw^2 + by + z, (x - z)^2 + (y - w)^2, 2dxw - 2byz\}$.
- $P3P = \{(1 - a)y^2 - ax^2 - py + arxy + 1, (1 - b)x^2 - by^2 - qx + brxy + 1\}$.

Parallel Reduction of Matrices in Gröbner Bases Computations

Severin Neumann

Fakultät für Mathematik und Informatik
Universität Passau, D-94030 Passau, Germany
neumans@fim.uni-passau.de

Abstract. In this paper we provide an parallelization for the reduction of matrices for Gröbner basis computations advancing the ideas of using the special structure of the reduction matrix [4]. First we decompose the matrix reduction in three steps allowing us to get a high parallelization for the reduction of the bigger part of the polynomials. In detail we do not need an analysis of the matrix to identify pivot columns, since they are obvious by construction and we give a rule set for the order of the reduction steps which optimizes the matrix transformation with respect to the parallelization. Finally we provide benchmarks for an implementation of our algorithm. This implementation is available as open source.

1 Introduction

Computing Gröbner bases is one of the major tasks of a computer algebra system, because they are required for several questions and tasks which one might like to solve. Since the complexity of computing Gröbner bases is double-exponential [8], there has been a lot of research on finding optimizations for Buchberger's algorithm and there are new algorithms like F4 [2], F5 [3] or GVW [5] being faster and more efficient. Another possibility of speeding up computations is parallelization meaning that many processor operations are carried out simultaneously. A sequential algorithm can be transformed into a parallel algorithm only if it is composed of independent steps and can be done in any order. In principle it is possible to write a parallel version of Buchberger's algorithm, but it has been shown that it is not speeding up in a suitable manner [11]. However Faugère's algorithm F4 is using matrix transformations which are known for being well parallelizable. Furthermore the special structures of the matrices occurring during reduction can be used to bring in even more parallelization [4]. Following we will present a method to compute an almost reduced row-echelon form of the matrix in parallel using this knowledge. Lastly we will show benchmarks of an implementation of our algorithm, available at https://github.com/svrnm/parallelGBC as open source.

2 Preliminaries

Without going into detail, computing Gröbner bases can be split up in three parts being called until all *critical pairs* are processed:

V.P. Gerdt et al. (Eds.): CASC 2012, LNCS 7442, pp. 260–270, 2012.
© Springer-Verlag Berlin Heidelberg 2012

1. Creating and updating the critical pairs of all elements of the basis using certain criteria. Usually the UPDATE function of Gebauer's and Möller's installation [6] does this best. Important for us is the property that there are no two critical pairs having the same least common multiple.
2. Selecting a subset of the critical pairs using a certain strategy. For this we prefer using the sugar cube strategy [7] allowing us the skip the SIMPLIFY heuristic of F4 and gain an equal or better performance [10].
3. Reducing the S-polynomials of the critical pairs until they are zero or top-irreducible. This means, none of the yet found elements of the Gröbner basis can reduce the leading term of the current S-polynomial. In the following we will only concentrate on this step.

So we assume that for a given input set $\{g_1, \ldots, g_m\} \subseteq P := K[x_1, \ldots, x_n]$, where K is a field, and a term ordering σ there is a set of critical pairs $B \subseteq \{(i, j) \mid i, j \in \{1, \ldots, m\}\}$ and a subset $B' \subseteq B$ being selected by the chosen strategy. In particular there is a set of polynomials $F := \{f_1, \ldots, f_s\} \subseteq P$, where s is a multiple of 2, with the following properties:

P1: Each two polynomials $f_k, f_{k+1} \in F$ and $k \in \{1, 3, \ldots, s-1\}$ are the minuend and subtrahend of the S-polynomial

$$S_{i,j} = t_{i,j} \cdot g_i - t_{j,i} \cdot g_j = f_k - f_{k+1}, \text{ with}$$
$$t_{i,j} = \frac{\mathrm{LT}_\sigma(g_j)}{\mathrm{LC}_\sigma(g_j) \cdot \gcd(\mathrm{LT}_\sigma(g_i), \mathrm{LT}_\sigma(g_j))}$$
$$t_{j,i} = \frac{\mathrm{LT}_\sigma(g_i)}{\mathrm{LC}_\sigma(g_i) \cdot \gcd(\mathrm{LT}_\sigma(g_i), \mathrm{LT}_\sigma(g_j))}$$

P2: Only those polynomials f_k, f_{k+1} have the same leading term with respect to the term ordering σ. As mentioned this is guaranteed by the UPDATE function.

P3: Each polynomial $f \in F$ has leading coefficient $LC_\sigma(f) = 1$. This is ensured by the division by the leading coefficients.

P4: Additionally we need reduction polynomials $R := \{r_1, \ldots, r_u\} \subseteq P$ having the property that for each term $t \in Support(F) \setminus LT_\sigma(F)$ there is exactly one polynomial $r \in R$ with $LT_\sigma(r) = t$. Thereby is ensured that by applying the reduction polynomials on F the resulting polynomials are zero or top-irreducible.

By writing the leading coefficients of each polynomial $h \in F' = F \cup R$ in a row in which each term is equivalent to a column, we obtain a reduction matrix. By transforming this matrix into row echelon form we reduce the S-polynomials of the current step into elements of the Gröbner basis – or zero. This is true, since the calculation of the S-polynomials f_k, f_{k+1} is done by subtracting the row k from the row $k+1$ and a reduction polynomial r is applied if $LT_\sigma(r) \in Support(h)$. This is equivalent to subtracting the row r from h. From now on we consider the following matrix M, where $\{t_1, \ldots, t_v\} := Support(F')$ and $c_{p,q}$ is the coefficient of t_q in the polynomial f_p or r_{p-s}:

$$
\begin{array}{c}
\begin{array}{cccc}
t_1 & t_2 & \cdots & t_v
\end{array} \\
\begin{array}{c}
f_1 \\ f_2 \\ \vdots \\ f_s \\ r_1 \\ r_2 \\ \vdots \\ r_u
\end{array}
\left(
\begin{array}{cccc}
c_{1,1} & c_{1,2} & \cdots & c_{1,v} \\
c_{2,1} & c_{2,2} & \cdots & c_{2,v} \\
\vdots & \vdots & \ddots & \vdots \\
c_{s,1} & c_{s,2} & \cdots & c_{s,v} \\
c_{s+1,1} & c_{s+1,2} & \cdots & c_{s+1,v} \\
c_{s+2,1} & c_{s+2,2} & \cdots & c_{s+2,v} \\
\vdots & \vdots & \ddots & \vdots \\
c_{s+u,1} & c_{s+u,2} & \cdots & c_{s+u,v}
\end{array}
\right)
\end{array}
$$

The following example demonstrates, how we can transform a set of polynomials into a reduction matrix.

Example 1. Let $K := \mathbb{F}_{32003}$ be the field with 32003 elements. It is chosen since our implementation of the parallelization is for finite fields and $\mathbb{F}_{32003}$ is also used for the benchmarks. Let $P := K[x_1, x_2, x_3]$ be a polynomial ring with three indeterminants over K. Let the three polynomials $g_1 = x_1^2 + x_2^2$, $g_2 = x_1 \cdot x_2 + x_2^2 + x_2 \cdot x_3$ and $g_3 = x_2^2 + x_3^2 + x_3$ form the Ideal $I := \langle g_1, g_2, g_3 \rangle$ for which a Gröbner basis with respect to the reverse degree lexicographic term-ordering (DegRevLex) should be computed. Using UPDATE we get the critical pairs $(2,3)$ and $(1,2)$ having the same sugar degree $sugar(g_2, g_3) = 3 = sugar(g_1, g_2)$, so we reduce the following polynomials in the first reduction:

$$
\begin{aligned}
f_{1,2} &= (x_1^2 + x_2^2) \cdot x_2 = x_1^2 \cdot x_2 + x_2^3 \\
f_{2,1} &= (x_1 \cdot x_2 + x_2^2 + x_2 \cdot x_3) \cdot x_1 = x_1^2 \cdot x_2 + x_1 \cdot x_2^2 + x_1 \cdot x_2 \cdot x_3 \\
f_{2,3} &= (x_1 \cdot x_2 + x_2^2 + x_2 \cdot x_3) \cdot x_2 = x_1 \cdot x_2^2 + x_2^3 + x_2^2 \cdot x_3 \\
f_{3,2} &= (x_2^2 + x_3^2 + x_3) \cdot x_1 = x_1 \cdot x_2^2 + x_1 \cdot x_3^2 + x_1 \cdot x_3
\end{aligned}
$$

and as set of reduction polynomials we obtain:

$$
\begin{aligned}
r_1 &= g_2 \cdot x_3 = x_1 \cdot x_2 \cdot x_3 + x_2^2 \cdot x_3 + x_2 \cdot x_3^2 \\
r_2 &= g_3 \cdot x_2 = x_2^3 + x_2 \cdot x_3^2 + x_2 \cdot x_3 \\
r_3 &= g_3 \cdot x_3 = x_2^2 \cdot x_3 + x_3^3 + x_3^2
\end{aligned}
$$

Finally we get the following matrix M:

	$x_1^2 x_2$	$x_1 x_2^2$	x_2^3	$x_1 x_2 x_3$	$x_2^2 x_3$	$x_1 x_3^2$	$x_2 x_3^2$	x_3^3	$x_1 x_3$	$x_2 x_3$	x_3^2
$f_{1,2}$	1	0	1	0	0	0	0	0	0	0	0
$f_{2,1}$	1	1	0	1	0	0	0	0	0	0	0
$f_{2,3}$	0	1	1	0	1	0	0	0	0	0	0
$f_{3,2}$	0	1	0	0	0	1	0	0	1	0	0
r_1	0	0	0	1	1	0	1	0	0	0	0
r_2	0	0	1	0	0	0	1	0	0	1	0
r_3	0	0	0	0	1	0	0	1	0	0	1

This matrix can be transformed into reduced row-echelon form and we receive the following matrix $\tilde{M}$:

$$
\begin{array}{c}
\\
\\
\\
\\
\\
g_4 = \\
g_5 =
\end{array}
\begin{pmatrix}
1 & 0 & 0 & 0 & 0 & 0 & 0 & 2 & 0 & 1 & 2 \\
0 & 1 & 0 & 0 & 0 & 0 & 0 & 1 & 0 & 1 & 1 \\
0 & 0 & 1 & 0 & 0 & 0 & 0 & -2 & 0 & -1 & -2 \\
0 & 0 & 0 & 1 & 0 & 0 & 0 & -3 & 0 & -2 & -3 \\
0 & 0 & 0 & 0 & 1 & 0 & 0 & 1 & 0 & 0 & 1 \\
0 & 0 & 0 & 0 & 0 & 1 & 0 & -1 & 1 & -1 & -1 \\
0 & 0 & 0 & 0 & 0 & 0 & 1 & 2 & 0 & 2 & 2
\end{pmatrix}
$$

The last rows of this matrix are elements of the Gröbner basis, since their leading terms are not contained in the leading term ideal of g_1, g_2, g_3. We set $g_4 = x_1 \cdot x_3^2 - x_3^3 + x_1 \cdot x_3 - x_2 \cdot x_3 - x_3^2$ and $g_5 = x_2 \cdot x_3^2 + 2 \cdot x_3^3 + 2 \cdot x_2 \cdot x_3 + 2 \cdot x_3^2$. In the following step we have to reduce the S-polynomials of the critical pairs $(1,4), (2,4), (3,5)$ and after computing the reduction polynomials we obtain the following new matrix M, which will be reduced in example 2 and 3 using the parallel method:

$$
\begin{pmatrix}
1 & 0 & 0 & -1 & 0 & 0 & 1 & -1 & 0 & -1 & 0 & 0 & 0 & 0 & 0 \\
1 & 0 & 1 & 0 & 0 & 0 & 0 & 0 & 0 & 0 & 0 & 0 & 0 & 0 & 0 \\
0 & 1 & 0 & 0 & -1 & 0 & 0 & 1 & -1 & 0 & -1 & 0 & 0 & 0 & 0 \\
0 & 1 & 1 & 0 & 1 & 0 & 0 & 0 & 0 & 0 & 0 & 0 & 0 & 0 & 0 \\
0 & 0 & 1 & 0 & 2 & 0 & 0 & 0 & 2 & 0 & 2 & 0 & 0 & 0 & 0 \\
0 & 0 & 1 & 0 & 0 & 1 & 0 & 0 & 0 & 0 & 0 & 1 & 0 & 0 & 0 \\
0 & 0 & 0 & 1 & 0 & -1 & 0 & 0 & 0 & 1 & -1 & -1 & 0 & 0 & 0 \\
0 & 0 & 0 & 0 & 1 & 2 & 0 & 0 & 0 & 0 & 2 & 2 & 0 & 0 & 0 \\
0 & 0 & 0 & 0 & 0 & 0 & 1 & 0 & 1 & 0 & 0 & 0 & 0 & 0 & 0 \\
0 & 0 & 0 & 0 & 0 & 0 & 1 & 1 & 1 & 0 & 1 & 0 & 0 & 0 & 0 \\
0 & 0 & 0 & 0 & 0 & 0 & 0 & 1 & 0 & 0 & 1 & 0 & 0 & 0 & 1 \\
0 & 0 & 0 & 0 & 0 & 0 & 0 & 0 & 1 & 0 & -1 & 1 & -1 & -1 \\
0 & 0 & 0 & 0 & 0 & 0 & 0 & 0 & 0 & 1 & 2 & 0 & 2 & 2
\end{pmatrix}
$$

3   Matrix Decomposition

As in [4] we take a detailed look on the special structure of the matrix M, before we will reduce the matrix in parallel. Reconsider property P2 and P4. If we look on one half of our S-polynomials, e.g. the minuends with odd index and on all reduction polynomials, we know that every polynomial of this selection has a distinct leading term. This means in the context of the matrix M that each row representing these polynomials has a distinct pivot column. After reordering the rows representing those polynomials $f_1, f_3, ..., f_{s-1}$ and $r_1, ..., r_u$ with respect to our term ordering σ we obtain an upper triangular sub-matrix A. In addition all entries have by property P3 in the pivot columns value 1. Putting the remaining rows of the S-polynomials in another sub-matrix B gives us the following matrix M', where $\underline{1}$ is the pivot of the column and $*$ a wild-card for any number:

$$
\begin{array}{c}
A \\
\\
\\
\\
\\
B \\
\\
\\
\end{array}
\left(
\begin{array}{ccccccccccccccc}
\underline{1} & * & \cdots & * & * & * & \cdots & * & * & \cdots & * & * & \cdots & * \\
0 & 0 & \cdots & 0 & \underline{1} & * & \cdots & * & * & \cdots & * & * & \cdots & * \\
0 & 0 & \cdots & 0 & 0 & 0 & \cdots & 0 & \underline{1} & \cdots & * & * & \cdots & * \\
\vdots & \vdots & \ddots & 0 & \vdots & \vdots & \ddots & \vdots & \vdots & \ddots & \vdots & \vdots & \ddots & \vdots \\
0 & 0 & \cdots & 0 & 0 & 0 & \cdots & 0 & 0 & \cdots & 0 & \underline{1} & \cdots & * \\
* & * & \cdots & * & * & * & \cdots & * & * & \cdots & * & * & \cdots & * \\
\vdots & \vdots & \ddots & \ddots & \vdots & \vdots & \ddots & * & \vdots & \ddots & \vdots & \vdots & \ddots & \vdots \\
* & * & \cdots & * & * & * & \cdots & * & * & \cdots & * & * & \cdots & *
\end{array}
\right)
$$

In the following section we will see, that this special structure of the matrix allows us to divide the reduction into separate steps which can be processed with more parallelism than with any parallel implementation of Gaussian elimination or other algorithm for generic matrices.

4   Parallel Reduction in Three Steps

The parallel reduction of the matrix can be split up in three steps. First we transform matrix A into reduced row echelon form. Reconsider that A is upper triangular. For this reason we do not need an analysis of the matrix to identify pivot columns, since they are obvious through the construction of A. Since this matrix contains one half of the polynomials F and all reduction polynomials R, this step processes the bigger part of the matrix and we will concentrate on the parallelization of this step. Secondly we will reduce the matrix B by the matrix A. Finally we will use a parallel version of the Gaussian elimination to bring the matrix B in reduced row echelon form. Those rows of B being not equal to zero and thus are irreducible with respect to the rows of A and the other rows of B are elements of the Gröbner basis. Remind that we use the sugar cube strategy and therefore we skip the SIMPLIFY step. This allows us to leave out an additional step which requires us to apply the reduced rows of B on the rows of A to obtain a reduced row echelon form of the whole reduction matrix M.

4.1   Step 1 – Reduction of the Upper Triangular Matrix

Reducing the matrix A can be done by the following sequential algorithm:

1. Set $i := n_A$, where n_A is index of the lowest row.
2. Set $j := i - 1$
3. If row j has an entry $a_{j,i} \neq 0$ in the pivot column of i, subtract row i $a_{j,i}$ times from row j.
4. Set $j := j - 1$ and if $j > 0$ go back to 3, if not go on.
5. Set $i := i - 1$ and if $i > 1$ go back to 2, if not return the reduced matrix $\tilde{A}$.

Step 3 ensures that after processing this algorithm there is no entry in the pivot column of every row in $\tilde{A}$ and so this matrix is in reduced row-echelon form.

For an implementation we will need the entries of all columns having a pivot only as an *operation* with the information that a row j has an entry $a_{j,i}$ in column i, which has to be reduced. Therefore these rows of our matrices A and B are stored in coordinate list format meaning that an entry is represented by a tuple $(j, i, a_{j,i})$.

The inner loop of this algorithm can be computed in parallel, since each j is distinct and $j \neq i$ in every step. We say all operations $(j_p, i, a_{j_p,i})$ with $k \neq i$ can be applied in parallel, after all operations $(j_q, i-1, a_{j_q,i})$ have been applied. We can rewrite the sequential algorithm to:

1. Set $i := n_A$, where n_A is index of the lowest row.
2. For all $j \in 1, \ldots, i-1$, if row j has an entry $a_{j,i} \neq 0$ in the pivot column of i, subtract row i $a_{j,i}$ times from row j.
3. Set $i := i-1$ and if $i > 1$ go back to 2, if not return the reduced matrix $\tilde{A}$.

If no entry above the diagonal is zero, every step of the outer loop has $n_A - i$ parallel steps. Mostly the matrix is not dense, so the number of parallel operations is clearly smaller. However the given set of rules can be modified by the following: two operations $(j, i-1, a_{j,i-1})$ and $(k, i, a_{k,i})$ can be applied in parallel, if there is no operation $(i, i-1, a_{i,i-1})$. This allows to apply an operation $(j, i, a_{j,i})$ just at the time the row i is reduced.

By this the ordering of all operations is not certain before runtime. Therefore an analysis of the matrix is required. This can be done during generation. The following example will show the difference between the two sets of rules.

Example 2. A part of matrix M of example 1 will be transformed in the following matrix A on which the explained step will be applied. Thereby the pivot columns are highlighted:

$$
\begin{pmatrix}
\mathbf{1} & 0 & 0 & -1 & 0 & 0 & 1 & -1 & 0 & -1 & 0 & 0 & 0 & 0 & 0 \\
0 & \mathbf{1} & 0 & 0 & -1 & 0 & 0 & 1 & -1 & 0 & -1 & 0 & 0 & 0 & 0 \\
0 & 0 & \mathbf{1} & 0 & 2 & 0 & 0 & 0 & 2 & 0 & 2 & 0 & 0 & 0 & 0 \\
0 & 0 & 0 & \mathbf{1} & 0 & -1 & 0 & 0 & 0 & 1 & -1 & -1 & 0 & 0 & 0 \\
0 & 0 & 0 & 0 & \mathbf{1} & 2 & 0 & 0 & 0 & 0 & 2 & 2 & 0 & 0 & 0 \\
0 & 0 & 0 & 0 & 0 & 0 & \mathbf{1} & 0 & 1 & 0 & 0 & 0 & 0 & 0 & 0 \\
0 & 0 & 0 & 0 & 0 & 0 & 0 & \mathbf{1} & 1 & 0 & 1 & 0 & 0 & 0 & 0 \\
0 & 0 & 0 & 0 & 0 & 0 & 0 & 0 & \mathbf{1} & 0 & 0 & 1 & 0 & 0 & 1 \\
0 & 0 & 0 & 0 & 0 & 0 & 0 & 0 & 0 & \mathbf{1} & 0 & -1 & 1 & -1 & -1 \\
0 & 0 & 0 & 0 & 0 & 0 & 0 & 0 & 0 & 0 & \mathbf{1} & 2 & 0 & 2 & 2
\end{pmatrix}
$$

To reduce this matrix into row-echelon form the following operations have to be applied, if the unmodified set of rules is used:

$$(2, 10, -1) \,\|\, (3, 10, 2) \,\|\, (4, 10, -1) \,\|\, (5, 10, 2)$$
$$(1, 9, -1) \,\|\, (4, 9, 1)$$
$$(2, 8, -1) \,\|\, (3, 8, 2) \,\|\, (6, 8, 1) \,\|\, (7, 8, 1)$$
$$(1, 7, -1) \,\|\, (2, 7, 1)$$
$$(1, 6, 1)$$
$$(2, 5, -1) \,\|\, (3, 5, 2)$$
$$(1, 4, -1)$$

In doing so each line stands for a list of operations which can be applied in parallel. If we use the modification, the following order will be used:

$$(1, 9, -1) \,\|\, (2, 10, -1) \,\|\, (3, 10, 2) \,\|\, (4, 10, -1) \,\|\, (5, 10, 2) \,\|\, (6, 8, 1) \,\|\, (7, 8, 1)$$
$$(1, 7, -1) \,\|\, (2, 8, -1) \,\|\, (3, 8, 2) \,\|\, (4, 9, 1)$$
$$(1, 6, 1) \,\|\, (2, 7, 1)$$
$$(1, 4, -1) \,\|\, (2, 5, -1) \,\|\, (3, 5, 2)$$

The maximal count of parallel operations is raised from four to seven and all operations are completed after four steps instead of seven. Afterwards we obtain the following reduced matrix $\tilde{A}$:

$$
\begin{pmatrix}
\mathbf{1} & 0 & 0 & 0 & 0 & -1 & 0 & 0 & 0 & 0 & 0 & -1 & 0 & 0 & 0 \\
0 & \mathbf{1} & 0 & 0 & 0 & 2 & 0 & 0 & 0 & 0 & 0 & 4 & 0 & 0 & 2 \\
0 & 0 & \mathbf{1} & 0 & 0 & -4 & 0 & 0 & 0 & 0 & 0 & -2 & 0 & 4 & 2 \\
0 & 0 & 0 & \mathbf{1} & 0 & -1 & 0 & 0 & 0 & 0 & 0 & 2 & -1 & 3 & 3 \\
0 & 0 & 0 & 0 & \mathbf{1} & 2 & 0 & 0 & 0 & 0 & 0 & -2 & 0 & -4 & -4 \\
0 & 0 & 0 & 0 & 0 & 0 & \mathbf{1} & 0 & 0 & 0 & 0 & -1 & 0 & 0 & -1 \\
0 & 0 & 0 & 0 & 0 & 0 & 0 & \mathbf{1} & 0 & 0 & 0 & -3 & 0 & -2 & -3 \\
0 & 0 & 0 & 0 & 0 & 0 & 0 & 0 & \mathbf{1} & 0 & 0 & 1 & 0 & 0 & 1 \\
0 & 0 & 0 & 0 & 0 & 0 & 0 & 0 & 0 & \mathbf{1} & 0 & -1 & 1 & -1 & -1 \\
0 & 0 & 0 & 0 & 0 & 0 & 0 & 0 & 0 & 0 & \mathbf{1} & 2 & 0 & 2 & 2
\end{pmatrix}
$$

4.2 Step 2 – Apply $\tilde{A}$ on B

At this point we have to reduce the matrix B by the matrix $\tilde{A}$. Because the rows of $\tilde{A}$ are independent, the reduction of matrix B can be done in parallel for all n_B rows of B:

1. For all $j \in 1, \ldots, n_B$, if row j of B has an entry $b_{j,i}$ in the pivot the column of row i of A, subtract row i from j $b_{j,i}$ times.
2. Return the reduced matrix $\tilde{B}$

Later we will show that we can combine the previous and this step. We split them up to be more comprehensible.

4.3   Step 3 – Parallel Gaussian Elimination of $\tilde{B}$

At this point matrix A can be discarded, because every row has a pivot column being equivalent to a leading term which is already contained in the leading term ideal of the Gröbner basis. So the reduction of B remains. Because this matrix has no specific structure, a parallel Gaussian elimination [12] computes the reductions. Finally the rows not equal to zero in all columns are elements of the Gröbner basis.

Example 3. At example 2 we finished with reducing matrix A to $\tilde{A}$. This matrix can now be applied to the remaining rows of B:

$$\begin{pmatrix} 1\,0\,1\,0\,0\,0\,0\,0\,0\,0\,0\,0\,0\,0\,0 \\ 0\,1\,1\,0\,1\,0\,0\,0\,0\,0\,0\,0\,0\,0\,0 \\ 0\,0\,1\,0\,0\,1\,0\,0\,0\,0\,0\,1\,0\,0\,0 \end{pmatrix}$$

The following operations are applied. The numbers 1 to 10 are representing the rows in $\tilde{A}$ and 11 to 13 the rows in B:

$$(11, 1, 1) \,\|\, (12, 2, 1) \,\|\, (13, 3, 1)$$
$$(11, 3, 1) \,\|\, (12, 3, 1)$$
$$(12, 5, 1)$$

Now we obtain the following matrix B':

$$\begin{pmatrix} 0\,0\,0\,0\,0\,5\,0\,0\,0\,0\,3\,0\,0\,-4\,-2 \\ 0\,0\,0\,0\,0\,0\,0\,0\,0\,0\,0\,0\,0\,\,0\,\,\,0 \\ 0\,0\,0\,0\,0\,5\,0\,0\,0\,0\,3\,0\,0\,-4\,-2 \end{pmatrix}$$

By using (parallel) Gaussian elimination the third row gets removed and the first row gets normalized. So finally the new element of the Gröbner basis is $g_6 = x_3^4 - 6400 \cdot x_3^3 - 12802 \cdot x_2 \cdot x_3 - 6401 \cdot x_3^2$. After one step more the algorithm does not find any further element and terminates.

4.4   Merging Step 1 and 2

As mentioned before step 1 and step 2 can be merged. It is not required to transform the whole matrix A, before step 2 can be processed, because a single row of A can already be applied on the matrix B if there are no further operations on it. Thus the algorithm does not need synchronization between step 1 and 2.

5   Benchmarks

The presented algorithm has been implemented using C++ as programming language and the application interface OpenMP as well as "Streaming SIMD

Extensions" (SSE) were used for parallelization. The latter can be used since the non-pivot columns of each row are stored in a continuous vector. The code is available as open source at `https://github.com/svrnm/parallelGBC`.

This project also includes all input files of the following tests. This gives everyone the possibility to reproduce the tests on a different platform and to compare the presented algorithm with other implementations. The tests were computed on a system with 48 AMD Opteron^{TM} 6172 processors and 64 gigabyte of main memory.

To compare our implementation with existing implementations of Gröbner basis algorithms we used the open source computer algebra systems Singular 3-1-3 and CoCoALib 0.9949.

As examples we took the commonly used polynomial systems Katsura-12, Katsura-13 [9], Cyclic-8 and Cyclic-9 [1] over the field with 32003 elements ($\mathbb{F}_{32003}$) and the degree reverse lexicographic term ordering (DegRevLex).

Table 1. Computation time of the given input system in seconds

Input	CoCoALib	Singular	Our implementation					
# of processors	1	1	1	2	4	8	16	32
Katsura-12	12401	1543	345	199	125	89	69	62
Katsura-13	202620	crashed	2468	1478	861	552	394	324
Cyclic-8	611	81	52	39	31	26	23	22
Cyclic-9	182888	23162	10663	5990	3437	2092	1410	1116

Table 1 shows that our implementation is a lot faster than existing comparable implementations of Gröbner basis algorithms in the sequential case. Beyond that

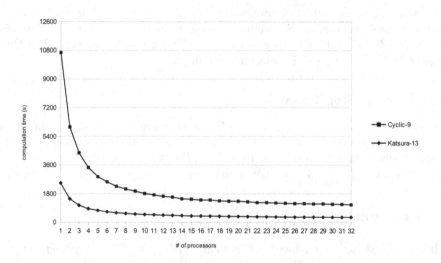

Fig. 1. Decreasing speedup of Cyclic-9 and Katsura-13 with many processors

the parallelization speeds up the computations of the examples by a factor of two to nine if 32 processors are used. Especially the harder problems can be solved a lot faster. This is due to the fact that in these cases the reduction takes up a larger part of the computation time.

Figure 1 illustrates that for each input there is a maximum amount of processors increasing the speedup in a measurable way. While Cyclic-9 can be computed still faster using 16 processors, the limit for Katsura-13 is reached with eight processors. One reason for that is that there are still the two other steps of the algorithm left which are consuming a relatively larger amount of computation time. Another reason is that to a certain amount the possibility of parallelization is exhausted. To speed up even more, further optimizations for all steps of the algorithm must be found.

6 Conclusion

We presented an parallel implementation of an Gröbner basis algorithm which is freely available for use. We compass to use it to solve further difficult problems. For this purpose we intend to further optimize the speedup and to decrease particularly the memory consumption which currently is the reason why we are unable to compute for example the Gröbner basis of Cyclic-10.

References

1. Björck, G., Haagerup, U.: All cyclic p-roots of index 3, found by symmetry-preserving calculations (2008)
2. Faugére, J.-C.: A new efficient algorithm for computing Gröbner bases (F4). Journal of Pure and Applied Algebra 139(1-3), 61–88 (1999)
3. Faugére, J.-C.: A new efficient algorithm for computing Gröbner bases without reduction to zero (F5). In: Proceedings of the 2002 International Symposium on Symbolic and Algebraic Computation, ISSAC 2002, New York, NY, USA, pp. 75–83 (2002)
4. Faugére, J.-C., Lachartre, S.: Parallel Gaussian Elimination for Gröbner bases computations in finite fields. In: Proceedings of the 4th International Workshop on Parallel and Symbolic Computation, PASCO 2010, New York, USA, pp. 89–97 (July 2010)
5. Gao, S., Volny IV, F., Wang, M.: A New Algorithm for Computing Gröbner Bases (2010)
6. Gebauer, R., Michael Möller, H.: On an installation of Buchberger's algorithm. Journal of Symbolic Computation 6, 275–286 (1988)
7. Giovini, A., Mora, T., Niesi, G., Robbiano, L., Traverso, C.: One sugar cube, please or selection strategies in the Buchberger algorithm. In: Proceedings of the 1991 International Symposium on Symbolic and Algebraic Computation, ISAAC 1991, New York, USA, pp. 49–54 (1991)
8. Huynh, D.T.: A superexponential lower bound for Gröbner bases and Church-Rosser Commutative Thue systems. Inf. Control 68, 196–206 (1986)
9. Katsura, S., Fukuda, W., Inawashiro, S., Fujiki, N., Gebauer, R.: Distribution of effective field in the ising spin glass of the $\pm J$ model at $T = 0$. Cell Biochemistry and Biophysics 11, 309–319 (1987)

10. McKay, C.E.: An analysis of improvements to Buchberger's algorithm for Gröbner basis computation. Master thesis, University of Maryland, USA (2004)
11. Ponder, C.G.: Evaluation of "performance enhancements" in algebraic manipulation systems. PhD thesis. University of California, USA (1988)
12. Quinn, M.J.: Parallel Programming in C with MPI and OpenMP. McGraw-Hill Education Group (September 2003)

Real and Complex Polynomial Root-Finding
by Means of Eigen-Solving

Victor Y. Pan[1,2,*], Guoliang Qian[2], and Ai-Long Zheng[2]

[1] Department of Mathematics and Computer Science
Lehman College of the City University of New York
Bronx, NY 10468 USA
victor.pan@lehman.cuny.edu,
http://comet.lehman.cuny.edu/vpan/
[2] Ph.D. Programs in Mathematics and Computer Science
The Graduate Center of the City University of New York
New York, NY 10036 USA
gqian@gc.cuny.edu, azheng-1999@yahoo.com

Abstract. Our new numerical algorithms approximate real and complex roots of a univariate polynomial lying near a selected point of the complex plane, all its real roots, and all its roots lying in a fixed half-plane or in a fixed rectangular region. The algorithms seek the roots of a polynomial as the eigenvalues of the associated companion matrix. Our analysis and experiments show their efficiency. We employ some advanced machinery available for matrix eigen-solving, exploit the structure of the companion matrix, and apply randomized matrix algorithms, repeated squaring, matrix sign iteration and subdivision of the complex plane. Some of our techniques can be of independent interest.

Keywords: Root-finding, Eigen-solving, Randomization, Matrix sign function.

1 Introduction

Univariate polynomial root-finding is a classical problem of mathematics and computational mathematics, having important applications to modern computing. It is a fundamental problem of computer algebra. New effective solution algorithms, in particular, numerical iterations are welcome by users.

Frequently only the real roots of a polynomial are required, being much less numerous than all its roots, but the best numerical subroutines (such as MP-Solve) approximate all real roots about as fast (and as slow) as all complex roots. The preceding work on numerical real polynomial root-finding, however, seems to be limited to [10] and two sections of [11], whose techniques are not used in our present paper.

A number of recent successful root-finders for a polynomial $p = p(x)$ of a degree n rely on the reduction of the root-finding task to eigen-solving for the

* Supported by NSF Grant CCF-1116736 and PSC CUNY Award 64512–0042.

V.P. Gerdt et al. (Eds.): CASC 2012, LNCS 7442, pp. 271–282, 2012.
© Springer-Verlag Berlin Heidelberg 2012

associated $n \times n$ companion matrix C_p. This enriches root-finding methods with the highly developed numerical techniques for matrix eigen-solving. E.g., the modified Rayleigh Quotient iterations of [11] quadratically converge to an *eigenpair* made up of a simple eigenvalue λ of C_p and the associated eigenvector provided the ratios $|\lambda^{(0)} - \lambda| / |\lambda^{(0)} - \mu|$ are substantially less than one for an initial approximation $\lambda^{(0)}$ to λ and for all other eigenvalues μ of C_p.

Hereafter we use the acronym RQ for Rayleigh Quotient.

In fact the RQ iteration can be extended to the case where an eigenvalue λ of C_p has multiplicity $r > 1$ or belongs to an isolated cluster of r simple eigenvalues of C_p. E.g., the Inverse Orthogonal iterations [5, page 339] (resp. Orthogonal iterations) quadratically converge to the eigenspace defined by the m absolutely smallest (resp. largest) eigenvalues of a nonsingular matrix C_p provided that the other eigenvalues dominate (resp. are dominated), that is have substantially larger (resp. smaller) absolute values. If the domination is strong enough, close approximations to these eigenvalues are obtained already in single iteration. In Section 3 we detail some of such general eigen-solving techniques and specify them for the companion matrix C_p.

These techniques simplify eigen-solving and root-finding where we know that the matrix C_p has a small number r of dominant or dominated eigenvalues, or even if some polynomial or well defined rational matrix function $F = F(C_p)$ has this property. In the latter case we can approximate the r eigenvalues of the matrix F and the associated eigenspace of dimension r. This eigenspace is shared by the matrices F and C_p, and we can use its basis to compute a $r \times r$ auxiliary matrix that shares its r eigenvalues with C_p. The reduction of the problem size is dramatic where the ratio n/r is large.

If required one can approximate the remaining eigenvalues by using deflation or by reapplying the same techniques to the shifted matrices $C_p - sI$ and their inverses for proper scalars s and the identity matrix I.

Paper [11] proposes an algorithm that maps the real line onto the unit circle centered at the origin, then isolates this circle from the images of nonreal roots of $p(x)$, approximates the images of real roots by applying the RQ iterations or their amendment, and finally recovers the real eigenvalues of the input matrix.

As an immediate application of the above extension of the RQ iteration, we can simplify the last stage of the latter real root-finder of [11]. Instead of at least r invocations of RQ iterations used in [11] for computing the r real roots of $p(x)$, we can just once apply the eigen-solver to a single auxiliary $r \times r$ matrix. Besides observing this, we propose three alternatives to the algorithms of [11].

1. Apply repeated squaring of C_p instead of RQ iterations (see Remark 3 in Section 3). High powers $C_p^{2^h}$ tend to have small sets of dominant eigenvalues, and so we can first compute the respective eigenspace shared by the matrices $C_p^{2^h}$ for all h and then compute the associated eigenvalues of C. For repeated squaring we apply the algorithm of [9], extending the one of [3], but we simplify the recovery of the eigenvalues from the eigenspace by using the above recipe. Every RQ iteration uses $O(n)$ *ops* for an $n \times n$ companion matrix C_p (ops is our abbreviation for arithmetic operations), versus $O(n \log n)$ ops per each squaring, but the RQ

process outputs a single eigenvalue versus r eigenvalues of repeated squaring, and unlike the RQ iterations, repeated squaring needs no initial approximations to eigenvalues. By applying this approach to the inverse matrices $(C_p - sI)^{-1}$ we extend it to the approximation of the eigenvalues of C_p lying near the selected shift values s.

2. Like papers [3] and [2], we employ the matrix sign classical iterations [6] to polynomial root-finding by using $O(n \log n)$ ops per iteration, but unlike [3] and [2], we avoid numerical stability problems by combining the iterations with the subdivision techniques (see the end of Section 4).

3. As our another novelty, we enforce domination of the images of the real roots of C_p over all other images of the roots of $p(x)$. In this approach we use random shifts of the matrix C_p to avoid singularities.

Some of our techniques can be of independent interest, e.g., our extension of the matrix sign iteration to real eigen-solving. As our another technical novelty, we simplify the computation of the bases of dominant eigenspaces by applying randomization (see Remark 4 in Section 3)). This application is supported by our general estimates of [12] for the condition numbers of randomized matrix products. Our generic randomized computation of such a basis involves $O(rn \log n)$ ops versus order of rn^2 in standard algorithms. Furthermore in a heuristic variation supported empirically we use $O(n \log n)$ ops.

Our analysis and experiments show effectiveness of the proposed real and complex polynomial root-finders.

Our root-finders are numerical; they allow rounding errors. The respective error and perturbation analysis is quite involved but also well developed [5], [13]. We can substantially simplify the analysis of our algorithms where we deal with a small set of dominant or dominated eigenvalues (see Remark 2 in Section 3).

In the presence of rounding errors our real numerical root-finders treat together real and nearly real roots x_j such that $|\Re(x_j)/x_j| \leq \epsilon$ (for a fixed tolerance ϵ and $\Re(x)$ denoting the real part of a complex number x). Indeed we cannot distingush between these two classes of roots in the presence of rounding errors. So we approximate all of them and at the end readily select the real roots among them. Our resulting real root-finders are effective as long as both real and nearly real roots together are less numerous than the other roots.

Further advance may rely on more intricate maps of the complex plane and on the combination with other polynomial root-finders, e.g., the Rayleigh Quotient iteration [5], [11], Newton's iteration (both can be concurrently applied at distinct initial points where *no communication between the processors is required*), and nonnumerical real polynomial root-finders, namely, subdivision and continued fraction methods (see [4], [14], and the bibliography therein). These successful algorithms can supply auxiliary information for our computations (e.g., the number of real roots and their bounds) and can handle the inputs that are hard for our numerical treatment.

We organize our paper as follows. We recall some definitions and auxiliary results in the next section. In Section 3, we cover some methods for the approximation of dominant and dominated eigenvalues as well as the eigenvalues lying near a selected real or complex point. We discuss these methods for both general and companion matrices and comment on some numerical issues and randomization techniques. We also comment on repeated squaring for obtaining the desired domination of the eigenvalues of a companion matrix. In Section 4, we recall and slightly extend the classical definition of the matrix sign function. We use this concept to enable the computation of some selected sets of the eigenvalues of the companion matrix C_p by means of the techniques of Section 3. We stabilize these computations numerically by means of subdivision techniques. Section 5 covers some effective iterations for computing matrix sign function for the matrix C_p and specifies the application to real eigen-solving. Section 6 is the contribution of the second and the third authors; it is devoted to numerical tests.

2 Definitions and Preliminaries

Hereafter "op" stands for "arithmetic operation". We assume computations in the fields of complex and real numbers $\mathbb{C}$ and $\mathbb{R}$, respectively. $\Re(z)$ and $\Im(z)$ are real and imaginary parts of a complex number $z = \Re(z) + \Im(z)\sqrt{-1}$.

Matrix computations: fundamentals [5]. $(B_j)_{j=1}^s = (B_1 \mid \ldots \mid B_s)$ is the $1 \times s$ block matrix with blocks $B_1, \ldots, B_s$. $\mathrm{diag}(B_j)_{j=1}^s = \mathrm{diag}(B_1, \ldots, B_s)$ is the $s \times s$ block diagonal matrix with diagonal blocks $B_1, \ldots, B_s$. $I = I_n = (\mathbf{e}_1 \mid \ldots \mid \mathbf{e}_n)$ is the $n \times n$ identity matrix with columns $\mathbf{e}_1, \ldots, \mathbf{e}_n$. $J = J_n = (\mathbf{e}_n \mid \ldots \mid \mathbf{e}_1)$ is the $n \times n$ reflection matrix, $J^2 = I$. $O_{k,l}$ is the $k \times l$ matrix filled with zeros. M^T is the transpose of a matrix M. $\mathcal{R}(M)$ is its range. $\mathcal{N}(M) = \{\mathbf{v} : M\mathbf{v} = \mathbf{0}\}$ is its null space, $\mathrm{rank}(M) = \dim(\mathcal{R}(M))$. A matrix of full column rank is a *matrix basis* of its range. M^+ is the Moore–Penrose pseudo inverse of M. An $n \times m$ matrix $X = M^{(I)}$ is a left (resp. right) inverse of an $m \times n$ matrix M if $XM = I_n$ (resp. if $MY = I_m$). M^+ is an $M^{(I)}$ for a matrix M of full rank; $M^{(I)}$ is unique iff the matrix M is nonsingular, and then $M^{(I)} = M^{-1}$. We use the matrix norms $||\cdot||_h$ for $h = 1, 2, \infty$, write $||\cdot|| = ||\cdot||_2$, and recall that $||A||^2 \le ||A||_1 ||A||_\infty$ for any matrix A. A matrix U is *unitary* and *orthogonal* if $U^T U = I$.

Theorem 1. *[5, Theorem 5.2.2]. A matrix M of full column rank has unique QR factorization $M = QR$ where $Q = Q(M)$ is a unitary matrix and $R = R(M)$ is a square upper triangular matrix with positive diagonal entries.*

Matrix computations: eigenspaces [5], [13]. $\mathcal{S} \subseteq \mathbb{C}^{n \times n}$ is an *invariant subspace* or *eigenspace* of a matrix $M \in \mathbb{C}^{n \times n}$ if $M\mathbf{v} \in \mathcal{S}$ for all $\mathbf{v} \in \mathcal{S}$.

Theorem 2. *[13, Theorem 4.1.2]. For all matrix bases $U \in \mathbb{C}^{n \times r}$ of an eigenspace $\mathcal{U}$ of $M \in \mathbb{C}^{n \times n}$ we have $MU = UL$ for unique matrix $L = U^{(I)} MU$.*

The pairs $\{L,\mathcal{U}\}$ and $\{L,U\}$ for L, $\mathcal{U}$ and U above are *eigenpairs* of a matrix M, L is its *eigenblock* and $\mathcal{U}$ is the *associated eigenspace* of L [13]. If $L = \lambda I_n$, then $\{\lambda,\mathcal{U}\}$ and $\{L,U\}$ are also called eigenpairs of a matrix M. In this case $\det(\lambda I - M) = 0$ and $\mathcal{N}(M - \lambda I)$ is the eigenspace associated with the *eigenvalue* λ and made up of its *eigenvectors*. $\Lambda(M)$ is the set of all eigenvalues of M, called its *spectrum*. $\rho(M) = \max_{\lambda \in \Lambda(M)} |\lambda|$ is the *spectral radius* of M. Theorem 2 implies that $\Lambda(L) \subseteq \Lambda(M)$. For an eigenpair $\{\lambda,\mathcal{U}\}$ write $\psi = \min |\lambda/\mu|$ over $\lambda \in \Lambda(L)$ and $\mu \in \Lambda(M) - \Lambda(L)$; call the eigenspace $\mathcal{U}$ *dominant* if $\psi > 1$, *dominated* if $\psi < 1$, *strongly dominant* if $1/\psi \approx 0$, and *strongly dominated* if $\psi \approx 0$. Here and hereafter the notation $\approx$, $\ll$ and $\gg$ (meaning "approximately equals", "is much less" and "is much greater", respectively) and the concepts "strong", "large", "small", "near", "close" etc. are defined in context, as is customary in the field of numerical computations with rounding errors.
A scalar λ is *nearly real* (within $\epsilon > 0$) if $|\Im(\lambda)| \leq \epsilon |\lambda|$.

An $n \times n$ matrix M is called *diagonalizable or nondefective* if SMS^{-1} is a diagonal matrix for some matrix S, e.g., if M has n distinct eigenvalues. A random real or complex perturbation makes the matrix diagonalizable with probability 1. In all our algorithms, we assume diagonalizable input matrices.

Theorem 3. *[6, Theorem 1.13]. $\Lambda(F(M)) = F(\Lambda(M))$ for a square matrix M and a function $F(x)$ bounded on its spectrum and being either a rational function or a limit of such functions. Furthermore $(F(\lambda),\mathcal{U})$ is an eigenpair of $F(M)$ if M is diagonalizable and has an eigenpair $(\lambda,\mathcal{U})$.*

Corollary 1. *$1/(\lambda - s)$ is an eigenvalue of $(M - sI)^{-1}$ if λ is an eigenvalue of M and if the matrix $M - sI$ is nonsingular.*

Toeplitz and companion matrices. $T = (t_{i-j})_{i,j=1}^n$ is an $n \times n$ Toeplitz matrix. Its every entry is invariant in the shifts along its diagonal. On matrices having structure of Toeplitz type, see [8, Chs. 2 and 4]. Z is the $n \times n$ downshift matrix: $Z\mathbf{v} = (v_i)_{i=0}^{n-1}$ for $\mathbf{v} = (v_i)_{i=1}^n$ and $v_0 = 0$.
$C_p = Z - \frac{1}{p_n}\mathbf{e}_n^T\mathbf{p}$ for $\mathbf{p} = (p_j)_{j=0}^{n-1}$ and $C_{\text{prev}} = JC_pJ$ denote the $n \times n$ companion matrices of the polynomial $p(x) = \det(xI - C_p) = \sum_{i=0}^n p_i x^i = p_n \prod_{j=1}^n (x - \lambda_j)$ and its reverse polynomial $p_{\text{rev}}(x) = \det(xI - C_{\text{prev}}) = x^n p(\frac{1}{x}) = \sum_{i=0}^n p_i x^{n-i} = p_n \prod_{j=1}^n (1 - x\lambda_j)$, respectively.

Theorem 4. *[3], [9]. A companion matrix $C_p \in \mathbb{C}^{n \times n}$ of a polynomial $p(x)$ generates an algebra $\mathcal{A}$ of matrices having structure of Toeplitz type. One needs $O(n)$ ops to add in $\mathcal{A}$, $O(n \log n)$ ops to multiply in $\mathcal{A}$, $O(n \log^2 n)$ ops to invert in $\mathcal{A}$, and $O(n \log n)$ ops to multiply a matrix from $\mathcal{A}$ by a square Toeplitz matrix.*

3 Basic Eigen-Solving Steps

Theorem 5. *Suppose F is an $n \times n$ diagonalizable matrix, Λ_r and $\Lambda^{(n-r)}$ are the sets of its r dominant and $n - r$ dominated eigenvalues associated with the eigenspaces $\mathcal{U}$ and $\mathcal{W}$, respectively. Let $k \to \infty$. Then the matrices $F^k/\|F^k\|$*

and $F^{-k}/||F^{-k}||$ *(if defined) converge to the matrices* U *of rank* r *and* W *of rank* $n - r$, *respectively, such that* $\mathcal{R}(U) \approx \mathcal{U}$, *whereas* $\mathcal{R}(W) \approx \mathcal{W}$.

Proof. Represent every vector as $\mathbf{v} = \mathbf{u} + \mathbf{w}$ where $\mathbf{u} \in \mathcal{U}$ and $\mathbf{w} \in \mathcal{W}$. Then $F^k\mathbf{v} = F^k\mathbf{u} + F^k\mathbf{w}$ and $F^{-k}\mathbf{v} = F^{-k}\mathbf{u} + F^{-k}\mathbf{w}$, and clearly $||F^k\mathbf{u}|| \gg ||F^k\mathbf{w}||$, whereas $||F^{-k}\mathbf{u}|| \ll ||F^{-k}\mathbf{w}||$ for all sufficiently large integers k.

On the extension to non-diagonaizable matrices F see [13, Section 6.1.1].

To implement the computation of the high powers F^k for $k = 2^h$ we can repeatedly square the matrix $F_0 = F$ and orthogonalize the outputs,

$$F_i = Q(F_{i-1}^2), \ i = 1, 2, \ldots, h. \tag{1}$$

Given a matrix F_h lying close to a matrix that has an unknown rank r, we can obtain r and a nearby matrix $\tilde{F}$ of rank r by computing rank revealing QR or LU factorization of the matrix F_h (cf. [7]).

Now assume that an $n \times n$ matrix F has r dominant eigenvalues and approximate them by applying the following procedure.

Procedure 1. Computing dominant eigenvalues.

1. Compute the matrix F_h *of (1) for a sufficiently large integer* h.

2. Compute an $n \times r$ *nearby matrix* $\tilde{F}$ *of full rank* r. *(It approximates a matrix basis* U *for the eigenspace associated with the* r *dominant eigenvalues of* F_h.)

3. Compute the $r \times r$ *matrix* $\tilde{L} = \tilde{F}^{(I)}F\tilde{F}$ *(cf. Theorem 2).*

4. Compute and output the spectrum $\Lambda(\tilde{L})$, *that approximates the set of the* r *dominant eigenvalues of* F.

5. Refine these approximate eigenvalues of F *by applying the RQ iteration.*

REMARKS.

1. Extension to the approximation of dominated eigenvalues and the eigenvalues near a selected point. Having completed the procedure, one can deflate the matrix F (cf. [5], [13]) and reapply the same algorithm to the resulting $(n - r) \times (n - r)$ matrix or to its shifted inverse that has dominant eigenvalues. Furthermore we can apply the procedure to the matrix $(F - sI)^{-1}$ (provided the matrix $F - sI$ is nonsingular) to approximate the eigenvalues of F lying near the real or complex point s. For a fixed or random small positive s (where F is singular) or for $s = 0$, these are precisely the dominated eigenvalues of F.

2. Some numerical issues. In numerical implementation one should take into account rounding errors and estimate their impact on the output, in particular to decide when we should stop our repeated squaring. This delicate subject is well developed but quite involved [13, Section 6.1.1]. If the ratio r/n is small, however, then Stages 3 and 4 of the proceedure are dramatically simplified, and we can readily compute the relative residual norm $||F\tilde{F} - \tilde{F}\tilde{L}||/||\tilde{F}\tilde{L}||$ to stop where it is small enough (cf. Theorem 2).

3. Repeated squaring with no orthogonalization. Squaring preserves the structure of a companion matrix $F = C_p$ and the shifted matrices $F = C_p - sI$ as well as of any their positive or negative integer power and only requires $O(n \log n)$

ops (cf. [3] or [9]), but orthogonalization and pivoting at Stages 1 and 2 of the procedure destroy this structure. One does not need orthogonalization for a few initial squarings, and even a single squaring is sufficient where the domination of the eigenvalues of F is already strong enough. Furthermore for a fixed or random real or complex shift s, we can write $F = C_p - sI$ or $F = (C_p - sI)^{-1}$ and repeatedly square F with no orthogonalization,

$$F_{h+1} = a_h F_h^2, \ a_h \approx 1/||F_h||^2 \text{ for } h = 0, 1, \ldots \qquad (2)$$

to approximate a single real or complex eigenvalue or a pair of complex conjugate eigenvalues of C_p. Quite typically the matrix $(C_p - sI)^{-1}$ has such a dominant eigenvalue or a pair of eigenvalues, both in the presence of rounding errors and without them.

4. *Stage 2 for small subset of the spectrum, randomized compression, and the case of companion matrix F.* We can simplify Stage 2 if we have a sufficiently small upper bound r_+ on the number r of dominant eigenvalues. In this case we can replace the matrix F_k by $F_k G$ for $n \times r_+$ standard Gaussian random matrix G. Decompose every column $G e_j$ for $j = 1, \ldots, r_+$ of the matrix G into the sum $\mathbf{u}_j + \mathbf{w}_j$ where $\mathbf{u}_j \in \mathcal{U}$ and $\mathbf{w}_j \in \mathcal{W}$ as in the proof of Theorem 5. It follows that for sufficiently large integers k the ratio $||F_k \mathbf{w}_j||/||F_k \mathbf{u}_j||$ is close to 0 with probability close to 1, that is the U term of every column $F_k e_j$ dominates its W term. Suppose that the W terms have been dominated and deleted for all j. Then the resulting matrix has rank at most r and lies in $\mathcal{U}$ near $F_k G$. Furthermore, with probability close to 1, the matrix $F_k G$ cannot lie near a matrix of rank less than r by virtue of [12, Corollary 5.2]. Consequently, with probability near 1, rank revealing LU and QR factorization of the matrix $F_k G$ defines an $n \times r$ matrix lying near a matrix basis of $\mathcal{U}$. Overall the cost of the computations at Stage 2 is dominated by the order $r_+ n \log n$ ops applied for multiplication of F_k by G. The cost decreases to $O(n \log n)$ where G is a Toeplitz matrix. [12, Corollary 5.2] does not hold in the case where the multiplier G is a standard Gaussian random Toeplitz matrix, but the respective extension of this corollary is in good accordance with test results in [12].

4   Matrix Sign Function and Eigen-Solving

Definition 1. $\text{sign}(x + y\sqrt{-1})$ *(for real x and y) is 1 if $x > 0$ and is -1 if $x < 0$. Fix any $r \times r$ real diagonal matrix D_r, e.g., $D_r = O_{r,r}$, and let $A = ZJZ^{-1}$ be a Jordan canonical decomposition of a matrix $A \in \mathbb{C}^{n \times n}$, $J = \text{diag}(J_-, J_0, J_+)$, $J_- \in \mathbb{C}^{p \times p}$, $J_0 \in \mathbb{C}^{r \times r}$, $J_+ \in \mathbb{C}^{q \times q}$, $\Re(d_-) < 0$, $\Re(d_0) = 0$, $\Re(d_+) > 0$ for all diagonal entries d_-, d_0 and d_+ of the matrices J_-, J_0, and J_+, respectively, $n = p + q + r$. Then define a generalized matrix sign function $\text{sign}(A)$ by writing $\text{sign}(A) = Z \text{diag}(-I_p, D_r\sqrt{-1}, I_q)Z^{-1}$.*

For $r = 0$ this is the classical *matrix sign function* [6], equivalently defined as $\text{sign}(A) = A(A^2)^{-1/2}$ or $\text{sign}(A) = \frac{2}{\pi} A \int_0^\infty (t^2 I_n + A^2)^{-1} dt$.

Theorem 6. *Assume the generalized matrix sign function* $\text{sign}(A)$ *for an* $n \times n$ *matrix* $A = ZJZ^{-1}$. *Then for some real* $r \times r$ *diagonal matrix* D_r *we have*

$$I_n - \text{sign}(A) = Z^{-1} \text{diag}(2I_p, I_r - D_r\sqrt{-1}, O_{q,q})Z,$$

$$I_n + \text{sign}(A) = Z^{-1} \text{diag}(O_{p,p}, I_r + D_r\sqrt{-1}, 2I_q)Z,$$

$$I_n - \text{sign}(A)^2 = Z^{-1} \text{diag}(O_{p,p}, I_r + D_r^2, O_{q,q})Z.$$

Corollary 2. *Under the assumptions of Theorem 6 the matrix* $I_n - \text{sign}(A)^2$ *has dominant eigenspace of dimension* r *associated with the eigenvalues* λ *of the matrix* A *such that* $\Re(\lambda) = 0$, *whereas the matrices* $I_n - \text{sign}(A)$ *(resp.* $I_n + \text{sign}(A))$ *have dominant eigenspaces associated with the eigenvalues* λ *of* A *such that* $\Re(\lambda) \leq 0$ *(resp.* $\Re(\lambda) \leq 0$).

Given the matrix $F(A) = I_n - \text{sign}(A)^2$, Procedure 1 approximates the eigenvalues of A lying on the imaginary axis $\{\lambda : \Re(\lambda) = 0\}$. Likewise, having the matrices A and $F(A) = I_n - \text{sign}(A)$ (resp. $F(A) = I_n + \text{sign}(A))$ available, we can apply Procedure 1 to approximate all eigenvalues λ of A such that $\Re(\lambda) \leq 0$ (resp. $\Re(\lambda) \leq 0$). In these cases, the square matrices L in Procedure 1 have dimensions p_+ and q_+, respectively, where $p \leq p_+ \leq p+r$ and $q \leq q_+ \leq q+r$. For $M = C_p$ and a pair of large integers p_+ and $n - p_+$ or q_+ and $n - q_+$, we split the polynomial $p(x)$ into two high degree factors, whose coefficients can grow dramatically versus the ones of $p(x)$, e.g., where $p(x) = x^n + 1 = u(x)v(x)$, $v(x) = \prod_{j=1}^{n/2}(x - \lambda_j)$, $\Re(\lambda_j) > 0$ for all j. The *subdivision* techniques based on the following simple fact give us a universal remedy, however.

Theorem 7. *Suppose* $\mathcal{U}$ *and* $\mathcal{V}$ *are two eigenspaces of* A *and* $\Lambda(\mathcal{U})$ *and* $\Lambda(\mathcal{V})$ *are the sets of the associated eigenvalues. Then* $\Lambda(\mathcal{U}) \cap \Lambda(\mathcal{V})$ *is the set of the eigenvalues of* A *associated with the eigenspace* $\mathcal{U} \cap \mathcal{V}$.

By computing the matrix sign function of the matrices $\alpha A - \sigma I$ for various selected pairs of complex scalars α and σ, we can define the eigenspace of A associated with the eigenvalues lying in a selected region on the complex plane bounded by straight lines, e.g., in any fixed rectangle. By including matrix inversions into this game, we define the eigenvalue regions bounded by straight lines, their segments, circles and their arcs.

5 Iterative Algorithms for the Matrix Sign Computation

[6, equations (6.17)–(6.20)] define effective iterative algorithms for the square root function $B^{1/2}$, and one can readily extend them to $\text{sign}(A) = A(A^2)^{-1/2}$. We, however, employ two popular alternatives: Newton's iteration based on the Möbius transform $x \to (x + 1/x)/2$ and the $[2/0]$ Padé iteration [6, Chapter 5],

$$N_0 = A, \ N_{i+1} = (N_i + N_i^{-1})/2, \ i = 0, 1, \ldots, \tag{3}$$

$$N_0 = A, \ N_{i+1} = (15I_n - 10N_i^2 + 3N_i^4)N_i/8, \ i = 0, 1, \ldots \tag{4}$$

Theorem 3 implies the following simple corollary.

Corollary 3. *Assume iterations (3) and (4) where neither of the matrices N_i is singular. Let $\lambda = \lambda^{(0)}$ denote an eigenvalue of the matrix N_0 and define*

$$\lambda^{(i+1)} = (\lambda^{(i)} + (\lambda^{(i)})^{-1})/2, \ i = 0, 1, \ldots, \tag{5}$$

$$\lambda^{(i+1)} = \lambda^{(i)}(15 - 10(\lambda^{(i)})^2 + 3(\lambda^{(i)})^4)/8, \ i = 0, 1, \ldots \tag{6}$$

Then $\lambda^{(i)} \in \Lambda(N_i)$ for $i = 1, 2, \ldots$ provided the pairs $\{N_i, \lambda^{(i)}\}$ are defined by the pairs of equations (3) and (5) or (4) and (6).

Corollary 4. *In iterations (5) and (6) the images $\lambda^{(i)}$ of an eigenvalue λ of the matrix N_0 for all i lie on the imaginary axis $\{\lambda : \Re(\lambda) = 0\}$ if so does λ.*

By virtue of the following theorems the sequences $\{\lambda^{(0)}, \lambda^{(1)}, \ldots\}$, defined by (5) and (6) converge to ± 1 exponentially fast, right from the start.

Theorem 8. *(See [6], [2, page 500].) Write $\lambda = \lambda^{(0)}$, $\delta = \mathrm{sign}(\lambda)$ and $\gamma = |\frac{\lambda-\delta}{\lambda+\delta}|$. Assume (5) and $\Re(\lambda) \neq 0$. Then $|\lambda^{(i)} - \delta| \leq \frac{2\gamma^{2^i}}{\gamma^{2^i}+\delta}$ for $i = 0, 1, \ldots$.*

Theorem 9. *[2, Proposition 4.1]. Write $\delta_i = \mathrm{sign}(\lambda^{(i)})$ and $\gamma_i = |\lambda^{(i)} - \delta_i|$ for $i = 0, 1, \ldots$. Assume (6) and $\gamma_0 \leq 1/2$. Then $\gamma_i \leq \frac{32}{113}(\frac{113}{128})^{3^i}$ for $i = 1, 2, \ldots$*

Substitute $N_0 = M$ in lieu of $N_0 = A$ into the matrix sign iterations (3) and (4) and equivalently rewrite them to avoid involving nonreal values,

$$N_{i+1} = 0.5(N_i - N_i^{-1}) \text{ for } i = 0, 1, \ldots, \tag{7}$$

$$N_{i+1} = -(3N_i^5 + 10N_i^3 + 15N_i)/8 \text{ for } i = 0, 1, \ldots. \tag{8}$$

Now the matrices N_i and the images $\lambda^{(i)}$ of every real eigenvalue λ of M are real for all i, whereas the results of Theorems 8 and 9 are immediately extended. The images of every nonreal λ converge to $\mathrm{sign}(\Im(\lambda))\sqrt{-1}$ quadratically under (7) if $\Re(\lambda) \neq 0$ and cubically under (8) if $|\lambda - (\mathrm{sign}(\Im(\lambda))\sqrt{-1})| \leq 1/2$.

Under the maps $M \to I_n + N_i^2$ for N_i in the above iterations, the images $1 + (\lambda^{(i)})^2$ of nonreal eigenvalues λ of M lying in the respective basins of convergence converge to 0, whereas for real λ the images are real and are at least 1 for all i. Thus for sufficiently large integers i we yield strong domination of the eigenspace of N_i associated with the images of real eigenvalues of M.

Iteration (7) fails where for some i the matrix N_i is singular or nearly singular, that is has eigenvalue 0 or near 0, but then we can approximate it by applying the Rayleigh Quotient Iteration [5, Section 8.2.3], [1] or the Inverse Orthogonal Iteration [5, page 339]. If we seek other real eigenvalues as well, we can deflate the matrix M and apply Procedure 1 to the resulting matrix of a smaller size. Alternatively we can apply it to the matrix $N_i + \rho_i I_n$ for a random real shift ρ_i; with $|\rho_i|$ small enough for all i, the images of all nonreal eigenvalues of M would still rapidly converge to a small neighborhood of the points $\pm\sqrt{-1}$, ensuring their isolation from the images of real eigenvalues.

6   Numerical Tests

We tested our algorithms for the approximation of the eigenvalues of $n \times n$ companion matrix C_p and of the matrix $C_p - sI_n$ for polynomials $p(x)$ with

Table 1. Repeated Squaring

n	dimension/squarings	min	max	mean	std
64	dimension	1	10	5.31	2.79
128	dimension	1	10	3.69	2.51
256	dimension	1	10	4.25	2.67
64	squarings	6	10	7.33	0.83
128	squarings	5	10	7.37	1.16
256	squarings	5	11	7.13	1.17

Table 2. Newton's iteration (7)

n	min	max	mean	std
64	7	11	8.25	0.89
128	8	11	9.30	0.98
256	9	13	10.22	0.88

Table 3. 5 N-steps (7) + P-steps (8)

n	P-steps or %	min	max	mean	std
64	P-steps	1	4	2.17	0.67
128	P-steps	1	4	2.05	0.63
256	P-steps	1	3	1.99	0.58
64	% w/o RQ steps	0	100	64	28
128	% w/o RQ steps	0	100	39	24
256	% w/o RQ steps	0	100	35	20
64	% w/RQ steps	0	100	89	19
128	% w/RQ steps	0	100	74	26
256	% w/RQ steps	0	100	75	24

random real coefficients for $n = 64, 128, 256$ and for random real s. For each class of matrices, each input size and each iterative algorithm we generated 100 input instances and run 100 tests. Our tables show the minimum, maximum, and average (mean) numbers of iteration loops in these runs (until convergence) as well as the standard deviations in the columns marked by "**min**", "**max**", "**mean**", and "**std**", respectively.

We applied repeated squaring of equation (2) to the matrix $C_p - sI$ for shifts $s \neq 0$ because polynomials $p(x)$ with random real coefficients tend to have all roots near the unit circle $\{\lambda : |\lambda| = 1\}$, and for such inputs repeated squaring for C_p advances very slowly.

We first applied iteration (7) to approximate the matrix sign function for the matrix C_p and then Procedure 1 to approximate real eigenvalues.

In both groups of tests above we output roots with at least four correct decimals. In our next group of tests we output roots with at least three correct decimals. In these tests we applied real version (8) of Padé iteration without stabilization to the matrices produced by five Newton's steps (7).

Table 1 displays the results of testing repeated squaring of equation (2). The first three lines show the dimension of the output subspace and the matrix L. The next three lines show the number of squarings performed until convergence. Table 2 displays the number of Newton's steps (7) performed until convergrence.

Table 3 covers the tests where we first performed five Newton's steps (7) followed by Padé steps, counted in the first three lines of the table. The next three lines display the percent of the real roots of the polynomials $p(x)$ which the algorithm computed with at least three correct decimals (compared to the overall number of the real eigenvalues of L). The next three lines show the increased percent of computed roots when Rayleigh Quotient iteration refined the crude approximations. The iteration rapidly converged from all these initial approximations but in many cases to the same roots from distinct initial points.

Acknowledgement. We are very grateful to the reviewers for helpful comments.

References

1. Bini, D.A., Gemignani, L., Pan, V.Y.: Inverse power and Durand/Kerner iteration for univariate polynomial root-finding. Computers and Math (with Applics.) 47(2/3), 447–459 (2004)
2. Bini, D., Pan, V.Y.: Graeffe's, Chebyshev, and Cardinal's processes for splitting a polynomial into factors. J. Complexity 12, 492–511 (1996)
3. Cardinal, J.P.: On two iterative methods for approximating the roots of a polynomial. Lectures in Applied Math 32, 165–188 (1996)
4. Emiris, I.Z., Mourrain, B., Tsigaridas, E.P.: Real Algebraic Numbers: Complexity Analysis and Experimentation. In: Hertling, P., Hoffmann, C.M., Luther, W., Revol, N. (eds.) Real Number Algorithms. LNCS, vol. 5045, pp. 57–82. Springer, Heidelberg (2008)
5. Golub, G.H., Van Loan, C.F.: Matrix Computations, 3rd edn. Johns Hopkins University Press, Baltimore (1996)

6. Higham, N.J.: Functions of Matrices: Theory and Computations. SIAM (2008)
7. Pan, C.-T.: On the existence and computation of Rrank-revealing LU factorization. Linear Algebra and Its Applications 316, 199–222 (2000)
8. Pan, V.Y.: Structured Matrices and Polynomials: Unified Superfast Algorithms. Birkhäuser, Boston, and Springer, NY (2001)
9. Pan, V.Y.: Amended DSeSC power method for polynomial root-finding. Computers and Math (with Applics.) 49(9-10), 1515–1524 (2005)
10. Pan, V.Y., Qian, G., Murphy, B., Rosholt, R.E., Tang, Y.: Real root-finding. In: Vershelde, J., Stephen Watt, S. (eds.) Proc. Third Int. Workshop on Symbolic–Numeric Computation (SNC 2007), London, Ontario, Canada, pp. 161–169. ACM Press, New York (2007)
11. Pan, V.Y., Zheng, A.: New progress in real and complex polynomial root-finding. Computers and Math (Also in Proc. ISSAC 2010) 61, 1305–1334 (2010)
12. Pan, V.Y., Qian, G., Zheng, A.: Randomized Matrix Computations II. Tech. Report TR 2012006, Ph.D. Program in Computer Science, Graduate Center, the City University of New York (2012), http://www.cs.gc.cuny.edu/tr/techreport.php?id=433
13. Stewart, G.W.: Matrix Algorithms, Vol II: Eigensystems, 2nd edn. SIAM, Philadelphia (2001)
14. Yap, C., Sagraloff, M.: A simple but exact and efficient algorithm for complex root isolation. In: Proc. ISSAC 2011, pp. 353–360 (2011)

Root-Refining for a Polynomial Equation

Victor Y. Pan

Department of Mathematics and Computer Science
Lehman College and the Graduate Center of the City University of New York
Bronx, NY 10468 USA
victor.pan@lehman.cuny.edu,
http://comet.lehman.cuny.edu/vpan/

Abstract. Polynomial root-finding usually consists of two stages. At first a crude approximation to a root is slowly computed; then it is much faster refined by means of the same or distinct iterations. The efficiency of computing an initial approximation resists formal study, and the users employ empirical data. In contrast, the efficiency of refinement is formally measured by the classical concept $q^{1/\alpha}$ where q is the convergence order and α is the number of function evaluations per iteration. To cover iterations not reduced to function evaluations alone, e.g., ones simultaneously refining n approximations to all n roots of a degree n polynomial, we let d denote the number of arithmetic operations involved in an iteration divided by $2n$ because we can evaluate such a polynomial at a point by using $2n$ operations. For this task we employ recursive polynomial factorization to yield refinement with the efficiency $2^{cn/\log^2 n}$ for a positive constant c. For large n this is a dramatic increase versus the record efficiency 2 of refining an approximation to a single root of a polynomial. The advance could motivate practical use of the proposed root-refiners.

Keywords: Root-refining, Efficiency, Polynomial factorization.

1 Introduction

1.1 Two Stages of Iterative Polynomial Root-Finding

The classical problem of polynomial root-finding is still a subject of intensive study because of its important applications to geometric modelling, financial mathematics, signal processing, control, and in particular to computer algebra, for which this is a fundamental task. We refer the reader to Bell (1940), Boyer (1968), and Pan (1997 and 1998) on the rich history of this subject and to McNamee (2002 and 2007) and McNamee and Pan (2012) on numerous old and new polynomial root-finders.

A typical iterative polynomial root-finder consists of two stages. At first substantial effort is invested into computing an initial point that lies much closer to one of the roots than to any other of them. Then the same or another iterative algorithm refines this approximation. Here is a formal support for this approach.

V.P. Gerdt et al. (Eds.): CASC 2012, LNCS 7442, pp. 283–293, 2012.
© Springer-Verlag Berlin Heidelberg 2012

Theorem 1. *(Corollary 4.5 from Renegar (1987).) Assume a polynomial*

$$p(x) = \sum_{i=0}^{n} p_i x^i = p_n(x - z_1) \cdots (x - z_n), \ p_n \neq 0, \tag{1}$$

and Newton's iterations

$$x_{i+1} = x_i - p(x_i)/p'(x_i), \ i = 0, 1, \ldots \tag{2}$$

where $5n^2|x_0 - z_1| \leq \min_{j>1}|x_0 - z_j|$. Then $|x_i - z_1| \leq 2^{3-2^i}|x_0 - z_1|$, that is the iterations converge quadratically from the initial point x_0.

On preceding works, variations and extensions, which cover Newton's processes in Banach spaces and other iterative root-finders, see Kantorovich and Akilov (1982), Theorem V.4.3; Kim (1985), Smale (1986), Renegar (1987), Curry (1989), Petkovic and Herceg (2001), and the bibliography therein.

1.2 Divide-and-Conquer Factorization

The issue of the computational complexity of polynomial root-finding has been raised in Smale (1981) and Schönhage (1982). The algorithms supporting the record upper estimates rely on recursive factorization of the polynomial $p(x)$. Schönhage (1982), Neff and Reif (1994), and Pan (1995, 1996 and 2002) numerically factorize a polynomial $p(x) = \sum_{i=0}^{n} p_i x^i$ into the product of two nonconstant factors and continue this splitting process recursively until factorization (1) of $p(x)$ into the product of n linear factors is closely approximated. Then the n approximate roots z_j are readily recovered such that $|x_j - z_j| \leq 1/2^b$ for a sufficiently large b and $j = 1, \ldots, n$. This process in Pan (1995) uses $\tilde{O}(n)$ ops with the precision $O(bn)$, translated into $\tilde{O}(n^2 b)$ Boolean (that is bitwise) operations. Here and hereafter "ops" stand for "arithmetic operations" and $\tilde{O}(f(b, n))$ for $O(f(b, n))$ up to polylog factors in $b + n$. The estimated complexity of root-finding for the worst case input polynomial is smaller than Schönhage's by a factor n and is still record low, but Pan (2002) decreases the Boolean cost and precision bounds by a factor n where one just seeks $x_1, \ldots, x_n$ such that

$$||p(x) - p_n(x - x_1) \cdots (x - x_n)||_1 \leq 2^b ||p(x)||_1, \ || \sum_i u_i x^i ||_1 = \sum_i |u_i|. \tag{3}$$

Computing such a polynomial factorization is important in its own right because of the applications to time series analysis, Weiner filtering, noise variance estimation, covariance matrix computation, the study of multi-channel systems (see Wilson (1969), Box and Jenkins (1976), Barnett (1983), Demeure and Mullis (1990), and Van Dooren (1994)), and the isolation of the roots of a polynomial $p(x)$ with n distinct roots and with integer coefficients, each of length at most l. (Isolation means computation of n disjoint discs, each containing a single root of $p(x)$.) For $b = \lceil (2n+1)(l+1+\log(n+1)) \rceil$ the extension is proved in Section 20 of Schönhage (1982) based on the gap theorem of Mahler (1964). Combination with the estimates of Pan (2002) yields the following result.

Theorem 2. *Let polynomial $p(x)$ of (1) have n distinct simple zeros and integer coefficients in the range $[-2^\tau, 2^\tau]$. Then one can isolate the n zeros of $p(x)$ from each other by using $\tilde{O}(n^2\tau)$ Boolean operations.*

The cited arithmetic and Boolean cost estimates from Pan (1995, 1996 and 2002) are optimal up to polylogarithmic factors, but the users prefer to employ functional iterative root-finders such as the Weierstrass–Durand–Kerner (hereafter WDK) and Ehrlich–Aberth algorithms (see Weierstrass (1903), Durand (1960), Kerner (1966), Ehrlich (1967), and Aberth (1973)). The known upper estimates for the complexity of these algorithms is no match to the ones of Pan (2002), but the gap disappears if we estimate the complexity of these functional iterations based on informal empirical data confirming their excellent global convergence, that is convergence right from the start (cf. the Appendix). Our next sections show, however, that just for the refinement of approximate roots even these highly recognized algorithms remain by far inferior to the recursive factorization algorithm. Thus the latter algorithm is definitely worth further study and may eventually become the method of choice for root-refining.

1.3 Efficiency of Refinement

Assume d function evaluations per iteration that refines an initial approximation and converges with order q. Since Ostrowski (1966), it is customary to measure the efficiency of the refinement by

$$\text{eff} = q^{1/\alpha} \tag{4}$$

or by $\log_{10} \text{eff} = (1/\alpha)\log_{10} q$ (cf. McNamee and Pan (2012)). For example, we have $q = \alpha = 2$ and $\text{eff} = \sqrt{2}$ for Newton's iterations $x_{i+1} = x_i - p(x_i)/p'(x_i)$, $i = 0, 1, \ldots$, whereas $\alpha = 1$ and $q = \text{eff} \approx 1.839$ for the root-refiner of Muller (1956). More generally, we write $\alpha = 0.5f/n$ provided the iteration uses f ops and the input polynomial has degree n and therefore can be evaluated at a point in $2n$ ops. The record efficiency of the known root-refiners for a single root of a polynomial is 2 (see McNamee and Pan (2012)); e.g., one can yield it by combining a linearly convergent root-refiner with the convergence acceleration by extrapolation of Aitken (1926), also called Δ^2 convergence acceleration. In the next sections, however, we refine all n roots of $p(x)$ with the much greater efficiency

$$\text{eff} = 2^{cn/\log^2 n} \text{ for a positive constant } c. \tag{5}$$

The supporting algorithms are numerically stable, their precision is controlled, and their Boolean cost stays nearly optimal. In particular, we avoid using polynomial evaluation at n points in $O(n\log^2 n)$ ops, which would have saved ops but is numerically unstable and has inferior Boolean complexity.

1.4 Some Technicalities

We employ the well developed techniques for polynomial root-finding based on recursive factorization, but show its dramatic simplification for the more restricted but still highly important task of root-refining. In particular, this simplification is quantified in our Theorem 3, based on divide-and-conquer process. In this process we incorporate the techniques from Schönhage (1982), Neff and Reif (1994), Pan (1995, 1996 and 2002), and Kirrinnis (1998) and accentuate their power by employing sufficiently close initial approximations to the roots, which we assume available for refinement. Our formal support for simplified refinement boils down to devising a fast algorithm that computes a reasonably wide root-free annulus on the complex plane that separates from one another two sets of the roots of $p(x)$, consisting of βn and $(1 - \beta)n$ roots, respectively, where β is independent of n and $0 < \beta < 1$. The techniques for computing such an annulus in Neff and Reif (1994) and Pan (1995, 1996 and 2002) are quite involved and are the bottleneck of the efficient implementation of the resulting root-finders, but based on our Procedure 1 and Theorem 5 we dramaticaly simplify them for root-refining. The resulting algorithms support our record efficiency estimate for root-refining and promise to be practically valuable.

1.5   Organization of the Paper

Our exposition incorporates many well known techniques developed for polynomial root-finding, and readily adjustable to root-refining. We recall them briefly, referring the reader to the original papers for details. In particular in the next section we first briefly recall the recursive divide-and-conquer factorization approach to polynomial root-finding and its specific version of Kirrinnis (1998) and then prove our main Theorem 3 subject to providing sufficiently wide root-free annuli that would separate the root sets of the factors in each factorization step, to support balanced recursive factorization. In Section 3 we cover computing such annuli. We briefly comment on some implementation issues in Section 4 and devote the Appendix to a conjecture on convergence of functional iterations.

2   Root-Refining Via Recursive Divide-And-Conquer Factorization and Kirrinnis' Algorithm

Suppose we are given the coefficients of a polynomial $p(x)$ of (1) and approximations $z_1, \ldots, z_n$ to its n simple roots $x_1 \ldots, x_n$. This defines an approximate factorization

$$p = p(x) \approx f(x) = p_n(x - z_1) \cdots (x - z_n) \tag{6}$$

(cf. (3)). Schönhage (1982) has extended Ostrowski (1940 and 1966) to bound the approximation errors $|x_j - z_j|$ for $|z_j| \leq 1$ and $|\frac{1}{x_j} - \frac{1}{z_j}|$ for $|z_j| \geq 1$, $j = 1, \ldots, n$ in terms of the norm $||p(x) - f(x)||_1$. Newton's multivariate iterations for refining (6) turn into the WDK algorithm (cf. Pan and Zheng (2011b)).

   Now suppose we are given an initial approximate factorization of the polynomial p into the product of two factors of comparable degrees,

$$p \approx f = f_1 f_2, \tag{7}$$

$$\deg f_1 < c' \deg f_2 < c'' \deg f_1 \tag{8}$$

for two positive constants c' and c''. Furthermore assume that the two root sets of these two factors are separated by a root-free annulus $A(z, r_+, r_-) = \{x : r_- \leq |x - z| \leq r_+\}$ bounded by two circles with a center z and radii r_- and r_+, respectively, such that $r_+/r_- > 1 + c/n^d$ for two constants $c > 0$ and d. We call the ratio r_+/r_- the *relative width* of the annulus and call $A(z, r_-, r_+)$ a (c, d) *annulus*. We call it a (c, d) *separating annulus* if (8) holds.

One can recursively refine the factors f_1 and f_2 by computing the polynomials $f_1^{\text{new}} = f_1 + t_1$ and $f_2^{\text{new}} = f_2 + t_2$ as well as Newton's correction polynomials t_1 and t_2 that satisfy

$$\frac{r}{f} = \frac{t_1}{f_1} + \frac{t_2}{f_2} \tag{9}$$

where $r = p - f$, $f = f_1 f_2$, $\deg t_1 < \deg f_1$ and $\deg t_2 < \deg f_2$ (cf. Schönhage (1982)). The well known algorithms compute such a partial fraction decomposition (hereafter referred to as PFD) by using $O(n \log^2 n)$ ops (cf. Bini and Pan (1994), Problem 4.2c (PART·FRAC), pages 30–31). The computation is prone to numerical stability problems, but the modification in the next section avoids them. Schönhage (1982) has ignored the numerical stability issue and instead directed his work to decreasing upper bounds on the Boolean complexity of polynomial root-finding (although, as we said, they are inferior to the ones of Pan (1995, 1996 and 2002) by a factor n). Kirrinnis (1998), however, has devised numerically stable quadratically convergent factorization algorithm of Newton type, provided sufficiently wide root-free (c, d) annulus (for $c > 0$) separating the roots of the factors f_1 and f_2 has been supplied (see the next section).

As soon as we closely approximate the factors f_1 and f_2, we recursively factorize both of them in the same fashion until we arrive at a refined sufficiently close complete approximate factorization (6) such that $z_1, \ldots, z_n$ approximate the n roots $x_1, \ldots, x_n$ with a desired accuracy.

Actually Kirrinnis (1998) has extended the above techniques to the refinement of an initial factorization $p \approx f = f_1 \cdots f_s$ into the product of s nonconstant factors for any integer s from 2 to n, that is he assumed $1 < s \leq n$, $\deg f_j > 0$ for all j, and $\deg f_1 + \cdots + \deg f_s = n$. Furthermore, he has proved quadratic convergence of the iterations as well as of its variant in which he improved the efficiency and numerical stability of the refinement. In this variant he confined the most expensive and numerically unstable stage of the PFD computation to the first iteration. At all subsequent iterations he updated the corrections t_j and new factors f_j as follows (we decrypt his formulas a little):

$$f = f_1 \cdots f_s, \tag{10}$$

$$t_j^{\text{new}} = (2 - t_j f/f_j) t_j \bmod f_j, \quad j = 1, \ldots, s, \tag{11}$$

$$f_j^{\text{new}} = f_j + (t_j^{\text{new}} p \bmod f_j), \quad j = 1, \ldots, s. \tag{12}$$

In the above variations polynomial multiplications replace the computation of PFDs; this improves numerical stability and decreases the number of ops per refinement iteration to $O(n \log n)$. Kirrinnis (1998) has also estimated the precision and the Boolean cost of these computations. Next assume $s = 2$ and balancing (8), extend splitting recursively, and arrive at bound (5).

Theorem 3. *Assume n close initial approximations to n distinct roots of a polynomial $p(x)$ of (1) and refine them by recursively applying equations (10)–(12) for $s = 2$. Further assume that (c, d) separating annuli for $c > 0$, e.g., for $c = d = 1$, are available throughout recursive factorization. Then the refinement has efficiency (5).*

Proof. Represent the above recursive refinement process by a binary tree whose root p has two children f_1 and f_2, each of them in turn has at most two children such that $f_1 \approx f_{11}f_{12}$ and $f_2 \approx f_{21}f_{22}$, and so on. At every level of the tree its nodes represent polynomials whose degrees sum to $n - l$ where l denotes the number of linear factors output at the previous levels. The tree has $O(\log n)$ levels because of recursive balancing (8). It follows that computing Newton's corrections for all factor polynomials at each level takes $O(n \log n)$ ops per iteration. This is translated into $\alpha = O(\frac{\log n}{n})$ per level, $\alpha = O(\frac{\log^2 n}{n})$ for all the $O(\log n)$ levels, and thus into (5) because $q = 2$ for Newton's iterations.

3   Computation of $(1, 1)$ Separating Annuli

Next we complete the refinement algorithms by computing $(1, 1)$ separating annuli; the first annulus defines the factors f_1 and f_2 satisfying (8), and the next annuli define recursive splittings of the polynomials f_1, f_2, and their factors.

We can compute the desired annuli by employing the techniques of Schönhage (1982), Neff and Reif (1994) and Pan (1995, 1996 and 2002), but we only partly reuse these rather complicated and expensive techniques. In particular, we avoid expensive computation of the roots of the higher order derivatives $p^{(h)}(x)$. We proceed more efficiently because we use approximate roots of $p(x)$, but technical challenge persists. Even if we had all these roots available with no error, still computing the $(1, 1)$ separating annuli would not be trivial. Next we supply the required details. We begin with recalling the following result.

Theorem 4. *(Cf. Schönhage (1982), Theorem 14.2.) Given polynomial $p(x)$ of (1), two real constants $c > 0$ and d, and a complex value v_0, we need $O(n \log^2 n)$ ops to approximate within relative errors of $1 + c/n^d$ the distances $|v_0 - x_j|$ between v_0 and all roots x_j of $p(x)$ for $j = 1, \dots, n$.*

By applying the theorem at a complex point v_0, one can find a desired separating $(1, 1)$ annulus unless most of the roots lie in a narrow annulus about a fixed circle $\{x : |v_0 - x| = R\}$ for some positive R. In the latter case we can reapply the theorem twice, for v_0 replaced by $v_0 + 2R$ and $v_0 + 2R\sqrt{-1}$. Then in these three applications we obtain either a desired $(1, 1)$ separating annulus or three narrow annuli, each containing almost all roots of $p(x)$. In the latter case the intersection of the three annuli is a small region with a large collection of roots. Namely we have the following corollary (cf. Neff and Reif (1994)).

Corollary 1. *Given a polynomial $p(x)$ of (1) with n distinct roots, a fixed real $u > 1$ and a complex v_0, we need $O(n \log^2 n)$ ops to compute either (i) a required wide separating annulus for $p(x)$ or (ii) a disc $D(v_1, \rho_1) = \{x : |x - v_1| \le \rho_1\}$ containing at least $n/12$ roots of $p(x)$ where $|v_0 - v_1| \ge u\rho_1$.*

In our current application, we decrease the estimates of the theorem and consequently of the corollary by a factor $\log^2 n$ because approximate roots $x_1, \ldots, x_n$ are assumed to be available. It remains to extend computations from case (ii) to arrive at case (i).

Apply the amended corollary for v_1 replacing v_0 to yield the desired case (i) or to obtain a disc $D(v_2, \rho_2) = \{x : |x - v_2| \le \rho_2\}$ with at least $n/12$ roots of $p(x)$ and such that $|v_2 - v_0|/\rho_2$ has order u^2 (cf. Pan (1995, 1996 and 2002)).

We reapply the corollary $t = O(\log n)$ times, by using $O(n \log n)$ ops overall, to ensure either the desired case (i) or the bound $|v_t - v_0|/\rho_t \ge s = cn^d$ for any fixed real $c > 0$ and d. It remains to treat the latter case where it is sufficient for us to choose c and d such that $s = cn^d > 6$ and thus

$$|v_t - v_0|/\rho_t > 6. \tag{13}$$

Let n_s denote the number of the roots of $p(x)$ in the disc $D(v_k, s\rho_k) = \{x : |x - v_k| \le s\rho_k\}$. By assumption (ii) we have $n_s \ge n/12$ for $s \ge 1$. Consider n_s for $s = s_h = (1 + 1/n)^h$ for $h = 0, 1, \ldots$. If $n_{s_h} = n_{s_{h-1}}$, then $A(v_k, s_{h-1}\rho_k, s_h\rho_k)$ is a root-free annulus with relative width $1 + 1/n$, that is a $(1, 1)$ annulus. Clearly, $n_{s_h} > n_{s_{h-1}}$ for at most $n - n/12$ integers h because at most $n - n/12$ roots lie outside the disc $D(v_t, \rho_t)$. Therefore we have a $(1, 1)$ separating annulus $A(v_t, s_{h-1}\rho_t, s_h\rho_t)$ for $h \le n - n/12$. It supports approximate factorization (7). Note that

$$s_h \le (1 + 1/n)^h \le (1 + 1/n)^n < 3 \text{ for } s \le n. \tag{14}$$

Now it remains to ensure the degree bound (8). We achieve this by applying the above construction to a proper complex v_0. In its computation we keep assuming that sufficiently close approximations z_j to the n distinct roots x_j of $p(x)$ are available for $j = 1, \ldots, n$. Furthermore assume that the n roots are distinct, $n = 4k$ is divisible by 4, $\Re(z_j) \le \Re(z_{2k}) < \Re(z_l)$ for $j < 2k$ and $l > 2k$, $\Im(z_j) \le \Im(z_k) < \Im(z_l)$ for $j < k$ and $k < l \le 2k$, and $\Im(z_j) \le \Im(z_{3k}) < \Im(z_l)$ for $2k < j < 3k$ and $3k < l$. We yield these bounds by properly enumerating the n roots having $2n$ distinct projections on the real and imaginary axes. The latter property holds with probability 1 under random rotation of the complex plane.

Now we proceed as follows.

Procedure 1. Computing a center v_0 in the search for $(1,1)$ annulus.

COMPUTATIONS:

    1. Compute the half-sum $a = 0.5(\Re(z_{2k}) + \Re(z_{2k+1}))$.

    2. Compute the three half-sums $a_1 = 0.5(\Im(z_k) + \Im(z_{k+1}))$,
        $a_2 = 0.5(\Im(z_{3k}) + \Im(z_{3k+1}))$ and $b = 0.5(a_1 + a_2)$.

OUTPUT $v_0 = a + b\sqrt{-1}$.

Theorem 5. *Suppose that Procedure 1 has output a complex v_0 and that t recursive applications of Corollary 1 have produced a complex value v_t and positive ρ_t such that the disc $D(v_t, \rho_t)$ contains at least $n/12$ roots of $p(x)$, whereas its 6-dilation $D(v_t, 6\rho_t)$ does not contain the point v_0 (cf. (13)). Then at least $n/4$ roots of $p(x)$ lie at the distance at least $2\rho_t$ from the disc $D(v_t, \rho_t)$.*

Proof. Partition the complex plane into the four domains D_1, D_2, D_3, and D_4 bounded by the straight line $L = \{x : \Re(x) = a\}$ and two half-lines $R_1 = \{x : \Re(x) \le a$ and $\Im(x) = a_1\}$ and $R_2 = \{x : \Re(x) \ge a$ and $\Im(x) = a_2\}$, both orthogonal to the line L. Let us show that the disc $D(v_t, \rho_t)$ cannot lie at the distances less than $2\rho_t$ from all these domains simultaneously.

Indeed otherwise its 3-dilation $D(v_t, 3\rho_t)$ would have intersected all four domains. Then the diameter $6\rho_t$ of this dilation would have exceeded the distance $|a_2 - a_1|$ between the points a_1 and a_2. Moreover, being convex, the dilation would have also intersected both half-lines R_1 and R_2, and consequently (invoke convexity again) the line interval $\{y : \Im(a_2) \le y/\sqrt{-1} \le \Im(a_1)\}$ as well. Therefore, the distance of v_0 from this 3-dilation $D(v_t, 3\rho_t)$ would have been less than $|a_1 - a_2|/2 \le 3\rho_t$. It would have followed that the 6-dilation $D(v_t, 6\rho_t)$ of the disc $D(v_t, \rho_t)$ contained v_0. This would have contradicted the assumption of the theorem, and therefore, the distance of $D(v_t, \rho_t)$ to at least one of the four domains D_1, D_2, D_3, and D_4 is at least $2\rho_t$. The theorem follows because by definition each of the four domains contains exactly $k = n/4$ roots of $p(x)$.

Recall that the disc $D(v_t, \rho_t)$ contains at least $n/12$ roots of $p(x)$, combine the latter theorem, Corollary 1 and the argument that we used for deducing (14), and obtain a desired $(1,1)$ separating annulus $A(v_0, \rho_t, r_+)$ for $r_+ < 3\rho_t$.

Remark 1. The paper McNamee and Pan (2012) has estimated the efficiency of great many known root-refiners. It also has pointed out that bound (5) can rely on recursive factorization, but has omitted the issue of computing (c, d) annuli for recursive splitting and relied on numerically unstable computation of PFDs described in our Section 2.

4   Some Implementation Issues

A polynomial root-finder based on factorization was implemented by X. Gourdon in the PARI Functions and was superseded by the subroutines of MPSolve (based on Ehrlich–Aberth's algorithm) and Eigensolve (based on combining the WDK and Rayleigh Quotient iterations). The prospects for factorization-based root-refining are much brighter because the hardest stage of initialization is skipped and our refinement algorithms achieve faster progress by balancing the degrees of the factors and dramatically simplifying the computation of the wide root-free separating annuli. For a large class of input polynomials, heuristic computation of these annuli can be further simplified. One can frequently find them already in a single, double or triple application of Theorem 4 without using Corollary 1 and can try other promising heuristic recipes. E.g., examine various regions on the complex plane between pairs of lines orthogonal to the real axis and passing through the consecutive values in the set $\{\Re(x_1), \ldots, \Re(x_n)\}$ of the projections of the given approximate roots onto this axis. Then transform the complex plane onto itself to move these lines into circles and examine the annuli between them.

Our root-refiners are amenable to parallel acceleration, but in that respect they can be superseded by the iterations directed to a single root such as Newton's root-finder or Rayleigh Quotient Iterations for eigenvalues of the companion matrix of the polynomial $p(x)$ (cf. Pan and Zheng (2011a)). Namely, assume that we concurrently apply such iterations at h crude but sufficiently close initial approximations to h distinct roots for $1 < h \leq n$ (see Pan and Zheng (2011b)). Part of the processes can diverge or can converge to the same root of $p(x)$, although by using sufficiently many distinct initial points one can ensure convergence of Newton's root-finder to all n roots (cf. Bollobàs, Lackmann, and Schleicher (2012)). As their very attractive but apparently yet unexploited feature, this particular parallel processing requires no data exchange among the h processors, thus allowing acceleration of the refinement by a factor h.

Appendix

A   On Expanding the Set of Constraints and Variables

Pan and Zheng (2011b) suggest that the reduction of root-finding for a univariate polynomial of a degree n to the multivariate polynomial system of n Viète's (Vieta's) polynomial equations with n unknowns can explain the empirical strength of global convergence of the WDK and Ehrlich–Aberth iterations, based on such a system. The authors argue that multiple additional constraints keep the iterative process on its course to convergence stronger than the single polynomial equation can do. If true, this suggests a more general recipe of properly expanding an original system of constraints to support more reliable convergence of its iterative solution. Empirical global convergence of iterative solution of a polynomial equation is also quite strong for some companion matrix algorithms (see Pan and Zheng (2011a)). Then again the algorithms compute the solution to n constraints defining n unknowns: namely, an eigenvalue and an eigenvector of dimension n (defined up to scaling), versus the single constraint defined by a univariate polynomial equation. Yet another example is the known effect of using the duality in linear and nonlinear programming and in the solution of a multivariate system of polynomial equations (see Mourrain and Pan (2000) and Faugère (2002)). Should these examples motivate further attempts of improving global convergence of iterative solution to a system of constraints (in particular a system of multivariate polynomial equations) by means of their proper expansion with additional constraints and variables? The idea has strong support from many proverbs such as "One's as good as none", "There's strength in numbers", "One man does not make a team" (see more in Pan and Zheng (2011b)), but does not seem to be yet explicitly proposed in sciences.

Acknowledgement. Our work has been supported by NSF Grant CCF-1116736 and PSC CUNY Award 64512-0042. We are also very grateful to the reviewers for helpful comments.

References

1. Aberth, O.: Iteration Methods For Finding All Zeros of a Polynomial Simultaneously. Mathematics of Computation 27(122), 339–344 (1973)
2. Aitken, A.C.: On Bernoulli's numerical solution of algebraic equations. Proc. Roy. Soc. Edin. 46, 289–305 (1926)
3. Barnett, S.: Polynomial and Linear Control Systems. Marcel Dekker, NY (1983)
4. Bell, E.T.: The Development of Mathematics. McGraw-Hill, New York (1940)
5. Bini, D., Pan, V.Y.: Polynomial and Matrix Computations, Fundamental Algorithms, vol. 1. Birkhäuser, Boston (1994)
6. Bollobàs, B., Lackmann, M., Schleicher, D.: A small probabilistic universal set of starting points for finding roots of complex polynomials by Newton's method. Math. of Computation (in press, 2012), arXiv:1009.1843
7. Box, G.E.P., Jenkins, G.M.: Time Series Analysis: Forecasting and Control. Holden-Day, San Francisco (1976)
8. Boyer, C.A.: A History of Mathematics. Wiley, New York (1968)
9. Curry, J.H.: On zero finding methods of higher order from data at one point. J. of Complexity 5, 219–237 (1989)
10. Durand, E.: Equations du type F(x) = 0: Racines d'un polynome, In Solutions numérique équation algébrique, Masson, Paris, vol. 1 (1960)
11. Demeure, C.J., Mullis, C.T.: A Newton–Raphson method for moving-average spectral factorization using the Euclid algorithm. IEEE Trans. Acoust., Speech, Signal Processing 38, 1697–1709 (1990)
12. Ehrlich, L.W.: A modified Newton method for polynomials. Comm. of ACM 10, 107–108 (1967)
13. Faugère, J.C.: A new efficient algorithm for computing Gröbner bases without reduction to zero (F5). In: Proc. ISSAC 2002, pp. 75–83. ACM Press, NY (2002)
14. Householder, A.S.: Dandelin, Lobachevskii, or Graeffe. American Mathematical Monthly 66, 464–466 (1959)
15. Kantorovich, L.V., Akilov, G.P.: Functional Analysis. Pergamon Press (1982)
16. Kerner, I.O.: Ein Gesamtschrittverfahren zur Berechnung der Nullstellen von Polynomen. Numerische Mathematik 8, 290–294 (1966)
17. Kim, M.-H.: Computational complexity of the Euler type algorithms for the roots of complex polynomials. PhD Thesis, City University of New York (1985)
18. Kirrinnis, P.: Polynomial factorization and partial fraction decomposition by simultaneous Newton's iteration. J. of Complexity 14, 378–444 (1998)
19. Mahler, K.: An Inequality for the Discriminant of a Polynomial. Michigan Math. Journal 11, 257–262 (1964)
20. McNamee, J.M.: A 2002 update of the supplementary bibliography on root of polynomials. J. Comput. Appl. Math. 142, 433–434 (2002)

21. McNamee, J.M.: Numerical Methods for Roots of Polynomials (Part 1). Elsevier, Amsterdam (2007)
22. McNamee, J.M., Pan, V.Y.: Efficient polynomial root-refiners: survey and new record estimates. Computers and Math. with Applics. 63, 239–254 (2012)
23. Mourrain, B., Pan, V.Y.: Multivariate polynomials, duality and structured matrices. J. of Complexity 16(1), 110–180 (2000)
24. Muller, D.E.: A method for solving algebraic equations using an automatic computer. Math. Tables Aids Comput. 10, 208–215 (1956)
25. Neff, C.A., Reif, J.H.: An $o(n^{1+\epsilon})$ algorithm for the complex root problem. In: Proc. STOC 1994, pp. 540–547. IEEE Computer Society Press (1994)
26. Ostrowski, A.M.: Recherches sur la méthode de Graeffe et les zéros des polynomes et des séries de Laurent. Acta Math 72, 99–257 (1940)
27. Ostrowski, A.M.: Solution of Equations and Systems of Equations, 2nd edn. Academic Press, New York (1966)
28. Pan, V.Y.: Optimal (up to polylog factors) sequential and parallel algorithms for approximating complex polynomial zeros. In: Proc. 27th Ann. ACM Symp. on Theory of Computing, pp. 741–750. ACM Press, New York (1995)
29. Pan, V.Y.: Optimal and nearly optimal algorithms for approximating polynomial zeros. Computers and Math. with Applications 31(12), 97–138 (1996)
30. Pan, V.Y.: Solving a polynomial equation: some history and recent progress. SIAM Review 39(2), 187–220 (1997)
31. Pan, V.Y.: Solving polynomials with computers. American Scientist 86 (January-February 1998)
32. Pan, V.Y.: Univariate polynomials: nearly optimal algorithms for factorization and rootfinding. J. Symbolic Computation 33(5), 701–733 (2002)
33. Pan, V.Y.: Amended DSeSC Power Method for polynomial root-finding. Computers and Math (with Applications) 49(9-10), 1515–1524 (2005)
34. Pan, V.Y., Zheng, A.–L.: New progress in real and complex polynomial rootfinding. Computers and Mathematics with Applications 61, 1305–1334 (2011a)
35. Pan, V.Y., Zheng, A.–L.: Root-finding by expansion with independent constraints. Computers and Mathematics with Applications 62, 3164–3182 (2011b)
36. Petkovic, M.S., Herceg, D.: Point estimation of simultaneous methods for solving polynomial equations: a survey. Computers Math. with Applics. 136, 183–207 (2001)
37. Renegar, J.: On the worst-case arithmetic complexity of approximating zeros of polynomials. J. of Complexity 3, 90–113 (1987)
38. Schönhage, A.: The fundamental theorem of algebra in terms of computational complexity. Department of Math., University of Tübingen, Germany (1982)
39. Smale, S.: The fundamental theorem of algebra and complexity theory. Bulletin of the American Mathematical Society 4, 1–36 (1981)
40. Smale, S.: Newton's method estimates from data at one point. In: Ewing, R.E., Cross, K.I., Martin, C.F. (eds.) The Merging Disciplines: New Directions in Pure, Applied and Computational Math., pp. 185–196. Springer (1986)
41. Van Dooren, P.M.: Some numerical challenges in control theory. Linear Algebra for Control Theory IMA Vol. Math. Appl (1994)
42. Weierstrass, K.: Neuer Beweis des Fundamentalsatzes der Algebra. Mathematische Werke, Band III, Mayer und Müller, Berlin, 251–269 (1903)
43. Wilson, G.T.: Factorization of the covariance generating function of a pure moving-average process. SIAM J. on Numerical Analysis 6, 1–7 (1969)

PoCaB: A Software Infrastructure to Explore Algebraic Methods for Bio-chemical Reaction Networks

Satya Swarup Samal[1], Hassan Errami[2], and Andreas Weber[3]

[1] Bonn-Aachen International Center for Information Technology, Universität Bonn, Dahlmannstraße 2, D-53113, Bonn, Germany
samal@cs.uni-bonn.de
[2] Institut für Mathematik, Universität Kassel, Heinrich-Plett-Straße 40, 34132 Kassel, Germany
errami@cs.uni-bonn.de
[3] Institut für Informatik II, Universität Bonn, Friedrich-Ebert-Allee 144, 53113 Bonn, Germany
weber@cs.uni-bonn.de

Abstract. Given a bio-chemical reaction network, we discuss the different algebraic entities e.g. stoichiometric matrix, polynomial system, deficiency and flux cones which are prerequisite for the application of various algebraic methods to qualitatively analyse them. We compute these entities on the examples obtained from two publicly available biodatabases called Biomodels and KEGG. The computations involve the use of computer algebra tools (e.g. polco, polymake). The results consisting of mostly matrices are arranged in form of a derived database called *PoCaB* (Platform of Chemical and Biological data). We also present a visualization program to visualize the extreme currents of the flux cone. We hope this will aid in the development of methods relevant for computational systems biology involving computer algebra. The database is publicly available at http://pocab.cg.cs.uni-bonn.de/

1 Introduction

In Systems Biology, the biological functions emerge from the interaction of several chemical species. So, it is desirable to study the set of components, which give rise to a discrete biological function. This leads to the formulation of *functional modules*. The interplay between these modules represent the diversity and richness of living systems [1]. To understand this, the first step is to analyse different networks responsible for biological functions. For this, there exists several publicly available databases which provide a diverse set of pathways and biological models e.g. Biomodels [2], KEGG [3], MetaCyc [4], etc. The data in such databases can be browsed using different biological functions, diseases, molecular complexes, etc. Moreover, the reactions and the chemical species can be downloaded in XML (Extensible Markup Language) [5], which is both computer readable and human legible.

V.P. Gerdt et al. (Eds.): CASC 2012, LNCS 7442, pp. 294–307, 2012.

In chemical systems and systems biology, reactions networks can be represented as a set of reactions. If it is assumed that they follow mass action kinetics [6], then the dynamics of these reactions can be represented by differential equations. Particularly, in complex systems it is difficult to estimate the values of the parameters of these equations, hence the simulation studies involving the kinetics is a daunting task. Nevertheless, quite a few things about the dynamics can be concluded from the structure of the reaction network itself. In this context there has been a surge of algebraic methods, which are based on the structure of the network and the associated stoichiometry of the chemical species. These methods provide a way to understand the qualitative behaviour (e.g. steady states, multistability, oscillations, stability) of the network. So, for a reaction network, one of the initial tasks is its mathematical representation and the subsequent tasks involve application of algebraic methods to gain insight into its qualitative behaviour. In this paper we describe a general framework for such an analysis on different biological models obtained from two publicly available databases. The purpose of this framework is twofold:

- It can serve as a manual for chemists and biologists to apply these algebraic methods in a systematic manner and interpret the results. It will also help us to formulate subsequent computational questions on the applicability, pros and cons of such methods for large and diverse datasets.
- By providing tools to extract relevant algebraic entities out of the network description such as stoichiometric matrices and their factorizations, flux cones, polynomial systems, deficiencies and differential equations, we generate a large *derived database* of examples that can be used by people working in computer algebra to benchmark their algorithms.

The main steps are as follows:

1. Firstly, we parse the reaction networks in public databases to generate a graph theoretic representation of reaction networks.
2. Secondly, we concentrate on the computation of extreme currents [7] and deficiency of the network [8].
3. Thirdly, as these extreme currents can be understood in terms of relevant pathway in a biological network so, one of the ways to interpret it is to visualize it, so we describe a Java based program to visualize the extreme currents.

2   The Building Blocks for the System

2.1   Data Source

Biomodels: We selected 275 biochemical reaction networks from Biomodels database in System Biology Markup Language (SBML). These models can be browsed by name of the disease, biological process and molecular complex.

KEGG: The KEGG database is another repository of biological pathways. The KEGG pathways can be downloaded in KGML format. For our analysis we downloaded a precompiled list of KEGG files in SBML format from http://www.systems-biology.org/resources/model-repositories/000275.html (Last accessed 22th March 2012) with specifications SBML Level-2 Ver-1 (without CellDesigner tag). We selected 103 models with organism code hsa (*Homo sapiens*). In addition, if downloaded in KGML format the files can be converted to SBML using KEGG translator [9].

However, our framework is not restricted to these databases but can be used for all sources that provide data in SBML form.

2.2   Representation of Reaction Networks in Databases

As discussed above, the reactions present in the biodatabases are present in XML based format. For the current analysis, we obtained data in SBML format [10] which is a XML based format to communicate biochemical reaction network consisting of metabolic pathways, signalling pathways, gene regulation pathways, etc and is also software independent. Further details about the various specifications and components can be found in the SBML tutorial [11].

2.3   Graph-Theoretic Representation of the Reaction Systems

Based on the notion that reaction networks can be modelled using differential equations, the rate of change of every chemical species (x) in such a network is denoted by $\dot{x}$ and can be represented as

$$\dot{x} = Y I_a I_K \Psi(x) \tag{1}$$

This representation has 3 matrices namely Y, I_a, I_K and one vector $\Psi(x)$. Let the reaction network has l reactions and m species. Here we follow the notation and terminology used in [12,13,14]. The above differential equation system can be represented by the use of two graphs, a weighted directed graph and a bipartite undirected graph. In addition to this, both sides of a reaction (i.e. products and reactants) are arranged in form of complexes (the complex may consist of a single species or a combination of species) as shown in Eq. 2.

$$C = \sum_{i=1}^{m} y^i x(i) \tag{2}$$

where y^i are numbers denoting the stoichiometry of the species $x(i) \ldots x(m)$. Let the network has n complexes. In the directed graph, there exists a directed edge between the two complexes describing a reaction. The edge weight is the rate constant of the reaction. From this graph two incidence matrices are defined I_a and I_K. The I_a is a n-by-l matrix and has the information whether the complex is present as a reactant (entry -1) or product (entry 1) vertex of the graph. The I_K is a l-by-n matrix that has non zero entries only for reactant vertices where the entry is the weight of the edge which is the rate constant of the reaction. The bipartite graph contains the set of complexes and the set of chemical species as vertices. If the complex contains a species then there exists an edge between them with the edge weight equals to the stoichiometry of the chemical species. The adjacency matrix of this graph is denoted by Y matrix (m-by-n). As per mass action kinetics every reaction has a certain velocity called the *reaction rate*, which is a monomial. These monomials can be obtained by the columns y_j of Y, assigning to each complex a monomial. This results in the mapping $\Psi_j(x) = x^{y_j}$, $j = 1 \ldots n$. These things are illustrated with an example (Eq. 4, 5, 6, 7). The product of $I_K \Psi(x)$ denotes the vector of monomials describing the flux of the reactions and the product $Y I_a$ denotes the stoichiometric matrix.

The matrix Y is an integer matrix with non-negative entries, I_a is an integer integer matrix with entries in the set $\{-1, 0, 1\}$, I_K a matrix consisting of symbolic parameters k_{ij} ranging over non-negative real values. The polynomial system is hence one over the fraction field of the polynomials in variables $x_1, \ldots, x_m$ and parameters $k_0, \ldots, k_{l-1}$ over the field of real algebraic numbers. Moreover, all variables and parameters are restricted to non-negative values. The Eq. 1 can be also represented as:

$$\dot{x} = Sv \qquad (3)$$

where $S = Y I_a$ and $v = I_K \Psi(x)$. In addition another matrix of interest is the kinetic order matrix [15] or Y_L matrix (m-by-l) (Eq. 8), which describes the exponents of monomials in $I_K \Psi(x)$. To illustrate further, a biochemical reaction from [16] representing fructose-2,6- bisphosphate cycle is presented in Fig. 1(a), 1(b), 1(c). Here S1 is fructose-6-phosphate and S2 is fructose- 2,6-bisphosphate and $x1, x2$ denote their respective concentrations.

$$I_a = \begin{bmatrix} -1 & 1 & -1 & -1 & 1 \\ 1 & -1 & 1 & 0 & 0 \\ 0 & 0 & 0 & 1 & -1 \end{bmatrix} \qquad (4)$$

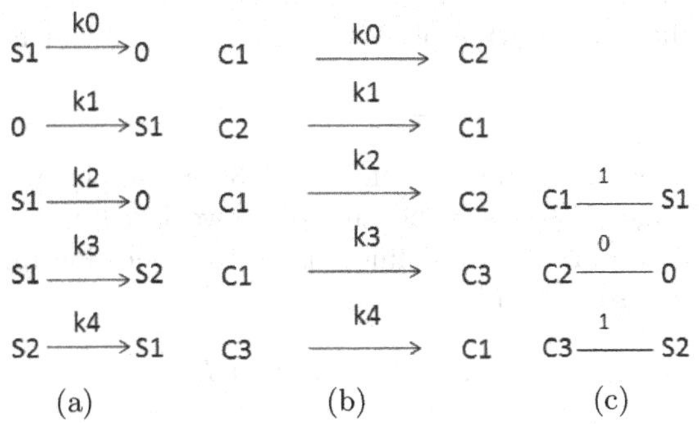

(a)   (b)   (c)

Fig. 1. Graph-theoretic representation (a) Chemical reaction system (b) Weighted directed graph (c) Bipartite undirected graph

$$I_K = \begin{bmatrix} k_0 & 0 & 0 \\ 0 & k_1 & 0 \\ k_2 & 0 & 0 \\ k_3 & 0 & 0 \\ 0 & 0 & k_4 \end{bmatrix} \tag{5}$$

$$Y = \begin{bmatrix} 1 & 0 & 0 \\ 0 & 0 & 1 \end{bmatrix} \tag{6}$$

$$\Psi(x) = \begin{bmatrix} x_1 \\ 1 \\ x_2 \end{bmatrix} \tag{7}$$

$$Y_L = \begin{bmatrix} 1 & 0 & 1 & 1 & 0 \\ 0 & 0 & 0 & 0 & 1 \end{bmatrix} \tag{8}$$

The corresponding differential equations are:

$$\dot{x}_1 = -1 \cdot k_0 \cdot x_1^1 + 1 \cdot k_1 - 1 \cdot k_2 \cdot x_1^1 - 1 \cdot k_3 \cdot x_1^1 + 1 \cdot k_4 \cdot x_2^1$$
$$\dot{x}_2 = 1 \cdot k_3 \cdot x_1^1 - 1 \cdot k_4 \cdot x_2^1$$

Additionally, the number of linkage classes, which will be utilized in Sect. 2.4, can be found from the weighted directed graph, i.e. it is the set of connected complexes. So, in this case there is one linkage class. Further details concerning complexes and linkage classes can be found in [8]. The graph-theoretic approach leads to solutions having a graph theoretic meaning [13] and also helpful for stability analysis [14].

2.4   Deficiency Value of the Reaction Network

Deficiency is a non negative integer for a reaction network which is an invariant of the network. In this context, two well known theorems are available which are Deficiency Zero and Deficiency One theorems respectively [8]. The first step in this direction is the computation of this deficiency value which is given by the following formula:

$$\delta = n - t - s \tag{9}$$

where n is the number of complexes (cf. Sect. 2.3 for complexes) in the network, t is number of linkage classes, s is the rank of network. Alternatively and equivalently the deficiency can also be defined using the following formula using the graph theoretic representation [12]:

$$\delta = \text{rank}(I_a) - \text{rank}(YI_a) \tag{10}$$

For the example in Fig. 1(a) the deficiency value is 0. This deficiency value enables to classify the reaction networks into the kind of dynamics they can possibly exhibit. The exact application of the deficiency zero and one theorems require some additional information which is mentioned in Sect. 3.

2.5   Extreme Currents of the Flux Cone

In steady state condition i.e. $\dot{x} = 0$, the Eq. 3 reduces to a system of homogenous linear equations (with unknown being v i.e. the flux of the reactions). The solution set of this system can be described as a polyhedron(or flux cone), given by the following formalism:

$$P = \{v \in \mathbb{R}^n : Sv = 0, Dv \geq 0\} \tag{11}$$

where D is a $l \times l$ diagonal matrix. $D_{i,i} = 1$ if the flux for ith reaction is irreversible, and 0 otherwise. Two types of constraints are considered while defining the cone. The first one is the steady state or equality constraint ($YI_a = 0$) and the other one is irreversibility or inequality constraint ($I_K \Psi(x) \geq 0$). As shown above, the flux cone will be present in non-negative orthant of the vector space spanned by fluxes and in order to achieve this, reversible reactions are split into forward and backward reactions. So, the irreversibility constraint is maintained. The edges of this flux cone are called as *extreme currents*. This is represented by a f-by-l matrix, where f denotes the number of extreme currents. This is also unique for a network, except for permutation of rows and arbitrary scaling factor to each row. Any possible steady state of the system is a non negative combination of these extreme currents. Mathematically this means any steady state of the network can be denoted by $v = Ej$, where E is the extreme current matrix and the j is the positive weighting. For the example in Fig. 1(a) the extreme current matrix is the following:

$$E = \begin{bmatrix} 1\,1\,0\,0\,0 \\ 0\,0\,0\,1\,1 \\ 0\,1\,1\,0\,0 \end{bmatrix} \tag{12}$$

In such a matrix, the first row is the first extreme current and the non-zero entry implies the reactions that carry the flux (active ones). This co-ordinate transformation of network to the flux cone can be also used for stability analysis [14].

The reversibility of reactions in the network can be tackled by different ways. In extreme currents mentioned above the reversible reactions are split but in elementary flux modes [17] they are not split and in extreme pathways [18] some may be split. The different ways to split the reactions affects the construction of stoichiometric matrix and hence different methods describe the cone in different vector spaces due to presence of reversible reactions[19,20]. It also affects the construction of cone, for extreme currents computation the cone is always pointed whereas for elementary flux modes and extreme pathways it may neither be pointed nor remain in non-negative orthant [20].

3   Software Workflow and Components

Pre-processing Step: From the data source (as described in Sect. 2.1) the networks were downloaded. Although the models in these databases are annotated and curated, still as a part of general framework we have a possibility to balance the reactions. This removes stoichiometric inconsistencies present in the model but works only when the annotation of model is correct and the chemical formula of species can be found. This works on the principle of mixed integer linear programming (MILP) and was done automatically using the Subliminal toolbox [21]. We present the results of this analysis only for Biomodels database. However, this step is optional and we report the results with and without balancing.

Main Steps:

1. The files were parsed by a Java based program to generate the Y, I_a , stoichiometric (YI_a) and Y_L matrices along with the basic information about the model concerning the number of species, reactions, complexes, rank of stoichiometric matrix and nullity of stoichiometric matrix respectively. While computing the various matrices the reversible reactions were split into forward and backward reactions and it increased the dimension of the stoichiometric matrix by one for every reversible reaction. To parse the SBML program using Java, the JSBML library was used [22]. The graph theoretic representation of the network(cf. Sect. 2.3) was done using the JGraphT Java Library [23].

2. The deficiencies of the networks were computed using the *ERNEST* library [24], which is a Matlab based program. This program also tests whether the Deficiency zero or one theorems are applicable and presents the result. We report only the deficiency value and for additional conditions for deficiency theorems, this tool can be used e.g. for computation of weak reversibility, linkage classes. We also computed the deficiency from Eq. 10 and it was equivalent to *ERNEST*.

3. The stoichiometric matrix acts as an input to Java based *polco* program available at http://www.csb.ethz.ch/tools/polco implementing double description algorithm [25] to compute the extreme currents. Apart from polco, there exists a matlab based library called metatool [26] to compute these entities. These two programs mainly concern to biochemical systems. In computational geometry these problems are solved with the help of libraries like polymake[27]. In polymake there are three different algorithms to compute the convex hull. We tried to solve all the examples of our database using all these four different approaches and summarize the results in Table 1. The different running times are reported. It can be seen the current implementation (i.e. *polco*) outperformed the other three approaches but we found five examples for which neither our current implementation nor the other three approaches succeeded. So, wherever polco failed none of the other approaches succeeded. Here failed means either there is out of memory exception or the computation time exceeded 1800 seconds and was forcefully terminated. These five models were not included in our final database.

4. The reaction network, can be also represented by inequalities and equations. One benefit of this representation is that different constraints can be put on individual reactions. In biology, the networks operate under different constraints[28], one important constraint is the effect of gene regulation

in which some genes are differentially expressed [29]. The changes in gene expression levels affect the reaction rates as the reactions are governed by enzymes which are gene products. One of the ways to model this phenomenon is to use inequalities [30]. The following formalism illustrates the above points:

$$YI_aI_K\Psi(x) = 0 \tag{13}$$

$$I_K\Psi(x) \geq 0 \tag{14}$$

$$I_K\Psi(x) \leq \beta_i \quad (i = 0, 1, \ldots, l-1) \tag{15}$$

where l denotes the number of reactions. Eq. (15) denotes the constraint on the flux of a reaction and hence is optional. In the current analysis we have not accounted for any constraints, so in a way the flux cone computed is maximal where all reactions occur at their maximal rate. We systematically generated this type of file for all the examples (also a part of our derived database)and was used for the computation of extreme currents using polymake. Apart from extreme current computation polymake offers to deduce different properties of the flux cone. Further details can be found at http://www.polymake.org/doku.php. One of the advantages of using this tool is the different choice of algorithms for convex hull computation. Further theoretical exploration into different properties of polyhedron can be found in [31].

5. The extreme currents are analogous to the pathways in reaction networks, so it is often desirable to visualize them with respect to their position in the network. This will enable some visual analytics to discover the distribution of extreme currents. We implemented a Java based program to dynamically visualize these extreme currents. We used *JUNG* Java library for this purpose [32], which offers different layout algorithms for visualization (Fig. 3). This results in a Java applet with pan and zoom functionality.

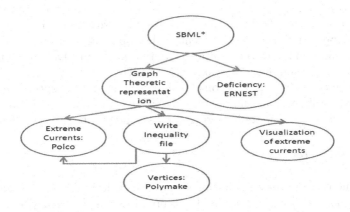

Fig. 2. Flowchart outlining the main steps. * denotes either SBML file directly from database or SBML file after balancing (cf. Sect. 3).

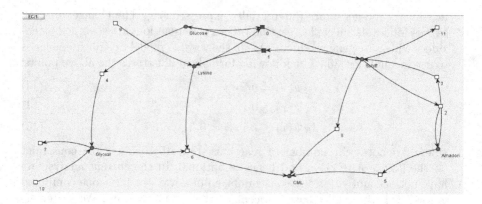

Fig. 3. Output from visualization program with FRLayout showing reaction network as a directed graph. One extreme current is shown at one time and it can be changed using the button at top left corner denoted by EC. The square nodes denote the reactions. The dark coloured square nodes show the active reactions (or carrying flux) in the current extreme current, while the other nodes are inactive reactions. The species are the oval shaped nodes and dark coloured. The directed edge's head points to product(s) while the tail to reactant(s) of the corresponding reaction.

4   Results

4.1   Database of Algebraic Entities

1. All the matrices including the extreme current matrix along with the polynomial system and the differential equation files are stored as text files using a delimiter (,).
2. In the above files mainly in the I_K, $I_K\Psi(x)$, polynomial system and differential equation files, the species name are mapped to certain variable. The rate constants are also mapped to corresponding reactions where they occur. This mapping information is present as a Mapping file. This file also contains the information about the reactions and the species involved in the network.
3. The statistical summary for Biomodels and KEGG is represented in Table 2.
4. A summary of the computations involving the number of species, reactions, complexes, dimensions, nullity and rank of stoichiometric matrix, nullity of Y and deficiency information is presented as a spreadsheet inside the database.

5   Discussions

We constructed a database having algebraic entities derived from biological reaction networks and using it computed two properties i.e. deficiency and extreme currents. From the results it can be seen that the number of extreme currents does not always correspond to the size of the network as seen in Biomodels database. Also the size of network doesn't relate to the deficiency also pointed

Table 1. Extreme current computations

	Biomodels (UnBalanced) (Time in seconds)
Number of Models	275
Average time with lrs(polymake)*	0.635
Average time with cdd(polymake)*	0.408
Average time with beneath_beyond(polymake)*	0.433
Average time with polco	0.055
Average time with metatool	0.203
# Models failed with lrs(polymake)*	12
# Models failed with cdd(polymake)*	5
# Models failed with beneath_beyond(polymake)*	5
# Models failed with polco(polymake)*	5
# Models failed with metatool(polymake)*	28
Models failed for all the approaches	BIOMOD(019,183,255,256,268)

*denotes the convex hull computation algorithms in polymake. The average time reported are calculated based on those examples where all the four algorithms were successful (in this case 247 out of 275 models).

Table 2. Summary of results in Biomodels and KEGG database

	UnBalanced Biomodels	KEGG	Balanced Biomodels
Number of Models	270	103	236
Maximum number of reactions* in a model	194	132	194
Maximum number of species in a model	120	139	120
Number of models with reactions upto 10	94	48	117
Number of models with reactions upto 50	237	97	213
Maximum number of EC* in a model	5130	282	5130
Dimension of SM* with maximum EC	17×48	139×132	19×48
Models with deficiency = 0	159 (58.8%)	80 (77.6%)	154 (65.2%)
Models with deficiency = 1	33 (12.2%)	11 (10.6%)	28 (11.8%)
Highest Deficiency	63	24	63
SM with highest deficiency	36×94	139×132	39×94
Maximum rank of SM	94	67	94
Maximum nullity of SM	100	65	100
Number of models with zero EC	24 (8.88%)	38 (36.8%)	50 (21.1%)

*The reactions here refer to columns of SM. SM = Stoichiometric matrix. EC=Extreme Currents.

out in [8]. But there are also some models displaying high deficiency with large dimensions and there is a need of improved algorithms or new approaches to address such systems. It can be seen around 71.1% (unbalanced) and 77.1 % (balanced) models in Biomodels correspond to deficiency one or zero. Similarly 88 % (unbalanced) of KEGG models correspond to this criteria, this implies the existing deficiency theorems are applicable to a large extent. The effect of balancing the reactions can also be seen in Table 2, there may be changes in the number of species and reactions during the balancing and this affects the number of extreme currents and deficiency computation. As our derived database contain diverse examples, it provides a corpus to test and benchmark different algebraic methods and designate the methods working for a particular class of examples. This will eventually lead to partitioning the database into classes which may be suitable for some methods and unsuitable for others. A natural partitioning occurs for examples with deficiency zero or one and there exists theorems to apply on such examples. Additional type of partitioning can be based on the dimension of various matrices, number of extreme currents and one such possibility is presented in Table 2 which is based on the number of reactions up to 10 and 50. There also exist tools e.g. CellDesigner [33] with graphical interface to encode reaction networks and export them to SBML format.

6   Conclusion and Future Work

Using the automation pipeline and the different conversion and computation tools described in this paper novel algebraic algorithms can be used to compute various structural information of reaction networks. For single examples one can perform algebraic algorithms without the need of a *database*. However, storing the information in a database we also have benchmarking collections for algebraic algorithms. Especially, we can extract benchmarks obeying certain biological and algebraic properties, e.g. retrieving all models of metabolic networks of humans that have deficiency higher than 1, with rank of stoichiometric matrix bigger than say 50 but with dimension of the nullspace of the stoichiometric matrix being small. Without storing such information in a database, one had to compute the relevant entities for all examples first, instead of being able to query them using standard relational database techniques.

Another aspect in the use of a database is to store information that might have required considerable computation time to come up with. Some examples of those are the following:

- Is the network one with *toric steady states* [34]? Not all reaction networks have toric steady states, but many it has been shown that many non-trivial examples have this property. Unfortunately, up to now only algorithms with at least exponential complexity are known to test for this property (see e.g. [35]), so storing the information once it has been obtained yields a significant benefit over recomputing it again and again.
- Similarly, more detailed information about the number of equilibria per set of reaction constants should be stored. General theorems about their uniqueness are only available in the context of deficiency zero and deficiency one theorems, but this property presumably holds for much larger classes (as was already conjectured by Clarke in 1981 [36]).
- There are some theorems with respect to the *stability* of the fixed points of the flow [15,37], but storing it in the database for any example, for which it can be computed (by which method it be) gives not only information with

respect to the biological applications, but might also help to establish new mathematical theorems by means of experimental mathematics.

- All of the above statements also apply to more complex questions on the dynamics of the reaction network, such as existence of Hopf bifurcation fixed points [14,38,39] or even globally the question for existence or absence of oscillations [40,41,42].

Our database structure is open to include these information and will be extended by it for any example once it has been computed.

Acknowledgement. This research was supported in part by *Deutsche Forschungsgemeinschaft* within SPP 1489.

References

1. Hartwell, L.H., Hopfield, J.J., Leibler, S., Murray, A.W.: From molecular to modular cell biology. Nature 402(6761 suppl.), C47–C52 (1999)
2. Bornstein, B., Broicher, A., Nove, N.L., Donizelli, M., Dharuri, H., Li, L., Sauro, H., Schilstra, M., Shapiro, B., Snoep, J.L., Hucka, M.: BioModels Database: a free, centralized database of curated, published, quantitative kinetic models of biochemical and cellular systems. Database 34, 689–691 (2006)
3. Kanehisa, M., Goto, S., Sato, Y., Furumichi, M., Tanabe, M.: KEGG for integration and interpretation of large-scale molecular data sets. Nucleic Acids Research 40(Database issue), D109–D114 (January 2012)
4. Caspi, R., Altman, T., Dreher, K., Fulcher, C.A., Subhraveti, P., Keseler, I.M., Kothari, A., Krummenacker, M., Latendresse, M., Mueller, L.A., Ong, Q., Paley, S., Pujar, A., Shearer, A.G., Travers, M., Weerasinghe, D., Zhang, P., Karp, P.D.: The MetaCyc database of metabolic pathways and enzymes and the BioCyc collection of pathway/genome databases. Nucleic Acids Research 40(Database issue), D742–D753 (2012)
5. Bray, T., Paoli, J., Sperberg-McQueen, C.M., Maler, E., Yergeau, F.: Extensible markup language (xml) 1.0 (fifth edition). Language (2008)
6. Horn, F., Jackson, R.: General mass action kinetics. Archive for Rational Mechanics and Analysis 47, 81–116 (1972), 10.1007/BF00251225
7. Clarke, B.: Stoichiometric network analysis. Cell Biochemistry and Biophysics 12, 237–253 (1988), 10.1007/BF02918360
8. Feinberg, M.: Review Article Number 25 Stability of Complex Isothermal Reactors–I. Chemical Engineering 42(10), 2229–2268 (1987)
9. Wrzodek, C., Dräger, A., Zell, A.: KEGGtranslator: visualizing and converting the KEGG PATHWAY database to various formats. Bioinformatics (Oxford, England) 27(16), 2314–2315 (2011)
10. Hucka, M., Finney, A., Sauro, H.M., Bolouri, H., Doyle, J.C., Kitano, H.: The rest of the SBML Forum: Arkin, A.P., Bornstein, B.J., Bray, D., Cornish-Bowden, A., Cuellar, A.A., Dronov, S., Gilles, E.D., Ginkel, M., Gor, V., Goryanin, I.I., Hedley, W.J., Hodgman, T.C., Hofmeyr, J.H., Hunter, P.J., Juty, N.S., Kasberger, J.L., Kremling, A., Kummer, U., Le Novère, N., Loew, L.M., Lucio, D., Mendes, P., Minch, E., Mjolsness, E.D., Nakayama, Y., Nelson, M.R., Nielsen, P.F., Sakurada, T., Schaff, J.C., Shapiro, B.E., Shimizu, T.S., Spence, H.D., Stelling, J., Takahashi, K., Tomita, M., Wagner, J., Wang, J.: The systems biology markup language (SBML): a medium for representation and exchange of biochemical network models. Bioinformatics 19(4), 524–531 (2003)

11. Hucka, M., Smith, L., Wilkinson, D., Bergmann, F., Hoops, S., Keating, S., Sahle, S., Schaff, J.: The Systems Biology Markup Language (SBML): Language Specification for Level 3 Version 1 Core. Nature Precedings (October 2010)
12. Gatermann, K.: Counting stable solutions of sparse polynomial systems in chemistry. In: Green, E., et al. (eds.) Symbolic Computation: Solving Equations in Algebra, Geometry and Engineering, vol. 286, pp. 53–69. American Mathematical Society, Providence (2001)
13. Gatermann, K., Huber, B.: A family of sparse polynomial systems arising in chemical reaction systems. Journal of Symbolic Computation 33(3), 275–305 (2002)
14. Gatermann, K., Eiswirth, M., Sensse, A.: Toric ideals and graph theory to analyze Hopf bifurcations in mass action systems. Journal of Symbolic Computation 40(6), 1361–1382 (2005)
15. Clarke, B.L.: Stability of complex reaction networks. Advances In Chemical Physics, vol. 43 (1980)
16. Schuster, S., Dandekar, T., Fell, D.A.: Detection of elementary flux modes in biochemical networks: a promising tool for pathway analysis and metabolic engineering. Trends in Biotechnology 17(2), 53–60 (1999)
17. Schuster, S., Hlgetag, C.: On elementary flux modes in biochemical reaction systems at steady state. Journal of Biological Systems 2(2), 165–182 (1994)
18. Schilling, C.H., Letscher, D., Palsson, B.O.: Theory for the systemic definition of metabolic pathways and their use in interpreting metabolic function from a pathway-oriented perspective. Journal of Theoretical Biology 203(3), 229–248 (2000)
19. Wagner, C., Urbanczik, R.: The geometry of the flux cone of a metabolic network. Biophysical Journal 89(6), 3837–3845 (2005)
20. Llaneras, F., Picó, J.: Which metabolic pathways generate and characterize the flux space? A comparison among elementary modes, extreme pathways and minimal generators. Journal of Biomedicine & Biotechnology 2010, 753904 (2010)
21. Swainston, N., Smallbone, K., Mendes, P., Kell, D., Paton, N.: The SuBliMinaL Toolbox: automating steps in the reconstruction of metabolic networks. Journal of Integrative Bioinformatics 8(2), 186 (2011)
22. Dräger, A., Rodriguez, N., Dumousseau, M., Dörr, A., Wrzodek, C., Keller, R., Fröhlich, S., Novère, N.L., Zell, A., Hucka, M.: JSBML: a flexible and entirely Java-based library for working with SBML. Bioinformatics, 4 (2011)
23. JGraphT: A free Java graph library (2009), http://jgrapht.sourceforge.net
24. Soranzo, N., Altafini, C.: Ernest: a toolbox for chemical reaction network theory. Bioinformatics 25(21), 2853–2854 (2009)
25. Terzer, M.: Large Scale Methods to Enumerate Extreme Rays and Elementary Modes (18538) (2009)
26. Kamp, A.V., Schuster, S.: Metatool 5.0: fast and flexible elementary modes analysis. Bioinformatics 22(15), 1930–1931 (2006)
27. Gawrilow, E., Joswig, M.: Polymake: a framework for analyzing convex polytopes. In: Kalai, G., Ziegler, G.M. (eds.) Polytopes—Combinatorics and Computation. Oberwolfach Seminars, vol. 29, pp. 43–73. Birkhäuser, Basel (2000), 10.1007/978-3-0348-8438-9_2
28. Palsson, B.O.: The challenges of in silico biology Moving from a reductionist paradigm to one that views cells as systems will necessitate. Nature Biotechnology 18, 1147–1150 (2000)
29. Covert, M.W., Schilling, C.H., Palsson, B.O.: Regulation of gene expression in flux balance models of metabolism. Journal of Theoretical Biology 213(1), 73–88 (2001)

30. Urbanczik, R.: Enumerating constrained elementary flux vectors of metabolic networks. IET Systems Biology 1(5), 274–279 (2007)
31. Ziegler, G.M.: Lectures on Polytopes. Graduate Texts in Mathematics. Springer (July 2001)
32. O'Madadhain, J., Fisher, D., White, S., Boey, Y.: The JUNG (Java Universal Network/Graph) Framework. Technical report, UCI-ICS (October 2003)
33. Funahashi, A., Morohashi, M., Kitano, H., Tanimura, N.: Celldesigner: a process diagram editor for gene-regulatory and biochemical networks. BIOSILICO 1(5), 159–162 (2003)
34. Pérez Millán, M., Dickenstein, A., Shiu, A., Conradi, C.: Chemical reaction systems with toric steady states. Bulletin of Mathematical Biology, 1–29 (October 2011)
35. Grigoriev, D., Weber, A.: Complexity of solving systems with few independent monomials and applications to mass-action kinetics. In: These Proceedings (2012)
36. Clarke, B.L.: Complete set of steady states for the general stoichiometric dynamical system. The Journal of Chemical Physics 75(10), 4970–4979 (1981)
37. Anderson, D.: A proof of the global attractor conjecture in the single linkage class case (2011)
38. Domijan, M., Kirkilionis, M.: Bistability and oscillations in chemical reaction networks. Journal of Mathematical Biology 59(4), 467–501 (2009)
39. Errami, H., Seiler, W.M., Eiswirth, M., Weber, A.: Computing Hopf bifurcations in chemical reaction networks using reaction coordinates. In: These Proceedings (2012)
40. Weber, A., Sturm, T., Abdel-Rahman, E.O.: Algorithmic global criteria for excluding oscillations. Bulletin of Mathematical Biology 73(4), 899–917 (2011)
41. Weber, A., Sturm, T., Seiler, W.M., Abdel-Rahman, E.O.: Parametric Qualitative Analysis of Ordinary Differential Equations: Computer Algebra Methods for Excluding Oscillations (Extended Abstract) (Invited Talk). In: Gerdt, V.P., Koepf, W., Mayr, E.W., Vorozhtsov, E.V. (eds.) CASC 2010. LNCS, vol. 6244, pp. 267–279. Springer, Heidelberg (2010)
42. Errami, H., Seiler, W.M., Sturm, T., Weber, A.: On Muldowney's Criteria for Polynomial Vector Fields with Constraints. In: Gerdt, V.P., Koepf, W., Mayr, E.W., Vorozhtsov, E.V. (eds.) CASC 2011. LNCS, vol. 6885, pp. 135–143. Springer, Heidelberg (2011)

Approximately Singular Systems and Ill-Conditioned Polynomial Systems[*]

Tateaki Sasaki[1] and Daiju Inaba[2]

[1] Professor emeritus, University of Tsukuba,
Tsukuba-city, Ibaraki 305-8571, Japan
sasaki@math.tsukuba.ac.jp
[2] Japanese Association of Mathematics Certification,
Ueno 5-1-1, Tokyo 110-0005, Japan
d.inaba@su-gaku.net

Abstract. By "approximately singular system" we mean a system of multivariate polynomials the dimension of whose variety is increased by small amounts of perturbations. First, we give a necessary condition that the given system is approximately singular. Then, we classify polynomial systems which seems ill-conditioned to solve numerically into four types. Among these, the third one is approximately singular type. We give a simple well-conditioning method for the third type. We test the third type and its well-conditioned systems by various examples, from viewpoints of "global convergence", "local convergence" and detail of individual computation. The results of experiments show that our well-conditioning method improves the global convergence largely.

Keywords: approximate ideal, approximately linear-dependent relation, approximately singular system, ill-conditioned polynomial system, multivariate Newton's method, well-conditioning.

1 Introduction

In [8], one of the authors (T.S), collaborated with Ochi and Noda, investigated solving multivariate polynomial systems having approximate GCD, showed that such systems are ill-conditioned for Newton's method, and presented a method of well-conditioning. The method is to transform the given system algebraically. Since then, he considered to generalize the method of [8]. He thought that an approximate Gröbner basis is a key for the generalization, but even the concept of approximate Gröbner basis has been unclear until recently. In [9], he has succeeded in constructing a theory of approximate Gröbner bases, and this is the first paper which follows [9].

The key concept in [9] is the "approximate ideal". Let $\mathbb{F}$ be a set of fixed-precision floating-oint numbers, to be abbreviated to *floats*. If a float f contains an error which begins at the $(k+1)$-st bit then we say the accuracy of f is 2^{-k} and express as $\mathrm{acc}(f) = 2^{-k}$. The accuracy of the polynomial is the minimum of

[*] Work supported by Japan Society for the Promotion of Science under Grants 23500003 and 08039686.

V.P. Gerdt et al. (Eds.): CASC 2012, LNCS 7442, pp. 308–320, 2012.
© Springer-Verlag Berlin Heidelberg 2012

accuracies of its coefficients. Let $(\boldsymbol{x}) = (x_1, \cdots, x_n)$ be a set of variables, and let $F_1, \cdots, F_m$ be given polynomials in $\mathbb{F}[\boldsymbol{x}]$, with initial accuracy $\varepsilon_{\text{init}}$. Consider the set $\{F = a_1 F_1 + \cdots + a_m F_m \mid \forall a_i \in \mathbb{C}[\boldsymbol{x}]\}$. This set seems to be an ideal, however, we may have the case that big cancellations occur in the sum $a_1 F_1 + \cdots + a_m F_m$ and the coefficients of F become fully erroneous. Therefore, we must modify the concept of ideal in $\mathbb{F}[\boldsymbol{x}]$. In [9], the *approximate ideal of tolerance* $\varepsilon_{\text{init}}$ was defined to be the set of only polynomials of the form F, that contain significant terms of accuracies < 1. If F is such that (for $\| \cdot \|$, see Sect. 2)

$$\|F\| = \varepsilon \max\{\|a_1 F_1\|, \cdots, \|a_m F_m\|\}, \qquad \varepsilon \ll 1,$$

then we call F *approximately linear-dependent relation*, to be abbreviated to *appLD-rel*, of tolerance ε. (We define appLD-rel more carefully in Sect. 2.) The existence of appLD-rel(s) makes the approximate ideal characteristic; if there exists no appLD-rel of tolerance which is significantly smaller than 1 then the approximate ideal is not much different from the exact ideal.

For the sake of later use, we explain the algorithm of computing *approximate Gröbner basis of tolerance* ε_{app}, $\varepsilon_{\text{init}} \leq \varepsilon_{\text{app}} \ll 1$, proposed in [9]. The algorithm is based essentially on two points: 1) introduce an "accuracy-guarding reduction" and use it instead of the conventional term-reduction, 2) employ Buchberger's procedure basically but discard polynomials whose accuracies are lost by amounts $\geq 1/\varepsilon_{\text{app}}$ due to the appLD-rels. Therefore, the appLD-rels play a crucial role not only in theory but also for algorithm. The appLD-rels lead us naturally to a concept of "approximately singular system".

In Sect. 2, we define approximately singular system and prove a necessary condition that the given system is approximately singular. In Sect. 3, we classify the ill-conditioned multivariate polynomial systems into four types. The reader will see that small deviations of input coefficients cause infinite changes of some solutions in one type. Another type is such that the deviations increase the number of solutions to infinite; we call this *approximately singular type*. In Sect. 4, we propose a simple method of transforming ill-conditioned systems of approximately singular type to well-conditioned ones. In Sect. 5, we test our well-conditioning method by multivariate Newton's method which has a fault in that some solutions are quite difficult to obtain in many cases (poor global convergence). Experiments show that our well-conditioning method improves the global convergence largely.

2   Approximately Singular Polynomial Systems

In this paper, we use symbols F and G for multivariate polynomials in variables $\boldsymbol{x} = x_1, \ldots, x_n$, $n \geq 2$. By $\|F\|$ we denote the norm of F; in this paper we employ the infinity norm, i.e., the maximum of absolute values of the numerical coefficients of F. By $\text{lt}(F)$ we denote the *leading term (monomial)* of F, with respect to a given *term order* $\succ$. Let $\Phi = \{F_1, \cdots, F_m\}$ be a given set of polynomials. By $\text{LT}(\Phi)$ we denote *leading-term set* of Φ, i.e., $\text{LT}(\Phi) = \{\text{lt}(F_1), \cdots, \text{lt}(F_m)\}$. By $\text{Var}(\Phi)$ we denote the *variety* of Φ, the set of all the solutions of coupled equations $F_1 = 0, \cdots, F_m = 0$.

In this section, we consider the case that the number of solutions is either finite or infinite, and we are interested in the dimension of variety, dim(variety). We can compute dim(Var(Φ)) over $\mathbb{C}$ practically, as follows: compute a Gröbner basis Γ of Φ, with respect to the total-degree order, then the dimension is given by dim(Var(LT(Γ))); see Sect. 9 of [2]. We assume that the given polynomials are normalized as $\|F_1\| = \cdots = \|F_m\| = 1$, and express an appLD-rel of tolerance ε, among $F_1, \ldots, F_m$ as follows.

$$\delta F = A_1 F_1 + \cdots + A_m F_m, \qquad \|\delta F\| = \varepsilon \max\{\|A_1 F_1\|, \cdots, \|A_m F_m\|\}. \qquad (2.1)$$

Speaking rigorously, the *syzygy* $(A_1, \cdots, A_m)$ must satisfy two conditions. 1) There is no "nearby" syzygy $(A_1', \cdots, A_m')$ such that $A_i' \simeq A_i$ $(i = 1, \ldots, m)$ and satisfies $A_1' F_1 + \cdots + A_m' F_m = 0$. 2) The syzygy must be *minimal*, that is, it does not include $(a_1, \cdots, a_m)$ which satisfies $a_1 F_1 + \cdots + a_m F_m = 0$. For more details of and how to compute appLD-rels, see [10].

Definition 1 (approximately singular system). *Let $\delta F_1, \ldots, \delta F_m$ be polynomials in $\mathbb{F}[\boldsymbol{x}]$, of small norms, satisfying $F_i \succ \delta F_i$ $(i = 1, \ldots, m)$. Let*

$$\Phi_\varepsilon = \{F_1 + \delta F_1, \cdots, F_m + \delta F_m\}, \qquad \max\{\|\delta F_1\|, \cdots, \|\delta F_m\|\} = \varepsilon \ll 1, \quad (2.2)$$

be an ε-perturbed system. If dim(Var(Φ_ε)) $>$ dim(Var(Φ)) *for some $\delta F_1, \ldots, \delta F_m$ then Φ is called an* approximately singular system *of tolerance ε.* □

Given an approximately singular system Φ, we consider in this section an important problem: find a condition for that Φ is approximately singular.

Lemma 1. *Let $\Phi' = \{F_2, \cdots, F_m\}$. If we have a non-trivial linear-dependent relation $A_1 F_1 + A_2 F_2 + \cdots + A_m F_m = 0$, with $A_1 \neq 0$, then we have*

$$\text{Var}(\Phi) = \text{Var}(\Phi') \setminus \text{Var}(\{A_1, F_2, \cdots, F_m\}). \qquad (2.3)$$

Proof. Since Var(Φ') = Var($\{A_1 F_1 + \cdots + A_m F_m, F_2, \cdots, F_m\}$) = Var($\{A_1 F_1, F_2, \cdots, F_m\}$) = Var($\{A_1, F_2, \cdots, F_m\}$) $\cup$ Var(Φ), we obtain (2.3). □

This lemma tells that the existence of an exactly linear-dependent relation among $F_1, \ldots, F_m$ is, roughly speaking, equivalent to deleting a polynomial from Φ. Deleting a polynomial from Φ increases usually the dimension of the variety.

Theorem 1. *Assume that, in the computation of approximate Gröbner basis of tolerance ε, of Φ w.r.t. the total-degree order, the leading terms of polynomials appearing in the computation do not vanish by the perturbations of norms $\leq \varepsilon$, except that some polynomials vanish due to accuracy loss of amounts $> 1/\varepsilon$. Then, existence of appLD-rel(s) of tolerance $> \varepsilon$ among $F_1, \ldots, F_m$ is a necessary condition for that Φ is an approximately singular system of tolerance $O(\varepsilon)$.*

Proof. Let Γ and Γ_ε be approximate Gröbner bases of Φ and Φ_ε, respectively, of tolerance ε, w.r.t. the total-degree order, hence intermediate polynomials whose accuracies are lost by more than $1/\varepsilon$ are discarded. Due to the assumption on the leading terms, if there is no appLD-rel of tolerance $\geq \varepsilon$, then we have $\mathrm{LT}(\Gamma_\varepsilon) = \mathrm{LT}(\Gamma)$, which means that $\dim(\mathrm{Var}(\Phi_\varepsilon)) = \dim(\mathrm{Var}(\Phi))$. The theorem is the contraposition of this. $\qquad\qquad\qquad\qquad\qquad\qquad\qquad\qquad\qquad\qquad\quad\square$

Note that the existence of appLD-rel(s) is not a sufficient condition. In [10], a simple example is given to show this.

3   Classification of Ill-Conditioned Polynomial Systems

In this section, we assume that $m = n$ and $\dim(\mathrm{Var}(\Phi)) = 0$, that is, the n coupled equations in n variables have a finite number of solutions.

Let Φ_ε be an ε-perturbed system of Φ, $\varepsilon \ll 1$. Assume that both Φ and Φ_ε have n solutions $\{s_1, \ldots, s_n\}$ and $\{s'_1, \ldots, s'_n\}$, respectively, such that $s'_i \to s_i$ $(i = 1, \ldots, n)$ as $\varepsilon \to 0$. Put $\delta = \max\{|s'_1 - s_1|, \cdots, |s'_n - s_n|\}$. If δ/ε is extremely larger than 1 then Φ is called *ill-conditioned* (w.r.t. solution finding).

In the univariate case, roughly speaking, we have two types of ill-conditioned polynomials. *Close-root type*: polynomials having close roots (the multiple roots in $\mathbb{C}$ become close roots in $\mathbb{F}$). *Wilkinson type*: polynomials which oscillate very largely if one traces them from a root to its neighboring one. As for close-root type polynomials, Sasaki and Noda [11] proposed to separate the factors having close roots as approximately multiple factors by the approximate squarefree decomposition and compute the close roots by the scale transformation. As for Wilkinson type polynomials, Fortune [4] presented a wonderful method which converts such polynomials to fully well-conditioned ones.

We classify the ill-conditioned multivariate systems from viewpoints of parametric and approximate Gröbner bases, as follows.

Close-solution type: The $\mathrm{Var}(\Phi)$ contains close solutions.

Tiny leading-term type: A tiny change of some input coefficients makes some leading terms vanish in the computation of Gröbner basis w.r.t. the lexicographic order and causes large changes of some solutions.

Approximately singular type: The input system is approximately singular, where any tiny change of input coefficients does not make any leading term vanish in the computation of Gröbner basis w.r.t. the total-degree order.

Other types: The types which we have not clearly recognized so far, such as "multivariate Wilkinson-type polynomials", i.e., rapidly oscillating polynomials with huge coefficients.

The system of the second type seems similar to univariate Wilkinson polynomial, but the origin of the solution change is completely different.

Example 1. (ill-conditioned system of tiny leading-term type)   Consider

$$\Phi' : \begin{cases} F'_1 = x^2 y + 101/100xy + 101x + 200, \\ F'_2 = xy^2 + 100/100y^2 + 100y - 201. \end{cases} \tag{3.1}$$

$$\Phi : \begin{cases} F_1 = x^2 y + 100/100xy + 101x + 201, \\ F_2 = xy^2 + 100/100y^2 + 101y - 201. \end{cases} \tag{3.2}$$

$$\Phi'' : \begin{cases} F''_1 = x^2 y + 100/100xy + 100x + 201, \\ F''_2 = xy^2 + 101/100y^2 + 101y - 200. \end{cases} \tag{3.3}$$

Although systems Φ', Φ and Φ'' are different only a little in their coefficients, their Gröbner bases w.r.t. lexicographic order are very different, as follows; we show only polynomials in variable y:

$\Gamma' : G_1' = y^4 - 19800y^3 + 29899y^2 + 1989900y - 4040100$,

$\Gamma : G_1 = y^3 - y^2 - 101y + 201$,

$\Gamma'' : G_1'' = y^4 + 2030200/101y^3 - 1030000/101y^2 - 204000000/101y + 400000000/101$.

Some roots of G_1', G_1 and G_1'' are also very different, as follows:

$$G_1' : [2.0530\cdots, \ 9.7043\cdots, \ -10.242\cdots, \ \underline{19798.4\cdots}],$$
$$G_1 : [2.0323\cdots, \ 9.4421\cdots, \ -10.474\cdots],$$
$$G_1'' : [2.0224\cdots, \ 9.1390\cdots, \ -10.659\cdots, \ \underline{-20101.4\cdots}].$$

The fourth roots go to ∞ and $-\infty$, respectively, as $\Phi' \to \Phi \leftarrow \Phi''$.   □

This example shows clearly that, in some cases, polynomial structure of the Gröbner basis is changed largely by a tiny change of coefficients. The comprehensive Gröbner basis (i.e. Gröbner basis of the system with parametric coefficients) by Weispfenning [12] tells us in which case the Gröbner basis changes: *polynomial structure of the resulting Gröbner basis changes only in the case where, during the execution of Buchberger's procedure, at least one leading term of polynomial vanishes by the substitution of suitable set of numbers to the parameters.* The following theorem is obvious, so we omit the proof.

Theorem 2. *Appearance of tiny leading term(s) in computing a Gröbner basis of Φ w.r.t. the lexicographic order is a necessary condition that Φ is an ill-conditioned system of tiny leading-term type.*   □

Even if the resulting Gröbner basis is changed largely, one will often obtain the variety which is perturbed only a little. We can confirm this by changing the coefficients of y of F_1 and F_2 in Example 1 variously.

We explain the close-solution type briefly, which will help the reader to understand the third type deeply. Assume that $\boldsymbol{x}$ is close to a solution $\boldsymbol{s} = (s_1, \ldots, s_n)$. Put $s_i = x_i + \delta_i$, $|\delta_i| = O(\delta)$ $(i=1, \ldots, n)$, where δ is a small positive number. Expanding $F_i(\boldsymbol{x}+\boldsymbol{\delta})$ into Taylor series at $\boldsymbol{x}$, we obtain

$$F_i(\boldsymbol{s}) = 0 = F_i(\boldsymbol{x} + \boldsymbol{\delta}) = F_i(\boldsymbol{x}) + \partial F_i/\partial x_1 \delta_1 + \cdots + \partial F_i/\partial x_n \delta_n + O(\delta^2).$$

Solving $F_i(\boldsymbol{x} + \boldsymbol{\delta}) = 0$ $(i=1, \ldots, n)$ w.r.t. $\delta_1, \ldots, \delta_n$ by neglecting $O(\delta^2)$ terms, we obtain

$$\begin{pmatrix} \delta_1 \\ \vdots \\ \delta_n \end{pmatrix} \approx -J^{-1}(\boldsymbol{x}) \begin{pmatrix} F_1(\boldsymbol{x}) \\ \vdots \\ F_n(\boldsymbol{x}) \end{pmatrix}, \quad \text{where } J(\boldsymbol{x}) = \begin{pmatrix} \partial F_1/\partial x_1 & \cdots & \partial F_1/\partial x_n \\ \vdots & \ddots & \vdots \\ \partial F_n/\partial x_1 & \cdots & \partial F_n/\partial x_n \end{pmatrix}.$$

$$(3.4)$$

Let $\boldsymbol{s}$ be a multiple solution, hence $F_i(\boldsymbol{x}) = f_i(s_1 - \delta_1)^{\mu_1} \cdots (s_n - \delta_n)^{\mu_n} + O(\delta^2)$ for some $i \in \{1, \cdots, n\}$, where $f_i \in \mathbb{C}$ and at least one of $\mu_1, \ldots, \mu_n$ is greater than 1. Then, the i-th row of $J(\boldsymbol{x})$ becomes very small near $\boldsymbol{s}$. The same is true if s is a close solution. Therefore, we may set a criterion of ill-conditionedness to be that the Jacobian becomes quite small near some solution(s).

We next consider the systems of approximately singular type. If $\Phi = \{F_1, F_2\}$ then the existence of appLD-rel $A_1 F_1 + A_2 F_2 = \delta F$ means that F_1 and F_2 have an approximate GCD. Hence, in this case, the systems of the third type are nothing but the systems treated in [8]. However, if Φ contains three or more polynomials, the systems we are treating are more general than those treated in [8]; for example, no pair (F_i, F_j) with $i \neq j$ may have approximate GCD.

We will show that, for systems of approximately singular type, $J(x)$ becomes approximately singular at points on a curve or even a surface. Let Φ be approximately singular. Then, by Theorem 1, there exists an appLD-rel and a nearby singular system $\Phi_\varepsilon = \{F_1 + \delta F_1, \cdots, F_n + \delta F_n\}$, satisfying

$$A_1(F_1 + \delta F_1) + \cdots + A_n(F_n + \delta F_n) = 0. \tag{3.5}$$

If a solution s of Φ is close to continuous solutions of Φ_ε then we say s is *critical*.

Theorem 3. *Let Φ be of approximately singular type. Then, we have $|J(x)| = O(\varepsilon)$ or less on or near continuous zero-points of the singular system Φ_ε.*

Proof. By (3.5), we have the following relation for each $i \in \{1, \cdots, n\}$.

$$A_1 \frac{\partial F_1}{\partial x_i} + \cdots + A_n \frac{\partial F_n}{\partial x_i} = -\left(F_1 \frac{\partial A_1}{\partial x_i} + \cdots + F_n \frac{\partial A_n}{\partial x_i}\right) - \frac{\partial(A_1 \delta F_1 + \cdots + A_n \delta F_n)}{\partial x_i}.$$

Consider this relation on $z \in \mathrm{Var}(\Phi_\varepsilon)$ (or z' which is close to z). Since $F_i(z) + \delta F_i(z) = 0$ for each $i \in \{1, \cdots, n\}$, we have $|F_i(z)| = O(\varepsilon)$, hence

$$A_1(z)\partial F_1(z)/\partial x_i + \cdots + A_n(z)\partial F_n(z)/\partial x_i = O(\varepsilon) \quad \text{for any } i.$$

This means that n rows of $J(z)$ are approximately linear-dependent. □

Remark 1. *The Gröbner basis computation w.r.t. the total-degree order is used for defining systems of approximately singular type, while the computation w.r.t. the lexicographic order is used for tiny leading-term type. Therefore, these two types are not exclusive to each other.* □

4 Well-Conditioning of Systems of Approximately Singular Type

So far, many researchers have investigated ill-conditioned systems of close- solution type; see, for example, [1,3,5,6,13]. On the other hand, ill-conditioned systems of other types were scarcely investigated so far; as far as the authors know, only [8] (see also [7]) treated a special case of systems of the third type. Probably, the existence itself of ill-conditioned systems of types other than close-solution type will be not well recognized by researchers. For the systems of tiny leading-term type, we need to compute Gröbner bases carefully by checking the accuracies of the leading coefficients, and we must treat some coefficients to be 0

if their accuracies are lost fully. In this subsection, we consider well-conditioning
of only the systems of approximately singular type.

We assume that the given system Φ is approximately singular, hence we have
appLD-rels as in (2.1). We can compute such relations either by approximate
Gröbner basis algorithm [9] or better by eliminating a matrix the rows of which
are coefficient vectors of $S_j F_i$ ($i = 1, \ldots, n$; $j = 1, 2, \ldots$), where S_j is a monomial
bounded suitably; see [10] for details.

In approximately singular system Φ, the δF prevents $\mathrm{Var}(\Phi)$ from being an
infinite set. Therefore, if we convert Φ to another system Φ' in which the δF is
a main player, then Φ' will be well-conditioned. Following this idea, we convert
the input system as follows, where we assume that $A_i \neq 0$ in (2.1).

$$\Phi = \{F_1, \cdots, F_i, \cdots, F_n\} \implies \Phi' = \{F_1, \cdots, \delta F, \cdots, F_n\}. \tag{4.1}$$

Theorem 4. *We have* $\mathrm{Var}(\Phi) \subseteq \mathrm{Var}(\Phi')$.

Proof. We have $\mathrm{Var}(\Phi') = \mathrm{Var}(\{F_1, \cdots, A_1 F_1 + \cdots + A_n F_n, \cdots, F_n\}) =$
$\mathrm{Var}(\{F_1, \cdots, A_i F_i, \cdots, F_n\}) = \mathrm{Var}(\Phi) \cup \mathrm{Var}(\{F_1, \cdots, A_i, \cdots, F_n\})$. By this, we
obtain the theorem at once. $\qquad \square$

Remark 2. *The Φ' may not always be well-conditioned. Such a case occurs if,
for example, we have another appLD-rel $A_1' F_1 + \cdots + A_n' F_n = \delta F'$, $A_j' \neq 0$
($j \neq i$). In this case, we replace F_i and F_j by δF and $\delta F'$, respectively.* $\qquad \square$

5   Numerical Experiments

We have tested various approximately singular systems and their well-conditioned
ones by multivariate Newton's method which is an iterative method based on
(3.4); let $s^{(k)}$ be the k-th approximate solution, then the $(k+1)$-st one $s^{(k+1)}$ is
determined as

$$s^{(k+1)} = s^{(k)} + \delta, \qquad \delta = -J^{-1}(s^{(k)}) \cdot (F_1(s^{(k)}), \cdots, F_n(s^{(k)}))^{\mathrm{t}}. \tag{5.1}$$

We have implemented the above method with no sophistication. For easiness
of data processing, we have restricted ourselves to treat only real solutions of
systems in three variables, having $5 \sim 12$ real solutions.

In our experiments, we observed the performance of Newton's method from
three viewpoints: *global convergence, local convergence* and *detail of individual
computation*. That is, we performed the following three different computations.

Global convergence: Generate N_{try} points $(s_z^{(0)}, s_y^{(0)}, s_x^{(0)})$'s randomly on a real
   sphere of radius R, located at the origin, and apply Newton's method for
   each initial value $(s_z^{(0)}, s_y^{(0)}, s_x^{(0)})$. We count the percentages of convergence to
   each solution, with the stopping condition $\max\{|\delta_z|, |\delta_y|, |\delta_x|\} < 10^{-3}$. (In
   Examples A and B below, we also generate 8000 initial values distributed
   uniformly in the cube $[-R', R']^3$; the corresponding percentages are given in
   the **grid**-row.) The computation is forced to stop when the iteration reaches
   at 100. Percentages of convergence to non-solutions and non-convergence are
   given in the columns **nonS** and **nonC**, respectively.

Local convergence: Specify an approximate solution S (if $S_{i_1}, \ldots, S_{i_k}$ are similar to each other, we choose only several of them). Generate $N_{\mathbf{try}}$ points $(s_z^{(0)}, s_y^{(0)}, s_x^{(0)})$'s randomly on a real sphere of radius r (we set $r = 0.01$), located at S, and apply Newton's method for each initial value $(s_z^{(0)}, s_y^{(0)}, s_x^{(0)})$. We count the average number of iterations $\mathbf{It_{avr}}$ by which the computation converges to S or another solution S', with the stopping condition $\max\{|\delta_z|, |\delta_y|, |\delta_x|\} < \varepsilon$ (we set $\varepsilon = 10^{-10}$). In Tables, $\boldsymbol{\Phi}_i(\mathbf{j})$ means that the system $\boldsymbol{\Phi}_i$ is solved with initial points around $\mathbf{Solj}$.

Detail of computation: Choosing several events which show typical or slow local convergence, we check the following quantities: the number of iterations $N_{\mathbf{itr}}$ and values of the Jacobian, the initial one $J_{\mathbf{ini}}$, the maximum one $J_{\mathbf{mx}}$ and the final one $J_{\mathbf{fin}}$. If the computation converges to the specified solution S then $\mathbf{toS?}$ is YES, if converges to another solution S' then we write S'.

We show three examples and evaluate the experimental results.

Example A. (approximate-GCD type). We consider the following system Φ; we show some of its real solutions (**Sols**) and apparent appLD-rels (**apR**).

$$\Phi := \begin{cases} F_1 = (z^2+2y+z)*(y^2+x^2-y+x-3) - (y-z)/1000, \\ F_2 = (z^2+2x+z)*(y^2+x^2-y+x-3) - (z-x)/1000, \\ F_3 = (z^2+2y-x)*(y^2+x^2-y+x-3) - (x+y)/1000. \end{cases} \quad (5.2)$$

$$\text{Sols}: \begin{cases} \text{Sol0} = (0,\ 0,\ 0), \qquad \text{Sol0}_+ = (0,\ 2.30291\cdots, 0), \\ \text{Sol1} = (\ 1.12415\cdots, -1.26820\cdots, -1.12415\cdots), \\ \qquad\qquad \vdots \qquad\qquad \vdots \qquad\qquad \vdots \\ \text{Sol5} = (\ 0.99802\cdots, -0.99605\cdots, -0.99802\cdots). \end{cases} \quad (5.3)$$

$$\text{apR}: \begin{cases} G_{12} = 1000*((z^2+2x+z)*F_1 - (z^2+2y+z)*F_2), \\ G_{23} = 1000*((z^2+2y-x)*F_2 - (z^2+2x+z)*F_3), \\ G_{31} = 1000*((z^2+2y+z)*F_3 - (z^2+2y-x)*F_1). \end{cases} \quad (5.4)$$

Let Φ_ε be the singularized system in which the last terms $(y-z)/1000$ etc. in (5.2) are removed, then we have $\dim(\text{Var}(\Phi_\varepsilon)) = 2$. Among the 8 solutions, only Sol0, Sol1 and Sol5 are not close to $y^2+x^2-y+x-3 = 0$, hence they are *non-critical*. In addition to Φ, we investigate the following systems; note that $\text{Var}(\Phi_2)$ and $\text{Var}(\Phi_3)$ are of dimension 0 but $\dim(\text{Var}(\Phi_1)) = 1$. As we mentioned above, the grid-row shows the global convergence by initial points on grid.

$$\begin{cases} \Phi_1 = \{G_{23}, F_2, F_3\}, \quad \Phi_2 = \{F_1, G_{23}, F_3\}, \quad \Phi_3 = \{F_1, F_2, G_{23}\}, \\ \Phi_4 = \{F_1, G_{12}, G_{31}\}, \quad \Phi_5 = \{G_{12}, F_2, G_{23}\}, \quad \Phi_6 = \{G_{31}, G_{23}, F_3\}. \end{cases} \quad (5.5)$$

Tables A1, A2 and A3 show the global convergence, the local convergence, and the detail of individual computation, respectively. In Table A1, $N_{\text{try}} = 10000$ for Φ and 200 for others. In Table A2, $N_{\text{try}} = 200$ for each system.

Table A1. Global convergence ($R = 5$, $R' = 2$, $\varepsilon = 10^{-3}$)

	(0)	Sol0$_+$	Sol0$_-$	Sol1	Sol2	Sol3	Sol4	Sol5	nonS	nonC
Φ		42.6	13.2	40.7	2.2	0.5	.01			0.8
grid	6.6	32.7	20.9	31.7	2.1	0.2		4.9	0.1	0.6
Φ_4	2.0	38.5	10.5	4.5	2.5	1.0	2.5	0.5	37.5	0.5
Φ_5		23.0	3.5	11.5	36.0	1.5	5.5	0.5	17.5	1.0
Φ_6	1.0	43.0	8.5	2.0	2.0	8.0	8.5	1.0	25.0	1.0

Table A2. Local convergence ($r = 0.01$, $\varepsilon = 10^{-10}$)

	$\Phi(1)$	$\Phi(2)$	$\Phi(3)$	$\Phi(4)$	$\Phi(5)$	$\Phi_4(1)$	$\Phi_4(2)$	$\Phi_4(3)$	$\Phi_4(4)$	$\Phi_4(5)$
It$_{avr}$	5.7	7.2	13.5	9.6	4.0	4.1	4.6	4.1	4.0	4.0
toS	100	99.0	82.5	81.5	100	100	100	100	100	100
nonS		1.0	17.5	18.5						
nonC										

Table A3. Detail of computation ($r = 0.01$, $\varepsilon = 10^{-10}$)

	$\Phi(3)$	$\Phi(3)$	$\Phi(3)$	$\Phi_4(3)$	$\Phi_5(3)$	$\Phi_6(3)$	$\Phi_3(3)$	$\Phi_3(3)$
N_{itr}	10	19	35	4	4	4	29	96
J_{ini}	3.7e-2	7.9e-2	3.1e-2	7.27	56.7	18.0	1.11	5.5e-1
J_{mx}	''	97.8	1.3e+7	''	''	19.3	9.9e+5	3.e+13
J_{fin}	1.4e-5	1.9e-6	6.2e-5	6.26	56.3	19.3	8100	0.100
toS?	YES	Sol2	Sol0$_+$	YES	YES	YES	nonS	Sol1

Table A1 shows that some solutions of Φ are very hard to obtain, but all the solutions can be obtained by the well-conditioning. We note that some well-conditioned system is good for some solutions and another system is good for some other solutions. Table A2 shows that well-conditioning improves the local convergence, too. The nonS events there show that the local convergence is bad for critical solutions. The most drastic improvement is the value of Jacobian, shown in Table A3, where two columns for $\Phi(3)$ show that the iterative computation jumps from Sol3 to another solution. The last two columns there show that the convergence is very poor if "unsuitable system" is chosen.   □

Example B. ($\{F, G, aF + bG\}$-type) We consider the following system Φ; we show its real solutions and an apparent appLD-rel.

$$\Phi := \begin{cases} F_1 = (x + 1001/1000) * (z^2 + 1001/1000x^2 - 3) + 2/1000, \\ F_2 = (x - 999/1000) * (y^2 - 999/1000x^2 - 3) - 1/1000, \quad (5.6) \\ F_3 = (x + 1) * (x - 1) * ((z^2 + x^2 - 3) + (y^2 - x^2 - 3)). \end{cases}$$

$$\text{Sols :} \begin{cases} \text{Sol1} = (\ 1.41350 \cdots, \quad 2.23584 \cdots, \quad 1.00000 \cdots), \\ \text{Sol5} = (\ 1.19942 \cdots, \quad 2.13573 \cdots, \quad 1.24856 \cdots), \\ \text{SolA} = (\ 0.96391 \cdots, \quad 2.25185 \cdots, -1.43991 \cdots), \quad (5.7) \\ \qquad\quad \vdots \qquad\qquad\quad \vdots \qquad\qquad \vdots \end{cases}$$

$$\text{apR :}\ G = 1000 * (F_3 - (x - 1) * F_1 - (x + 1) * F_2). \quad (5.8)$$

Let Sol1 $= (a, b, c)$ then Sol2 $= (-a, b, c)$, Sol3 $= (a, -b, c)$ and Sol4 $= (-a, -b, c)$. The same is true for Sol5 ~ 8 and SolA $\sim$ D. In addition to Φ, we investigate the following systems.

$$\Phi_1 = \{G, F_2, F_3\}, \quad \Phi_2 = \{F_1, G, F_3\}, \quad \Phi_3 = \{F_1, F_2, G\}. \tag{5.9}$$

Tables B1, B2 and B3 show the global convergence, the local convergence, and the detail of individual computation, respectively. In Table B1, $N_{\text{try}} = 10000$ for Φ, Φ_3 and 100 for others. In Table B2, $N_{\text{try}} = 100$ for each system.

Table B1. Global convergence ($R = 5$, $R' = 3$, $\varepsilon = 10^{-3}$)

	Sol1, 2	Sol3, 4	Sol5, 6	Sol7, 8	SolA	SolB	SolC	SolD	nonS	nonC
Φ	0.4	0.4	1.2	1.1	1.3	1.3	1.4	1.5		91.4
grid	1.2	1.2	2.2	2.2	2.8	2.8	2.8	2.8		82.0
Φ_3	6.9	5.9	21.0	20.8	10.9	10.8	10.5	11.6		1.3
Φ_1	1.0	1.0	23.0	20.0	12.0	11.0	5.0	13.0	12.0	2.0
Φ_2	17.0	10.0	12.0	10.0	14.0	12.0	10.0	14.0		1.0

Table B2. Local convergence ($r = 0.01$, $\varepsilon = 10^{-10}$)

	$\Phi(1)$	$\Phi(5)$	$\Phi(A)$	$\Phi(D)$	$\Phi_3(1)$	$\Phi_3(5)$	$\Phi_3(A)$	$\Phi_3(D)$
It$_{\text{avr}}$	4.9	5.8	4.9	4.9	4.0	4.0	4.0	4.0
toS	100	93	100	100	100	100	100	100
nonS		1						
nonC		6						

Table B3. Detail of computation ($r = 0.01$, $\varepsilon = 10^{-10}$)

	$\Phi(1)$	$\Phi(5)$	$\Phi(5)$	$\Phi(A)$	$\Phi_3(1)$	$\Phi_3(5)$	$\Phi_3(5)$	$\Phi_3(A)$
N_{itr}	5	5	100	5	4	4	4	4
J_{ini}	6.0e-2	0.248	9.2e-4	0.191	53.5	60.9	69.6	166
J_{mx}	"	"	6.3e+6	"	"	66.6	"	"
J_{fin}	5.0e-2	6.6e-2	2.4e+4	0.159	50.4	66.4	66.4	159
toS?	YES	YES	nonC	YES	YES	YES	YES	YES

Table B1 shows that the convergence to every solution of Φ is quite bad, and the reason is "sub-system trapping" which we will explain at the end of this section. The non-convergent event $\Phi(5)$ in Table B3 is also due to sub-system trapping. We again see from Tables B1$\sim$B3 that the global convergence and the value of Jacobian are improved largely by well-conditioning.   □

Example C. ($\{AB, BC, ABC\}$-type) We consider the following system Φ; we show its real solutions and an apparent appLD-rel.

$$\Phi := \begin{cases} F_1 = (x - y - 1) * (y - z + 2) - 1/100000(x - 1), \\ F_2 = (y - z + 2) * (z - x - 1) - 2/100000(y - 1), \\ F_3 = (x - y - 1) * (y - z + 2) * (z - x - 1) + 3/200000(z^2 - 1). \end{cases} \tag{5.10}$$

$$\text{Sols}: \begin{cases} \text{Sol1} = (\ \ 0.99812\cdots, -0.99481\cdots, \ \ 0.00377\cdots), \\ \text{Sol2} = (\ \ 1.00189\cdots, -1.00518\cdots, -0.00377\cdots), \\ \qquad \vdots \qquad\qquad\qquad \vdots \qquad\qquad\qquad \vdots \\ \text{Sol6} = (\ \ 5.58652\cdots, \ \ 3.58652\cdots, -4.17306\cdots). \end{cases} \tag{5.11}$$

$$\text{apR}: G = 100000 * ((z - x - 1) * F_1 + (x - y - 1) * F_2 - 2 * F_3). \tag{5.12}$$

We note that Sol1 and Sol2 are rather close (not so close) to each other, and Sol3 and Sol4 are also so. In addition to Φ, we investigate the following systems.

$$\Phi_1 = \{G, F_2, F_3\}, \quad \Phi_2 = \{F_1, G, F_3\}, \quad \Phi_3 = \{F_1, F_2, G\}. \tag{5.13}$$

Tables C1, C2 and C3 show the global convergence, the local convergence, and the detail of individual computation, respectively. In Table C1, $N_{\text{try}} = 1000$ for Φ, Φ_3, and 100 for others. In Table C2, $N_{\text{try}} = 100$ for each system.

Table C1. Global convergence ($R = 10$, $\varepsilon = 10^{-3}$)

	Sol1	Sol2	Sol3	Sol4	Sol5	Sol6	nonS	nonC
Φ	0.2	0.2	31.7	12.4	20.3	22.0		13.2
Φ_3	19.8	33.7	39.3	7.1		0.1		
Φ_1	8.0	15.0	10.0	17.0	5.0	12.0	33.0	
Φ_2	6.0	14.0	6.0	6.0	14.0	15.0	39.0	

Table C2. Local convergence ($r = 0.01$, $\varepsilon = 10^{-10}$)

	$\Phi(1)$	$\Phi(2)$	$\Phi(5)$	$\Phi(6)$	$\Phi_3(1)$	$\Phi_3(2)$	$\Phi_3(5)$	$\Phi_3(6)$
It$_{\text{avr}}$	6.2	6.1	9.0	15.6	5.8	5.8	6.3	8.6
toS	72	69	94	40	79	70	34	32
nonS	23	27	2	3	21	30	64	50
nonC	5	4	4	57			2	18

Table C3. Detail of computation ($r = 0.01$, $\varepsilon = 10^{-10}$)

	$\Phi(1)$	$\Phi(2)$	$\Phi(5)$	$\Phi(6)$	$\Phi_3(1)$	$\Phi_3(2)$	$\Phi_3(5)$	$\Phi_3(6)$
N_{itr}	6	13	10	12	6	7	6	9
J_{ini}	1.4e-8	1.1e-10	1.1e-6	1.9e-5	2.8e-3	2.4e-5	1.3e-3	5.1e-4
J_{mx}	''	1.3e-5	''	''	''	4.0e-3	''	0.138
J_{fin}	3.0e-9	6.5e-9	3.4e-9	3.4e-9	6.0e-4	6.0e-4	6.0e-4	1.3e-3
toS?	YES	Sol3	YES	Sol5	YES	Sol1	Sol1	Sol3

Table C1 shows that Sol1 and Sol2 of Φ are not easy to obtain but are easy after the well-conditioning. The nonC events for Φ in Tables C1 and C2 are due to accuracy losing. In Table C2, nonS events for $\Phi(6)$ and $\Phi_3(6)$ are jumping to other solutions. $\square$

We explain the sub-system trapping. Suppose, w.l.o.g., that $F_1(x)$ does not contain variable x_1, as F_1 and F_2 in Example B, and that $s' = (s_2, \cdots, s_n)$ is an approximate solution of sub-system $\{F_1(x)=0\}$ (which may contain two or more polynomials). Then, $(1,1)$-element of $J(x)$ is 0. We further assume that $(n,1)$-element of $J(s)$ is not zero. Then, using the n-th row, we can eliminate $(2,1)$-, $\ldots$, $(n-1,1)$-elements of $J(s)$, and let the matrix eliminated be $J'(s)$. This transforms the system into $J'(s)(\delta_1, \delta_2, \cdots, \delta_n)^{\mathrm{t}} = -(F_1(s), F_2'(s), \cdots, F_n(s))^{\mathrm{t}}$, where the first column of $J'(x)$ is $(0, \cdots, 0, \partial F_n/\partial x_1)^{\mathrm{t}}$, and $F_i'(s) = F_i(s)-(J_{i1}/J_{n1})F_n(s)$ $(2 \leq i \leq n-1)$. This means that $\delta_2, \ldots, \delta_{n-1}$ are determined by only the first to $(n-1)$-st transformed equations, and δ_1 is determined by the n-th equation and $\delta_2, \cdots, \delta_n$. Although $F_2'(x), \ldots, F_{n-1}'(x)$ change from iteration to iteration, the $n-1$ equations for $\delta_2, \ldots, \delta_n$ are of the same form as Newton's formula. Hence $s'' = (s_2 + \delta_2, \cdots, s_n + \delta_n)$ will be an approximate solution of $\{F_1(x)= 0, F_2'(x)=0, \cdots, F_{n-1}'(x)=0\}$, showing that s'' is trapped to $F_1(x)=0$.

6   Concluding Remarks

Before performing the numerical experiments described in Sect. 5, the authors thought that the local convergence is damaged much by approximate singularness of the system. After the experiments, we were surprised at that our well-conditioning improved the global convergence largely. We, furthermore, got an impression that algebraic varieties determined by ill-conditioned polynomial systems are very interesting. For example, drawing the graphs of F_1, F_2 etc. in Example 1, we found that the graphs show complicated but very interesting behaviors, and such behaviors can be understood easily from the viewpoints of approximately singular systems and approximate Gröbner bases.

We should investigate the approximately singular systems variously. In [10], we have performed such a study from the viewpoint of "singularization", i.e., to perturb the given system so that the variety of the perturbed system is of higher dimension.

References

1. Corless, R.M., Gianni, P.M., Trager, B.M.: A reordered Schur factorization method for zero-dimensional polynomial systems with multiple roots. In: Proceedings of ISSAC 1997 (Intn'l Symposium on Symbolic and Algebraic Computation), pp. 133–140. ACM Press (1997)
2. Cox, D., Little, J., O'Shea, D.: Ideals, Varieties, and Algorithms. Springer, New York (1997)
3. Dayton, B.H., Zeng, Z.: Computing the multiplicity structure in solving polynomial systems. In: Proceedings of ISSAC 2005, pp. 116–123. ACM Press (2005)

4. Fortune, S.: Polynomial root finding using iterated eigenvalue computation. In: Proceedings of ISSAC 2001, pp. 121–128. ACM Press (2001)
5. Janovitz-Freireich, I., Rónyai, L., Szántó, A.: Approximate radical of ideals with clusters of roots. In: Proceedings of ISSAC 1997, pp. 146–153. ACM Press (2006)
6. Mantzaflaris, A., Mourrain, B.: Deflation and certified isolation of singular zeros of polynomial systems. In: Proceedings of ISSAC 2011, pp. 249–256. ACM Press (2011)
7. Noda, M.-T., Sasaki, T.: Approximate GCD and its application to ill-conditioned algebraic equations. J. Comput. App. Math. 38, 335–351 (1991)
8. Ochi, M., Noda, M.-T., Sasaki, T.: Approximate GCD of multivariate polynomials and application to ill-conditioned system of algebraic equations. J. Inf. Proces. 14, 292–300 (1991)
9. Sasaki, T.: A theory and an algorithm of approximate Gröbner bases. In: Proceedings of SYNASC 2011 (Symbolic and Numeric Algorithms for Scientific Computing), pp. 23–30. IEEE Computer Society Press (2012)
10. Sasaki, T.: Proposal of singularization of approximately singular systems. Preprint of Univ. Tsukuba, 14 pages (May 2012)
11. Sasaki, T., Noda, M.-T.: Approximate square-free decomposition and root-finding of ill-conditioned algebraic equations. J. Inf. Proces. 12, 159–168 (1989)
12. Weispfenning, V.: Comprehensive Gröbner bases. J. Symb. Comp. 14, 1–29 (1992)
13. Wu, X., Zhi, L.: Determining singular solutions of polynomial systems via symbolic-numeric reduction to geometric involutive forms. J. Symb. Comput. 47, 227–238 (2012)

Symbolic-Numeric Implementation
of the Method of Collocations and Least Squares
for 3D Navier–Stokes Equations

Vasily P. Shapeev and Evgenii V. Vorozhtsov

Khristianovich Institute of Theoretical and Applied Mechanics,
Russian Academy of Sciences, Novosibirsk 630090, Russia
{shapeev,vorozh}@itam.nsc.ru

Abstract. The method of collocations and least squares, which was
previously proposed for the numerical solution of the two-dimensional
Navier–Stokes equations governing steady incompressible viscous flows,
is extended here for the three-dimensional case. The derivation of the
collocation and matching conditions is carried out in symbolic form using the CAS *Mathematica*. The numerical stages are implemented in a
Fortran code, into which the left-hand sides of the collocation and matching equations have been imported from the *Mathematica* program. The
results of numerical tests confirm the second order of convergence of the
presented method.

1 Introduction

It is well known (see, for example, [19]) that the influence of the compressibility
of a gas or a liquid may be neglected if the flow Mach number does not exceed
the value of 0.3. In such cases, it is reasonable to use the Navier–Stokes equations
governing the viscous incompressible fluid flows. These equations are somewhat
simpler than the system of Navier–Stokes equations for compressible media.

The numerical solution of Navier–Stokes equations is simplified greatly if they
are discretized on a uniform rectangular spatial grid in Cartesian coordinates. It
is natural and convenient to use such grids at the solution of problems in regions
of rectangular shape. Many applied problems are, however, characterized by the
presence of curved boundaries. In such cases, other grid types are often used:
curvilinear grids, structured and unstructured triangular and polygonal grids.
Although such grids simplify the implementation of boundary conditions, their
use leads to new difficulties, such as the extra (metric) terms in equations, extra
interpolations, larger computational molecules, etc. [13].

During the last decade, a new method for numerical solution of the Navier-Stokes equations in regions with complex geometry has enjoyed a powerful development: the immersed boundary method (IBM). In this method, the computation
of gas motion is carried out on a rectangular grid, and the curved boundary
is interpreted as an interface. A survey of different recent realizations of the
IBM may be found in [15,24,17]. The immersed boundary method has extended

V.P. Gerdt et al. (Eds.): CASC 2012, LNCS 7442, pp. 321–333, 2012.

significantly the scope of applicability of the rectangular Cartesian grids at the numerical solution of applied problems of the incompressible fluid dynamics.

The projection finite difference methods [12,27,4] have gained widespread acceptance at the numerical solution of the incompressible Navier–Stokes equations. A recurring difficulty in these methods is the proper choice of boundary conditions for the auxiliary variables in order to obtain at least second order accuracy in the computed solution. A further issue is the formula for the pressure correction at each time step [3].

In the method of *artificial compressibility* (AC) [6], there is no need in the solution of the Poisson equation for the pressure correction because the time derivative of pressure is introduced in the left-hand side of the continuity equation. As a result, the entire system of the Navier–Stokes equations acquires the evolutionary character, and one can then apply for its numerical integration the methods, which were developed for the equations of compressible gas dynamics. The AC method has a significant drawback in the difficulty of choosing the AC parameter, improper choice of which leads either to slow convergence or divergence [14,16].

The use of computer algebra systems (CASs) is very useful at the development and investigation of new numerical methods. The approaches to the application of CASs for constructing and investigating the numerical methods of solving the boundary-value problems for partial differential equations (PDEs) may be found in [25,7]. The method of collocation and least squares (CLS) for the PDE systems was first implemented in [21] at the solution of the boundary-value problems for Stokes equations. The CLS method was extended in [20] for the numerical solution of the 2D incompressible Navier–Stokes equations. The advantage of the CLS method over the fractional-step difference methods and the artificial compressibility methods is that it does not require the numerical solution of the continuity equation because this equation is satisfied identically at any spatial point at the expense of an appropriate choice of the basis functions. As was mentioned in [20], the approximation order of the CLS method can be increased just by adding several further basis functions. But huge analytic work is needed when constructing the formulas of the CLS method for Navier–Stokes equations. This work was done in [20] with the aid of CASs REDUCE and Maple. The speed of numerical computations by the CLS method was optimized in [20] by optimizing the sequence of arithmetic computations at first in the CAS, and then by inputting the optimized formulas of the CLS method to the program written in the C programming language.

The application of CASs for the construction of difference schemes for the Navier–Stokes equations was considered in the works [22,9].

The CLS method was implemented in [20] for the 2D Navier–Stokes equations in the case of square cells of a spatial computational grid. In the works [10,11], the CLS method was extended for the case of rectangular grid cells in the plane of two spatial variables. The variants of the CLS method were developed in [11], which had the accuracy orders from 2 to 8. The authors of [11] were able to obtain with their CLS method the chain of the Moffatt vortices in the lower corners of the square cavity in the 2D lid-driven cavity flow problem for the Reynolds number Re = 1000 and to solve the given benchmark problem with the accuracy, which is still at the level of the best accuracies. In the works [10,11], a detailed comparison of the results of the numerical solution of the two-dimensional lid-driven cavity problem was carried out, and it was shown that the obtained results are among the most accurate ones obtained by other researchers (see [2,23,8,5]) with the aid of different high-accuracy numerical algorithms.

At present, the numerical solutions of the 3D lid-driven cavity problem are very scarce. Some comparison was done with [8], which showed that the results obtained by the CLS method and by the method of [8] are very close to one another in terms of the accuracy. This comparison is, however, omitted in the text of the present paper because of the shortage of the place and also in view of the fact that the main objective of the present paper is to show the efficiency of the application of symbolic computations at the development of new methods for solving the partial differential equations, in particular, the Navier–Stokes equations.

In the present work, we extend the CLS method of [20] for the case of three spatial variables. All symbolic computations needed for the derivation of the collocation and matching conditions were implemented with *Mathematica.*

We show the usefulness and efficiency of using the CASs for the construction, analysis, and numerical implementation of new complex numerical algorithms by the example of constructing the work formulas and their realization for solving the boundary-value problems for the 3D Navier–Stokes equations. In view of the specifics of the CASC Workshops themes, we discuss here to a lesser extent the numerical results; this discussion has been presented in sufficient detail in the foregoing publications of one of the authors [10,11]. In comparison with the 2D case [10,11], the amount of symbolic computations needed for the construction, analysis, and transfer of the work formulas to the code (written in Fortran,C, Pascal, etc.) for the numerical computation increases considerably in the 3D case. Therefore, the application of computer algebra systems becomes much more useful in the 3D case than in the 2D case.

The numerical computations needed for obtaining the final numerical solution in each cell were implemented by us in a Fortran code.

2   Description of the CLS Method

2.1   Problem Statement

Consider a boundary-value problem for the system of stationary Navier–Stokes equations

$$(\mathbf{V} \cdot \nabla)\mathbf{V} + \nabla p = \frac{1}{\text{Re}} \Delta \mathbf{V} - \mathbf{f}, \tag{1}$$

$$\text{div } \mathbf{V} = 0, \quad (x_1, x_2, x_3) \in \Omega, \tag{2}$$

which govern the flows of a viscous, non-heat-conducting, incompressible fluid in the cube

$$\Omega = \{(x_1, x_2, x_3), 0 \le x_i \le X, i = 1, 2, 3\} \tag{3}$$

with the boundary $\partial\Omega$, where $X > 0$ is the user-specified length of the rib of the cubic region Ω, and x_1, x_2, x_3 are the Cartesian spatial coordinates. In equations(1) and (2), $\mathbf{V} = \mathbf{V}(x_1, x_2, x_3)$ is the velocity vector having the components $v_1(x_1, x_2, x_3), v_2(x_1, x_2, x_3)$, and $v_3(x_1, x_2, x_3)$ along the axes x_1, x_2, and x_3, respectively; $p = p(x_1, x_2, x_3)$ is the pressure, $\mathbf{f} = (f_1, f_2, f_3)$ is a given vector function. The positive constant Re is the Reynolds number, $\Delta = \frac{\partial^2}{\partial x_1^2} + \frac{\partial^2}{\partial x_2^2} + \frac{\partial^2}{\partial x_3^2}$, $(\mathbf{V} \cdot \nabla) = v_1\frac{\partial}{\partial x_1} + v_2\frac{\partial}{\partial x_2} + v_3\frac{\partial}{\partial x_3}$.

The system of four equations (1) and (2) is solved under the Dirichlet boundary condition $\mathbf{V}\big|_{\partial\Omega} = \mathbf{g}$, where $\mathbf{g} = \mathbf{g}(x_1, x_2, x_3) = (g_1, g_2, g_3)$ is a given vector function.

2.2 Local Coordinates and Basis Functions

In the CLS method, the spatial computational region (3) is discretized by a grid with cubic cells $\Omega_{i,j,k}$. We search for the solution in the form of a piecewise smooth function on this grid. To write the formulas of the CLS method it is convenient to introduce the local coordinates y_1, y_2, y_3 in each cell $\Omega_{i,j,k}$. The dependence of local coordinates on the global spatial variables x_1, x_2, x_3 is determined by the relations $y_m = (x_m - x_{m,i,j,k})/h$, $m = 1, 2, 3$, where $x_{m,i,j,k}$ is the value of the coordinate x_m at the geometric center of the cell $\Omega_{i,j,k}$, and h is the halved length of the rib of the cubic cell $\Omega_{i,j,k}$. The local coordinates then belong to the interval $y_m \in [-1, 1]$. We now introduce the notations $\mathbf{u}(y_1, y_2, y_3) = \mathbf{V}(hy_1 + x_{1,i,j,k}, hy_2 + x_{2,i,j,k}, hy_3 + x_{3,i,j,k})$, $p(y_1, y_2, y_3) = p(hy_1 + x_{1,i,j,k}, hy_2 + x_{2,i,j,k}, hy_3 + x_{3,i,j,k})$. After the above change of variables the Navier–Stokes equations take the following form:

$$\frac{\Delta u_m}{\text{Re}h} - \left(u_1\frac{\partial u_m}{\partial y_1} + u_2\frac{\partial u_m}{\partial y_2} + u_3\frac{\partial u_m}{\partial y_3} + \frac{\partial p}{\partial y_m}\right) = hf_m, \quad m = 1, 2, 3; \quad (4)$$

$$\frac{1}{h}\left(\frac{\partial u_1}{\partial y_1} + \frac{\partial u_2}{\partial y_2} + \frac{\partial u_3}{\partial y_3}\right) = 0, \quad (5)$$

where $\Delta = \frac{\partial^2}{\partial y_1^2} + \frac{\partial^2}{\partial y_2^2} + \frac{\partial^2}{\partial y_3^2}$.

The basic idea of the method is to use the collocation method in combination with the least-squares method to obtain numerical solution. We call such a combined method the "collocation and least-squares" (CLS) method. One of the reasons for using this combination was that the application of the least-squares method often improves the properties of a numerical method. In turn, the collocation method is simple in implementation and gives good results when solving boundary-value problems for ODEs, both linear and nonlinear [1].

Following [20] we now linearize the Navier–Stokes equations(4). To this end, we present the desired improved solution in the form: $u_m = U_m + \bar{u}_m$ ($m = 1, 2, 3$), $p = P + \bar{p}$, where (U_1, U_2, U_3, P) is some known approximate solution, and $\bar{u}_m, \bar{p}$ are the corrections to the solution, they are to be found. Substituting this representation in equations (4) and neglecting the terms of the second order of smallness $\bar{u}_l \bar{u}_{m,y_l}$, $m, l = 1, 2, 3$, we obtain the linearized equations:

$$\Delta \bar{u}_m / (\text{Re} \cdot h) - (U_1 \bar{u}_{m,y_1} + U_2 \bar{u}_{m,y_2} + U_3 \bar{u}_{m,y_3} + \bar{u}_1 U_{m,y_1} + \bar{u}_2 U_{m,y_2} +$$
$$\bar{u}_3 U_{m,y_3} + \bar{p}_{y_m}) - F_m / (\text{Re} \cdot h) = 0, \ m = 1, 2, 3, \tag{6}$$

where $F_m = \text{Re} \left[h^2 f_m + h \left(U_1 U_{m,y_1} + U_2 U_{m,y_2} + U_3 U_{m,y_3} + P_{y_m} \right) \right] - \Delta U_m$,

$$\bar{u}_{m,y_l} = \partial \bar{u}_m / \partial y_l, \ U_{m,y_l} = \partial U_m / \partial y_l, \ \bar{p}_{y_m} = \partial \bar{p} / \partial y_m, \ l, m = 1, 2, 3.$$

We now present the approximate solution in each cell $\Omega_{i,j,k}$ as a linear combination of the basis vector functions φ_l

$$\begin{pmatrix} U_1 \\ U_2 \\ U_3 \\ P \end{pmatrix} = \sum_l a_{i,j,k,l} \varphi_l, \quad \begin{pmatrix} \bar{u}_1 \\ \bar{u}_2 \\ \bar{u}_3 \\ \bar{p} \end{pmatrix} = \sum_l b_{i,j,k,l} \varphi_l. \tag{7}$$

The basis functions φ_l are presented in Table 1. To approximate the velocity components we use the second-order polynomials in variables y_1, y_2, y_3, and the first-order polynomials are used for the approximation of the pressure. The basis functions for the velocity components are solenoidal, that is $\text{div} \, \varphi_l = 0$. This enables the identical satisfaction of the continuity equation in each cell.

Table 1. The form of basis functions φ_l

l	1	2	3	4	5	6	7	8	9	10
	1	0	0	y_1	0	0	y_2	0	0	y_3
φ_l	0	1	0	$-y_2$	y_1	0	0	y_2	0	0
	0	0	1	0	0	y_1	0	$-y_3$	y_2	0
	0	0	0	0	0	0	0	0	0	0
l	11	12	13	14	15	16	17	18	19	20
	0	y_1^2	0	0	$-2y_1y_2$	0	$-2y_1y_3$	0	y_1^2	y_2^2
φ_l	y_3	$-2y_1y_2$	y_1^2	0	y_2^2	0	0	y_1y_3	0	0
	0	0	0	y_1^2	0	y_1y_2	y_3^2	0	$-2y_1y_3$	0
	0	0	0	0	0	0	0	0	0	0
l	21	22	23	24	25	26	27	28	29	30
	0	0	y_2y_3	0	y_3^2	0	0	0	0	0
φ_l	y_2^2	0	0	$-2y_2y_3$	0	y_3^2	0	0	0	0
	$-2y_2y_3$	y_2^2	0	y_3^2	0	0	0	0	0	0
	0	0	0	0	0	0	1	y_1	y_2	y_3

The use of the solenoidal basis reduces here the number of the unknowns in a cell from 34 to 30. Such a reduction of the number of unknowns enables significant savings in the CPU time.

2.3   Derivation of the Overdetermined System from Collocation and Matching Conditions

Let us take the cell $\Omega_{i,j,k}$ and specify the collocation points inside it. Let ω be a user-specified value in the interval $0 < \omega < 1$, for example, $\omega = 1/2$. Let us specify eight collocation points in each cell as the points with local coordinates $(\pm\omega, \pm\omega, \pm\omega)$. Substituting these coordinates in the left-hand sides of equations (6) we obtain 24 equations of collocations.

Let us now proceed to the derivation of the equations following from the matching conditions. Similarly to [20] we use on the cell faces the matching conditions ensuring the only piecewise polynomial solution. These conditions of the solution continuity are the linear combinations of the form

$$
\frac{\partial (u^+)^n}{\partial n} + \eta (u^+)^n = \frac{\partial (u^-)^n}{\partial n} + \eta (u^-)^n;
$$
$$
\frac{\partial (u^+)^{\tau_1}}{\partial n} + \eta (u^+)^{\tau_1} = \frac{\partial (u^-)^{\tau_1}}{\partial n} + \eta (u^-)^{\tau_1};
$$
$$
\frac{\partial (u^+)^{\tau_2}}{\partial n} + \eta (u^+)^{\tau_2} = \frac{\partial (u^-)^{\tau_2}}{\partial n} + \eta (u^-)^{\tau_2},
$$
$$
p^+ = p^-.
$$

(8)

Here n is the external normal to the cell face, $(\cdot)^n$, $(\cdot)^{\tau_1}$, and $(\cdot)^{\tau_2}$ are the normal and tangential components of the velocity vector with respect to the face between the cells, u^+ and u^- are the limits of the function u at its tending to the cell face from within and from outside the cell; η is a positive parameter, which can affect the conditionality of the obtained system of linear algebraic equations and the convergence rate. The points at which equations (8) are written are called the matching points. Let us exemplify the way, in which the coordinates of these points are specified, by the example of the faces $y_1 = \pm 1$. On the face $y_1 = 1$, two matching points are specified as $(1, \zeta, \zeta)$ and $(1, -\zeta, -\zeta)$, and on the face $y_1 = -1$, two matching points are specified as $(-1, -\zeta, \zeta)$ and $(-1, \zeta, -\zeta)$, where $0 \le \zeta \le 1$, and ζ is specified by the user, for example, $\zeta = 1/2$.

Thus, we specify two matching points for the velocity components on each of the six faces of the cell. As a result, we obtain 12 matching points. We then substitute the coordinates of these points in each of the first three matching conditions in (8) and obtain 36 matching conditions for velocity components. The matching conditions for the pressure are specified at six points $(\pm 1, 0, 0)$, $(0, \pm 1, 0)$, $(0, 0, \pm 1)$.

If the cell face coincides with the boundary of region Ω, then we use the boundary conditions instead of the matching conditions: $u_m = g_m$, $m = 1, 2, 3$.

To identify the unique solution we specified the pressure at the coordinate origin. Let $\Omega_{1,1,1}$ be the cell one of the vertices of which coincides with the coordinate origin $(0,0,0)$. The local coordinates $y_1 = y_2 = y_3 = -1$ correspond to this point. We have set $p(-1, -1, -1) = p_0$, where p_0 is an arbitrary constant; it was specified from the numerical solution of the test problem below in Section 3.1. And in the computations of the lid-driven cavity flow, we used the value $p_0 = 1$ following [18].

Thus, the set of 24 collocation equations, 36 equations of matching conditions for velocity components, and six matching conditions for the pressure give, in the total, the overdetermined system containing 66 or 67 equations. These equations involve 30 unknowns $b_{i,j,k,l}$, $l = 1, 2, \ldots, 30$. All these equations were derived on computer in Fortran form by using symbolic computations with *Mathematica*.

At the obtaining of the final form of the formulas for the coefficients of the equations, it is useful to perform the simplifications of the arithmetic expressions of polynomial form to reduce the number of the arithmetic operations needed for their numerical computation. To this end, we employed standard functions of the *Mathematica* system, such as `Simplify` and `FullSimplify` for the simplification of complex symbolic expressions arising at the symbolic stages of the construction of the formulae of the method. Their application enabled a two-three-fold reduction of the length of polynomial expressions.

Let us denote the obtained collocation equations by

$$COL(s) = 0, \quad s = 1, \ldots, 24, \tag{9}$$

and let us denote the equations obtained from the matching conditions by

$$MATCH(s) = 0, \quad s = 1, \ldots, 42. \tag{10}$$

It would be impossible to present all these equations here in view of their very bulky form.

The system of equations (9) and (10) is overdetermined. It was solved numerically by the method of rotations with column pivoting [26].

Let us assume that the length of the resulting computer code in Fortran 90 language, which is measured in the number of lines of the code text, is equal to 100 %. Then the collocation equations take 26 %, and the matching conditions take 33 % of the entire code so that 59 % of the entire Fortran code were generated with *Mathematica*. It should be noted that the corresponding Fortran subroutines are of crucial importance for the implementation of the 3D CLS method under consideration because they enable the user the obtaining of the complete set of the equations of the overdetermined algebraic system (9), (10). Furthermore, the CAS *Mathematica* enables various check-ups of new methods directly at the stage of their development, including the convergence of the method. It is very important that the CAS performs the job on the development of new formulas, which need much attention and many efforts of the mathematician and saves him from many possible errors.

The remaining 41 % of the Fortran code implement the method of rotations, the iterations in nonlinearity, and the export of the final converged solution to the external files in the form of the *Mathematica* lists for the subsequent graphical plotting and analysis of the results with the aid of the corresponding *Mathematica* program.

3   Numerical Results

This section presents the results of computations of two problems to assess the capabilities of the proposed method. The first problem is a test problem with the known exact analytic solution, which has enabled us to obtain the exact error of the approximate solution and, consequently, to obtain the information about the convergence order of the numerical solution.

The second problem is the viscous flow in a lid-driven cavity, for which there is no exact solution. In this case, one can judge about the correctness of obtained results from the numerical experiments on a sequence of grids and comparisons with experimental and numerical results obtained by other researchers.

3.1   Test with Exact Analytic Solution

Let us consider the following exact solution of the Navier–Stokes equations (1) and (2) in the cubic region (3):

$$
\begin{aligned}
u_1 &= -\cos(x_1)\sin(x_2)\sin(x_3), \quad u_2 = 0.5\sin(x_1)\cos(x_2)\sin(x_3), \\
u_3 &= 0.5\sin(x_1)\sin(x_2)\cos(x_3), \\
p &= \cos(x_1)+\cos(x_2)+\cos(x_3)-(3/X)\sin(X).
\end{aligned}
\tag{11}
$$

It is to be noted here that the solution (11) satisfies the continuity equation (2). We now write down the right-hand sides f_1, f_2, f_3 of equations (1):

$$
\begin{aligned}
f_1 &= (3\cos(x_1)\sin(x_2)\sin(x_3) + \mathrm{Re}\sin(x_1)(1+\cos(x_1)(0.5\cos^2(x_3)\sin^2(x_2) \\
&\quad + (0.5\cos^2(x_2)+\sin^2(x_2))\sin^2(x_3))))/\mathrm{Re}, \\
f_2 &= (-1.5\cos(x_2)\sin(x_1)\sin(x_3) + \mathrm{Re}\sin(x_2)(1+\cos(x_2)\times \\
&\quad (-0.25\cos^2(x_3)\sin^2(x_1) + (0.5\cos^2(x_1)+0.25\sin^2(x_1))\sin^2(x_3))))/\mathrm{Re}, \\
f_3 &= (\mathrm{Re}\sin(x_3)+\cos(x_3)(-1.5\sin(x_1)\sin(x_2)+0.5\mathrm{Re}\cos^2(x_1)\sin^2(x_2)\sin(x_3) \\
&\quad + \mathrm{Re}\sin^2(x_1)(-0.25\cos^2(x_2)+0.25\sin^2(x_2))\sin(x_3)))/\mathrm{Re}.
\end{aligned}
$$

The test solution (11) has no singularities, therefore, one can observe the convergence order of the numerical solution already on rough grids. The region (3) was discretized by a uniform grid of cubic cells. The half-size h of the cell rib was equal to $h = X/(2M)$, where M is the number of cells along each coordinate direction. Let us assign the indices i, j, k of a cell to the geometric center of the cell, where i, j, k vary along the axes x_1, x_2, and x_3, respectively.

Since the present method uses the linearization of the Navier–Stokes equations, the iterations in nonlinearity are necessary. The zero initial guess for the quantities U_i and P was used.

The numerical results presented below in Table 2 were obtained by using the value $\omega = 1/2$ at the specification of collocation points.

To determine the absolute numerical errors of the method on a specific uniform grid with half-step h we have computed the following root mean square errors:

$$
\delta\mathbf{u}(h) = \left[\frac{1}{3M^3}\sum_{i=1}^{M}\sum_{j=1}^{M}\sum_{k=1}^{M}\sum_{\nu=1}^{3}(u_{\nu,i,j,k}-u_{\nu,i,j,k}^{ex})^2\right]^{0.5},
$$

$$
\delta p(h) = \left[\frac{1}{M^3}\sum_{i=1}^{M}\sum_{j=1}^{M}\sum_{k=1}^{M}(p_{i,j,k}-p_{i,j,k}^{ex})^2\right]^{0.5},
$$

where $\mathbf{u}^{ex}_{i,j,k}$ and $p^{ex}_{i,j,k}$ are the velocity vector and the pressure calculated in accordance with the exact solution (11). The quantities $\mathbf{u}_{i,j,k}$ and $p_{i,j,k}$ denote the numerical solution obtained by the method described in the foregoing sections. Let us denote by ν_u and ν_p the convergence orders calculated from the numerical solutions for the velocity vector $\mathbf{U}$ and for the pressure p. The quantities ν_u and ν_p were calculated by the following well-known formulas:

$$\nu_u = \frac{\log[\delta\mathbf{u}(h_{m-1})] - \log[\delta\mathbf{u}(\mathbf{h_m})]}{\log(h_{m-1}) - \log(h_m)}, \quad \nu_p = \frac{\log[\delta p(h_{m-1})] - \log[\delta p(h_m)]}{\log(h_{m-1}) - \log(h_m)}, \quad (12)$$

where h_m, $m = 2, 3, \ldots$ are some values of the step h such that $h_{m-1} \neq h_m$. Let us denote the value of the coefficient $b_{i,j,k,l}$ in (7) at the nth iteration by $b^n_{i,j,k,l}$, $n = 0, 1, \ldots$. The following condition was used for termination of the iterations in nonlinearity: $\delta b^{n+1} < \varepsilon$, where

$$\delta b^{n+1} = \max_{i,j,k} \left(\max_{1 \leq l \leq 30} \left| b^{n+1}_{i,j,k,l} - b^n_{i,j,k,l} \right| \right) \quad (13)$$

and ε is a small positive user-specified number, $\varepsilon \leq h^2$. We will term the quantity (13) the pseudo-error in the following.

The graphs of the errors δb^n, $\delta\mathbf{u}$ and δp as the functions of the number n of the iterations in nonlinearity were plotted. We do not present these graphs for the sake of brevity.

Table 2. The errors $\delta\mathbf{u}, \delta p$ and the convergence orders ν_u, ν_p on a sequence of grids, Re = 100

η	$M \times M \times M$	$\delta\mathbf{u}$	δp	ν_u	ν_p
2.0	$10 \times 10 \times 10$	$1.01 \cdot 10^{-2}$	$7.27 \cdot 10^{-3}$		
4.45	$20 \times 20 \times 20$	$2.13 \cdot 10^{-3}$	$2.53 \cdot 10^{-3}$	2.25	1.52
3.0	$30 \times 30 \times 30$	$7.74 \cdot 10^{-4}$	$1.47 \cdot 10^{-3}$	2.50	1.34

The values of the parameter η, which are nearly optimal for each specific grid, were determined as follows: twenty iterations were done on a specific grid for some chosen value of η. Then the value of η was deviated arbitrarily from the initial guess $\eta = 1$, which was used in [11]. If the pseudo-error after 20 iterations was smaller than in the foregoing variant, then the variation of η was continued in the same direction of the decrease or increase in η. Usually six or seven such runs were done to find a nearly optimal value of η. Similar investigations concerning the choice of η were done previously in [20,10,11] in the two-dimensional case.

It is to be noted that the specific size of the errors $\delta\mathbf{u}$ and δp also depends on the value of the collocation parameter ω so that one can reduce these errors in comparison with the case of $\omega = 0.5$. This value of ω was used as the initial guess similarly to [20,10,11]. It was found by numerical experiments similar to the above ones used for determining a quasi optimal value of η that the values $\omega = 0.6$

and $\zeta = 0.6$ were generally better in terms of accelerating the convergence of the iterative procedure than the values $\omega = 0.5$ and $\zeta = 0.5$.

It can be seen from Table 2 that in the case of the Reynolds number Re = 100, the convergence order ν_u lies in the interval $2.25 \leq \nu_u \leq 2.50$, that is it is somewhat higher than the second order.

The fact that the numerical solution converges to the exact solution with the expected (second) order is the additional evidence of correctness of formulas derived with the aid of *Mathematica*. It can be seen from Table 2 that the absolute errors in the velocity are less than the absolute errors in the pressure. This is explained by the fact that the first-degree polynomials are used in the cells to approximate the pressure, whereas the velocity components are approximated by the second-degree polynomials, see Table 1. The use of the quadratic approximation for the pressure was presented in [20,11] in the two-dimensional case, and it showed an increase in the accuracy of the pressure calculation in comparison with the linear approximation. The quadratic approximation of the pressure can, of course, be extended for the 3D case, but it will involve an increase in the CPU time expenses because of the appearance of several further expansion coefficients $b_{i,j,k,l}$ in the local representation of the pressure.

It is to be noted that at the application of the method of least squares for solving any problems, it is important that the equations in the overdetermined system, which play the equal role in the approximate solution, have equal weight coefficients. For example, in the given case, the solution accuracy can deteriorate if some equations of collocations are included in the system with the weight coefficients different from the weight coefficients of other equations.

3.2   Flow in the Lid-Driven Cavity

Consider the flow of a viscous incompressible fluid in a three-dimensional cavity whose lid moves in the given direction at a constant speed. The computational region is the cube (3). The coordinate origin lies in one of the cube corners, and the Ox_3 axis is directed upwards. The cube upper face moves in dimensionless coordinates at the unit velocity in the positive direction of the Ox_1 axis. The remaining faces of cube (3) are at rest. The no-slip conditions are specified on all cube faces: $v_1 = 1$, $v_2 = v_3 = 0$, if $x_3 = X$, and $v_m = 0$, $m = 1,2,3$, at the remaining boundary faces.

The lid-driven cavity flow has singularities in the corners of the region for which $x_3 = X$, which manifest themselves more strongly with the increasing Reynolds number. Therefore, the increase of Re needs the application of fine grids in the neighborhood of singularities to obtain a more accurate solution. In the present work, we have used only uniform grids whose size did not exceed $40 \times 40 \times 40$ cells. And, as a result, an increase in the Reynolds number resulted in an error increase.

Some results of the numerical computations of the viscous incompressible flow in a cubic lid-driven cavity are presented in Fig. 1 for the Reynolds number Re = 100. The arrows indicate the local directions of the motion of fluid particles. One can see the following features of the flow under consideration: there is a line

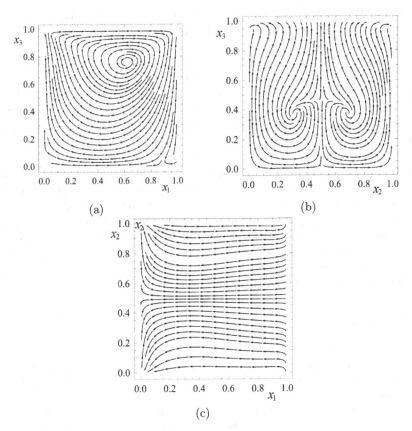

Fig. 1. Streamlines in different sections of the cubic cavity at Re = 100: (a) section $x_2 = 0.5$; (b) section $x_1 = 0.5$; (c) section $x_3 = 0.5$

of flow divergence in section $x_3 = 0.5$ (see Fig. 1, (c)); the equation of this line has the form $x_2 = 0.5$. Near the wall $x_1 = 0$, the fluid particles move at $x_2 < 0.5$ in the direction of the wall $x_2 = 0$, and at $x_2 > 0.5$, they move in the direction of the wall $x_2 = 1$. In section $x_1 = 0.5$ (Fig. 1, (b)), the flow is symmetric with respect to the line $x_2 = 0.5$. In section $x_2 = 0.5$ (Fig. 1, (a)), a vortex is observed, which is similar to the one obtained at the flow computation in the two-dimensional cavity [20]. Figures 1, (b) and (c) show that in the region of the fluid flow, there arise new types of singularities, which are different from the two-dimensional case. For example, the points of the bifurcation of streamlines are observed, and the corresponding flow patterns differ substantially from the flow pattern in the two-dimensional case.

4   Conclusions

A new symbolic-numeric method of collocations and least squares has been presented for the numerical solution of three-dimensional stationary Navier–Stokes equations. The method has been verified on a test problem having the exact analytic solution. The application of the method is given for the flow in a 3D lid-driven cavity.

The presented method can be generalized for the case of unsteady flow problems. The application of symbolic computations for the construction of the method for unsteady governing equations will be very useful also for deriving the basic formulae of the method for the unsteady case. Furthermore, the authors have an experience of implementing the method of collocations and least squares with the use of symbolic computations for a simpler scalar unsteady equation, namely, the one-dimensional heat equation.

It is also to be noted that the numerical stage of the presented symbolic-numeric method suits very well for parallelization because the computations in each cell of the spatial grid can be carried out at each iteration independently of other cells. The computational region may be partitioned into any number of subregions, which is equal to the number of available processors, in such a way that the computations in each subregion are carried out independently of each other, and their interaction with one another is realized by the refinement of the matching conditions between them after each iteration.

References

1. Ascher, U., Christiansen, J., Russell, R.D.: A collocation solver for mixed order systems of boundary value problems. Math. Comput. 33, 659–679 (1979)
2. Botella, O., Peyret, R.: Benchmark spectral results on the lid-driven cavity flow. Comput. Fluids 27, 421–433 (1998)
3. Brown, D.L., Cortez, R., Minion, M.L.: Accurate projection methods for the incompressible Navier–Stokes equations. J. Comp. Phys. 168, 464–499 (2001)
4. Chibisov, D., Ganzha, V.G., Mayr, E.W., Vorozhtsov, E.V.: Stability Investigation of a Difference Scheme for Incompressible Navier-Stokes Equations. In: Ganzha, V.G., Mayr, E.W., Vorozhtsov, E.V. (eds.) CASC 2007. LNCS, vol. 4770, pp. 102–117. Springer, Heidelberg (2007)
5. Erturk, E., Gokcol, C.: Fourth order compact formulation of Navier–Stokes equations and driven cavity flow at high Reynolds numbers. Int. J. Numer. Methods Fluids 50, 421–436 (2006)
6. Ferziger, J.H., Peric, M.: Computational Methods for Fluid Dynamics, 3rd edn. Springer, Heidelberg (2002)
7. Ganzha, V.G., Mazurik, S.I., Shapeev, V.P.: Symbolic Manipulations on a Computer and their Application to Generation and Investigation of Difference Schemes. In: Caviness, B.F. (ed.) EUROCAL 1985. LNCS, vol. 204, pp. 335–347. Springer, Heidelberg (1985)
8. Garanzha, V.A., Kon'shin, V.N.: Numerical algorithms for viscous fluid flows based on high-order accurate conservative compact schemes. Comput. Math. Math. Phys. 39, 1321–1334 (1999)
9. Gerdt, V.P., Blinkov, Y.A.: Involution and Difference Schemes for the Navier–Stokes Equations. In: Gerdt, V.P., Mayr, E.W., Vorozhtsov, E.V. (eds.) CASC 2009. LNCS, vol. 5743, pp. 94–105. Springer, Heidelberg (2009)
10. Isaev, V.I., Shapeev, V.P.: Development of the collocations and least squares method. Proc. Inst. Math. Mech. 261(suppl. 1), 87–106 (2008)

11. Isaev, V.I., Shapeev, V.P.: High-accuracy versions of the collocations and least squares method for the numerical solution of the Navier–Stokes equations. Computat. Math. and Math. Phys. 50, 1758–1770 (2010)
12. Kim, J., Moin, P.: Application of a fractional-step method to incompressible Navier–Stokes equations. J. Comp. Phys. 59, 308–323 (1985)
13. Kirkpatrick, M.P., Armfield, S.W., Kent, J.H.: A representation of curved boundaries for the solution of the Navier–Stokes equations on a staggered three-dimensional Cartesian grid. J. Comp. Phys. 184, 1–36 (2003)
14. Malan, A.G., Lewis, R.W., Nithiarasu, P.: An improved unsteady, unstructured artificial compressibility, finite volume scheme for viscous incompressible flows: Part I. Theory and implementation. Int. J. Numer. Meth. Engng. 54, 695–714 (2002)
15. Marella, S., Krishnan, S., Liu, H., Udaykumar, H.S.: Sharp interface Cartesian grid method I: An easily implemented technique for 3D moving boundary computations. J. Comp. Phys. 210, 1–31 (2005)
16. Muldoon, F., Acharya, S.: A modification of the artificial compressibility algorithm with improved convergence characteristics, Int. J. Numer. Meth. Fluids 55, 307–345 (2007)
17. Pinelli, A., Naqavi, I.Z., Piomelli, U., Favier, J.: Immersed-boundary methods for general finite-difference and finite-volume Navier–Stokes solvers. J. Comp. Phys. 229, 9073–9091 (2010)
18. Roache, P.J.: Computational Fluid Dynamics, Hermosa, Albuquerque, N.M (1976)
19. Schlichting, H., Truckenbrodt, E.: Aerodynamics of the Airplane. McGraw-Hill, New York (1979)
20. Semin, L., Shapeev, V.: Constructing the Numerical Method for Navier – Stokes Equations Using Computer Algebra System. In: Ganzha, V.G., Mayr, E.W., Vorozhtsov, E.V. (eds.) CASC 2005. LNCS, vol. 3718, pp. 367–378. Springer, Heidelberg (2005)
21. Semin, L.G., Sleptsov, A.G., Shapeev, V.P.: Collocation and least -squares method for Stokes equations. Computational Technologies 1(2), 90–98 (1996) (in Russian)
22. Shapeev, A.V.: Application of computer algebra systems to construct high-order difference schemes. In: 6th IMACS Int. IMACS Conf. on Applications of Computer Algebra, June, 25-28, pp. 92–93. Univ. of St. Petersburg, St. Petersburg (2000)
23. Shapeev, A.V., Lin, P.: An asymptotic fitting finite element method with exponential mesh refinement for accurate computation of corner eddies in viscous flows. SIAM J. Sci. Comput. 31, 1874–1900 (2009)
24. Uhlmann, M.: An immersed boundary method with direct forcing for the simulation of particulate flows. J. Comp. Phys. 209, 448–476 (2005)
25. Valiullin, A.N., Ganzha, V.G., Meleshko, S.V., Murzin, F.A., Shapeev, V.P., Yanenko, N.N.: Application of Symbolic Manipulations on a Computer for Generation and Analysis of Difference Schemes. Preprint Inst. Theor. Appl. Mech. Siberian Branch of the USSR Acad. Sci., Novosibirsk (7) (1981)
26. Voevodin, V.V.: Computational Foundations of Linear Algebra. Nauka, Moscow (1977) (in Russian)
27. Wesseling, P.: Principles of Computational Fluid Dynamics. Springer, Heidelberg (2001)

Verifiable Conditions on Asymptotic Stabilisability for a Class of Planar Switched Linear Systems[*]

Zhikun She and Haoyang Li

SKLSDE, LMIB and School of Mathematics and Systems Science
Beihang University, Beijing, China
zhikun.she@buaa.edu.cn

Abstract. In this paper, we propose a computer algebra based approach for analyzing asymptotic stabilisability of a class of planar switched linear systems, where subsystems are assumed to be alternatively active. We start with an algebraizable sufficient condition on the existence of stabilizing switching lines and a multiple Lyapunov function. Then, we apply a real root classification based method to under-approximate this condition such that the under-approximation only involves the parametric coefficients. Afterward, we additionally use quantifier elimination to eliminate parameters in the multiple Lyapunov function, arriving at a quantifier-free formula over parameters in the switching lines. According to our intermediate under-approximation as well as our final quantifier-free formula, we can easily design explicit stabilizing switching laws. Moreover, based on a prototypical implementation, we use an illustrating example to show the applicability of our approach. Finally, the advantages of our approach are demonstrated by the comparisons with some related works in the literature.

1 Introduction

A switched system [24,13] is a hybrid system which comprises a collection of dynamic systems and a switching law that specifies switching instances between two dynamic systems. Many real-world processes and systems, for example, gyroscopic systems, power systems and biological systems, can be modeled as switched systems. Among the various research topics on switched systems, stability and stabilisability of switched systems have attracted most of the attention [18,32,12,29,8,33,19,20]. For more references, the reader may refer to the survey papers [15,17] and the books [14,13].

On the analysis, design and synthesis of stable switched systems, there are two categories of stabilization strategies. One is the feedback stabilization, where state or output feedback control laws are designed based on a given class of switching laws. Usually, the following switching laws are considered: arbitrary switching [7], slow switching [6] and restricted switching induced by partitions of the state space [23], etc.

The other one is the switching stabilization, where the critical problem is the design of stabilizing switching laws by time domain restrictions [12,28] and state space restrictions [10,1,20,16,11]. For this critical problem, one challenge is to derive sufficient and necessary conditions for the existence of stabilizing switching laws. Some necessary and sufficient conditions have been derived for certain classes of switched systems,

---

[*] This work was partly supported by NSFC-61003021 and SKLSDE-2011ZX-16.

V.P. Gerdt et al. (Eds.): CASC 2012, LNCS 7442, pp. 334–348, 2012.

for instances, continuous-time linear switched systems in static state feedback [16] and generic switched linear systems with a pair of planar subsystems [11], etc. But, few results on general switched systems have been obtained. Thus, it has been becoming more practical to only find sufficient conditions [10,12,21,20]. However, in the literature, most of these found sufficient conditions are not easily verified [12,21,20]. Moreover, they are often used to prove the existence of stabilizing switching laws [10,12,16] instead of explicitly providing them.

It is well known that one sufficient stabilisability condition is the existence of a common Lyapunov function (CLF) or a multiple Lyapunov function (MLF) [3,20]. Thus, following our earlier work on computing quadratic MLFs [26], we will in this paper provide an easily verifiable condition on stabilizing switching laws to guarantee asymptotic stabilisability of a class of planar switched linear systems (PSLS), where subsystems are assumed to be alternatively active.

We first construct an algebraizable sufficient condition for the existence of stabilizing switching lines and a multiple Lyapunov function in quadratic form. Then, in order to make computation efficient, we reduce the number of parameters and classify our discussion on its asymptotic stabilisability into eight cases. For each case, we apply a real root classification (RRC) based method to under-approximate this sufficient condition such that each under-approximation only involves the coefficients of the pre-assumed quadratic Lyapunov functions and pre-assumed stabilizing switching lines. Furthermore, we additionally formulate each under-approximation as an existentially quantified constraint and use quantifier elimination (QE) to eliminate the coefficients of the pre-assumed Lyapunov functions, arriving at a quantifier-free formula over parameters appearing in the switching lines. Easily, we can design explicit stabilizing switching laws by using sample points satisfying our intermediate under-approximation as well as our final quantifier-free formula. Finally, based on a prototypical implementation, we use an illustrating example to show the verifiability and applicability of our approach.

Moreover, we compare our current approach with some related works in the literature [33,11]. Especially, we use a simple example to show that we can design stabilizing switching laws that do not satisfy the collinear condition in [11].

To our knowledge, this is the first time to apply computer algebra to provide explicit stabilizing switching laws for analyzing asymptotic stabilisability.

Distinguishingly different from [27], this paper aims to design stabilizing switching laws while [27] aims to calculate a MLF with given switching laws. Therefore, the switching lines as well as the state space of each subsystem are undetermined parameters in our sufficient condition and thus we additionally require QE to generate stabilizing conditions. Moreover, since the system is planar, our current conditions are simpler than the corresponding ones in [27] and thus their under-approximations or equivalences can be flexibly computed due to the classification.

2 Preliminaries and Problem Definitions

A switched system is a system that consists of a finite number of subsystems of form $\dot{x} = f_i(x)$ and a switching law $\mathbb{S}$, where $i \in \mathcal{M}$, $\mathcal{M} = \{1, \cdots, N\}$ is the set of discrete modes, $x \in X_i$, $X_i \subset \mathbb{R}^n$ is the continuous state space of mode i, $f_i(x)$ is a vector field describing the dynamics of mode i, and S determines switches between different modes (i.e., switches between different subsystems). Moreover, we assume that the time interval between two successive switching instances must be positive.

A trajectory of a switched system is a finite or an infinite sequence $r_0(t), r_1(t)$, $\cdots, r_p(t), \cdots$ of flows with a sequence $T_0, T_1, \cdots, T_p, \cdots$, such that for all $t \in [T_p, T_{p+1}]$, the system is active in a certain mode i and evolves according to $\dot{r}_p(t) = f_i(r_p(t))$. For simplicity, we denote a trajectory $r_0(t), r_1(t), \ldots, r_p(t), \ldots$ by $x(t)$ satisfying

1. $x(0) = r_0(0)$ and for all $p \geq 0$, $x(\sum_{i=0}^{p} T_i) = r_p(T_p) = r_{p+1}(0)$;
2. for all $t \in (\sum_{i=0}^{p-1} T_i, \sum_{i=0}^{p} T_i)$, $x(t) = r_p(t - \sum_{i=0}^{p-1} T_i)$.

Definition 1. *(Asymptotically Stable) A switched system is called stable if*

$$\forall \epsilon > 0 \exists \delta > 0 \forall t > 0 \, [\|x(0)\| < \delta \Rightarrow \|x(t)\| < \epsilon],$$

and attractive if for all trajectories $x(t)$, $\lim_{t \to +\infty} x(t) = 0$, where 0 is its equilibrium point. A switched system is called asymptotically stable if it is both stable and attractive.

Due to the existence of the switching law, the asymptotic stability of every subsystems is neither necessary nor sufficient for the asymptotic stability of the switched system, which can be seen from the following example.

Example 1. [27] Consider a pair of planar linear subsystems of form $\dot{x} = A_i x$, where

$$A_1 = \begin{pmatrix} -1 & -10 \\ 100 & -1 \end{pmatrix} \quad and \quad A_2 = \begin{pmatrix} -1 & -100 \\ 10 & -1 \end{pmatrix}.$$

Clearly, each subsystem $\dot{x} = A_i x$ is globally asymptotically stable. However, the switched system using $A_1 x$ in the second and fourth quadrants and $A_2 x$ in the first and third quadrants is unstable and the switched system using $A_1 x$ in the first and third quadrants and $A_2 x$ in the second and fourth quadrants is asymptotically stable, which can be easily seen from Fig. 1 and Fig. 2, respectively.

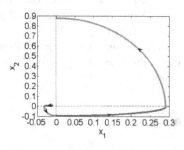

Fig. 1. An unstable switched system

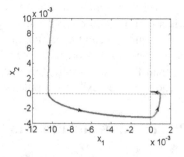

Fig. 2. A stable switched system

Thus, we need to concern about such a question: what restriction should we consider for the switching law in order to guarantee asymptotic stability of a switched system. This question is connected with asymptotic stabilisability defined as follows.

Definition 2. *(Asymptotic stabilisability) A switched system is called asymptotically stabilisable if there exists a switching law such that it is asymptotically stable.*

In general, stability analysis of switched systems is undecidable [2] and finding a switching law for guaranteeing stability is at least NP-hard [1]. Thus, in this paper, we will focus on a class of planar switched linear systems of form $\dot{x} = A_i x$, where $i = 1, 2$, $x \in \mathbb{R}^2$ and subsystems must be alternatively active. In addition, we assume that

1. each subsystem evolves anti-clockwisely, i.e., by letting $A_i = \begin{pmatrix} a_{i,11} & a_{i,12} \\ a_{i,21} & a_{i,22} \end{pmatrix}$, A_i should satisfy $a_{i,12} < 0 < a_{i,21}$ and $(a_{i,22} - a_{i,11})^2 + 4a_{i,12}a_{i,21} < 0$, and
2. switches must occur at two different candidate lines L_1 and L_2 through the origin.

We denote such a switched system as PSLS.

Therefore, given a switched system PSLS, we in this paper attempt to generate easily verifiable conditions on L_1 and L_2 such that switching laws can be directly derived to guarantee asymptotic stability, which will be discussed in Section 3 by using real root classification (RRC) [5,26,27] and quantifier elimination (QE) [4,9].

Remark 1. Our approach proposed in Section 3 can also be adapted to linear systems with subsystems in any form (e.g., a subsystem has a proper node or a saddle). Such an adaption to more general switched linear systems will be our future work. Moreover, our approach can also be adapted to systems where L_1 and L_2 are the same.

3   RRC and QE Based Approach for Generating Switching Laws

In this section, based on the algebraization of the existence condition on multiple Lyapunov functions, we will use real root classification (RRC) and quantifier elimination (QE) to generate switching laws for a given system PSLS such that PSLS using these switching laws is asymptotically stable.

3.1   Algebraic Analysis on Asymptotic Stabilisability

In this subsection, we will construct an algebraizable sufficient condition for asymptotic stability analysis of a given PSLS. To start with, we will introduce a classical theorem on multiple Lyapunov functions for a switched system as follows.

Theorem 1.   *[13,27] For a given switched system, if there exist a neighborhood $\mathbb{U}$ of the origin and continuously differentiable functions $V_i(x) : X_i \to \mathbb{R}, i \in M$, such that*

1. *for each $i \in M$, $V_i(0) = 0$ and $\frac{d}{dt}V_i(0) = (\nabla V_i(0))^T \cdot f_i(0) = 0$;*
2. *for each $i \in M$ and each $x \in X_i \cap \mathbb{U}$, if $x \neq 0$, then $V_i(x) > 0$;*
3. *for each $i \in M$ and each $x \in X_i \cap \mathbb{U}$, if $x \neq 0$, then $\frac{d}{dt}V_i(x) = (\nabla V_i(x))^T \cdot f_i(x) < 0$;*

4. *for all $i, j \in M$, if $x \in \{x \in U : \phi_{i,j}(x)\}$, then $V_i(x) \geq V_j(x)$, where $\phi_{i,j}(x)$ is a constraint describing that if $\phi_{i,j}(x)$ holds, a switching instance from mode i to mode j at the state x must occur,*

then the system is asymptotically stable. Here, the family $\{V_i(x) : i \in M\}$ is called as a *multiple Lyapunov function.*

Moreover, in order to apply the above MLF theorem for a given PSLS, we can without loss of generality provide some assumptions and notations as follows.

1. The candidate switching conditions (i.e., the two candidate lines) can be written as $g_i^T x = 0$, where $i \in \{1, 2\}$, $x = \begin{pmatrix} x_1 \\ x_2 \end{pmatrix}$, $g_i = \begin{pmatrix} k_{i1} \\ k_{i2} \end{pmatrix}$, $k_{i2} \geq 0$ and k_{i1}, k_{i2} are parameters;

2. A switch from mode 1 to mode 2 occurs at the state x if x satisfies $g_1^T x = 0$ and a switch from mode 2 to mode 1 occurs at the state x if x satisfies $g_2^T x = 0$;

3. X_1 is the region defined by an anti-clockwise rotation around the origin from the line $g_2^T x = 0$ to $g_1^T x = 0$ and X_2 is the region defined by an anti-clockwise rotation around the origin from the line $g_1^T x = 0$ to $g_2^T x = 0$. Thus, for each i, $X_i = X_{i1} \cup X_{i2}$, where for each $j \in \{1, 2\}$, $X_{ij} = \{x : E_{ij}x \geq 0\}$, $E_{ij} \in \mathbb{R}^{2 \times 2}$ is defined directly after classification in Subsection 3.2, and for two vectors $a = (a_1, a_2)$ and $b = (b_1, b_2)$, $a \geq b$ means $a_1 \geq b_1$ and $a_2 \geq b_2$;

4. For each $i \in \{1, 2\}$, let $V_i(x) = x^T P_i x$, where $P_i = \begin{pmatrix} a_i & \frac{b_i}{2} \\ \frac{b_i}{2} & c_i \end{pmatrix} \in \mathbb{R}^{2 \times 2}$ is a symmetric matrix and a_i, b_i, c_i are real parameters.

Thus, based on above assumptions and notations, we can construct an algebraizable sufficient condition for analyzing asymptotic stabilisability of a given PSLS as follows.

Theorem 2. *For a given system PSLS, if there exist functions $V_i(x) = x^T P_i x$ and two different switching lines defined by $g_i^T x = 0$, where $i \in \{1, 2\}$, such that:*

1. *for each $i, j \in \{1, 2\}$,*
   (1a) $\forall x[x \in X_{ij} \wedge x \neq 0 \Rightarrow V_i(x) \neq 0]$, and
   (1b) $\exists x_{ij,1}[E_{ij}x_{ij,1} > 0 \wedge V_i(x_{ij,1}) > 0]$;
2. *for each $i, j \in \{1, 2\}$,*
   (2a) $\forall x[x \in X_i \wedge x \neq 0 \Rightarrow \dot{V}_i(x) \neq 0]$, and
   (2b) $\exists x_{ij,2}[E_{ij}x_{ij,2} > 0 \wedge \dot{V}_i(x_{ij,2}) < 0]$;
3. *$\forall x\left[\left[g_1^T x = 0 \Rightarrow V_2(x) \leq V_1(x)\right] \wedge \left[g_2^T x = 0 \Rightarrow V_1(x) \leq V_2(x)\right]\right]$,*

where $\dot{V}_i(x) = (\nabla V_i(x))^T \cdot f_i(x)$, then the PSLS is asymptotically stabilisable.

Proof. For each $i \in \{1, 2\}$, since $V_i(x) = x^T P_i x$, it is clear that $V_i(0) = 0$ and $\dot{V}_i(0) = 0$.

Due to the condition (1a), for each $i \in \{1, 2\}$, 0 is the unique solution of $V_i(x) = 0$ in X_i. We want to prove that for each $i \in \{1, 2\}$ and each $x \in X_i$, if $x \neq 0$, then $V_i(x) > 0$.

Suppose that for each $i \in \{1, 2\}$, there is a point $x_i' \in X_i$ such that $x_i' \neq 0$ and $V_i(x_i') < 0$. Clearly, either $x_i' \in X_{i1}$ or $x_i' \in X_{i2}$.

1. If $x_i' \in X_{i1}$, according to the convexity of the set $\{x : E_{i1}x > 0\}$ and the continuity of $V_i(x)$, there exists a point $x_{ij,1}'' \in X_{i1}$ such that $x_{ij,1}'' \neq 0$ and $V_i(x_{ij,1}'') = 0$, contradicting the condition (1a).

2. If $x_i' \in X_{i2}$, we can similarly derive a contradiction.

Thus, for all $x \in \{x \in X_i | x \neq \mathbf{0}\}$, $V_i(x) > 0$.

Similarly, according to the condition (2a) and (2b), we can prove that for all $x \in \{x \in X_i | x \neq \mathbf{0}\}$, $\dot{V}_i(x) < 0$.

Due to Theorem 1, for the system PSLS, where the switching law is determined by $\phi_{12}(x) = [\mathbf{g_1}^T x = 0]$ and $\phi_{21}(x) = [\mathbf{g_2}^T x = 0]$, the family $\{V_i(x), i = 1, 2\}$ is a multiple Lyapunov function, implying that the PSLS is asymptotically stabilisable. $\square$

Remark 2. Different from Theorem 3 in [27], Theorem 2 in this paper additionally considers the parameters k_{i1} and k_{i2} $(i = 1, 2)$ in our sufficient condition for guaranteeing the asymptotic stabilisability of a given PSLS. Moreover, for each X_{ij}, we only require one interior point instead of two non-collinear points, which makes the conditions (1b) and (2b) in Theorem 2 simpler than the corresponding ones in [27]. This simpleness results from the definition of $E_{i,j}$s, which are all matrices of rank 2.

3.2   RRC and QE Based Approach for Guaranteeing Asymptotic Stabilisability

In this subsection, we will propose a RRC and QE based approach for analyzing asymptotic stabilisability of a given PSLS. That is, we will

1. apply a real root classification based approach to under-approximate the constraints in Theorem 2 respectively in the sense that every solution of the under-approximation is also a solution of the original constraint;
2. make use of these under-approximations to formulate an existentially quantified constraint, where the existential quantifiers are the parametric coefficients of the pre-assumed MLF;
3. solve this constraint by quantifier elimination, arriving at a quantifier-free formula over k_{i1} and k_{i2}, $i = 1, 2$, which provides switching laws for guaranteeing asymptotic stabilisability.

First, in order to reduce the number of parameters and make computation efficient, we assume that $V_1(x) = V_2(x) = x_1^2 + bx_1x_2 + cx_2^2$. Due to this assumption, the conditions (1a) and (1b) in Theorem 2 can simply be replaced by $b^2 - 4c < 0$ and the condition (3) is always true. So, we only need to under-approximate the conditions (2a) and (2b).

Remark 3. If we use $V_i(x) = a_i x_1^2 + b_i x_1 x_2 + c_i x_2^2$ for computation, the condition (1) can be under-approximated in a similar way. Moreover, we can directly get an equivalent formula for the condition (3) since switches in our planar systems occur at lines.

Second, for under-approximating the conditions (2a) and (2b), we will classify the discussion on its asymptotic stabilisability into the following two cases.

(1) One of the two candidate lines is the x_2-axis. Thus, we have two subcases: (1a) $k_{11} = k_1, k_{12} = 1, k_{21} = 1$ and $k_{22} = 0$; (1b) $k_{11} = 1, k_{12} = 0, k_{21} = k_1$ and $k_{22} = 1$.
(2) Both of the two candidate lines are not the x_2-axis. Then, by introducing k_1 and k_2, we can let $k_{11} = k_1, k_{12} = 1, k_{21} = k_2$, and $k_{22} = 1$. Thus, we have the following six subcases: (2a) $k_1 \leq 0$ and $k_2 > 0$; (2b) $k_2 \leq 0$ and $k_1 > 0$; (2c) $k_1 < k_2 \leq 0$; (2d) $k_2 < k_1 \leq 0$; (2e) $k_1 > k_2 \leq 0$; (2f) $k_2 > k_1 \leq 0$.

Due to the above classification, we can fix some notations as follows: for the subcases (1a) and (1b), $E_{11} = \begin{pmatrix} -k_1 & -1 \\ 1 & 0 \end{pmatrix}$, $E_{12} = \begin{pmatrix} k_1 & 1 \\ -1 & 0 \end{pmatrix}$, $E_{21} = \begin{pmatrix} k_1 & 1 \\ 1 & 0 \end{pmatrix}$, $E_{22} = \begin{pmatrix} -k_1 & -1 \\ -1 & 0 \end{pmatrix}$; for the other subcases, $E_{11} = \begin{pmatrix} -k_1 & -1 \\ k_2 & 1 \end{pmatrix}$, $E_{12} = \begin{pmatrix} k_1 & 1 \\ -k_2 & -1 \end{pmatrix}$, $E_{21} = \begin{pmatrix} k_1 & 1 \\ k_2 & 1 \end{pmatrix}$, $E_{22} = \begin{pmatrix} -k_1 & -1 \\ -k_2 & -1 \end{pmatrix}$.

Without loss of generality, we mainly analyze the subcases (1a), (2a), (2c) and (2e) for Theorem 2 and the other subcases can be discussed in a similar way.

3.2.1 $k_{11} = k_1$, $k_{12} = 1$, $k_{21} = 1$ and $k_{22} = 0$.

We observe that X_{11} and X_{12} (X_{21} and X_{22}) are symmetric with respect to the origin. Moreover, in Theorem 2, by denoting $\dot{V}_i(x)$ as $\dot{L}_i(x, b, c)$ for $i = 1, 2$, we also observe that $\dot{L}_i(x, b, c)$ is a symmetric function with respect to the origin. So, it is sufficient to only consider the condition (2) for $i \in \{1, 2\}$ and $j = 1$.

In details, we will under-approximatively solve the corresponding condition (2) in a conservative way such that every solution of the under-approximations is also a solution of the original conditions as follow.

For each $i \in \{1, 2\}$ and $j = 1$, we first equivalently transform the condition (2a) to $\bigwedge_{l=1}^{2} \phi_{i,1,l}(k_1, b, c)$, where $\phi_{i,1,1}(k_1, b, c) = [\forall x [E_{i1} x \geq 0 \wedge x_1 \neq 0 \Rightarrow \dot{L}_i(x, b, c) \neq 0]]$ and $\phi_{i,1,2}(k_1, b, c) = [\forall x [E_{i1} x \geq 0 \wedge x_2 \neq 0 \Rightarrow \dot{L}_i((0, x_2), b, c) \neq 0]]$.

1. For $\phi_{i,1,2}(k_1, b, c)$, we can easily obtain its equivalent constraint $\Lambda_{2a,i,1,2}(k_1, b, c)$.
2. For solving $\phi_{i,1,1}(k_1, b, c)$, we first use real root classification to get the sufficient and necessary condition $\delta_{i,1,1}$ over k_1, b, c, x_1, denoted as $\text{rrc}(\dot{L}_i(x, b, c) = 0, E_{i1} x \geq 0, x_1 \neq 0)$ or $\bigvee_s \bigwedge_l g_{s,l}(k_1, b, c, x_1) \Delta 0$, where $\Delta \in \{=, >, \geq, \neq\}$, such that the semi-algebraic system $\{\dot{L}_i(x, b, c) = 0, E_{i1} x \geq 0, x_1 \neq 0\}$ has no solution about x_2. Second, we can use Algorithm 1 to under-approximate $\forall x_1 [x_1 > 0 \Rightarrow \delta_{i,1,1}(k_1, b, c, x_1)]$, whose correctness can be assured by Lemma 2 in [27]. Denote these obtained under-approximations as $\Lambda_{2a,i,1,1}(k_1, b, c)$, $i \in \{1, 2\}$. Then, according to Lemma 4 in [27], $\Lambda_{2a,i,1,1}(k_1, b, c)$ is also an under-approximation for $\phi_{i,1,1}(k_1, b, c)$.

Algorithm 1. Computing an under-approximation

Input: $\forall x_1 \in \mathbb{R}[[ax_1 \geq 0 \wedge x_1 \neq 0] \Rightarrow \bigvee_{s=1}^{m} \bigwedge_{l=1}^{n_s} g_{s,l}(k_1, b, c, x_1) \Delta 0]$, where each $g_{s,l}$ is of form $\sum_{d=0}^{d_{s,l}} g_{s,l,d}(k_1, b, c) x_1^d$.
Output: an under-approximative constraint for the input.
1: **for** $s = 1 : 1 : m$ **do**
2:    **for** $l = 1 : 1 : n_s$ **do**
3:       **if** $d_{s,l} = 0$ **then** let $\delta_{s,l} = [g_{s,l}(k_1, b, c) \Delta 0]$;
4:       **else if** Δ in $g_{s,l} \Delta 0$ is "=" **then** let $\delta_{s,l} = [\bigwedge_{d=0}^{d_{s,l}} g_{s,l,d}(k_1, b, c) = 0]$;
5:          **else if** Δ is "$\neq$" **then** let $\delta_{s,l} = \text{rrc}(g_{s,l}(k_1, b, c, x_1) = 0, ax_1 \geq 0, x_1 \neq 0)$;
6:             **else if** $a \neq 0$ **then** let $\delta_{s,l} = \text{rrc}(ax_1 > 0, g_{s,l} = 0) \wedge g_{s,l}(k_1, b, c, sgn(a)) > 0$;
7:                **else** let $\delta_{s,l} = \text{rrc}(x_1 \neq 0, g_{s,l} = 0) \wedge g_{s,l}(k_1, b, c, 1) > 0 \wedge g_{s,l}(k_1, b, c, -1) > 0$;
8:    **end for**
9: **end for**
10: **return** $\bigvee_{s=1}^{m} \bigwedge_{l=1}^{n_s} \delta_{s,l}$.

Thus, $\Lambda_{2a,i,1,1}(k_1, b, c) \wedge \Lambda_{2a,i,1,2}(k_1, b, c)$ is an under-approximation for the condition (2a), denoted as $\Lambda_{2a,i,1}(k_1, b, c)$. Let $\Lambda_{2a}(k_1, b, c) = \wedge_{i=1}^{2} \Lambda_{2a,i,1}(k_1, b, c)$.

For solving the condition (1b), for each $i \in \{1, 2\}$, taking $x_{11,1} = (1, -k_1 - 1)$ and $x_{21,1} = (1, -k_1 + 1)$, we can simply replace the condition (1b) by the constraint $[\dot{L}(x_{i1,1}, b, c) < 0]$, denoted as $\Lambda_{2b,i,1}(k_1, b, c)$. Such a simple replacement will not lose any information, which can be proven as follows: letting $\Omega_{i1}^1 = \{(k_1, b, c) : \forall x[E_{i1}x \geq 0 \wedge x \neq 0 \Rightarrow \dot{L}_i(x, b, c) \neq 0]\}$, $\Omega_{i1}^2 = \{(k_1, b, c) : \exists x[E_{i1}x > 0 \wedge \dot{L}_i(x, b, c) < 0]\}$, $\Omega_{i1}^3 = \{(k_1, b, c) : (k_1, b, c) \in \Omega_1 \wedge (k_1, b, c) \in \Omega_2\}$ and $\Omega_{i1}^4 = \{(k_1, b, c) : (k_1, b, c) \in \Omega_1 \wedge \dot{L}_i(x_{i1,1}, b, c) < 0\}$, we have the following proposition.

Proposition 1. $\Omega_{i1}^3 = \Omega_{i1}^4$.

Proof. From the proof of Theorem 2, if $(k_1, b, c) \in \Omega_{i1}^3$, then $\forall x[x \neq 0 \wedge E_{i1}x > 0 \Rightarrow \dot{L}(x, b, c) < 0]$, implying that $\dot{L}(x_{i1,1}, b, c) < 0$. Thus $(k_1, b, c) \in \Omega_{i1}^4$ and then $\Omega_{i1}^3 \subset \Omega_{i1}^4$.

Similarly, if $(k_1, b, c) \in \Omega_4$, then $\forall x[x \neq 0 \wedge E_{i1}x > 0 \Rightarrow \dot{L}(x, b, c) < 0]$, implying that $(k_1, b, c) \in \Omega_{i1}^2$ and then $\Omega_{i1}^4 \subset \Omega_{i1}^3$. Thus, $\Omega_{i1}^3 = \Omega_{i1}^4$. □

Let $\Lambda_{2b}(k_1, b, c) = \wedge_{i=1}^{2} \Lambda_{2b,i,1}(k_1, b, c)$ and $\text{CASE}_{1a}(k_1, b, c) = [b^2 - 4c < 0 \wedge \Lambda_{2a} \wedge \Lambda_{2b}]$. Combining above discussions, it is straightforward to have:

Proposition 2. *If* (k_{10}, b_0, c_0) *makes* $\text{CASE}_{1a}(k_1, b, c)$ *hold, then* $g_1 = (k_{10}, 1)$, $g_2 = (1, 0)$ *and* $V_1(x) = V_2(x) = V(x, b, c)$ *satisfy the conditions of Theorem 2. Thus, PSLS is asymptoticallly stabilisable if its switching condition from mode 1 to mode 2 is* $k_{10}x_1 + x_2 = 0$ *and its switching condition from mode 2 to mode 1 is* $x_1 = 0$.

In addition, in order to generate conditions only involving switching laws, we can apply QE to the constraint: $\exists b \exists c [\text{CASE}_{1a}(k_1, b, c)]$, arriving at an equivalent quantifier-free formula over k_1, denoted as $\text{QE}_{1a}(k_1)$.

Thus, it is also straightforward to have:

Proposition 3. *If* k_{10} *makes* $\text{QE}_{1a}(k_1)$ *hold, then PSLS is asymptotically stabilisable if its switching condition from mode 1 to mode 2 is* $k_{10}x_1 + x_2 = 0$ *and its switching condition from mode 2 to mode 1 is* $x_1 = 0$.

Remark 4. For convenience, we can first manually optimize $\text{CASE}_{1a}(k_1, b, c)$ and then formulate it as a disjunction of conjunctions. Thus, $\text{QE}_{1a}(k_1)$ can be alternatively obtained by applying QE to each conjunction.

Remark 5. According to Propositions 2 and 3, stabilizing switching laws can be designed by sample points satisfying $\text{CASE}_{1a}(k_1, b, c)$ (or, $\text{QE}_{1a}(k_1)$), which can be found by an adaptive CAD based solver [25]. Moreover, for a given k_{10} (i.e., a given switching law), we can verify whether PSLS is asymptotically stable by checking whether there exist b and c such that $\text{CASE}_{1a}(k_{10}, b, c)$ holds (or, directly checking whether $\text{QE}_{1a}(k_{10})$ holds).

3.2.2 $k_1 \leq 0$ and $k_2 > 0$

Similarly, for under-approximatively solving the condition (2), it is also sufficient to only consider the condition (2) for $i \in \{1, 2\}$ and $j = 1$.

In details, for each $i \in \{1, 2\}$ and $j = 1$, we first equivalently transform the condition (2a) to $\wedge_{l=1}^{2} \phi_{i,1,l}(k_1, k_2, b, c)$, where $\phi_{i,1,1}(k_1, k_2, b, c) = [\forall \boldsymbol{x}[E_{i1}\boldsymbol{x} \geq 0 \wedge x_1 \neq 0 \Rightarrow \dot{L}_i(\boldsymbol{x}, b, c) \neq 0]]$ and $\phi_{i,1,2}(k_1, k_2, b, c) = [\forall \boldsymbol{x}[E_{i1}\boldsymbol{x} \geq 0 \wedge x_2 \neq 0 \Rightarrow \dot{L}_i((0, x_2), b, c) \neq 0]]$.

1. For $\phi_{i,1,2}(k_1, b, c)$, we can easily obtain its equivalent constraint $\Lambda_{2a,i,1,2}(k_1, b, c)$.

2. For solving $\phi_{1,1,1}(k_1, k_2, b, c)$, we first use real root classification to get the sufficient and necessary condition $\delta_{1,1,1}$ over k_1, k_2, b, c, x_1, such that the semi-algebraic system $\{\dot{L}_i(\boldsymbol{x}, b, c) = 0, E_{11}\boldsymbol{x} \geq 0, x_1 \neq 0\}$ has no solution about x_2; for solving $\phi_{2,1,1}(\boldsymbol{p}_i)$, we first use real root classification to get the sufficient and necessary condition $\delta_{2,1,1}$ over k_1, k_2, b, c, x_2, such that the semi-algebraic system $\{\dot{L}_i(\boldsymbol{x}, b, c) = 0, E_{21}\boldsymbol{x} \geq 0, x_1 \neq 0\}$ has no solution about x_1. Then, we similarly use Algorithm 1 to under-approximate $\forall x_1[x_1 > 0 \Rightarrow \delta_{1,1,1}(k_1, k_2, b, c, x_1)]$ and $\forall x_2[x_2 > 0 \Rightarrow \delta_{2,1,1}(k_1, k_2, b, c, x_2)]$, respectively. Denote these obtained under-approximations as $\Lambda_{2a,i,1,1}(k_1, k_2, b, c), i \in \{1, 2\}$.

Thus, $\Lambda_{2a,i,1,1}(k_1, k_2, b, c) \bigwedge \Lambda_{2a,i,1,2}(k_1, k_2, b, c)$ is an under-approximation for the condition (2a), denoted as $\Lambda_{2a,i,1}(k_1, k_2, b, c)$. Let $\Lambda_{2a}(k_1, k_2, b, c) = \wedge_{i=1}^{2} \Lambda_{2a,i,1}(k_1, k_2, b, c)$.

For solving the condition (2b), for each $i \in \{1, 2\}$, taking $\boldsymbol{x}_{11,1} = (1, -\frac{k_1+k_2}{2})$ and $\boldsymbol{x}_{21,1} = (-\frac{k_1+k_2}{2}, 1)$, we can simply replace the condition (2b) by its equivalent constraint $[\dot{L}(\boldsymbol{x}_{i1,1}, b, c) < 0]$, denoted as $\Lambda_{2b,i,1}(k_1, k_2, b, c)$.

Let $\Lambda_{2b}(k_1, k_2, b, c) = \wedge_{i=1}^{2} \Lambda_{2b,i,1}(k_1, k_2, b, c)$ and $\text{CASE}_{2a}(k_1, k_2, b, c) = [b^2 - 4c < 0 \bigwedge \Lambda_{2a} \bigwedge \Lambda_{2b}]$. Similarly, we can apply QE to the constraint $\exists b \exists c [\text{CASE}_{2a}(k_1, k_2, b, c)]$ and obtain an equivalent quantifier-free formula over k_1, denoted as $\text{QE}_{2a}(k_1, k_2)$.

3.2.3 $k_1 < k_2 \leq 0$

Similarly, for under-approximatively solving condition (2), it is also sufficient to only consider the condition (2) for $i \in \{1, 2\}$ and $j = 1$.

In details, for each $i \in \{1, 2\}$ and $j = 1$, we first equivalently transform the condition (2a) to $\wedge_{l=1}^{2} \phi_{i,1,l}(k_1, k_2, b, c)$, where $\phi_{i,1,1}(k_1, k_2, b, c) = [\forall \boldsymbol{x}[E_{i1}\boldsymbol{x} \geq 0 \wedge x_1 \neq 0 \Rightarrow \dot{L}_i(\boldsymbol{x}, b, c) \neq 0]]$ and $\phi_{i,1,2}(k_1, k_2, b, c) = [\forall \boldsymbol{x}[E_{i1}\boldsymbol{x} \geq 0 \wedge x_2 \neq 0 \Rightarrow \dot{L}_i((0, x_2), b, c) \neq 0]]$.

1. For $\phi_{i,1,2}(k_1, b, c)$, we can easily obtain its equivalent constraint $\Lambda_{2a,i,1,2}(k_1, b, c)$.

2. For solving $\phi_{i,1,1}(k_1, k_2, b, c)$, we first use real root classification to get the sufficient and necessary condition $\delta_{i,1,1}$ over k_1, k_2, b, c, x_1, such that the semi-algebraic system $\{\dot{L}_i(\boldsymbol{x}, b, c) = 0, E_{i1}\boldsymbol{x} \geq 0, x_1 \neq 0\}$ has no solution about x_2. Then, we similarly use Algorithm 1 to under-approximate $\forall x_1[x_1 > 0 \Rightarrow \delta_{1,1,1}(k_1, k_2, b, c, x_1)]$, and $\forall x_1[x_1 \neq 0 \Rightarrow \delta_{2,1,1}(k_1, k_2, b, c, x_1)]$, respectively. Denote these obtained under-approximations as $\Lambda_{2a,i,1,1}(k_1, k_2, b, c), i \in \{1, 2\}$.

Thus, $\Lambda_{2a,i,1,1}(k_1, k_2, b, c) \bigwedge \Lambda_{2a,i,1,2}(k_1, k_2, b, c)$ is an under-approximation for the condition (2a), denoted as $\Lambda_{2a,i,1}(k_1, k_2, b, c)$. Let $\Lambda_{2a}(k_1, k_2, b, c) = \wedge_{i=1}^{2} \Lambda_{2a,i,1}(k_1, k_2, b, c)$.

For solving the condition (2b), for each $i \in \{1, 2\}$, taking $\boldsymbol{x}_{11,1} = (1, -\frac{k_1+k_2}{2})$ and $\boldsymbol{x}_{21,1} = (0, 1)$, we can simply replace the condition (2b) by its equivalent constraint $[\dot{L}(\boldsymbol{x}_{i1,1}, b, c) < 0]$, denoted as $\Lambda_{2b,i,1}(k_1, k_2, b, c)$.

Let $\Lambda_{2b}(k_1, k_2, b, c) = \wedge_{i=1}^{2} \Lambda_{2b,i,1}(k_1, k_2, b, c)$ and $\text{CASE}_{2c}(k_1, k_2, b, c) = [b^2 - 4c < 0 \bigwedge \Lambda_{2a} \bigwedge \Lambda_{2b}]$. Similarly, we can apply QE to the constraint $\exists b \exists c [\text{CASE}_{2c}(k_1, k_2, b, c)]$ and obtain an equivalent quantifier-free formula over k_1, denoted as $\text{QE}_{2c}(k_1, k_2)$.

3.2.4 $k_1 > k_2 \geq 0$

Similarly, for under-approximatively solving the condition (2), it is also sufficient to only consider the condition (2) $i \in \{1, 2\}$ and $j = 1$.

In details, for each $i \in \{1, 2\}$ and $j = 1$, we first equivalently transform the condition (2a) to $\wedge_{l=1}^{2} \phi_{i,1,l}(k_1, k_2, b, c)$, where $\phi_{i,1,1}(k_1, k_2, b, c) = [\forall x[E_{i1}x \geq 0 \wedge x_1 \neq 0 \Rightarrow \dot{L}_i(x, b, c) \neq 0]]$ and $\phi_{i,1,2}(k_1, k_2, b, c) = [\forall x[E_{i1}x \geq 0 \wedge x_2 \neq 0 \Rightarrow \dot{L}_i((0, x_2), b, c) \neq 0]]$.

1. For $\phi_{i,1,2}(k_1, b, c)$, we can easily obtain its equivalent constraint $\Lambda_{2a,i,1,2}(k_1, b, c)$.
2. For solving $\phi_{i,1,1}(k_1, k_2, b, c)$, we first use real root classification to get the sufficient and necessary condition $\delta_{i,1,1}$ over k_1, k_2, b, c, x_1, such that the semi-algebraic system $\{\dot{L}_i(x, b, c) = 0, E_{i1}x \geq 0, x_1 \neq 0\}$ has no solution about x_2. Second, we similarly use Algorithm 1 to under-approximate $\forall x_1[x_1 < 0 \Rightarrow \delta_{1,1,1}(k_1, k_2, b, c, x_1)]$ and $\forall x_1[x_1 \neq 0 \Rightarrow \delta_{2,1,1}(k_1, k_2, b, c, x_1)]$, respectively. Denote these obtained under-approximations as $\Lambda_{2a,i,1,1}(k_1, k_2, b, c)$, $i \in \{1, 2\}$.

Thus, $\Lambda_{2a,i,1,1}(k_1, k_2, b, c) \wedge \Lambda_{2a,i,1,2}(k_1, k_2, b, c)$ is an under-approximation for the condition (2a), denoted as $\Lambda_{2a,i,1}(k_1, k_2, b, c)$. Let $\Lambda_{2a}(k_1, k_2, b, c) = \wedge_{i=1}^{2} \Lambda_{2a,i,1}(k_1, k_2, b, c)$.

For solving the condition (2b), for each $i \in \{1, 2\}$, taking $x_{11,1} = (0, -1)$ and $x_{21,1} = (1, -\frac{k_1+k_2}{2})$, we can simply replace the condition (2b) by its equivalent constraint $[\dot{L}(x_{i1,1}, b, c) < 0]$, denoted as $\Lambda_{2b,i,1}(k_1, k_2, b, c)$.

Let $\Lambda_{2b}(k_1, k_2, b, c) = \wedge_{i=1}^{2} \Lambda_{2b,i,1}(k_1, k_2, b, c)$ and $\text{CASE}_{2e}(k_1, k_2, b, c) = [b^2 - 4c < 0 \wedge \Lambda_{2a} \wedge \Lambda_{2b}]$. Similarly, we can apply QE to the constraint $\exists b \exists c [\text{CASE}_{2e}(k_1, k_2, b, c)]$ and obtain an equivalent quantifier-free formula over k_1, denoted as $\text{QE}_{2e}(k_1, k_2)$.

3.2.5   Summary

Similarly, for the subcase (1b), we have $\text{CASE}_{1b}(k_1, b, c)$ and $\text{QE}_{1b}(k_1)$; for the subcases (2b), (2d) and (2f), we have $\text{CASE}_{2b}(k_1, k_2, b, c)$, $\text{QE}_{2b}(k_1, k_2)$, $\text{CASE}_{2d}(k_1, k_2, b, c)$, $\text{QE}_{2d}(k_1, k_2)$, $\text{CASE}_{2f}(k_1, k_2, b, c)$ and $\text{QE}_{2f}(k_1, k_2)$, respectively.

Thus, it is straightforward to have:

Theorem 3. *If* $\text{CASE}_{1a}(k_{10}, b_0, c_0)$ *(*$\text{CASE}_{1b}(k_{10}, b_0, c_0)$*) or* $\text{QE}_{1a}(k_{10})$ *(*$\text{QE}_{1b}(k_{10})$*) holds, then* PSLS *is asymptotically stable if its switching condition from mode 1 to mode 2 is* $k_{10}x_1 + x_2 = 0$ *(*$x_1 = 0$*) and its switching condition from mode 2 to mode 1 is* $x_1 = 0$ *(*$k_{10}x_1 + x_2 = 0$*); If* $[k_{10} \leq 0 \wedge k_{20} > 0 \wedge \text{CASE}_{2a}(k_{10}, k_{20}, b_0, c_0)] \vee [k_{10} < k_{20} \leq 0 \wedge \text{CASE}_{2c}(k_{10}, k_{20}, b_0, c_0)] \vee [k_{10} > k_{20} \geq 0 \wedge \text{CASE}_{2e}(k_{10}, k_{20}, b_0, c_0)] \vee [k_{20} \leq 0 \wedge k_{10} > 0 \wedge \text{CASE}_{2b}(k_{10}, k_{20}, b_0, c_0)] \vee [k_{20} < k_{10} \leq 0 \wedge \text{CASE}_{2d}(k_{10}, k_{20}, b_0, c_0)] \vee [k_{20} > k_{10} \geq 0 \wedge \text{CASE}_{2f}(k_{10}, k_{20}, b_0, c_0)]$ *or* $[k_{10} \leq 0 \wedge k_{20} > 0 \wedge \text{QE}_{2a}(k_{10}, k_{20})] \vee [k_{10} < k_{20} \leq 0 \wedge \text{QE}_{2c}(k_{10}, k_{20})] \vee [k_{10} > k_{20} \geq 0 \wedge \text{QE}_{2e}(k_{10}, k_{20})] \vee [k_{20} \leq 0 \wedge k_{10} > 0 \wedge \text{QE}_{2b}(k_{10}, k_{20})] \vee [k_{20} < k_{10} \leq 0 \wedge \text{QE}_{2d}(k_{10}, k_{20})] \vee [k_{20} > k_{10} \geq 0 \wedge \text{QE}_{2f}(k_{10}, k_{20})]$ *holds, then the* PSLS *is asymptotically stable if its switching condition from mode 1 to mode 2 is* $k_{10}x_1 + x_2 = 0$ *and its switching condition from mode 2 to mode 1 is* $k_{20}x_1 + x_2 = 0$.

Remark 6. For under-approximating the condition (2a), we orderly eliminate x_2 and x_1 in [27] while in this paper, due to the definition of E_{ij}, we for certain subcases first eliminate x_1, e.g., the subcase (2a).

3.3 Real Root Classification

A *semi-algebraic system* (or short, SAS) is a system of form

$$\{p_1(x, u) = 0, \cdots, p_s(x, u) = 0, g_1(x, u) \geq 0, \cdots, g_r(x, u) \geq 0,$$
$$g_{r+1}(x, u) > 0, \cdots, g_t(x, u) > 0, h_1(x, u) \neq 0, \cdots, h_m(x, u) \neq 0\},$$

where variable x ranges in $\mathbb{R}^n$, parameter u ranges in $\mathbb{R}^d$, and p_i, g_j, h_k are polynomials in $Q[x, u]$. We can write the system as $[P, G_1, G_2, H]$, where P, G_1, G_2 and H stand for $[p_1, \cdots, p_s]$, $[g_1, \cdots, g_r]$, $[g_{r+1}, \cdots, g_t]$ and $[h_1, \cdots, h_m]$, respectively.

An SAS is called a *constant semi-algebraic system* if it contains no parameters, *i.e.*, $d = 0$; otherwise, a *parametric semi-algebraic system*.

For a parametric SAS S, one problem is to determine the sufficient and necessary condition on the parameters such that S has N distinct real solutions. The Maple package RegularChains [5] has the features for providing such a condition. To use these features, one may first type in the following commands orderly:

```
> with(RegularChains):
> with(ParametricSystemTools):
> with(SemiAlgebraicSetTools):
> infolevel[RegularChains]:=1:
```

Then, for a parametric SAS S of form $[P, G_1, G_2, H]$ and a given non-negative integer N, to determine the necessary and sufficient conditions on u such that the number of distinct real solutions of S equals N, one can orderly type in

```
> R := PolynomialRing([x,u]);
> rrc := RealRootClassification(P, G1, G2, H, d, N, R);
```

where R is a list of variables and parameters, and d indicates the last d elements of R are parameters. The output of `RealRootClassification` is a quantifier-free formula Φ in parameters and a *border polynomial* $\mathrm{BP}(u)$ which mean that, provided $\mathrm{BP}(u) \neq 0$, the necessary and sufficient condition for S to have exactly N distinct real solutions is Φ holds. Then, letting $P' = [p_1, ..., p_s, \mathrm{BP}]$, one can call

```
> rrc1 := RealRootClassification(P', G1, G2, H, d, N, R);
```

to find conditions on u when the parameters are on the "border" $\mathrm{BP}(u) = 0$.

4 An Illustrating Example

In this section, we continue to consider Example 1, where each subsystem is globally asymptotically stable. In details, by using a prototypical implementation, we first have:

$$\begin{aligned}
\Lambda_{2a,1,1}(k_1, b, c) = & [5b + c > 0 \wedge 1001b^2 + 10000c^2 - 2004c + 100 < 0] \\
& \vee [5b + c > 0 \wedge 1001b^2 + 10000c^2 - 2004c + 100 > 0 \\
& \quad \wedge 10bk_1 + 2ck_1 - b + 100c - 10 \geq 0 \\
& \quad \wedge 5bk_1^2 + ck_1^2 - bk_1 + 100ck_1 - 50b - 10k_1 + 1 > 0] \\
& \vee [5b + c > 0 \wedge 1001b^2 + 10000c^2 - 2004c + 100 = 0 \\
& \quad \wedge 10bk_1 + 2ck_1 - b + 100c - 10 > 0];
\end{aligned}$$

$$\Lambda_{2a,2,1}(k_1, b, c) = [50b + c > 0 \land 1001b^2 + 100c^2 - 2004c + 10000 < 0]$$
$$\lor [50b + c > 0 \land 1001b^2 + 100c^2 - 2004c + 10000 > 0$$
$$\land\ 100bk_1 + 2ck_1 - b + 10c - 100 \le 0$$
$$\land\ 50bk_1^2 + ck_1^2 - bk_1 + 10ck_1 - 5b - 100k_1 + 1 > 0]$$
$$\lor [50b + c > 0 \land 1001b^2 + 100c^2 - 2004c + 10000 = 0$$
$$\land\ 100bk_1 + 2ck_1 - b + 10c - 100 < 0];$$
$$\Lambda_{2b,1,1}(k_1, b, c) = [(5b + c)(k_1 + 1)^2 + (10 + b - 100c)(-k_1 - 1) + (1 - 50b) > 0];$$
$$\Lambda_{2b,2,1}(k_1, b, c) = [(50b + c)(k_1 - 1)^2 + (100 + b - 10c)(-k_1 + 1) + (1 - 5b) > 0].$$

Then, $CASE_{1a} = [b^2 - 4c < 0 \land \Lambda_{2a,1,1} \land \Lambda_{2a,2,1} \land \Lambda_{2b,1,1} \land \Lambda_{2b,2,1}]$. And we can in a similar way get $CASE_{1b}$, $CASE_{2a}$, $CASE_{2b}$, $CASE_{2c}$, $CASE_{2d}$, $CASE_{2e}$ and $CASE_{2f}$, respectively. Due to the space limit, we omit the corresponding computation results here.

We have also reformulated $CASE_{1a}$ as a disjunction of conjunctions and applied both REDLOG [9] and QEPCAD [4] to each conjunction. However, the corresponding results are very large. So we also omit the corresponding computation results here.

Moreover, although certain conjunctions of $CASE_{1a}$ do not hold, we can still find lots of sample points (e.g., $(k_1, b, c) = (-\frac{10}{9}, \frac{1}{1024}, \frac{7}{64})$) in semi-algebraic sets defined by other conjunctions, implying that there are explicit stabilizing switching laws and thus this planar system is asymptotically stabilisable, which can be seen from Fig. 3.

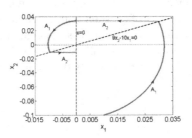

Fig. 3. A stable switched system

Further, since there exist sample points b and c satisfying $CASE_{1a}(0, b, c)$ (e.g., $(\frac{1}{64}, 9)$), we certify that the switched system using A_1x in the first and third quadrants and A_2x in the second and fourth quadrants is asymptotically stable.

5   Related Works with Comparisons

In the Subsection 3.1 of [33], the authors discussed the stabilization of two planar subsystems with unstable foci. They first defined several regions for providing a fixed switching law. Then, in their Theorem 1, by using this switching law, the asymptotic stability is assured if and only if $\|x^*\| < \|x_0\|$, where $x_0 \neq 0$ is an initial point in the switching line and x^* is the point obtained by following the trajectory of the switched system from x_0 around the origin for 2π. However, the inequality $\|x^*\| < \|x_0\|$ is not easily checked since in order to obtain x^*, one needs to solve both exponential functions and trigonometric functions, which is still a difficult problem.

In [11], the authors investigated the regional stabilization of two planar subsystems that are not asymptotically stable. They derived an verifiable, necessary and sufficient condition for regional stabilisability of the switched system. The *advantage* is that the existence of two independent vectors, along which the trajectories of the two subsystems are collinear, is a necessary condition for the switched system to be stabilisable. However, it is still hard to check the satisfiability of this sufficient condition when the spiralling case is considered, since for such a case, one still needs to follow the trajectory of the switched system around the origin for π and thus needs to solve trigonometric and matrix exponential functions and integrate rational trigonometric functions.

In this current paper, we analyze the global asymptotic stabilisability of planar linear switched systems, where each subsystem has a stable or unstable focus and the two subsystems are assumed to be alternatively active. Simply speaking, after regarding the switching lines to be parametric, we use a multiple Lyapunov functions based approach to produce a verifiable and sufficient condition. Especially, by using RRC and QE, we not only can avoid integrating rational trigonometric functions and solving trigonometric functions and matrix exponential functions, but also can *provide an explicit set of stabilizing switching laws* due to the theory of real-closed fields [31].

Consider a simple system $\dot{x} = A_i x$, where $A_1 = \begin{pmatrix} 0.1 & \sqrt{5} \\ -\sqrt{5} & 0.1 \end{pmatrix}$ and $A_2 = \begin{pmatrix} 0.1 & \sqrt{5}/5 \\ -5\sqrt{5} & 0.1 \end{pmatrix}$.
Although each subsystem evolves clockwisely, we can similarly apply our RRC and QE based approach and obtain a set of constant semi-algebraic systems over the slopes. Especially, the switched system using $A_1 x$ in the second and fourth quadrants and $A_2 x$ in the first and third quadrants is asymptotically stable, which can be seen from Fig. 4. Clearly, $v = (1, 0)^T$ and $w = (0, 1)^T$ do not satisfy the collinear condition in [11].

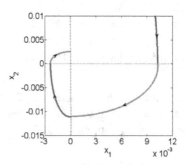

Fig. 4. A stable switched system

Thus, in addition to the verifiability, an advantage of our approach over [33,11] is that our condition can provide more selectable stabilizing switching laws that do not satisfy the collinear condition although the trajectory of the system using these stabilizing switching laws will converge to the origin with lower speeds.

6   Conclusion

In this paper, we use a RRC and QE based algebraic approach to analyze asymptotic stabilisability of a class of planar switched linear systems, where subsystems are assumed to be alternatively active, arriving at an intermediate under-approximation over all parameters and a final quantifier-free formula over the switching lines. Based on our intermediate under-approximation as well as our final formula, we can easily design explicit stabilizing switching laws. Moreover, we use an illustrating example to show the applicability of our approach.

Our short-term goal is to simplify our classification on stabilizing switching lines and make our analysis fully automatic. Our long-term goal is to analyze the stabilisability of more general switched systems [13], e.g., planar linear switched systems in any form and planar nonlinear switched systems. Moreover, we would also like to analyze the practical stabilisability [34] by computing multiple Lyapunov-like functions [22].

References

1. Blondel, V.D., Bournez, O., Koiran, P., Tsitsiklis, J.N.: NP-hardness of Some Linear Control Design Problems. SIAM J. Control and Optimization 35(8), 2118–2127 (1997)
2. Blondel, V.D., Tsitsiklis, J.N.: The Stability of Saturated Linear Dynamical Systems is Undecidable. Journal of Computer and System Sciences 62(3), 442–462 (2001)
3. Branicky, M.S.: Multiple Lyapunov Functions and other Analysis Tools for Switched and Hybrid Systems. IEEE Transactions on Automatic Control 43(4), 751–760 (1998)
4. Brown, C.W.: QEPCAD B: A System for Computing with Semi-algebraic Sets via Cylindrical Algebraic Decomposition. SIGSAM Bull 38(1), 23–24 (2004)
5. Chen, C., Lemaire, F., Li, L., Maza, M.M., Pan, W., Xie, Y.: The ConstructibleSetTools and ParametricSystemsTools Modules of the RegularChains library in Maple. In: Proc. of the International Conf. on Computational Science and Its Applications, pp. 342–352 (2008)
6. Cheng, D., Guo, L., Lin, Y., Wang, Y.: Stabilization of Switched Linear Systems. IEEE Transactions on Automatic Control 50(5), 661–666 (2005)
7. Daafouz, J., Riedinger, R., Iung, C.: Stability Analysis and Control Synthesis for Switched Systems: A Switched Lyapunov Function Approach. IEEE Transactions on Automatic Control 47, 1883–1887 (2002)
8. Decarlo, R.A., Braanicky, M.S., Pettersson, S., Lennartson, B.: Perspectives and Results on the Stability and Stabilisability of Hybrid Systems. Proceedings of the IEEE 88(7), 1069–1082 (2000)
9. Dolzmann, A., Sturm, T.: REDLOG: Computer Algebra Meets Computer Logic. SIGSAM Bull 31(2), 2–9 (1997)
10. Feron, E., Apkarian, P., Gahinet, P.: Analysis and Synthesis of Robust Control Systems via Parameter-dependent Lyapunov Functions. IEEE Transactions on Automatic Control 41(7), 1041–1046 (1996)
11. Huang, Z.H., Xiang, C., Lin, H., Lee, T.H.: Necessary and Sufficient Conditions for Regional Stabilisability of Generic Switched Linear Systems with a Pair of Planar Subsystems. International Journal of Control 83(4), 694–715 (2010)
12. Hespanha, J.P., Morse, A.S.: Stability of Switched Systems with Average Dwell-time. In: Proceedings of the 38th Conference on Decision and Control, pp. 2655–2660 (1999)
13. Lunze, J., Lamnabhi-Lagarrigue, F.: Handbook of Hybrid Systems Control: Theory, Tools, Applications. Cambridge University Press (2009)

14. Liberzon, D.: Switching in Systems and Control. Birkhauser, Boston (2003)
15. Liberzon, D., Morse, A.S.: Basic Problems in Stability and Design of Switched Systems. IEEE Control Systems Magazine 19(5), 59–70 (1999)
16. Lin, H., Antasklis, P.J.: Switching Stabilisability for Continuous-time Uncertain Switched Linear Systems. IEEE Transactions on Automatic Control 52(4), 633–646 (2007)
17. Lin, H., Antasklis, P.J.: Stability and Stabilizability of Switched Linear Systems: A Survey of Recent Results. IEEE Transactions on Automatic Control 52(2), 308–322 (2009)
18. Narendra, K.S., Balakrishnan, J.: Adaptive Control using Multiple Models and Switching. IEEE Transactions on Automatic Control 42(2), 171–187 (1997)
19. Narendra, K.S., Xiang, C.: Adaptive Control of Discrete-time Systems using Multiple Models. IEEE Transactions on Automatic Control 45, 1669–1686 (2000)
20. Pettersson, S.: Synthesis of Switched Linear Systems. In: Proceedings of the 42nd Conference on Decision and Control, pp. 5283–5288 (2003)
21. Pettersson, S., Lennartson, B.: Stabilization of Hybrid Systems Using a Min-Projection Stategy. In: Proceedings of the American Control Conference, pp. 223–288 (2001)
22. Ratschan, S., She, Z.: Providing a basin of attraction to a target region of polynomial systems by computation of Lyapunov-like functions. SIAM Journal on Control and Optimization 48(7), 4377–4394 (2010)
23. Rodrigues, L., Hassibi, A., How, J.P.: Observer-based Control of Piecewise-affine Systems. International Journal of Control 76, 459–477 (2003)
24. van der Schaft, A., Schumacher, H.: An Introduction to Hybrid Dynamical Systems. Springer, London (2000)
25. She, Z., Xia, B., Xiao, R., Zheng, Z.: A semi-algebraic approach for asymptotic stability analysis. Nonlinear Analysis: Hybrid System 3(4), 588–596 (2009)
26. She, Z., Xue, B., Zheng, Z.: Algebraic Analysis on Asymptotic Stability of Continuous Dynamical Systems. In: Proceedings of the 36th International Symposium on Symbolic and Algebraic Computation, pp. 313–320 (2011)
27. She, Z., Xue, B.: Algebraic Analysis on Asymptotic Stability of Switched Hybrid Systems. In: Proceedings of the 15th International Conference on Hybrid Systems: Computation and Control, pp. 197–196 (2012)
28. She, Z., Yu, J., Xue, B.: Controllable Laws for Stability Analysis of Switched Linear Systems. In: Proceedings of the 3rd IEEE International Conference on Computer and Network Technology, vol. 13, pp. 127–131 (2011)
29. Shorten, R.N., Narendra, K.S.: Necessary and Sufficient Conditions for the Existence of a Common Quadratic Lyapunov Function for Two Stable Second Order Linear Time-invariant Systems. In: Proceedings of the American Control Conference, pp. 1410–1414 (1999)
30. Skafidas, E., Evans, R.J., Savkin, A.V., Petersen, I.R.: Stability Results for Switched Controller Systems. IEEE Transactions on Automatic Control 35(4), 553–564 (1999)
31. Tarski, A.: A Decision Method for Elementary Algebra and Geometry. Univ. of California Press (1951)
32. Wicks, M.A., Peleties, P., DeCarlo, R.A.: Switched Controller Design for the Quadratic Stabilization of a Pair of Unstable Systems. European J. of Control 4(2), 140–147 (1998)
33. Xu, X., Antasklis, P.J.: Switching Stabilisability for Continuous-time Uncertain Switched Linear Systems. International Journal of Control 73(14), 1261–1279 (2000)
34. Xu, X., Zhai, G.: On Practical Stability and Stabilization of Hybrid and Switched Systems. In: Alur, R., Pappas, G.J. (eds.) HSCC 2004. LNCS, vol. 2993, pp. 615–630. Springer, Heidelberg (2004)

Improving Angular Speed Uniformity by Optimal C^0 Piecewise Reparameterization

Jing Yang[1], Dongming Wang[2], and Hoon Hong[3]

[1] LMIB – School of Mathematics and Systems Science, Beihang University,
Beijing 100191, China
yangjing@smss.buaa.edu.cn

[2] Laboratoire d'Informatique de Paris 6, CNRS – Université Pierre et Marie Curie,
4 place Jussieu – BP 169, 75252 Paris cedex 05, France
Dongming.Wang@lip6.fr

[3] Department of Mathematics, North Carolina State University,
Box 8205, Raleigh, NC 27695, USA
hong@ncsu.edu

Abstract. We adapt the C^0 piecewise Möbius transformation to compute a C^0 piecewise-rational reparameterization of any plane curve that approximates to the arc-angle parameterization of the curve. The method proposed on the basis of this transformation can achieve highly accurate approximation to the arc-angle parameterization. A mechanism is developed to optimize the transformation using locally optimal partitioning of the unit interval. Experimental results are provided to show the effectiveness and efficiency of the reparameterization method.

Keywords: Parametric plane curve, angular speed uniformity, optimal piecewise Möbius transformation, locally optimal partition.

1 Introduction

There are two kinds of representations, implicit and parametric, which are used for curves and surfaces in computer aided geometric design and related areas. Representations of each kind have their own advantages in applications. In this paper, we are concerned with rational parametric plane curves and their reparameterization.

A parametric curve may have infinitely many different parameterizations. One may need to choose a suitable one out of them for a concrete application. For example, when the speed of a point moving along the curve has to be controlled (e.g., for plotting and numerical control machining), one typically wishes to choose an arc-length parameterization [2,4,5,6,7,10] or an arc-angle parameterization [8,12]. Our investigations will be focused on the latter, which was first proposed by Patterson and Bajaj [8] as *curvature parameterization*.

The problem of arc-angle parameterization has been studied by the authors in [12], where an optimality criterion is introduced as a quality measure for parameterizations of plane curves and a method is proposed to compute arc-angle

V.P. Gerdt et al. (Eds.): CASC 2012, LNCS 7442, pp. 349–360, 2012.

reparameterizations of given parametric curves. We proved that among all the rational curves, only straight lines have rational arc-angle parameterizations. For any curve other than lines, a rational reparameterization that has the maximum uniformity among all the parameterizations of the same degree may be found by using the Möbius transformation. Since there is only one parameter in the Möbius transformation, the possibility of finding an admissible transformation with special parametric value is limited and sometimes one cannot obtain a sufficiently accurate approximation to the arc-angle parameterization.

In this paper, we extend our study of the Möbius transformation by exploring the C^0 piecewise Möbius transformation suggested by Costantini and others [3], aiming at finding C^0 piecewise-rational reparameterizations whose uniformities are close to 1. We also apply Zoutendijk's method of feasible directions [13] from optimization theory to optimize the partitioning of the unit interval, resulting in highly accurate approximations to arc-angle parameterizations. To overcome the difficulty of computing partial derivatives in each iteration step, we propose to do so using an explicit formula.

The paper is organized as follows. In Section 2, we recall the measure of angular speed uniformity for the quality of parameterizations of plane curves. In Section 3, we derive formulas for the optimal C^0 piecewise Möbius transformation for any given or undetermined partition of the unit interval. In Section 4, an optimization mechanism is developed for computing locally optimal partitions of the unit interval. We summarize the results into an algorithm in Section 5. Experimental results are provided and the performance of the method is analyzed in Section 6.

2   Review of Angular Speed Uniformity

Let

$$p = (x(t), y(t)) : [0, 1] \to \mathbb{R}^2$$

be a regular parameterization of a plane curve and

$$\theta_p = \arctan \frac{y'}{x'}, \quad \omega_p = |\theta_p'|, \quad \mu_p = \int_0^1 \omega_p(t) \, dt, \quad \sigma_p^2 = \int_0^1 (\omega_p(t) - \mu_p)^2 \, dt.$$

Definition 1 (Angular Speed Uniformity). *The* angular speed uniformity u_p *of* p *is defined as*

$$u_p = \frac{1}{1 + \sigma_p^2 / \mu_p^2}$$

when $\mu_p \neq 0$. *Otherwise*, $u_p = 1$.

The value of u_p ranges over $(0, 1]$. The more uniform the angular speed ω_p is, the closer to 1 u_p is. When the angular speed is uniform, $u_p = 1$.

Let r be a proper parameter transformation which maps $[0, 1]$ onto itself with $r(0) = 0$ and $r(1) = 1$. It is shown in [12] that the uniformity of the reparameterization $p \circ r$ is

$$u_{por} = \frac{\mu_p^2}{\eta_{p,r}}, \quad \text{where} \quad \eta_{p,r} = \int_0^1 \frac{\omega_p^2}{(r^{-1})'}(t)\,dt \qquad (1)$$

and r^{-1} is the inverse function of r. It is easy to show that $p \circ r$ has a uniform angular speed when

$$r^{-1} = \int_0^t \omega_p(s)\,ds/\mu_p.$$

We call such r a uniformizing transformation. The uniformizing transformation is an irrational function in most cases. One natural question is how to compute a rational function which is close enough to the uniformizing transformation. There are two typical approaches: one is to use polynomials of higher degree (e.g., the Weierstrass approximation approach) and the other is to use piecewise functions with each piece having a low degree. Between the two approaches, the latter is preferable in applications. In the following sections, we will show how to use the C^0 piecewise Möbius transformation to approximate the uniformizing transformation.

3 Optimal C^0 Piecewise Möbius Transformation

We begin by recalling the standard definition of a C^0 piecewise Möbius transformation. Let

$$\alpha = (\alpha_0, \ldots, \alpha_{N-1}), \quad T = (t_0, \ldots, t_N), \quad S = (s_0 \ldots, s_N)$$

be three sequences such that $0 \leq \alpha_0, \ldots, \alpha_{N-1} \leq 1$, $0 = t_0 < \cdots < t_N = 1$, and $0 = s_0 < \cdots < s_N = 1$.

Definition 2 (C^0 Piecewise Möbius Transformation). *A map m is called a C^0 piecewise Möbius transformation if it has the following form*

$$m(s) = \begin{cases} \vdots \\ m_i(s), \quad \text{if } s \in [s_i, s_{i+1}]; \\ \vdots \end{cases}$$

where

$$m_i(s) = t_i + \Delta t_i \frac{(1 - \alpha_i)\tilde{s}}{(1 - \alpha_i)\tilde{s} + (1 - \tilde{s})\alpha_i}$$

and $\Delta t_i = t_{i+1} - t_i$, $\Delta s_i = s_{i+1} - s_i$, $\tilde{s} = (s - s_i)/\Delta s_i$.

It is easy to verify that $m(s_i) = t_i$. With different choices of the three sequences α, T, and S, one may get different C^0 piecewise Möbius transformations.

Let p be a regular rational parameterization of a plane curve. From now on, we assume that the curve is not a straight line and that the angular speed of p is nonzero over $[0, 1]$. Let the sequence T be arbitrary but fixed. We would like to find the two sequences α and S that maximize the uniformity $u_{p \circ m}$ of the reparameterization of p by m.

Theorem 1 (Optimal Transformation). *The uniformity* u_{pom} *is maximum if and only if*

$$\alpha_i = \alpha_i^* = \frac{1}{1 + \sqrt{C_i/A_i}} \quad and \quad s_i = s_i^* = \frac{\sum_{k=0}^{i-1} \sqrt{M_k}}{\sum_{k=0}^{N-1} \sqrt{M_k}}, \tag{2}$$

where

$$A_i = \int_{t_i}^{t_{i+1}} \omega_p^2(t) \cdot (1 - \tilde{t})^2 \, dt, \quad B_i = \int_{t_i}^{t_{i+1}} \omega_p^2(t) \cdot 2\tilde{t}(1 - \tilde{t}) \, dt, \tag{3}$$

$$C_i = \int_{t_i}^{t_{i+1}} \omega_p^2(t) \cdot \tilde{t}^2 \, dt, \quad M_k = \Delta t_k \left(2\sqrt{A_k C_k} + B_k \right),$$

$$\tilde{t} = (t - t_i)/\Delta t_i.$$

Proof. Recall (1). Since μ_p is a constant, the problem of maximizing u_{pom} is equivalent to that of minimizing

$$\eta_{p,m} = \int_0^1 \frac{\omega_p^2}{(m^{-1})'}(t) \, dt = \sum_{i=0}^{N-1} \int_{t_i}^{t_{i+1}} \frac{\omega_p^2(t)}{(m_i^{-1})'(t)} \, dt.$$

First we rewrite the above integrals as simply as possible. Note that

$$\frac{1}{(m_i^{-1})'(t)} = m_i'(s) = \frac{\Delta t_i}{\Delta s_i} \cdot \frac{\alpha_i(1 - \alpha_i)}{[(1 - \alpha_i)\tilde{s} + (1 - \tilde{s})\alpha_i]^2}. \tag{4}$$

By solving the equation $t = m_i(s)$ for $\tilde{s}$, we get

$$\tilde{s} = \frac{\alpha_i \tilde{t}}{\alpha_i \tilde{t} + (1 - \alpha_i)(1 - \tilde{t})}. \tag{5}$$

Substitution of (5) into (4) yields

$$\frac{1}{(m_i^{-1})'(t)} = \frac{\Delta t_i}{\Delta s_i} \cdot \frac{[(1 - \alpha_i)(1 - \tilde{t}) + \alpha_i \tilde{t}]^2}{(1 - \alpha_i)\alpha_i}.$$

Thus

$$\eta_{p,m} = \sum_{i=0}^{N-1} \frac{\Delta t_i}{\Delta s_i} \left(A_i \frac{1 - \alpha_i}{\alpha_i} + B_i + C_i \frac{\alpha_i}{1 - \alpha_i} \right) \tag{6}$$

using the notation (3). Clearly $\eta_{p,m} \geq 0$. When $(\alpha_0, \ldots, \alpha_{N-1}, s_1, \ldots, s_{N-1})$ approaches the boundary of the admissible set, the value of $\eta_{p,m}$ goes to $+\infty$. Therefore, the global minimum of $\eta_{p,m}$ is reached at an internal critical point, satisfying the equations

$$\frac{\partial \eta_{p,m}}{\partial \alpha_i} = 0, \quad \frac{\partial \eta_{p,m}}{\partial s_i} = 0.$$

The above equations can be written explicitly as

$$-\frac{A_i}{\alpha_i^2} + \frac{C_i}{(1-\alpha_i)^2} = 0, \tag{7}$$

$$\frac{\Delta t_i}{\Delta s_i^2}\left(A_i\frac{1-\alpha_i}{\alpha_i} + B_i + C_i\frac{\alpha_i}{1-\alpha_i}\right)$$
$$-\frac{\Delta t_{i-1}}{\Delta s_{i-1}^2}\left(A_{i-1}\frac{1-\alpha_{i-1}}{\alpha_{i-1}} + B_{i-1} + C_{i-1}\frac{\alpha_{i-1}}{1-\alpha_{i-1}}\right) = 0, \tag{8}$$

Solving the system of equations (7) for α_i, we obtain the optimal value

$$\alpha_i^* = 1/\left(1 \pm \sqrt{C_i/A_i}\right).$$

Observing that

$$1/\left(1 + \sqrt{C_i/A_i}\right) \in (0,1), \quad 1/\left(1 - \sqrt{C_i/A_i}\right) \notin (0,1),$$

we choose $\alpha_i^* = 1/\left(1 + \sqrt{C_i/A_i}\right)$. Substituting $\alpha_i = \alpha_i^*$ into (8) and solving the resulting equations for Δs_i, we obtain the optimal value

$$\Delta s_i^* = \Delta s_0^*\sqrt{M_i/M_0}$$

using the notation (3). Noting that $\sum_{i=0}^{N-1}\Delta s_i^* = 1$, we have

$$\Delta s_0^* = \left(\sum_{k=0}^{N-1}\sqrt{\frac{M_k}{M_0}}\right)^{-1}.$$

Thus

$$s_i^* = \sum_{k=0}^{i-1}\Delta s_k^* = \frac{\sum_{k=0}^{i-1}\sqrt{M_k/M_0}}{\sum_{k=0}^{N-1}\sqrt{M_k/M_0}} = \frac{\sum_{k=0}^{i-1}\sqrt{M_k}}{\sum_{k=0}^{N-1}\sqrt{M_k}}.$$

Recall that only $\alpha_0, \ldots, \alpha_{N-1}$ and $s_1, \ldots, s_{N-1}$ are the free parameters. Thus there is only one critical point and that critical point must be the unique point for the global minimum of $\eta_{p,m}$, and equivalently for the global maximum of u_{pom}. □

From now on, let m^* denote the optimal transformation obtained by Theorem 1. Note that it depends on the choice of the sequence T. The following theorem ensures that m^* converges, as expected, to the uniformizing transformation as Δt_i approaches zero.

Theorem 2 (Convergence). *If the sequence T satisfies*

$$\max_{0 \le i \le N-1} \Delta t_i \le \epsilon < 1,$$

then

$$u_{pom^*} \ge 1 - O(\epsilon^4).$$

Proof. It is easy to show that

$$\eta_{p,m} = \sigma_{pom}^2 + \mu_{pom}^2 \geq \mu_{pom}^2 = \mu_p^2.$$

Hence $\eta_{p,m^*} \geq \mu_p^2$. Substituting (2) into (6) and simplifying the result, we have

$$\eta_{p,m^*} = \left(\sum_{i=0}^{N-1} \sqrt{M_i} \right)^2. \tag{9}$$

Carrying out Taylor expansion of $\sqrt{M_i}$ at $t_{i+1} = t_i$ with Lagrange remainder [1] (and t_{i+1} regarded as variable) using Maple, we obtain

$$\sqrt{M_i} = \sum_{k=1}^{4} \frac{\omega_p^{(k-1)}(t_i)}{k!}(\Delta t_i)^k + \frac{F_i(t_i^*)}{120}(\Delta t_i)^5,$$

where $F_i(t_{i+1})$ is the fifth derivative of $\sqrt{M_i}$ with respect to t_{i+1} and $t^* \in [t_i, t_{i+1}]$. One can verify that $F_i(t_i) < +\infty$ and see easily that $F_i(t)$ is finite for any $t \in (t_i, t_{i+1}]$. Similarly,

$$\int_{t_i}^{t_{i+1}} \omega_p(t)\, dt = \sum_{k=1}^{4} \frac{\omega_p^{(k-1)}(t_i)}{k!}(\Delta t_i)^k + \frac{\omega_p^{(4)}(t_i^*)}{120}(\Delta t_i)^5,$$

where $t_i^* \in [t_i, t_{i+1}]$ and $\omega_p^{(4)}(t)$ is also finite for $t \in [t_i, t_{i+1}]$. Hence there is a positive constant c_i such that $\left| F_i(t_i^*) - \omega_p^{(4)}(t_i^*) \right|/120 \leq c_i$. It follows that

$$\left| \sqrt{M_i} - \int_{t_i}^{t_{i+1}} \omega_p(t)\, dt \right| = \frac{\left| F_i(t_i^*) - \omega_p^{(4)}(t_i^*) \right|}{120}(\Delta t_i)^5 \leq c_i(\Delta t_i)^5.$$

Thus

$$\left| \sqrt{\eta_{p,m^*}} - \mu_p \right| = \left| \sum_{i=0}^{N-1} \sqrt{M_i} - \sum_{i=0}^{N-1} \int_{t_i}^{t_{i+1}} \omega_p(t)\, dt \right|$$

$$\leq \sum_{i=1}^{N} \left| \sqrt{M_i} - \int_{t_i}^{t_{i+1}} \omega_p(t)\, dt \right| \leq \sum_{i=0}^{N-1} c_i(\Delta t_i)^5$$

$$\leq \max(c_0, \ldots, c_{N-1}) \cdot \max_{0 \leq i \leq N-1}(\Delta t_i)^4 \cdot \sum_{0 \leq i \leq N-1} \Delta t_i$$

$$= \max(c_0, \ldots, c_{N-1}) \cdot \max_{0 \leq i \leq N-1}(\Delta t_i)^4 \leq O(\epsilon^4).$$

This implies that

$$\eta_{p,m^*} \leq (\mu_p + O(\epsilon^4))^2 = \mu_p^2 + O(\epsilon^4).$$

Therefore we can conclude that $u_{pom^*} \geq 1 - O(\epsilon^4)$. □

4   Locally Optimal Partition of $[0, 1]$

We want to find an optimal partition of $[0, 1]$, that is, a sequence T that maximizes u_{pom^*}. Referring to (1) and (9), this problem is equivalent to minimizing η_{p,m^*} and, therefore, equivalent to minimizing

$$\phi = \sqrt{\eta_{p,m^*}}, \tag{10}$$

which can be solved by using Zoutendijk's method of feasible directions (linear constraints) [13] from optimization theory. Zoutendijk's optimization method is well known and we will not reproduce its steps here. We only note that this method involves the calculation of $\partial\phi/\partial t_i$, which can be very daunting because ϕ has a highly nonlinear dependence on t_i. We provide the following theorem in the hope to alleviate the pain.

Theorem 3 (Partial Derivatives). *For $1 \leq i \leq N - 1$,*

$$\frac{\partial\phi}{\partial t_i} = \frac{\frac{B_{i-1}C_{i-1}+A_{i-1}\omega_p^2(t_i)\Delta t_{i-1}}{\sqrt{A_{i-1}C_{i-1}}} + 2\,C_{i-1}}{2\sqrt{\Delta t_{i-1}\left(2\sqrt{A_{i-1}C_{i-1}} + B_{i-1}\right)}} - \frac{\frac{A_iB_i+C_i\omega_p^2(t_i)\Delta t_i}{\sqrt{A_iC_i}} + 2\,A_i}{2\sqrt{\Delta t_i\left(2\sqrt{A_iC_i} + B_i\right)}}.$$

Proof. For each i, only M_i and M_{i-1} contribute to $\partial\phi/\partial t_i$, so

$$\frac{\partial\phi}{\partial t_i} = \frac{\partial M_{i-1}/\partial t_i}{2\sqrt{M_{i-1}}} + \frac{\partial M_i/\partial t_i}{2\sqrt{M_i}}$$

$$= \frac{\left(\Delta t_{i-1}\left(2\sqrt{A_{i-1}C_{i-1}} + B_{i-1}\right)\right)'}{2\sqrt{M_{i-1}}} + \frac{\left(\Delta t_i\left(2\sqrt{A_iC_i} + B_i\right)\right)'}{2\sqrt{M_i}}$$

$$= \frac{\Delta t_{i-1}\left(\frac{A'_{i-1}C_{i-1}+A_{i-1}C'_{i-1}}{\sqrt{A_{i-1}C_{i-1}}} + B'_{i-1}\right) + \left(2\sqrt{A_{i-1}C_{i-1}} + B_{i-1}\right)}{2\sqrt{M_{i-1}}}$$

$$+ \frac{\Delta t_i\left(\frac{A'_iC_i+A_iC'_i}{\sqrt{A_iC_i}} + B'_i\right) - \left(2\sqrt{A_iC_i} + B_i\right)}{2\sqrt{M_i}}, \tag{11}$$

where all the derivatives are with respect to t_i. We want to compute $A'_{i-1}, B'_{i-1}, C'_{i-1}$ and A'_i, B'_i, C'_i. First,

$$A_{i-1} = \int_{t_{i-1}}^{t_i} \omega_p^2(t)(1-\tilde{t})^2 \, dt = \int_{t_{i-1}}^{t_i} \omega_p^2(t)\left(\frac{t_i - t}{t_i - t_{i-1}}\right)^2 dt$$

$$= \frac{1}{\Delta t_{i-1}^2}\left[t_i^2 \int_{t_{i-1}}^{t_i} \omega_p^2(t)\, dt - 2\,t_i \int_{t_{i-1}}^{t_i} \omega_p^2(t)t\, dt + \int_{t_{i-1}}^{t_i} \omega_p^2(t)t^2\, dt\right].$$

Thus

$$A'_{i-1} = -\frac{2}{\Delta t_{i-1}}A_{i-1} + \frac{1}{\Delta t_{i-1}^2}\left[2\,t_i \int_{t_{i-1}}^{t_i} \omega_p^2(t)\, dt - 2\int_{t_{i-1}}^{t_i} \omega_p^2(t)t\, dt\right]$$

$$= -\frac{2}{\Delta t_{i-1}}A_{i-1} + \frac{1}{\Delta t_{i-1}}(2\,A_{i-1} + B_{i-1}) = \frac{B_{i-1}}{\Delta t_{i-1}}. \tag{12}$$

Similarly, we have

$$B'_{i-1} = \frac{2\,C_{i-1} - B_{i-1}}{\Delta t_{i-1}}, \quad C'_{i-1} = \omega_p^2(t_i) - \frac{2\,C_{i-1}}{\Delta t_{i-1}}, \tag{13}$$

$$A'_i = -\omega_p^2(t_i) + \frac{2\,A_i}{\Delta t_i}, \quad B'_i = \frac{B_i - 2\,A_i}{\Delta t_i}, \quad C'_i = -\frac{B_i}{\Delta t_i}. \tag{14}$$

Substituting (12)–(14) into (11), we arrive at the conclusion.   □

5   Algorithm

We summarize the results from the previous two sections into an algorithm.

Input: p, a rational parameterization of a plane curve;
           N, a number of pieces.

Output: p^*, a more uniform reparameterization via a C^0 piecewise Möbius transformation (which is locally optimal with respect to the sequence T and globally optimal with respect to the sequences α and S).

1. Construct the expression ϕ in terms of $t_1, \ldots, t_{N-1}$ using (10), (9), and (3).

2. Compute the value of T which locally minimizes ϕ using Zoutendijk's method with an initial value $(\frac{1}{N}, \ldots, \frac{N-1}{N})$, where the derivatives of ϕ with respect to t_i are calculated according to Theorem 3.

3. Compute the values of the sequences α and S which globally maximize u_{pom}, using Theorem 1.

4. Construct the piecewise map m using the three computed sequences T, α, and S, according to Definition 2.

5. Return $p^* = p \circ m$.

The following example illustrates this algorithm.

Example 1. Consider the parameterization

$$p = \left(\frac{t^3 - 6\,t^2 + 9\,t - 2}{2\,t^4 - 16\,t^3 + 40\,t^2 - 32\,t + 9}, \frac{t^2 - 4\,t + 4}{2\,t^4 - 16\,t^3 + 40\,t^2 - 32\,t + 9} \right).$$

For $N = 2$, after steps 1 and 2, we get a locally optimal partition of $[0, 1]$ which is $[0, 0.5869], [0.5869, 1]$. Under this partition, we calculate

$$\alpha_0^* \approx 0.1398, \quad \alpha_1^* \approx 0.8505, \quad s_1^* \approx 0.5726,$$

$$m(s) \approx \begin{cases} \dfrac{0.8817\,s}{1.2580\,s + 0.1398}, & s \in [0, 0.5726]; \\[2mm] \dfrac{0.8184\,s - 0.9679}{1.6405\,s - 1.7899}, & s \in [0.5726, 1]. \end{cases}$$

Then $p^* = p \circ m$ is a reparameterization satisfying the conditions in the output of the above algorithm. It may be verified that

$$u_p \approx 0.4029 \quad \text{and} \quad u_{p^*} \approx 0.9727.$$

The curve of Example 1 with parameterization p, together with its optimal reparameterizations and their corresponding angular speeds using locally optimal partitions for $N = 2, 5, 10$, is plotted in Figure 1. In practice, the plotting is satisfactory when the uniformity of the curve parameterization is greater than 0.99. This uniformity can often be achieved by a locally optimal partition for $N = 5$. For two parameterizations of the same curve with uniformity greater than 0.99 (as in Figure 1), their plots look almost the same.

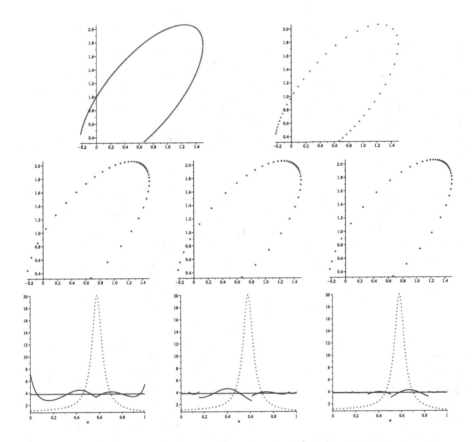

Fig. 1. Curve p and its reparameterizations using the C^0 piecewise Möbius transformation under locally optimal N-partitions: the first row shows the test curve and its dot plotting using p, the second row shows the dot plottings of the curve for $N = 2, 5, 10$, and the third row displays the angular speeds (where the dot ones are the original angular speed)

6   Experimental Results and Performance Analysis

In this section, we present some experimental results with a preliminary implementation of the proposed method in Maple. The experiments were performed on a PC Intel(R) Core(TM)2 Quad CPU Q9500 @2.83GHz with 3G of RAM. Five parametric curves[1] C1–C5 are selected from [9,11] for our experiments. For curves C1–C3, application of the α-Möbius transformation does not improve the uniformity significantly. By reparameterization using the C^0 piecewise Möbius transformation, the uniformities for C1–C3 can be improved dramatically and those for C4 and C5 may become very close to 1.

Table 1. Experimental results with locally optimal N-partitions of $[0, 1]$

Curve	Original u	$N = 1$		$N = 2$		$N = 5$		$N = 10$	
		Optimal u	Time (sec)	Optimal u	Time (sec)	Optimal u	Time (sec)	Optimal u	Time (sec)
C1	0.5518	0.5575	0.063	0.9914	0.670	0.9993	0.459	0.9999	1.240
C2	0.4034	0.6427	0.140	0.9380	1.281	0.9909	1.029	0.9996	1.218
C3	0.4029	0.4121	0.141	0.9727	1.052	0.9938	1.253	0.9996	1.469
C4	0.9259	0.9725	0.078	0.9987	0.292	0.9999	0.450	1.0000	0.378
C5	0.9607	0.9757	3.265	0.9983	11.422	0.9999	6.932	1.0000	9.391

Table 1 shows the uniformities of C^0 piecewise-rational reparameterizations of the five curves using locally optimal partitions and the times consumed for computing such partitions. When $N = 1$, the C^0 piecewise Möbius transformation becomes the α-Möbius transformation studied in [12]. From this table, one sees that reparameterizations with uniformity very close to 1 may be obtained by using locally optimal partitions even with a very small N (e.g., the uniformity is greater than 0.9 when $N = 2$ and greater than 0.99 when $N = 5$). For curves C1, C2, and C5, computation of the locally optimal partitions takes more time for $N = 2$ than for $N = 5$. This is because for $N = 2$, more iterations are needed for the optimization method to achieve the optimal partition.

Note that one may use a given (uniform) partition of the unit interval instead of computing a locally optimal partition. In this case, the reparameterization is not optimized with respect to the sequence T, so its uniformity may be not so close to 1, but some computing time can be saved.

Table 2 shows the uniformities of C^0 piecewise-rational reparameterizations of the five curves obtained under locally and *quasi-globally* optimal partitions of the interval $[0, 1]$, where quasi-globally (or *q-globally* for short) optimal partitions[2] are carried out as follows: first divide $[0, 1]$ into 20 equidistant subintervals; then choose optimal t_i ($1 \leq i \leq N - 1$) by direct enumeration under the constraints $0 < t_i < t_{i+1} < 1$ ($1 \leq i \leq N - 2$) and $t_i \in \{k/20 : k = 1, \ldots, 19\}$. Under the q-globally optimal partition the uniformity of the obtained reparameterization may be smaller than that under a locally optimal partition (e.g., for C1–C4 with $N = 2, 3$, when the locally optimal partition is very likely to be q-globally optimal).

---

[1] The parametric equations for these curves are available from the authors upon request.

[2] A q-globally optimal partition is not necessarily globally optimal, but it is likely closer to the globally optimal partition than a locally optimal one is.

Table 2. Comparison of optimal u with locally optimal partitions and q-globally optimal partitions

Curve	$N = 2$		$N = 3$		$N = 6$	
	u (locally optimal)	u (q-globally optimal)	u (locally optimal)	u (q-globally optimal)	u (locally optimal)	u (q-globally optimal)
C1	0.9914	0.9906	0.9977	0.9969	0.9989	0.9997
C2	0.9380	0.9362	0.9905	0.9793	0.9986	0.9991
C3	0.9727	0.9683	0.9916	0.9826	0.9991	0.9987
C4	0.9987	0.9987	0.9998	0.9998	1.0000	1.0000
C5	0.9983	0.9990	0.9996	0.9998	1.0000	1.0000

In some cases (e.g., C5 with $N = 2, 3$) when the locally optimal partition is not globally optimal, the q-globally optimal partition leads to better uniformity. For C5 with $N = 3$ the q-globally optimal partition is approximately $[0, 0.15, 0.65, 1]$, whereas the locally optimal partition (with initial value $[0, 1/3, 2/3, 1]$) ends at $[0, 0.317, 0.650, 1]$.

7 Conclusion

We have adapted the C^0 piecewise Möbius transformation to compute C^0 piecewise-rational reparameterizations of plane curves with high uniformity. Using partitions of the unit interval with moderate or even small numbers, we can produce reparameterizations which closely approximate to the uniform ones. The reparameterization method based on the C^0 piecewise Möbius transformation, even with given interval partitions, is practically more powerful than that based on the α-Möbius transformation. We have further enhanced the power of the method by developing an optimization mechanism for locally optimal partitioning of the unit interval. This mechanism is based on Zoutendijk's method of feasible directions and a formula we have derived for the calculation of derivatives of a certain type of functions dependent nonlinearly on their variables. Experimental results have shown the applicability and high efficiency of our new method, in particular with locally optimal interval partitioning.

The piecewise Möbius transformation and piecewise-rational reparameterizations studied in this paper are only C^0 continuous with respect to the new parameter. Optimal C^1 piecewise reparameterization is clearly an interesting problem that remains for further investigation.

Acknowledgements. Part of this work has been supported by the ANR-NSFC Project ANR-09-BLAN-0371-01/60911130369 (EXACTA) and the Open Fund SKLSDE-2011KF-02.

References

1. Abramowitz, M., Stegun, I.A. (eds.): Handbook of Mathematical Functions with Formulas, Graphs, and Mathematical Tables (10th printing). United States Government Printing, Washington, D.C (1972)

2. Cattiaux-Huillard, I., Albrecht, G., Hernández-Mederos, V.: Optimal parameterization of rational quadratic curves. Computer Aided Geometric Design 26(7), 725–732 (2009)
3. Costantini, P., Farouki, R., Manni, C., Sestini, A.: Computation of optimal composite re-parameterizations. Computer Aided Geometric Design 18(9), 875–897 (2001)
4. Farouki, R.: Optimal parameterizations. Computer Aided Geometric Design 14(2), 153–168 (1997)
5. Farouki, R., Sakkalis, T.: Real rational curves are not unit speed. Computer Aided Geometric Design 8(2), 151–157 (1991)
6. Gil, J., Keren, D.: New approach to the arc length parameterization problem. In: Straßer, W. (ed.) Prodeedings of the 13th Spring Conference on Computer Graphics, Budmerice, Slovakia, June 5–8, pp. 27–34. Comenius University, Slovakia (1997)
7. Jüttler, B.: A vegetarian approach to optimal parameterizations. Computer Aided Geometric Design 14(9), 887–890 (1997)
8. Patterson, R., Bajaj, C.: Curvature adjusted parameterization of curves. Computer Science Technical Report CSD-TR-907, Paper 773, Purdue University, USA (1989)
9. Sendra, J.R., Winkler, F., Pérez-Díaz, S.: Rational Algebraic Curves: A Computer Algebra Approach. Algorithms and Computation in Mathematics, vol. 22. Springer, Heidelberg (2008)
10. Walter, M., Fournier, A.: Approximate arc length parameterization. In: Velho, L., Albuquerque, A., Lotufo, R. (eds.) Prodeedings of the 9th Brazilian Symposiun on Computer Graphics and Image Processing, Fortaleza-CE, Brazil, October 29-November 1, pp. 143–150. Caxambu, SBC/UFMG (1996)
11. Wang, D. (ed.): Selected Lectures in Symbolic Computation. Tsinghua University Press, Beijing (2003) (in Chinese)
12. Yang, J., Wang, D., Hong, H.: Improving angular speed uniformity by reparameterization (preprint, submitted for publication, January 2012)
13. Zoutendijk, G.: Methods of Feasible Directions: A Study in Linear and Nonlinear Programming. Elsevier Publishing Company, Amsterdam (1960)

Usage of Modular Techniques for Efficient Computation of Ideal Operations
(Invited Talk)

Kazuhiro Yokoyama

Department of Mathematics, Rikkyo University
3-34-1 Nishi Ikebukuro, Toshima-ku, Tokyo, 171-8501, Japan
kazuhiro@rikkyo.ac.jp

Modular techniques are widely applied to various algebraic computations. (See [5] for basic modular techniques applied to polynomial computations.) In this talk, we discuss how modular techniques are efficiently applied to computation of various ideal operations such as Gröbner base computation and ideal decompositions. Here, by *modular techniques* we mean techniques using certain projections for improving the efficiency of the total computation, and by *modular computations*, we mean corresponding computations applied to projected images. The usage of modular techniques for computation of ideal operations might be very roughly classified into the following:

(1) For the computation of the target ideal operation, we use its corresponding modular computations. In this computation, we need methods for recovering the true result from its modular images (the results of modular computations).

(2) We use modular computations partly. For example, for some part of the computation of the target ideal operation, we apply modular techniques. Also we make good use of certain information derived from modular computations for improving the total efficiency.

Here we give several examples: Concerning (1), modular techniques for Gröbner bases computation have been proposed by several authors ([15,13,11,6,9,1]), where projections $\mathbb{Z}$ to $\mathbb{F}_p$ are mainly considered and *Chinese remainder algorithm* or *Hensel lifting* are applied to recovering the true results. As for triangular set, modular techniques for triangular decomposition of 0-dimensional case are proposed in [3] based on theoretical estimations on coefficient bound [4] and techniques based on *interpolation*, where a projection from a polynomial ring to its coefficient field is considered, are used for theoretical estimations on coefficient bound for positive dimensional case in [2]. In [12,8] practical combination of two different modular techniques, interpolation by *modular zeros* and Chinese remainder algorithm on integers, is proposed with help of the Galois group of target algebraic structure. As to (2), modular techniques are applied to computation of minimal polynomials, which are important objects for ideal decompositions, in [9,10,7]. Using useful informations derived from modular computations can be seen in [14], where they are used for avoiding unnecessary polynomial

V.P. Gerdt et al. (Eds.): CASC 2012, LNCS 7442, pp. 361–362, 2012.

reductions among Gröbner basis computation. Also, for the problem of prime (radical) decomposition, *quick tests* for primality (square-freeness) can be provided by modular techniques. (See [9,7].)

The most important points in (1) are to provide recovering methods from modular images and to guarantee the correctness of the computed results. To this end, notions on *luckiness* or *goodness* of projections have been introduced by several authors. Here we explain techniques for guaranteeing the correctness, which are mainly based on *ideal inclusion* or theoretical estimation on the number of *unlucky* projections. Also, we refer effective combination between modular techniques and informations derived from the mathematical structure of the target object such as the action of the Galois group or a Gröbner basis with respect to another ordering. Moreover, concerning (2), we give some discussion on efficiently computable quick tests, which have good effects on decomposition algorithms in practice.

References

1. Arnold, E.: Modular algorithms for computing Gröbner bases. J. Symb. Comp. 35, 403–419 (2003)
2. Dahan, X., Kadri, A., Schost, É.: Bit-size estimates for triangular sets in positive dimension. J. Complexity 28, 109–135 (2012)
3. Dahan, X., Moreno Maza, M., Schost, É., Wu, W., Xie, Y.: Lifting techniques for triangular decompositions. In: Proc. ISSAC 2005, pp. 108–115. ACM Press, New York (2005)
4. Dahan, X., Schost, É.: Sharp estimates for triangular sets. In: Proc. ISSAC 2004, pp. 103–110. ACM Press, New York (2004)
5. von zur Gathen, J., Gerhard, J.: Modern Computer Algebra. Cambridge University Press, Cambridge (1999)
6. Gräbe, H.: On lucky primes. J. Symb. Comp. 15, 199–209 (1993)
7. Idrees, N., Pfister, G., Steidel, S.: Parallelization of modular algorithms. J. Symb. Comp. 46, 672–684 (2011)
8. Orange, S., Renault, G., Yokoyama, K.: Efficient arithmetic in successive algebraic extension fields using symmetries. Math. Comput. Sci. (to appear)
9. Noro, M., Yokoyama, K.: A modular method to compute the rational univariate representation of zero-dimensional ideals. J. Symb. Comp. 28, 243–263 (1999)
10. Noro, M., Yokoyama, K.: Implementation of prime decomposition of polynomial ideals over small finite fields. J. Symb. Comp. 38, 1227–1246 (2004)
11. Pauer, F.: On lucky ideals for Gröbner bases computations. J. Symb. Comp. 14, 471–482 (1992)
12. Renault, G., Yokoyama, K.: Multi-modular algorithm for computing the splitting field of a polynomial. In: Proceedings of ISSAC 2008, pp. 247–254. ACM Press, New York (2008)
13. Sasaki, T., Takeshima, T.: A modular method for Gröbner-bases construction over ℚ and solving system of algebraic equations. J. Inform. Process. 12, 371–379 (1989)
14. Traverso, C.: Gröbner Trace Algorithms. In: Gianni, P. (ed.) ISSAC 1988. LNCS, vol. 358, pp. 125–138. Springer, Heidelberg (1989)
15. Winkler, F.: A p-adic approach to the computation of Gröbner bases. J. Symb. Comp. 6, 287–304 (1988)

Author Index